Human Adaptation in the Asian Palaeolithic

This book examines the first human colonization of Asia and particularly the tropical environments of Southeast Asia during the Upper Pleistocene. In studying the unique character of the Asian archaeological record, it reassesses long-held views about the development of human 'modernity'. Ryan J. Rabett proposes that an evolutionary relationship exists between the process of colonization, the challenges encountered during its pursuit – especially in relation to climatic and environmental change – and the forms of human behaviour that emerge as a result. The book argues that human modernity was not something achieved in the remote past in one part of the world but rather that it has developed out of a diverse, flexible and locally contingent process of adaptation; one that continues to this day.

Ryan J. Rabett is a Research Fellow at the McDonald Institute for Archaeological Research at the University of Cambridge. His research focuses on archaeological sites in Asia, and since 2007, he has been the director of a major project in northern Vietnam. He is the author of more than 40 articles, which primarily focus on prehistoric subsistence and technological strategies in Asia.

Human Adaptation in the Asian Palaeolithic

Hominin Dispersal and Behaviour during the Late Quaternary

Ryan J. Rabett
McDonald Institute, Cambridge University

CAMBRIDGE
UNIVERSITY PRESS

CAMBRIDGE UNIVERSITY PRESS
Cambridge, New York, Melbourne, Madrid, Cape Town,
Singapore, São Paulo, Delhi, Mexico City

Cambridge University Press
32 Avenue of the Americas, New York, NY 10013-2473, USA

www.cambridge.org
Information on this title: www.cambridge.org/9781107018297

First published 2012

Printed in the United States of America

A catalog record for this publication is available from the British Library.

Library of Congress Cataloging in Publication Data

Rabett, Ryan J., 1970–
Human adaptation in the Asian palaeolithic : hominin dispersal and behaviour during the late quaternary /
Ryan J. Rabett.
p. cm.
Includes bibliographical references and index.
ISBN 978-1-107-01829-7 (hardback)
1. Paleolithic period – Asia. 2. Fossil hominids – Asia. 3. Human evolution – Asia.
4. Social evolution – Asia. 5. Asia – Antiquities. I. Title.
GN772.3.A1R33 2012
950'.1–dc23 2011053166

ISBN 978-1-107-01829-7 Hardback

Contents

List of Figures and Tables — *page* vii
Acknowledgements — xi

Chapter 1: **The Journey East** — 1
1.1 Introduction — 1
1.2 The Palaeolithic of Asia — 3
1.3 Human Adaptation in the Asian Palaeolithic — 5

Chapter 2: **The Pleistocene Planet** — 8
2.1 Introduction — 8
2.2 The Quaternary Period — 9
2.3 Rapid Climate Change — 18
2.4 Conclusion — 24

Chapter 3: **Hominin Dispersal beyond Africa during the Lower and Middle Pleistocene** — 25
3.1 Introduction — 25
3.2 Lower Pleistocene (2.58–0.775 Ma) Dispersals of Anatomically Archaic Hominins — 25
3.3 Middle Pleistocene (775–127 KBP) Dispersals of Anatomically Archaic *Homo* — 46
3.4 Dispersal Mechanisms: Faunal Turnovers, Carnivory and Habitat Specificity — 62
3.5 Conclusion — 66

Chapter 4: **Regional Trajectories in Modern Human Behaviour** — 68
4.1 Introduction — 68
4.2 The 'Human Revolution' in Europe — 68
4.3 Modern Human Behaviour Beyond Europe — 80
4.4 Conclusion — 84

Chapter 5: **The Initial Upper Pleistocene Dispersal of *Homo sapiens* Out of Africa** — 85
5.1 Introduction — 85
5.2 MIS-5e (126–116 KBP) — 85
5.3 MIS-5d to the Last Termination (116-22 KBP) — 97
5.4 Conclusion — 139

Chapter 6: **Climate, Dispersal and Technological Change during the Last Termination and Early Holocene in Southeast Asia** 142
6.1 Introduction 142
6.2 Regional Climate and Conditions across the Last Termination (22–11.7 KBP) 142
6.3 Genetic Evidence of Human Dispersal during the Last Termination 149
6.4 Technological Change 150
6.5 Conclusion 204

Chapter 7: **Tropical Subsistence Strategies at the End of the Last Glacial** 208
7.1 Introduction 208
7.2 Northern Borneo: Niah Caves 210
7.3 Northern Vietnam: Tràng An 232
7.4 Eastern Peninsular Malaysia: Gua Sagu and Gua Tenggek 251
7.5 Inter-Site Patterns of Subsistence through the Last Termination and Early Holocene 261
7.6 Conclusion 264

Chapter 8: ***Ex Levis Terra*** 266
8.1 Introduction 266
8.2 The Colonization of New Environments and the Recolonization of Changing Ones 267
8.3 The Last Termination in Southeast Asia 281
8.4 Discussion: Hominin Dispersal, Climate Change and Behavioural Evolution 284
8.5 Conclusion 288

Appendix 291
Bibliography 307
Index 361

List of Figures and Tables

Figures

2.1 Chronostratigraphic correlations for the Quaternary period. *page* 11
2.2 Fox Glacier, South Island, New Zealand. 13
2.3 Map showing the locations of the Greenland GRIP, GISP2, NGRIP and NEEM ice-coring operations and the Antarctic locations of the Vostok and EPICA Kohnen Station (EDML) and Dome C ice core drilling sites. 15
2.4 The drill hole at the NEEM ice core drilling camp. 17
2.5 The $\delta^{18}O$ GISP2 ice core curve and tropical Atlantic sea surface temperature curve. GIS stages back to *c.* 57 KBP and GS stages back to *c.* 40 KBP, with the most recent named stadial and interstadial phases from the European sequence. 19
2.6 An ice floe from the Breiðamerkurjökull glacier, Jökulsárlón lagoon, southeast coast of Iceland. 22
2.7 Schematic of the global thermohaline circulation. 23
3.1 Map showing the East Asian Lower and Middle Pleistocene sites, together with the location of the Chinese Loess Plateau and key stratigraphic reference sections. 27
3.2 Map showing the locations of the Toba and Tambora volcanoes and the distribution of the Australasian tektite strewn field. 43
3.3 Excavations during the 1997 field season at the Gesher Benot Ya'aqov site on the River Jordan, Israel. 47
3.4 Gunung Temporung, Perak, West Malaysia. 57
3.5 Two bifaces from Bukit Bunuh, West Malaysia. 59
4.1 A portrayal of early anatomically and behaviourally modern *H. sapiens* during our species' Late Pleistocene dispersal through Europe. 69
4.2 Map showing northwest Eurasian sites. 71
4.3 A portrayal of body ornamentation and pigment use (here, on the scalp) among Late Pleistocene *H. sapiens* around the time of their entry into Europe. 72
4.4 A miniature horse pendant (47 mm in length, 25 mm in height) sculpted from mammoth ivory (perforated between the front legs), dated to the Aurignacian, 30–36 kbp, Vogelherd, Germany. 73
4.5 The panel of horses from the Grotte Chauvet, Ardèche, France. 75
4.6 Franchthi Cave, southeastern Argolid, Greece. 79
4.7 Two *N. gibbosulus* shell beads from Es-Skhūl, Israel. 81
4.8 A selection of MSA bone points from Blombos Cave, South Africa. 83
5.1 MIS-5 (74–126 KBP) oxygen isotope curve. 87
5.2 Map showing the trans-Sahara watershed, together with key African and Middle Eastern sites. 89
5.3 The Mount Carmel caves viewed from the northwest. 90

5.4 The small north-facing Es-Skhūl cave and terrace, Mount Carmel. 91
5.5 Map of South, Central and northern Asia, showing the locations of sites discussed in the text. 95
5.6 Lang Rongrien rock-shelter under excavation 1984–5. 111
5.7 Photograph of the immediate environs of the Bukit Bunuh island site from a palaeolake, Lenggong Valley, West Malaysia. 113
5.8 Looking out through the West Mouth entrance to the Niah Caves complex. 116
5.9 Looking north across the 'Hell' trench system as preserved today towards the rock overhang and the Deep Skull still *in situ*. 117
5.10 A fragment of hominin cranium with pigment adhering to the interior surface. 118
5.11 Pigmented Geoemydidae plastron from Area A of the West Mouth, Niah Caves, dated to *c.* 41 cal. KBP. 119
5.12 A selection of struck flakes from the Hell and lower E/B1 trenches, Niah Caves. 120
5.13 Photographs of the earliest bone tool from the Hell trench, Niah Caves. 121
5.14 Inter-cutting pit features in the Pleistocene deposits of the West Mouth during the 1959 excavations of this part of the cave and the section sampled by the Niah Caves Project. 123
5.15 Indonesian–Australian excavations at Liang Bua, Flores and deep excavations in Sectors VII and XI. 124
5.16 Excavating LB1 in Sector VII. 125
5.17 Map of Pleistocene Southeast Asia, showing the exposed Sunda Shelf to −116 m, prevailing ocean currents and possible human migration routes through Wallacea. 133
6.1 Map of Pleistocene Southeast Asia, showing the locations of terrestrial and ocean sediment core records. 144
6.2 Photograph of the Situ Bayongbong location during coring. 145
6.3 Map of Pleistocene Southeast Asia, showing the distribution of bone tool assemblages across the Sunda Shelf, Wallacea and northwest Sahul. 153
6.4 Examples of bone points from the Phase II occupation at Lobang Hangus, classified as likely to have been projectile armatures. 154
6.5 A 'self-barbed' bone point from the Lobang Hangus, Phase II assemblage, compared with a bone-barbed spear from Australia. 155
6.6 An example of a longitudinally split short convex point, or 'gouge', with significant rounding and polishing to the utilised surface, Phase II, Lobang Hangus, and an example of a more complete specimen of this kind of implement from the Late Holocene levels at nearby Kain Hitam Cave. 156
6.7 Stingray barb points (probably *Pastinachus sephen*) excavated by the Niah Caves Project in Area D of the West Mouth. 157
6.8 Evidence of extensively imposed form to create a hafting surface, Ban Lum Khao. 165
6.9 Examples of scrapers from the Early Holocene horizons at Lie Siri. 169
6.10 Elongated, blade-like scrapers from horizons I and II at Bui Ceri Uato, East Timor. 170
6.11 Tanged and other retouched blades from Uai Bobo 1, horizons V to IIIc. 171
6.12 Fish hook made from *Trochus niloticus*, Lene Hara. 172
6.13 Ulu Leang 1 at the end of excavations, July 1969. 178
6.14 Examples of Maros points. 179
6.15 An example of a unifacial, centripally flaked pebble tool or 'Sumatra type', 16 × 84 mm, from Gunong Pondok, layer G, 3.56 m below datum. 187

6.16 A pebble tool still *in situ* in the Deep Sounding of the Hang Boi excavation. 189
6.17 Limestone blade-flake and core face rejuvenation flake from Hang Trống, Tràng An, Vietnam. 191
6.18 An early stage in excavations at Gua Kerbau in 1926. 195
7.1 Photograph showing excavation work beneath the rock overhang (Area A, Niah Caves Project), probably 1965. 211
7.2 The archaeological compound of the West Mouth, Niah Caves. 221
7.3 Fragments of Cercopithecidae long bone constituting manufacturing debris from bone tool production. 227
7.4 Examples of cut-marked distal humeri from Cercopithecidae from the Lobang Hangus, Phase II occupation. 229
7.5 The NISP values and frequency of occurrence of butchery traces to the distal humerus of *Arctictis binturong* compared to other key taxonomic groups from the Phase II cut-marked assemblage from Lobang Hangus. 231
7.6 A view of the Upper Cave at Hang Boi looking east towards the archaeological compound and access to the Lower Cave, with the opening to the cave mouth visible. 232
7.7 Contexts and sample columns in the east and south-facing sections of Trench 1 at Hang Boi. 233
7.8 The southeast-facing section of excavations at Hang Boi (2009 field season), illustrating the whole shell and sediment-rich crushed shell stratigraphy that has been revealed in the Upper Cave shell midden. 235
7.9 The remains of a hunter's small fire found on the open floor of the Upper Cave at Hang Boi in 2010. 238
7.10 Three pierced Neritidae shells recovered from midden contexts at Hang Boi. 239
7.11 *Villopotamon* sp. Raffles Museum of Biodiversity Research at the University of Singapore and specimens of archaeological chelae from Hang Boi. 240
7.12 *Panthera* cf. *pardus* specimen from Hang Boi compared against *Panthera tigris*. 241
7.13 Archaeological (L) maxilla fragment with PM^1, M^1 and M^3 *in situ*, compared alongside *Ratufa* (= *Sciurus*) *bicolor*. 242
7.14 Keeled box turtle (*Cuora mouhotii*) and a nuchal fragment identified to this species. 245
7.15 One of a population of *Macaca arctoides* (stump-tailed macaque) that currently inhabits Tràng An Park. 247
7.16 Asian leaf turtle (*Cyclemys dentata*) and specimens of third and fourth (L) peripheral dermal plates of this species from Gua Sagu. 257
7.17 Cut-marks to the inside of the hyo- or hypoplastron dermal plate of a chelonid, Gua Sagu, and cut-marks to a chelonid femur or humerus mid-shaft. 259
8.1 View of wet rice fields just prior to the transplantation of rice seedlings, near the village of Pa' Dalih in the Kelabit Highlands of Sarawak. 275

Tables

3.1 Identified Mammalian Fauna from Goudi/Majuangou III and Xiaochangliang, Northeast China 33
3.2 Identified Mammalian Fauna from Chenjiawo and Gongwangling, Central China 36
3.3 Identified Mammalian Fauna from the M3 and M4 Components of the 'Yuanmou Hominid Fauna', Yuanmou Basin, North Central China 38

7.1 Identified Taxa from the Early and Late Phase II Occupation of Area A, in the West Mouth of the Niah Caves (Excluding Human Remains) 213
7.2 Taxa Represented during the Early Holocene Occupation of the West Mouth, Area D 222
7.3 Taxa Represented during the GIS-1 Occupation of Lobang Hangus 224
7.4 Identified Invertebrate Macrofaunal Remains from the Upper Cave Shell Midden (2007–8 Samples), Hang Boi 237
7.5 Vertebrate Fauna Identified and Quantified from the Cave Mouth Midden at Hang Boi (2008–9 samples quantified) 243
7.6 Identified Vertebrate Fauna from Pleistocene and Early Holocene Levels at Gua Sagu (spits 5–11) and Gua Tenggek (spits 4–7) 255
8.1 The Broad Adaptive Circumstances and Strategies of Tropical Hunter-Gatherer Communities during most of the Last Termination, Southeast Asia 283

Acknowledgements

Human Adaptation in the Asian Palaeolithic draws together work from several research projects with which I have been involved over the last decade; as such, numerous people are owed my sincere thanks. Among these, I must thank first of all my colleagues and friends involved in the Niah Caves and Tràng An Archaeological projects: among many others, Jo Appleby, Graeme Barker, Huw Barton, Bori Borbala-Nyiri, Jasminda Ceron, the Earl of Cranbrook, John deVos, Lucy Farr, Dave Gilbertson, Jason Hawkes, Daphne Hills, Chris Hunt, John Krigbaum, Lindsay Lloyd-Smith, Natalie Ludgate, Lisa Marlow, John Meneely, Mike Morley, Nguyễn Cao Tấn, Nguyễn Van Son, Nguyễn Van Truong, Kirsty Penkman, Phil Piper, Peter Pritchard, Ian Reeds, Tim Reynolds, Richard Sabin, David Simpson, Chris Stimpson, Paul Storm, Kath Szabó, Marc Verhoeven and Hannah Zukswert. In Tràng An, I must also thank, in particular, all the park staff who have supported the project through six field seasons and have become key components of it as well as becoming valued friends. I am very grateful to the organisations and institutions that have made both this latter project and my wider research possible over the last few years: including the D. M. McDonald Grants and Awards Fund, the Templeton Foundation, the Evans Fund, the Xuan Truong Corporation, the Ninh Bình People's Committee, Pusat Penyelidikan Arkeologi Malaysia (Universiti Sains Malaysia), the Arts and Humanities Research Council, the University of Cambridge and the British Academy.

The original idea for this book came out of discussions with Preston Miracle, in the Department of Archaeology, Cambridge, following a year when I had been lecturing on the Upper Palaeolithic (2006–7). Without Preston's encouragement and my students' interest in the topic (Risa Carlson, Alice Clough, Rob Hedge, Jenny Lee and Madeline Steele), it may never have got off the ground. The manuscript ultimately took shape between late 2008 and mid-2009, with reviews and revision over the following months (and years). At Cambridge University Press, I must extend my thanks to Beatrice Rehl and Amanda Smith for taking on this project and seeing the book through its various stages of gestation. At Aptara, thanks also to Shana Meyer and her colleagues for the way they so quickly and efficiently shepherded it from manuscript to final proofs. Many thanks also to the anonymous reviewers, who both gave me thorough, critical, encouraging and above all helpful comments on the manuscript during both rounds of the review process.

The text is illustrated through a combination of my own images and figures and information generously provided by colleagues and friends, including Douglas Anderson, Graeme Barker, Peter Bellwood, Torsten Blanck, Katie Boyle, Emil Charles, David Cobbett, Nicholas Conard, Fiona Coward, Ipoi Datan, Frantz Delpla, Robin Dennell, Lucy Farr, Ian Glover, Naarma Goren-Inbar, Dora Kemp, Lindsay Lloyd-Smith, Matt Lowe, Tohru Naruse, Jessica Rippengal, Berenice Robinson, Mokhtar Saidin, Brian Stewart, Marian Vanhaeren, Michael Williams, Sibylle Wolf and Zuraina Majid. Some figures in the book were kindly provided by Wall to Wall Television, the journals *Antiquity* and *Journal of Archaeological Science*, Taylor and Francis Publishing and, via open access, the NEEM ice core drilling project. The research appearing in the book was conducted at the McDonald Institute for Archaeological Research, University of Cambridge; the Cambridge University Museum of Archaeology and Anthropology; the University Museum of Zoology, Cambridge; the Natural History Museum, London; the Sarawak Museum; the Pitt Rivers Museum, Oxford; Universiti Sains Malaysia, Penang, and the Chelonian Institute, Florida. At the McDonald Institute, special thanks must go to Dora Kemp who, despite her schedule, somehow still found the time to produce the brilliant series of maps used in this book; also to Brian Stewart, Chris Stimpson (who also appears on the cover), Jo Appleby, Katie Boyle, Sacha Jones and Becky Farbstein for their thoughts, ideas and support, all of which are greatly appreciated.

Among the many people who have influenced this work, three stand out especially. I would not have even embarked on the study of the Palaeolithic were it not for Paul Mellars – partly because he interviewed me when I first came up to Cambridge but mostly because of the inspiration I got out of his lectures and supervisions. While much of what is written herein is, in a way, a response to ideas and models advanced by Paul, none of the questions I have asked would have come to be without the huge contribution he has made to the study of the Palaeolithic. The second person is Graeme Barker. More than any other senior colleague, Graeme's influence has shaped the way I approach academic research, and he has been an ever-pragmatic guiding force behind my work. I owe him a great deal. Finally, warm thanks must go to my close colleague and friend Phil Piper. Phil and I struck up a great and productive professional partnership through our work on the Niah Caves Project – one that we have maintained in the years since.

Human Adaptation in the Asian Palaeolithic was mostly written during spells in Canada staying with family, either through the depths of −30 degree winters or during long, hot summers. I must thank my parents and sister; my girlfriend, Nellie; and the Verzijlenberg family for their wonderful support, tolerance and laughter during those times; as well, Carac, for telling me long ago that I might just get this far. Most of all, I must thank my nephew and niece (Alec and Mya), whom I was babysitting at the same time as writing. They were incredibly good and patient in allowing me 'my mornings' to work. This book is dedicated to them.

Sundridge, July 2012

Chapter 1

The Journey East

1.1 Introduction

The proposition that migration and dispersal played an important role in early human evolution was put forward by Henri Breuil a hundred years ago (Breuil, 1912) and marked a significant shift in the way Palaeolithic culture history would come to be portrayed (Boyle *et al.*, 1963; Chazan, 1995). Breuil inspired his students, including Miles Burkitt and, especially, Dorothy Garrod – who would go on to become pioneers in British Palaeolithic archaeology (Smith, 2009) – to look beyond the confines of mainland Europe in their search to understand the emergence of modern humanity. Nonetheless, despite Garrod's famous discoveries in Gibraltar (Garrod *et al.*, 1928) and the tradition of research she instigated in the Levant (Garrod and Bate, 1937), much of the scholarship generated on this subject in the time since has still been concerned with the first appearance, practices and subsequent spread of *Homo sapiens* in Europe. Regionally specific research programmes that examine modern human settlement of Africa and Asia have only begun to emerge recently in any quantity (e.g. Akazawa *et al.*, 1998; Barker *et al.*, 2007; Brantingham *et al.*, 2004; d'Errico and Blackwell, 2005; Gamble, 1993; Gamble and Soffer, 1989a, 1989b; Haidle and Pawlik, 2010; McBrearty and Brooks, 2000; McBrearty, 2007; Mellars, 1990). While there is no doubting the importance of the European record, it represents only a tiny portion of the lands colonized during the early dispersal of our species. As Breuil (1912: 238) remarked, and as Garrod demonstrated, Europe is but a small peninsula of Africa and Asia (Clark, 1999). To fully appreciate the character of our species, the call remains the same today as it did a hundred years ago: we must look further afield.

Until the beginning of this century, the position held by most scholars, based largely on the European data, was that a package of behaviours regarded as quintessentially those of modern humans (including symbolic representation, use of ornament, technological and subsistence innovation) did not appear until *c.* 40,000 years ago. Their arrival in the archaeological record was linked closely to the first sustained colonization of western Eurasia and the European peninsula by *H. sapiens* (Mellars, 2005). The process of adaptation to the harsh climate and environments encountered, as well as contact and competition with indigenous

H. neanderthalensis populations, became the principal sources of explanation for the appearance of these traits, which then through migration, were assumed to become established across much of the rest of the world.

Over the last decade, new discoveries, particularly from Africa, have been pushing back the genesis of many of these defining attributes to periods long before the colonization of Europe (e.g. McBrearty and Brooks, 2000), raising questions about the strength of previous interpretations and leaving us to re-examine the place of the European sequence in a global interpretation of early human development. Some archaeologists (e.g. Finlayson and Carrión, 2007) propose that it is perhaps more prudent to view the European sequence as the product of a population of *H. sapiens* adapting survival mechanisms they already possessed to meet a new set of regionally specific challenges and opportunities. The European record may give us an expression of how a population of African *H. sapiens* coped with resource structure and competition from other species in what was broadly a sub-Arctic landscape; but it represents only one of potentially several regional stories of early human dispersal and of developmental trajectories in early humanity.

Removing Europe from the centre of the Palaeolithic world has the effect of democratising pioneer settlement, with each enclave of humans less liable to be judged against the standards of cultural development in the European evidence. It also does something else: it challenges the notion that a single suite of behavioural traits can be applicable to all adaptive circumstances. As well as being European centred, under prevailing orthodoxy, humanity also had a firmly one-track mind that was out of step with one of the key principles of evolutionary biology: that adaptation is situational.

Once the characteristics of behavioural modernity had been attained, the assumption had been that they were fixed in early human populations to be carried around the world: a superior codex of practice that could be bent to the requirements of any situation. This perspective was underpinned by the long-standing tradition of comparative studies that has examined the behavioural similarities and differences between *H. sapiens* and *H. neanderthalensis*. Consequently, very little systematic attention has been paid to the possibility that there existed significant behavioural variation between different regional populations of early *H. sapiens*. Wherever variance in behaviour was observed between regions tended to be explained away by appealing to local restrictions in material resources that were 'inhibiting innovation' and development or else to cultural degeneration – in other words, almost always starting from the premise of how was complexity lost. If populations must acclimate to different ecological settings, behavioural adaptation needs to be dynamic, not fixed; it needs to be viewed as a process, not a faculty.

Evolution by natural selection and the emergence of new species are fundamentally predicated on the adaptive process of organisms to new habitats and the consequential and eventual emergence of reproductive isolation and speciation. Although there remains considerable uncertainty about the timing and relative contribution of the many forces which can cause speciation, the long-known but less-studied significance of ecological barriers is now being championed as a potentially dominant factor (e.g. Schemske, 2000, 2010). In the late 1940s, Mayr (1947: 266, 285–6) argued that the first step towards speciation involves the founding of a new intra-specific population whose broader ecological tolerance (possibly due to its existence near the periphery of a species' range) enables expansion and whose geographic situation is or becomes spatially segregated from the parent population. The differing ecological factors that come into play in a newly colonized habitat lead to evolutionary divergence by adaptation. In this way, adaptation may be defined as the processes of modification through which an organism converges on a closeness of fit to the affordances of an environmental setting through its survival and successful reproduction, and through its presence, also affects changes within that setting (i.e. niche construction; Laland and Brown, 2006). In this process the stability and heterogeneity of the colonized environment will be crucial to determining the extent of selection pressures acting on a population (Kirch 1980: 103). To assess this for human prehistory, we must turn to the palaeoclimatic record.

As well as being affected by long-term cycles of climate deterioration and amelioration, palaeoclimate research has determined that over at least the last 100,000 years, the Earth has been subject to much shorter term climatic fluctuations, operating over centuries or even decades – timescales that would have impacted directly on our ancestors' societies across and within generations. The effect this had not only on the routes and timing of hominin dispersals but also on the very processes of adaptation and the roots of human behaviour, therefore, needs to be considered much more closely. A better understanding of how groups adapted under different regional settings whilst subject to the capricious climates of the last 100,000 years not only has great importance for our understanding of how we have been shaped as a species but has profoundly important implications for how we will continue to adapt and survive in the future.

Few palaeo-anthropologists would argue that the behavioural and demographic traits visible in the Palaeolithic record – the evidence of human dispersal, settlement and cultural character – were strongly shaped by the way early humans responded as individuals and socially as groups to the opportunities and challenges of different environments. It is, therefore, unfortunate that investigation into the extent to which local conditions affected group survival and behavioural development has tended to remain somewhat obscured behind the dominance of European data sets and perceptions. The emerging and older archaeological evidence from Africa is now successfully challenging this (even if findings from this continent are still often defined according to European behavioural criteria). It is with the aim of further expanding such a revision that this book explores the inter-related impact of regionality, behavioural contingency and the effects of attenuated climate change on the localised processes of early human adaptation. It takes as its geographical area of focus that part of the northern and eastern hemispheres that has, despite established local traditions of archaeological investigation, received the least systematic attention with respect to early human behaviour, namely, Asia.

1.2 The Palaeolithic of Asia

Traditionally, Asia is recognised as comprising the vast eastern province of Eurasia. This landscape covers almost 30 per cent of the world's total land surface area, or 44,579,000 km^2. The *Encyclopaedia Britannica* defines Asia geographically as the area east of the Suez Canal and Ural Mountains. Asia is bounded by the Arctic, Pacific and Indian oceans and by the Mediterranean Sea to the west and by the Timor and Arafura seas to the southeast. As well as the mainland, Asia's geographical area also includes numerous islands. Of particular interest in this study is Southeast Asia. During periods of glaciation at higher latitudes and reduced sea level, many of the islands in this part of the region (including Borneo, Java and Sumatra) were incorporated into a single landmass that extended the Asian mainland into the southern hemisphere. Molengraaf and Weber (1921) made early reference to this as the Sunda Shelf and to the emergent continent as 'Sunda Land' (hereinafter 'Sundaland'). In extent, this landmass (and the modern continental shelf) covers *c.* 1,850,000 km^2 (Movius, 1948), or an area equivalent to that of the state of Queensland, Australia.

Prehistoric research in Asia has a long, if still comparatively unintegrated, history. Sections of the Palaeolithic record, in places such as China, Siberia, India, Indonesia and Malaysia (*see* e.g. Barker *et al.*, forthcoming; Bellwood, 2007; Davis and Ranov, 1999; Derev'anko *et al.*, 1998; De Terra and Movius, 1943; Haslam *et al.*, 2010; Heine-Geldern, 1945; Ranov and Davis, 1979; Tweedie, 1953; van Heekeren, 1957; Wu and Olsen, 1985), continue to be separated by expanses of land that are far less well-known, such as vast stretches of central Asia or the Tibetan Plateau. This situation is improving. For example, a rejuvenated programme of systematic Palaeolithic archaeology in the Philippines is making significant strides in prehistoric knowledge in that part of Island Southeast Asia; however, across Asia, local traditions of outstanding research have limited reach to wider audiences because of linguistic, financial and occasionally political constraints. Hard economic realities have also made archaeology a less than attractive or indeed practical career path for young Asians; a situation that leaves the ranks of prehistorians there considerably and consistently thinner than in the countries of Europe or North America. Interest in the value of archaeological heritage is, though, on the rise, and it is beginning to draw governments, business, the general public and academia to this sphere, heralding, it is hoped, a fundamental change in the subject's recognition and potential.

Prior to the second half of the twentieth century, colonial administrators featured prominently in the emergence of prehistoric research in several parts of South and Southeast Asia. Fortuitously, some of these individuals held certification in archaeological training as this was deemed a useful preparation for colonial duties (Dennell, 1990) and probably helped inspire curiosity. As a result, nascent investigations undertaken in these regions kept pace with discoveries and techniques being developed in Europe. For example, in 1883, Robert Bruce Foote, a superintendent of the Geological Survey of India, began pioneering Palaeolithic research at the Billa Surgam caves in the Kurnool District of Andhra Pradesh, India (*see* Haslam *et al.*, 2010); while in 1886 and 1891, the first curator of the Taiping Museum, Malaya (Leonard Wray Jr.), was also actively exploring cave sites of archaeological interest in the Malay Peninsula (Wray, 1897, 1905). Both initiatives were being conducted less than a decade after the first assertions had been published in North America about the Palaeolithic settlement of that continent (Grote, 1877) and slightly before General Pitt-Rivers, one of the pioneers in scientific excavation in Britain, had begun his seminal work at Cranborne Chase in Dorest (1887–98; *see* Renfrew and Bahn, 1991). In Southeast Asia, work was also spurred on by a certain amount of subtle rivalry between the colonial powers. Collings (1938: 575) wrote in the journal *Nature* when announcing the discovery of the first Pleistocene site on the Malay Peninsula – a river terrace site called Kota Tampan – that it was 'with regret' that it had taken so long to find anything, given how close Malaya was to Java, then a Dutch possession.

At the same time as early investigations in Malaya, in Java, the finding of the Wajak skull by van Rietschoten in 1889 (Jacob, 1973) – which would turn out to be that of an anatomically modern human (Storm, 1995) – was closely followed by the historic discovery between 1891 and 1892 of a skull cap, femora and teeth belonging to the archaic hominin *Pithocanthropus erectus* (now *Homo erectus*) by the Dutch anatomist Eugene Dubois (1896). On the mainland, in China, the finding of hominin fossils attributed to the species *Sinanthropus pekinensis* (also now considered to have been *Homo erectus*) at Choukoutien (hereinafter Zhoukoudian) added to the list of evidence for the presence of archaic humans in east Asia and, in conjunction with other finds of the time, including the Javanese evidence, a proposed central Asiatic origin for humanity (Black, 1926). Indeed the formation of the Chinese Cenozoic Research Laboratory – part of the Geological Survey of China and a cornerstone of Chinese Palaeolithic studies – came about as a direct result of the finds and the questions raised by the Zhoukoutien material (Black, 1934). It is worth remembering that at this point in time, all the major hominin fossils were being uncovered not in Africa, but in Eurasia, and most in these two isolated areas of its eastern provinces. Neither of the first classic discoveries made in these locales was met with resounding acceptance (*see* Dubois, 1896; G.E.S., 1934). It is perhaps to a certain degree ironic that one of the most vocal alternate prognoses for the *Pithocanthropus erectus* skull cap from Trinil was that it belonged to a microcephalic modern human (Dubois, 1896); a similar contention, rightly or wrongly, laid upon the recent diminutive hominin remains from Flores (e.g. Jacob *et al.*, 2006).

As well as pioneering Palaeolithic archaeology in Asia, investigations also tended to import foreign attitudes to social change and prehistoric classification, with direct (and sometimes unfavourable) comparisons to the European sequence, none of which particularly helped elevate the profile of Asian archaeological records, their continued investigation or indeed perceptions about their value and need for protection. The concluding comments of Hallam Movius in his 1948 report *Lower Palaeolithic Cultures of Southern and Eastern Asia* are well-known and unintentionally came to typify this position. More positively, though, this same passage also highlighted the uniqueness of the anthropological and archaeological legacy held inside this vast and varied landscape – a feature that continues to bring forth more than its fair share of surprises:

> Southern and Eastern Asia as a whole was a region of cultural retardation. Therefore, it seems very unlikely that this vast area could ever have played a vital and dynamic role in early human evolution, although very primitive forms of Early Man apparently persisted there long after types at a comparable stage of physical evolution had become extinct elsewhere. (Movius, 1948: 411)

While evidence of anatomically archaic hominins of Lower Pleistocene age (*c.* 2.85–0.77 million years ago) continues to be uncovered in Asia – with finds from the Georgian site of Dmanisi principal among the more recent of these – as the age of the Flores material testifies (*c.* 74–17,000 years ago), taxonomic puzzles here are also extending into the much more recent past. New discoveries in the Philippines (Mijares *et al.*, 2010) and Siberia (Krause *et al.*, 2010) point to the possibility of further additions to Asia's ledger of hominins, while the proposed antiquity of modern human remains in China at *c.* 100,000 years ago (Lui *et al.*, 2010; Shen *et al.*, 2002) even challenges, if controversially, the existing model of *H. sapiens* dispersal out of Africa. Much has been made historically of the relationship between *H. sapiens* and *H. neanderthalensis* in Europe; the potential for contemporaneity between hominin species in Asia looks to match, if not exceed, this, with late-surviving archaics, the possibility of precociously early moderns and *in situ* evolution. Although the importance of Africa is now firmly established (and mostly accepted) as the cradle of humanity through fossil and genetic records, whatever role Asia held in the unfolding history of our genus and our species, the steady accumulation of often quite sensational discoveries would attest that it was complex and vital – and anything but a 'backwater'.

1.3 Human Adaptation in the Asian Palaeolithic

The anthropological finds from Asia continue to stir both fascination and controversy in equal measure. As with the general picture of Palaeolithic hominins across this continent, the data, though, still only come from a comparative few islands of understanding amidst a sea of unknown possibilities. Detailed regional reviews of hominin dispersal into Asia continue to be exceedingly rare. Dennell's (2009) excellent *The Palaeolithic Settlement of Asia* and Norton and Braun's (2010) timely *Asian Paleoanthropology* volumes are the latest additions to a slim catalogue; though like almost all other works on this area of the Palaeolithic, their stated concern is with the migration of ancestral species of *Homo* rather than with *H. sapiens*. There is growing interest within archaeology towards our species' behavioural evolution in its regional and adaptive contexts (*see* Brantingham *et al.*, 2004; Burroughs, 2005; Gamble, 1993; Hetherington and Reid, 2010; Potts, 1996b; Rockman and Steele, 2003) and an exceedingly wide literature on traditional interpretations of modern human culture and its Palaeolithic ancestry. What there is not, however, is any concerted attention towards tackling directly the importance and distinctiveness of the Asian record when it comes to these issues. Equally, while there are well-established English-language texts on the prehistory of Southeast Asia specifically (most notably Bellwood's (2007) essential work, *The*

Prehistory of the Indo-Malaysian Archipelago, now in its third edition), the matter of how the first *H. sapiens* adapted to the tropical conditions and changing landscapes of Southeast Asia has not been the subject of a dedicated exposition. Through its coverage, the current volume encourages debate about both these under-represented spheres of Palaeolithic research.

In the first half of the book, discussion of the evidence progresses at its broadest scale to establish a thorough overview of the topic. In this way, Chapter 2 introduces basic features of the global palaeoclimatic record and of climate conditions during the glacial–interglacial phases of the Pleistocene epoch. It also presents evidence concerning the rapidity of climate change during this epoch and its potential impact on hominins at visible, inter- and intra-generational scales. Chapter 3 summarises current knowledge on the initial spread of early hominins from Africa into western and eastern Eurasia and the emergence of regionally specific trajectories of biological and cultural evolution. Chapter 4 expands upon this theme with specific reference to *Homo sapiens*, presenting an assessment of the Middle–Upper Palaeolithic transition in Europe and the evidence surrounding each of the principal components of the behavioural package displayed by modern humans in that record. This evidence is then placed into the context of emerging and independently distinct African and Asian sequences of modern human behavioural development. Chapter 5 considers the genetic and archaeological evidence for the Southern Route Hypothesis, which traces perhaps the most likely path of successful colonization beyond Africa. Whereas hominin dispersal and settlement during the Lower and Middle Pleistocene can be broadly tied to a climatic schedule, records of climate change during the Upper Pleistocene are much more extensive and fine grained. The archaeological record for this period is also chronologically much more precise, though confident correlations between the two streams of data remain, for the most part, only possible at a millennial scale. Nonetheless, increasingly unstable climate and palaeo-environmental change were demonstrably the circumstances under which the pioneering dispersal of *H. sapiens* occurred.

Beginning the second half of the book, the discussion in Chapter 6 becomes more narrowly focused on Southeast Asia and on one time period: the Last Termination to early post-glacial. It is by this point that archaeological chronologies more closely match the resolution of climatic records, making it possible to track and compare rapid climatic, environmental and human behavioural changes. The chapter presents details of those changes that affected Southeast Asia and the technological adaptations in organic and inorganic media that emerged at that time. Chapter 7 explores what is known about early tropical subsistence and settlement practices in this part of the world and compares faunal assemblages from a sample of five sites in Island and Mainland Southeast Asia to assess patterning in early foraging strategies (drawing on the author's own experience working in East and West Malaysia and Vietnam). In Chapter 8, the mechanisms of early human dispersal and adaptation to the local conditions in the region are studied through modelling the dynamics of human colonization. From this emerges a reasoning as to why there would have been distinctly regional trajectories in early human behaviour, how Palaeolithic cultures developed and even how they may have differed from those of the Holocene.

The book makes, ultimately, three propositions. Firstly, it advocates that modern human behaviour is evolutionarily emergent rather than attained – an on-going process, not a faculty restricted to our species' infancy – and that it was subject (though not exclusively) to strong external selective pressures prior to the Holocene and stronger internal selection pressures thereafter. This is not to suggest any kind of sudden shift in the way humanity has been affected by selective forces. There is, for example, growing evidence that climate change has continued to feature during the recent period, affecting society in different ways through the Holocene (e.g. Burroughs, 2005; Hetherington and Reid, 2010). Rather it is to highlight a trend towards the increasing impact of internal cultural dynamics in recent millennia. Secondly, the book proposes that the increasing pace of climatic instability was a premier driver in hominin dispersive and adaptive trajectories since at least the last inter-glacial and perhaps earlier. Thirdly, it argues that regionalism in early *H. sapiens* behaviour developed to a significant degree through the demographic and acclimation processes of colonizing new or remodelled environments under these conditions of climatic caprice.

The simplicity and perhaps overly deterministic nature of these propositions are acknowledged from the outset. The importance of environmental and climatic changes on the development of particularly early human society, though, should not be underestimated (*see* e.g. Finlayson, 2004; Kirch, 1980: 147). This book is one attempt at a synthesis of the growing body of evidence now requiring us to consider what may become a new global reality about our species' past. It argues that the place of Asia within that past can no longer be left ignored, despite the challenges this creates; to do otherwise would very likely result in a seriously biased if not spurious picture of the early phases of the human journey.

The consequences of these propositions are potentially far-reaching, for they also have the capacity to push Palaeolithic archaeology beyond its traditional confines within the discipline of archaeology. The unique perspective that the study of Palaeolithic societies offers on the behavioural evolution of our species is well placed to make a major contribution to the ecologically and climatically oriented paradigm that is emerging within Western science more generally and our need to adapt to modern climate change rather than simply either to ignore or try to somehow halt it. The arguments presented are also, therefore, offered as one tentative step towards extending the study of the human past into the human future; of demonstrating that our humanity arises out of a long, varied and crucially *on-going* journey.

Chapter 2

The Pleistocene Planet

2.1 Introduction

The Earth's climate system experiences many cycles of change. Long-term, stately shifts ultimately stemming from the Earth's orbital geometry and the movement, creation and subduction of the planet's tectonic plates have taken it from extremes of warmth with ice-free poles to extremes of cold with massive continental ice sheets and expanses of polar ice. The earliest and perhaps deepest of Earth's glaciations arose during the Precambrian eras of the Palaeoproterozoic (2.5–1.6 billion years ago (Ba)) and Neoproterozoic (1000–542 million years ago (Ma)). Geological evidence suggests that during these massive glaciations, ice sheets extended to low latitudes (Hoffman and Schrag, 2002) and created planetary conditions that have been described as either 'Snowball Earth' or 'Slushball Earth' scenarios, depending on the proposed completeness of global glaciation (Micheels and Montenari, 2008). Rather than restricting or extinguishing life, however, they may have been instrumental in assuring its diversity. Deglaciation from the second of these deep freezes is linked to the opening of myriad ecological niches and may have played a significant role in paving the way for the Cambrian 'explosion', an array of new multicellular life forms that appeared after *c.* 575 Ma (e.g. Hoffman and Schrag, 2000; Narbonne and Gehling, 2003).

By contrast, the high to mid-latitude Quaternary glaciations commencing after 2.58 Ma were neither as deep nor as extensive, but they too are coincidental with the emergence of a highly diverse life form that would eventually have such planet-enveloping proportions that its impact on the biosphere may warrant its own geological epoch – the 'Anthropocene' (Crutzen, 2002). The glacial and interglacial cycles that characterize the Quaternary emerged as the consequence of long-term cooling, thought to be partly attributable to the uplift of the Tibetan Plateau. The last phase of the Quaternary also marked the climax of a long period of increasingly pronounced global climatic instability. This chapter introduces the essential chronology, mechanisms and pace of deep-time planetary climate change and the accumulating evidence for abrupt climate shifting during the Upper Pleistocene – the context in which our species, *Homo sapiens*, the last of the hominin forms to arise in Africa and the only one to survive, would begin to colonize the planet.

2.2 The Quaternary Period

The glacial cycles that formed the context for the evolution of anatomically modern humans in Africa and their expansion across Asia, Greater Australia and Europe are shown in Figure 2.1 (*see* Gibbard *et al.*, 2005). The Quaternary period (or system) is currently subdivided into two epochs (or series), the Pleistocene and the (present-day) Holocene – though the latter, strictly speaking, represents just the most recent of many warm interludes that have appeared through the course of the Quaternary. It does, however, mark a major stabilization in climatic conditions (Petit *et al.*, 1999); if Crutzen's proposal is adopted formally, the Anthropocene will be the third epoch and the one in which we have been living since the late eighteenth century.

Until 2009, the Quaternary period was officially recognised as commencing *c.* 1.8 Ma. The then unofficial alternative base, favoured by many working in the field of Quaternary science, was more than half a million years earlier than this, at 2.58 Ma. The 1.8 Ma baseline was designated as a Global Stratotype Section and Point (GSSP) through reference to a type geological section in Vrica, Calabria, in Italy in 1984 and following the 27th International Geological Congress in Moscow thereafter. Authorities who favoured the much earlier base argued that the scale of changes observed *c.* 2.58 Ma exceeded those seen at *c.* 1.8 Ma, indicating this to have been, globally, the more important threshold (Anderson *et al.*, 2007). That transition coincides with the appearance of ice-rafted debris – material dropped to the ocean floor from melting icebergs (*see* Section 2.3.2) – at mid-latitudes as well as the start of pronounced, planet-wide changes in sedimentary and biological systems on land. The *c.* 2.58 Ma horizon signalled the intensification of a long-term cooling trend that stretches back deeper into the Neogene (now defined as the most recent period before the Quaternary and starting *c.* 23 Ma) within the current Cenozoic era. The horizon is also commensurate with global geomagnetic stratigraphy that registers a major shift in the Earth's magnetic poles at this time; the 1.8 Ma GSSP was situated close to the end of the Olduvai geomagnetic event – a much shorter-lived shift known as an *excursion*. Following a partition for reassessment of the status of the Quaternary (Gibbard and Head, 2008), as of June 2009, the Executive Committee of the International Union of Geological Sciences (IUGS) formally ratified the 2.58 Ma base (Gibbard *et al.*, 2010). The Quaternary is defined by this new standard herein. The Late Quaternary of this book's title refers to the last 126,000 years, i.e. the present interglacial, the last glaciation and the preceding interglacial (*see* Walker, 2005).

2.2.1 Pleistocene Chronology

The other major chronological divisions used in the present study are based primarily on the Global Chronostratigraphic Correlation Table (POSTERstratchart v2011). The Lower Pleistocene is taken to begin *c.* 2.58 Ma and is the approximate time of the Gauss–Matuyama reversal in the Earth's geomagnetic field (Gibbard *et al.*, 2005). This first part of the Pleistocene epoch lasted until around the time of the Matuyama–Brunhes reversal at *c.* 775 thousand years ago (KBP; Gibbard and van Kolfschoten, 2005). The following sub-epoch, the Middle Pleistocene, covers the interval from this time (*c.* 775 KBP) until *c.* 126 KBP, ending slightly at the Blake geomagnetic excursion (Gibbard, 2003; Gibbard and van Kolfschoten, 2005). For clarity, herein the *Upper* Pleistocene is defined as lasting from *c.* 126 KBP to 11,700 BP – the formally recognised base of the Holocene (Gibbard and van Kolfschoten, 2005; Walker *et al.*, 2009). The *Late* Pleistocene is informally defined as comprising the period *c.* 40 KBP to 11,700 cal. BP, commencing at the Laschamp geomagnetic excursion – ($^{40}Ar/^{39}Ar$) dated to 40 ± 2 KBP (Guillou *et al.*, 2004).

A sizable portion of this book is concerned with a sub-stage of the Late Pleistocene described as the Last Termination. The termination point in a glacial cycle was defined as the mid-point in the rapid warming phase after a final glacial maximum (or Pleniglacial; e.g. Fairbridge, 1972). The Last Termination is defined here more broadly to encompass the period from the height of the Last Glacial Maximum (LGM) to the Pleistocene–Holocene boundary: *c.* 22,000–11,700 cal. BP (*after* Björck *et al.*, 1998; Blockley *et al.*, 2006; *see also* Denton *et al.*, 2010). A standardized dating of the LGM itself remains problematic as different regional records did not necessarily reach a temperature minimum concurrently (Ehlers and Gibbard, 2007). Herein the LGM is taken to span the period *c.* 26,500–19,000 cal. BP, an estimation based on global ice-sheet extent and onset of retreat (*after* Clark *et al.*, 2009).

Calibrated radiocarbon (^{14}C) dates are presented here as cal. BP or cal. KBP (thousands of years) before a 'present' defined as 1950 and are used particularly in Chapters 6 and 7 to standardize comparison between the archaeological and climatic records. Unless otherwise stated, and to help maintain consistency, all quoted calibrated ^{14}C dates were obtained using the Fairbanks_0107 calibration curve to 1 standard deviation (*see* Fairbanks *et al.*, 2005; Appendix, this volume). Uncalibrated radiocarbon dates and estimated dates – including genetic relative dates (*see* Endicott *et al.*, 2009) are given as (lowercase) bp or kbp values. All other radiometric dates (e.g. thermoluminescence (TL) dates), annual increment (e.g. ice-core chronologies), other relative and age equivalent dates (e.g. oxygen isotope stratigraphy) appear capitalized and are quoted wherever appropriate and possible with lab numbers, error margins and the methods used to obtain them (*see* Walker, 2005, for a detailed discussion of different Quaternary dating methods). Finally, the term *Palaeolithic* is reserved exclusively for discussion about Pleistocene material cultural (archaeological) evidence; it is not used as a term with geochronological reference.

2.2.2 Milankovitch Cycles

On the premise that surface temperatures on the Earth should vary according to regular and predictable changes in the planet's orbit and axis of rotation, the concept of Astronomical Theory was published in a book by J. Croll in 1875 called *Climate and Time*. In the early 1920s, this theory was elaborated by the Serbian geophysicist M. Milankovich and has since become central to our understanding of the passage of the ice ages that have characterized the Quaternary.

The effect of solar and other gravitational forces means that the Earth's orbit goes through a cycle from being nearly circular to more elliptical and back again over the course of *c.* 96,000 years – this is referred to as its *eccentricity*. As well as passage through this cycle, the planet's rotational axis describes another: moving *c.* 3° (between 21°39′ and 24°36′) and back again over the course of *c.* 41,000 years – this is the *obliquity of the ecliptic* (the ecliptic being the plane of the Earth's movement around the Sun). Finally, because of the gravitational influence of the Sun and the Moon, the Earth also wobbles slightly on its axis. This has the effect of varying the season when the Earth is closest to the Sun (its perihelion). This cycle, which in fact is probably two closely interlocked ones, takes *c.* 19,000–23,000 years to complete and is called *precession*. In combination, these are known as *Milankovitch cycles*, and they have a profound effect on the amount of insolation received by the Earth and how it is distributed around the globe (Hewitt, 2003). The total amount of incoming solar radiation is largely determined by the planet's eccentricity; the interaction of the other cycles affects how insolation is distributed at different latitudes. Owing to the predictability of these cycles, Milankovitch was able to estimate the receipt of insolation and hence the likely prevailing temperature at the surface during different periods in the Earth's history.

Two obstacles, though, initially hampered the general acceptance of this hypothesis. The first of these was identifying which aspects of the insolation budget were the most critical to climate change. Depending on which latitude and season were adopted, different models could be produced, giving a variety of chronological scenarios. Estimates for the last

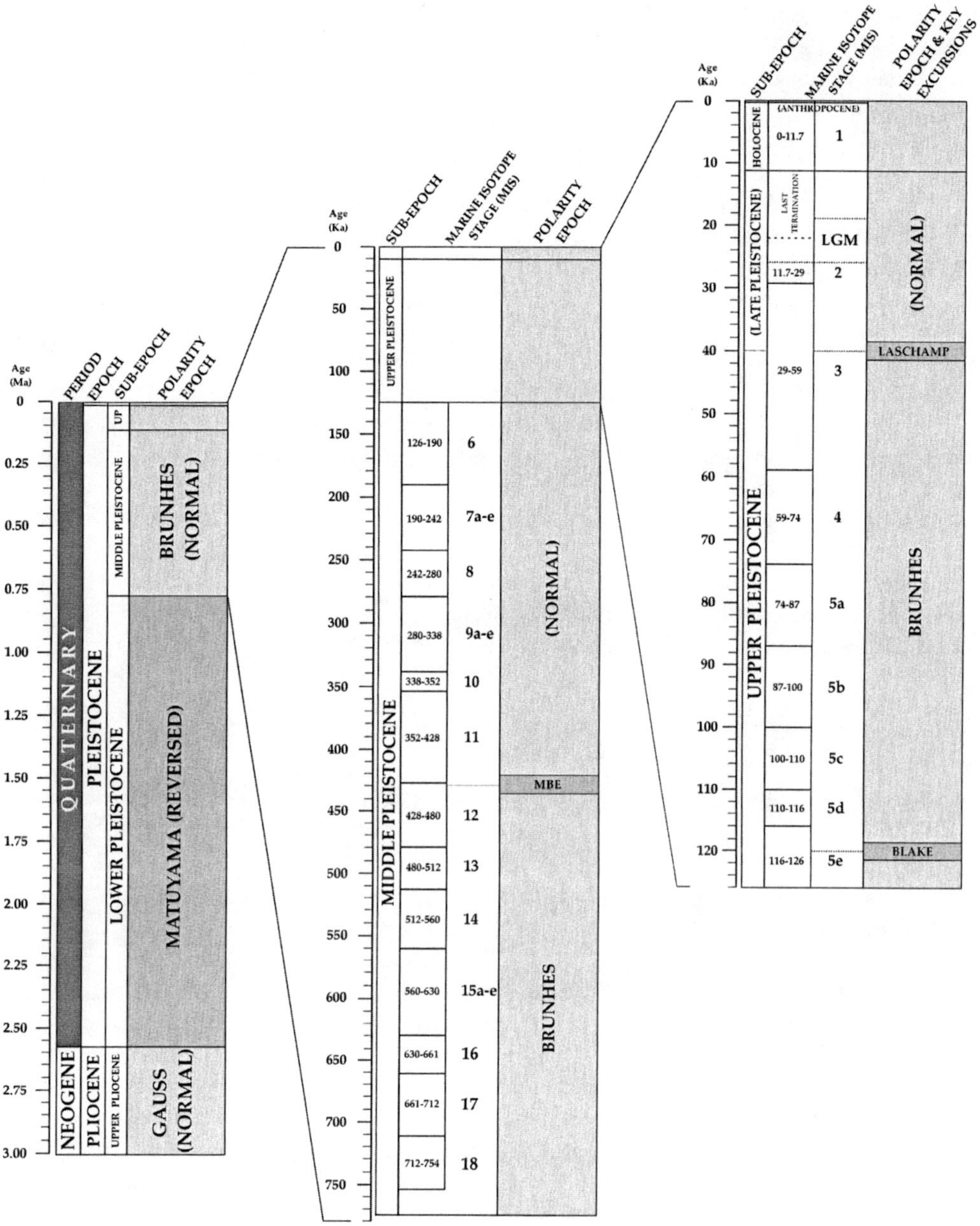

Figure 2.1. Chronostratigraphic correlations for the Quaternary period. Terms vary slightly from those presented on the POSTERstratchart v2011, e.g. 'Epoch' rather than 'Series'; also, the 'Late Pleistocene' is used to refer specifically to the period back to 40 KBP – the time span with which this book is most particularly concerned. 'UP' is 'Upper Pleistocene', 'LGM' is 'Last Glacial Maximum' and 'MBE' is the Mid-Brunhes palaeomagnetic event. In addition to POSTERstratchart v2011, other references used in the creation of this figure include data presented in Andersen *et al.* (2006, 2007), Gibbard *et al.* (2010), Jouzel *et al.* (2007), Prokopenko *et al.* (2002), Shackleton *et al.* (2004), Spahni *et al.* (2005), Spötl *et al.* (2008), Swann *et al.* (2005), Theissen *et al.* (2003), Walker (2005) and Walker *et al.* (2009).

interglacial, for example, varied from as early as 180 kbp to as late as 80 kbp. Secondly, when it came to testing the orbital theory, uncertainties persisted because of geological chronology itself, due principally to the error margins in the then available dating techniques (Hays *et al.*, 1976).

The recovery of continuous cores through deep-sea sediment that would help address these issues was made possible through the invention of the Kullenberg piston corer in 1947. While terrestrial archives of past climates are obtained through a range of proxies – such as pollen sequences or through water, wind or glacially accumulated deposits, ancient soils (palaeosols) or landforms (e.g. caves or fossil dunes) – that remain invaluable in characterising local or regional conditions, they are often discontinuous and can be subject to numerous complicating factors of burial or preservation. Though not immune to such disturbance themselves, marine sequences form under what are generally much less disrupted depositional environments. Marine sediment records can exhibit comparatively steady rates of accumulation and hence provide a reasonable picture of regional conditions. These, together with marine coral sequences, allowed researchers to bring clarity to climatic variability – an advance that provided the first unequivocal evidence of fluctuations that match the 100,000 year eccentricity cycle as well as the *c.* 41,000 year obliquity and 19,000–23,000 year precession cycles in the geological record (Broecker *et al.*, 1968; Hays *et al.*, 1976). Subsequent evidence obtained both though traditional and new proxy records – most notably the high precision of deep continuous ice cores (see later) – added strong support to the hypothesis that changes in the Earth's orbit (orbital forcing) were ultimately a key driver behind climatic fluctuations. Intersecting with these grand cycles are secondary (and indeed tertiary) processes, ranging from the constant tectonic activity of the Earth's crustal plates to very short term events such as iceberg discharges.

2.2.3 Glacial–Interglacial Cycles

The Quaternary is divided into glacial and interglacial stages. The former is typified by intensely cold and arid conditions with substantial ice sheet encroachment from polar to mid-latitudes and within mountain ranges. With the much larger ice build-up, sea level is correspondingly depressed in glacial stages. During the most recent of these, associated with the LGM, sea level dropped to between −116 and −120 m from its modern datum.

Quaternary interglacials, by comparison, represent shorter intervals of glacial retreat (e.g. Figure 2.2), when temperatures and sea levels have broadly paralleled or even exceeded those of today. The existence of interglacials was first recognised in the 1850s, where studies in the Alps identified soil–alluvial deposits sandwiched between two glacial till (unconsolidated debris left by a glacier) deposits. It was out of this relationship that the term itself arose (Fairbridge, 1972). Glaciations in the southern hemisphere appear to have been out of step with those in the northern hemisphere, beginning considerably earlier, principally in Antarctica but also in South America from early in the Neogene period (*c.* 23–2.58 Ma). Although the base of the Quaternary is marked by a shift towards colder conditions, from *c.* 2.6–0.8 Ma, climatic fluctuations were dominated by the 41,000 year precession cycle, and only 14 of the 41 cold stages actually progressed to full glacial conditions (Ehlers and Gibbard, 2007). The transition to the 100,000 year eccentricity cycle began *c.* 1.2 Ma but did not become fully established until *c.* 0.8 Ma. The first major cold events, where ice advanced to extents comparable to those that would typify the later Pleistocene of the northern hemisphere, occurred *c.* 880–870 KBP. While there is broad correspondence between the timing of these cycles around the northern hemisphere, there are local variations and potentially important differences in severity (*see* e.g. Prokopenko *et al.*, 2002; *see also* Chapter 3). The regional development of Quaternary research has meant that terminology is still not standardized globally, with numerous titles being used to describe glacial and

Figure 2.2. Fox Glacier, South Island, New Zealand. Lateral moraines are visible in the foreground. (Photograph by M. Williams and D. Cobbett, reproduced with kind permission.)

interglacial stages (e.g. *see* Anderson *et al.*, 2007, figure 3.3). In many cases preference is now given to using the much more standardized marine isotope stage (MIS) schema (*see* Section 2.2.4). Over the course of the last 800,000 years there have been around ten glacial and ten interglacial stages; over the entire span of the Quaternary, there are now known to have been more than 100 isotopic stages (glacial or interglacial; Walker, 2005). In view of the inter-regional nature of the discussion herein, the MIS standard is adopted throughout wherever feasible, but familiar European terrestrial terms (such as the Younger Dryas) are retained owing to their ubiquity in much of the English-language literature on this subject.

Although glacial and interglacial are still commonly used and remain valid terms of reference, they represent the largest of the climatic cycles within the Quaternary. Nested within these, further subdivisions have been identified. These represent a preferable scale of analysis if one is attempting to relate climate, environment and human-scale activities into a single coherent whole. Early attempts to create an authoritative picture of conditions during glacial and interglacial stages proved extremely difficult principally because of the often fragmentary nature of most long-term terrestrial sediment sequences – an important exception to this is the Chinese loess record (*see* Chapter 3) – and also localised variation between them. The drive to locate undisturbed stratigraphic profiles of environmental change led researchers to extract cores not only from the oceans but also from another area where they would be assured of continual, largely undisturbed and, as it turned out, exquisitely fine-grained deposits: from within the northern and southern hemispheres' ancient ice sheets. The particular attraction of the oxygen isotope record that has emerged from these extreme environments has been its geographic consistency; through which it has been possible to match portions of sequences from different records around the world and establish patterns of global climate change.

2.2.4 Marine Oxygen Isotope Record

Ocean sediments tend to be made up of very fine-grained particles containing a mixture of mineral and biological components (Anderson *et al.*, 2007). Although the relative proportion of each varies from one site on the ocean floor to another, the minerals will typically be formed of wind-borne (aeolian) materials, such as dust, ash and loess, as well as material transported into the ocean from terrestrial river systems. The biological component is dominated by the shell remains of microscopic planktonic (surface dwelling) and benthic (deep dwelling) species of foraminifera (tiny grazing or predatory organisms that belong to the Protozoa subkingdom) that formally lived in the ocean. The occurrence, species diversity and chemical structure of these shells provide insights into ocean circulation and temperature (and from it, atmospheric temperature) during the Quaternary. Among a range of different types of data that can be retrieved from proxy records within marine sediments – for example, aeolian dust can be used as a proxy for changes in terrestrial aridity – species of foraminifera (specifically the oxygen isotope signature within their shells) offers a good record of temperature change.

Given no direct contact with the atmosphere and more or less consistent temperature within the depositional environment, changes occurring in the wider ocean and, by extension, the atmosphere will register as variation in the isotopic character of the calcium carbonate in the shells. The ratio of oxygen isotopes $\delta^{18}O$ (heavy) and $\delta^{16}O$ (light) in the natural environment is, on average, 1:500, or *c.* 0.2 per cent. As foraminifera form their shells, oxygen is extracted from the surrounding water, and this creates the link between the isotopic signature in the shell and that prevailing in the ocean of the time.

Variation in the prevailing $\delta^{18}O$ to $\delta^{16}O$ ratio relates, basically, to the fact that evaporation at the water's surface draws off the lighter $\delta^{16}O$ in preference to the heavier $\delta^{18}O$. This process is temperature dependent: cold air masses are less capable of absorbing the heavier oxygen isotope; moisture deposited from glacial air currents will be markedly deficient in $\delta^{18}O$ compared to $\delta^{16}O$, which in turn becomes a characteristic of the glacial ice. Therefore, during glacial stages, with higher proportions of $\delta^{16}O$ locked into the ice sheets, and the oceans comparatively richer in $\delta^{18}O$, the ratio rises from 1:500 to between *c.* 2:500 and *c.* 6.9:500 (or *c.* 0.47%–1.37%). During interglacials, the trapped $\delta^{16}O$ reserves re-enter the ocean and return values to approximately their current ratio of 1:500.

Although these records have the advantage of providing long and often unbroken profiles, and collectively span the entire Quaternary and beyond, back to *c.* 65 Ma (e.g. Zachos *et al.*, 2001), this longevity must be tempered against a number of limiting factors. For the deep-sea cores, the rate of sediment accumulation is usually very low: approximately 1–3 cm per thousand years. Also bioturbation (e.g. burrowing by deep-sea creatures) has the effect of introducing varying degrees of mixing within the upper few centimetres of sediment at any one time. Over the long term, this does not affect the signature of major environmental and climatic changes in the composition of the sedimentation process, it does, however, tend to blur the small, more rapid changes. As a result, marine sediment cores tend to be most suited to tracking the kind of broader trends that proved so important in establishing the significance of orbital forcing and glacial–interglacial cycles.

Recent core drilling on continental shelves has been able to find a good compromise between time depth and resolution. While most of these sediment cores do not have the temporal span to reach beyond 200,000–300,000 years ago (*see* Denison *et al.*, 2006, for an exception), they remain continuous for that period, and with much higher rates of sedimentation. For example, cores off the coast of Portugal (Roucoux *et al.*, 2006) and Brazil (Arz *et al.*, 1998) have been drilled through sediments with deposition rates of 10 cm per thousand years and 15 cm per thousand years, respectively. With closely spaced sampling and attention to a broad spectrum of proxies along the length of the core, such depositional rates

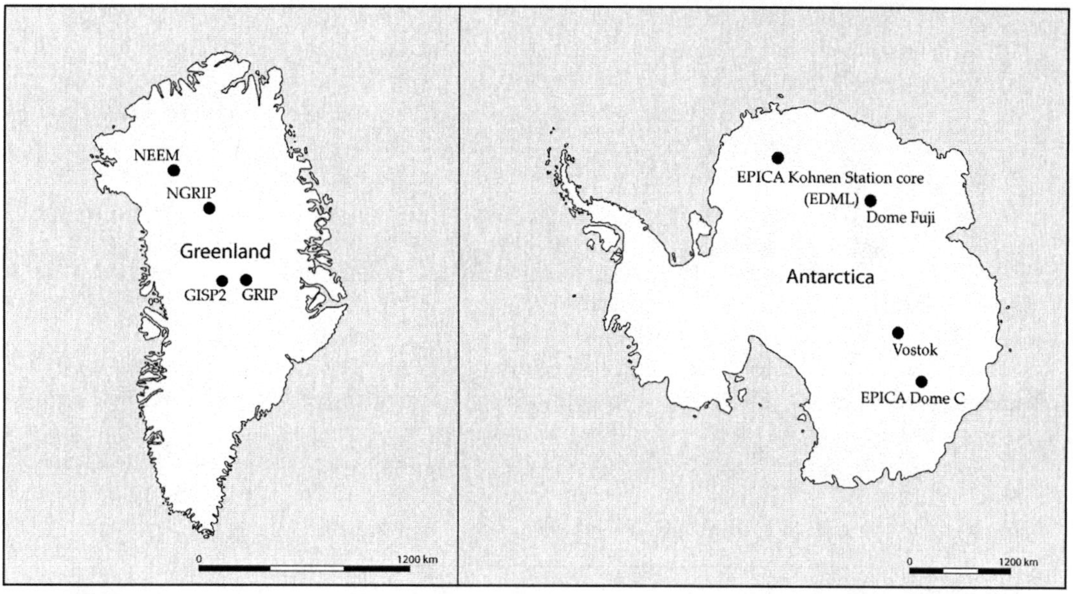

Figure 2.3. Map showing the locations of the Greenland GRIP, GISP2, NGRIP and NEEM ice-coring operations and the Antarctic locations of the Vostok, Dome Fuji, EPICA Kohnen Station (EDML) and Dome C ice core drilling sites. (Maps drawn and prepared by D. Kemp.)

permit millennial–centennial levels of resolution and accurate correlation between changes in ocean conditions and conditions on land.

The very close correlation in oxygen isotopic ratios between different cores around the world strongly suggests that the world's oceans were responding coherently and, over the course of the Quaternary, almost certainly in response to orbital forcing. Broad correspondence between the characteristic peaks and troughs that make up the saw-tooth pattern of individual MIS profiles (see later) allows global designation of isotopic stages and the creation of a principal oceanic-based palaeoclimatic stratigraphy.

In this way, the MIS record provides an independent age equivalent chronology for the Quaternary that can then be anchored using radiometric and astronomical dating techniques. As a result, this schema has been applied not only to marine sequences but also to cores obtained from the equally extreme setting of the polar ice sheets (Rousseau *et al.*, 2006).

2.2.5 Oxygen Isotope Record: Ice Cores

Starting in the 1960s, seven (of numerous) ice core drilling projects undertaken on the Greenland ice sheet reached bedrock, beginning with the Camp Century and Renland cores (Johnsen *et al.*, 2001; Langway, 2008). The more recent (and more well-known) of these are the Greenland Ice Sheet Program 2 (GISP2; Grootes *et al.*, 1993), Greenland Ice Core Project (GRIP; Summit; Dansgaard *et al.*, 1993) and North Greenland Ice Core Project (NGRIP) drilling operations (NGRIP Members, 2004; Figure 2.3). The most recent (2010) is the North Greenland Eemian Ice Drilling (NEEM) coring project.

The American-coordinated GISP2 and the European-coordinated GRIP concluded in 1993 (3027 m) and 1992 (3053 m), respectively. The general location of both these cores had been identified during the earlier 11-year (1970–81) GISP operation. The objective of GISP had been to core the entire ice sheet and included three intermediate-depth cores (Dye 3, Milcent and Crete) ahead of the principal core – drilled also at the Dye 3 location (*see* Dansgaard *et al.*, 1982). Bedrock was reached in 1981 (2035 m).

The NGRIP drilled between 1996 and 2003 (to 3085 m) with the aim of providing the first Greenland record spanning the whole of the last interglacial (MIS-5e) and the beginning of MIS-6. Ultimately, the core terminated at a depth-equivalent age of *c.* 123 KBP, *c.* 7000 years short of this objective, as pressure-induced melting of the ice layers below this prevented further coring (NGRIP Members, 2004). The data that have emerged (and will continue to emerge) from these ice records have transformed our understanding of cycles of climate change, of the low- and high-frequency events within it and of changes in atmospheric composition – knowledge that is now central not only to researchers across a raft scientific disciplines (including archaeology and biological anthropology) but also increasingly to government policy making (e.g. Intergovernmental Panel on Climate Change, 2007).

Analysis of a range of ice constituents – trapped gases, particulates, layer thickness and isotopic character – has revealed a great deal about the Earth's climate history and especially the high tempo of change during the Late Quaternary (Fuhrer *et al.*, 1993). In a similar fashion to that used in marine cores, temperature curves are extrapolated using isotopes of oxygen, hydrogen and deuterium. When water vapour from clouds condenses to form droplets, the ratio of $\delta^{18}O$ to $\delta^{16}O$ is affected by ambient temperature and hence latitude. Snow at higher latitudes carries a low ratio of $\delta^{18}O$ to $\delta^{16}O$ since the heavier $\delta^{18}O$ will already have been condensed out of the air. Glaciers and ice sheets become rich in $\delta^{16}O$ and poor in $\delta^{18}O$ compared to the oceans. Change in the ratio of these isotopes over time provides a good approximation of changes in ice volume and build-up and, therefore, changes in air temperature. It is chiefly because ice cores are not only used in close association with marine core sequences but are also often compared against them that the MIS chronology is also applied to describing ice core stratigraphy. Although the latter do not have the temporal reach of marine sequences, their resolution is in most cases considerably higher, and this has made it possible to identify the existence, amplitude and frequency of abrupt climatic shifts.

Incremental dating of the Greenland ice cores is achievable because many of the constituents found in the cores have marked annual cyclicity (Johnsen *et al.*, 2001). The most immediately discernible of these cycles relates to differences between winter snow and summer snow, which appear as banding in the core. Although errors do creep in and a degree of divergence between core sequences at depth remains an issue (*see* Alley *et al.*, 1997), counting these visible strata provides a very fine grained chronology that can be linked to the core's environmental signals.

Taking one of the principal ice cores, GISP2, as an example, the accuracy of increment dates along its *c.* 3 km length used to decrease from a ± 1 per cent margin for error over a century within the Holocene to a margin for error of ± 2 per cent per thousand years to *c.* 40 KBP and ± 5–10 per cent for periods beyond that (Meese *et al.*, 1997). A new generation of counting protocols are already reducing these margins considerably; they currently run at a *c.* 0.5 per cent error back to *c.* 50 KBP – a level that may eventually extend to the whole of the last glacial period (Shackleton *et al.*, 2004). For younger (particularly Holocene) periods, these chronologies can also be tied to other annual increment chronologies, such as tree ring dating (dendrochronology) and also events of known date such as volcanic eruptions. In a manner not dissimilar to the creation of a dendrochronological sequence, portions of ice cores can be matched in terms of their chemical and particulate (e.g. dust or volcanic ash) composition against other ice core sequences or against signatures in the generally longer marine cores, portions of which may be radiometrically dated (even though these are also subject to their own error margins). For example, Hughen *et al.* (2004) used 280 ^{14}C dates on foraminifera obtained from Ocean Drilling Program (ODP) cores (1002D and 1002E) in the Cariaco Basin, off the east coast of Venezuela, to calibrate the GISP2 core back to *c.* 50 KBP.

Inter-comparison between available ice cores from both polar regions is hampered by the fact that the individually established chronologies of each do not always neatly dovetail right

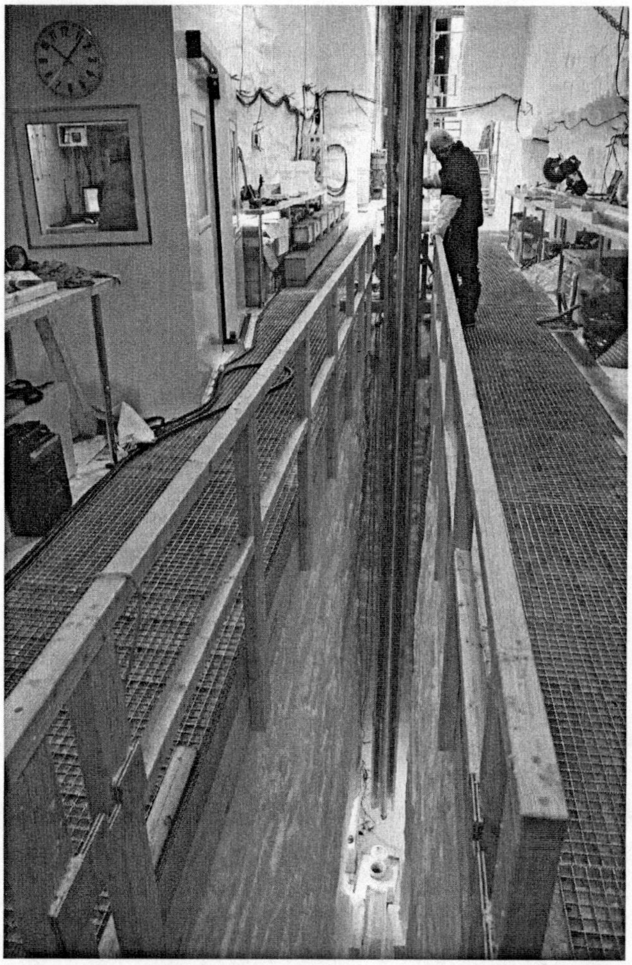

Figure 2.4. The drill hole at the NEEM ice core drilling camp. (NEEM ice core drilling project, http://www.neem.ku.dk; photograph by Sune Olander Rasmussen.)

along their respective sequences partly because of error factors, leaving the drive towards a common ice-core chronology an on-going one (*see* e.g. Rasmussen *et al.*, 2008; Shackleton *et al.*, 2004; Svenssen *et al.*, 2008). The reduction of error margins is just one of the things that is expected to emerge from the (2009–11) high-resolution NEEM project (77.45°N, 51.06°W; Svenssen *et al.*, 2008), which is seeking to establish the first clear picture of conditions during the last interglacial (Figure 2.4). Bedrock was reached on this project in July 2010 at 2537.36 m, and international publication of its results is pending.

While the Greenland cores have pushed back towards the base of the last interglacial, the deepest ice core records have been extracted on the other side of the world, through the Antarctic ice sheet. Where the GRIP and GISP2 cores hit bedrock at a little over 3 km, the Vostok core in Russian Antarctica (Figure 2.3) was halted at a depth of 3.5 km, 120 m above the ice-encased and pristine environment of Vostok Lake. Differences in local conditions and the cumulative effect of ice compaction with increasing depth mean a comparatively short difference in the respective maximum depths of cores can translate to a considerable difference in age. Preliminary data from the Vostok core indicate that it extends back to *c.* 420 KBP and has an accuracy of ± 5000 years for points within the last 110,000 years (Petit *et al.*, 1999). Answering concerns about regional variance between core sequences in Antarctica, Watanabe *et al.* (2003) reported on an isotope study of the Dome Fuji core (drilled 1995–6), 1500 km away from Vostok Station in a different sector of East Antarctica. Their results were in good correspondence and provided excellent support for the validity of the Vostok core; the maximum extent of the Dome Fuji core dates to MIS-9, *c.* 330 KBP.

The most recent and most ambitious work carried out in Antarctica has come through the European Project for Ice Coring in Antarctica (EPICA) initiative. This project has undertaken two explorative coring operations in the last decade. The first of these, the Kohnen Station core, is located on the Dronning Maud Land (EDML) sector of Antarctica facing the South Atlantic at 75°00′S, 00°04′E, 2892 m above sea level (asl). This site was chosen because of the high snow accumulation rates in this part of the continent, reaching 2–3 times higher than at other drilling sites in Antarctica. This permitted a very high level of record resolution, equivalent to that obtained from the Greenland cores, something that has not otherwise been available in Antarctica. Consequently, this core is capable of tracking the same type of millennial to decadal shifts in climate over the last glacial–interglacial cycle – the EDML extends to *c.* 150 KBP (EPICA Community Members, 2006). Drilling at EDML concluded in January 2006 at a depth of 2774.15 m.

The second core, Concordia Station, Dome C (EDC; 75°06′S, 123°21′E, 3233 m asl) was drilled through the central plateau of the East Antarctic ice sheet, *c.* 500 km away from Vostok and to a depth of 3270.02 m – 5 m above bedrock – concluding in December 2004. Its chronology was established through four control windows (EPICA Community Members, 2004: supplementary material). For the first of these windows *c.* 7130 BP, the EDC core was calibrated using volcanic events that had been dated within the Vostok core (themselves matched against dendrochronology records). The second window *c.* 12,400 BP was matched to methane records in the GRIP and GISP2 curves, via the upper part of the 2110 m Byrd Station core in Antarctica (*see* Johnsen *et al.*, 1972; Langway, 2008), whereas the third window was linked to a peak in ^{10}Be (Beryllium 10) of known date (41 ± 2 KBP). For depths greater than 800 m and older than *c.* 50,000 years, where modelling was used to estimate core age, chronological ties were established against the marine isotope curve of Bassinot *et al.* (2003) from the Indian Ocean core MD900963. Disturbed stratigraphy in the basal 80 m of ice compromised the use of this part of the core in palaeoclimatic reconstruction. The EDC core still extends back to *c.* 800 KBP (MIS-20) at a depth of 3190.56 m (Lüthi *et al.*, 2008), yielding by far the deepest ice core climate record currently available, with a resolution greater than that at Vostok because of differences in sampling protocols and a clarity sufficient to pick up the larger of the many rapid climatic cycles that are such a prominent feature of the Greenland cores.

2.3 Rapid Climate Change

The recognition of glacial–interglacial cycles originally carried with it expectations of comparative uniformity within each state and that transition from one to the other would occur gradually over a long period of time. As data have accumulated from ice core records and other sources, it has become increasingly evident that neither contention is correct. Glacial and sometimes even interglacial climate has turned out to be far less stable than was previously thought.

2.3.1 Stadials and Interstadials

It has been known for some time that within glacial cycles there exist periods of extreme cold – as evidenced through palaeo-environmental records and relic landforms indicating the advance of local glaciers. These downturns are defined as *stadials*. The LGM (*c.* 26.5–19 cal. KBP) and the Younger Dryas (12.8–11.7 cal. KBP) both represent examples of stadial

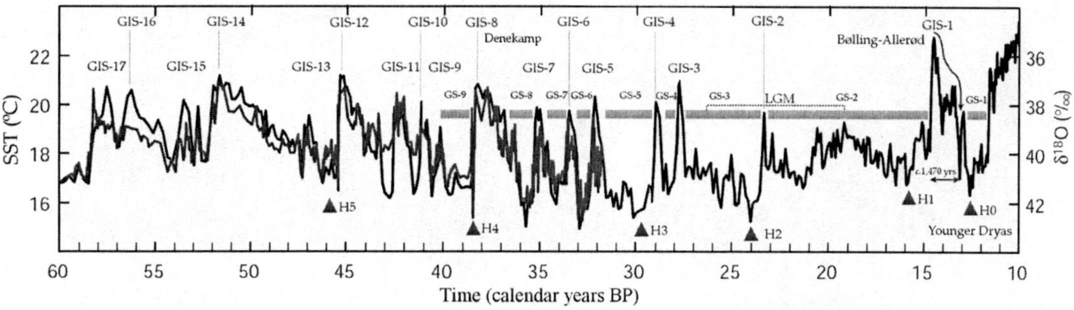

Figure 2.5. The $\delta^{18}O$ GISP2 ice core curve (black) and tropical Atlantic sea surface temperature curve (grey). GIS stages back to *c.* 57 KBP and GS stages back to *c.* 40 KBP, with the most recent named stadial and interstadial phases from the European sequence. One Dansgaard–Oeschger cycle (1470±500 years) is shown, covering the period from the rapid temperature rise of GIS-1 (Bølling-Allerød), steady, and then precipitous decline into GS-1 (Younger Dryas); Heinrich events 0–5 are also illustrated. (Redrawn after Rahmstorf, 2002, with reference to Lowe *et al.*, 2008, and Svensson *et al.*, 2008; LGM after Clark *et al.*, 2009.)

downturns during the latter part of the last glaciation. In the Greenland ice core record, deterioration towards the LGM probably begins within Greenland stadial 3(GS-3). This trend is interrupted by a brief spell of climatic amelioration, before conditions plunge to a final minima during the first half of GS-2, giving a preferred local definition for the LGM of *c.* 23-19 KBP (Mix *et al.*, 2001); a slightly narrower window than the global definition used herein. The Younger Dryas is represented by GS-1 (*see* Lowe *et al.*, 2008; Walker *et al.*, 1999). Some debate continues, though, with respect to the preferred numbering protocols for Greenland stages, so an authoritative sequence on this matter has probably yet to appear (*see* e.g. Wolff *et al.*, 2010).

As indicated, short-term warming trends are also known to have occurred during the last glacial period. They did not generally attain interglacial standing and as such are termed *interstadials*. Examples from the European record include the Denekamp (GIS-8) and Bølling–Allerød (GIS-1) interstadials (Figure 2.5). The character of these interludes in the Atlantic realm is recorded in changes to the structure of European habitats. For example, though warmer, interstadial lowland climax forest was still characteristically dominated by boreal species; compared to during interglacial periods, when large areas of the landscape were recolonized by mixed broadleaf woodlands. The terrestrial record, though, varies according to a number of factors such as altitude, latitude and the duration of the warming trend itself, and recent criticism has been levelled at the credibility of some traditionally pollen-derived interstadial markers in this part of the world (*see* e.g. d'Errico and Sánchez Goñi, 2003).

In the early 1970s, stadials and interstadials were recognised as characteristic features, subcycles, occurring within larger glacial cycles, and their frequency and amplitude became progressively more pronounced towards the end of each glaciation (Fairbridge, 1972). It has been one of the most striking findings to emerge from the Greenland ice core data that, while linked to cycles of glaciological activity, these attenuated fluctuations in temperature do not follow simple, mechanistic feedback loops during each glacial period (*see* Dansgaard *et al.*, 1993). A total of 24 warming episodes have been identified in the Greenland cores, spanning a period *c.* 14–110 KBP. In an effort to standardize chronology (and terminology), most researchers now favour using the high and low-amplitude extremes (the GIS and GS peaks and troughs) in the ice cores' jagged profile of Late Quaternary climate to mark these short-term changes. While the complexities of the mechanisms behind these fluctuations remain to be completely defined, their consequences and periodicity are now reasonably well understood. Two phenomena are central: Dansgaard–Oeschger (D-O) cycles and Heinrich events.

2.3.2 Dansgaard–Oeschger Cycles

The high-resolution ice core records that began to appear in the late 1980s and early 1990s showed that shifts between stadial and interstadial phases during the Late Pleistocene, and across the Pleistocene–Holocene transition, occurred very abruptly. Dansgaard *et al.* (1989) reported that the latter shift was marked in almost all parameters used to study the deep Dye 3 core (i.e. isotope, chemical and particulate proxies), describing a curve that could be matched closely to lake sediment sequences from Alpine Europe. Significantly, the rate of climate change was out of step with and happened more quickly than the rate at which the ice sheets were melting, taking no more than a decade or two. The same kind of delayed (in this case by *c.* 3000 years) response is noted between the climatic minimum of the LGM and the maximum extent of glacial ice build-up (Johnsen *et al.*, 1992). On the basis of the available proxy evidence, the Pleistocene–Holocene transition was likely accompanied by a 50 per cent increase in the rate of precipitation and a 7°C rise in temperature over a time interval of *c.* 50 years – or approximately two human generations.

In an effort to bring an independent line of evidence to bear on this emerging picture of punctuated change, electrical conductivity measurements (EDM) were taken along the upper part of the GISP2 core (40.5–10 cal. KBP; Taylor *et al.*, 1993). EDM values alter according to the balance of acid (e.g. from volcanic, terrestrial and oceanic sources; *see* Bates *et al.*, 1992) and alkaline (e.g. terrestrial dust) compounds within the ice. The method's close correspondence to transitions on the isotope curve from the GISP2 core underscored its reliability as a measure of changing climate conditions. It proved effective at revealing the more prominent millennial-scale shifts, such as the Younger Dryas (GS-1) and Bølling–Allerød (GIS-1), and century-scale fluctuations, including changes taking place within the latter. At finer scales, divergences appear between the two curves due to the fact that each is a measure of a different aspect of climate variability (temperature and particulates, respectively). Nonetheless, the EDM record revealed decadal and sub-decadal fluctuations when the transitional phases between these larger-scale events were analysed closely. Rather than being rapid but smooth transitions, these major climatic reversals were marked by high-frequency oscillations. The record also revealed continuous sharp fluctuations during the course of interstadials themselves, a picture that contrasts with the more stable conditions of the Holocene. During the late glacial, that part of the climate system controlling the proxies that the EDM was measuring, was acting effectively like a 'flickering switch', oscillating between climatic states (Taylor *et al.*, 1993: 435). At this scale, the visibility of such shifts to human groups alive at the time cannot be doubted.

The rapidity of change between stadial and interstadial (and interglacial) conditions was not, however, confined to the very end of the Pleistocene; evidence has come to light that these two climatic conditions are tied to a repetitive, but asymmetric, cycle of colder and warmer phases that stretch at least as far back as the last interglacial. This cycle may even persist in a more muted form, but at the same average pace of 1470 ± 500 years, within the Holocene (*see* Bond *et al.*, 1997). The resolution available in the GRIP, GISP2 and NGRIP ice cores makes it possible to track the course of the Pleistocene component of these cycles. Each commenced with a rapid warming phase of 8°–16°C from previous glacial temperatures over a period of ≤30 years. This was followed by a plateau-like phase of several centuries with only slow cooling, before a more precipitous drop back into cold stadial conditions, which then also lasted *c.* 1500 before the onset of the next cycle. Other, 3000–4500 year intervals also appear to be part of this mechanism, but are less prominent (Ganopolski and Rahmstorf, 2001; Huber *et al.*, 2006; Rahmstorf, 2002). Currently, 24 D-O cycles (*see* Dansgaard *et al.*, 1989; Johnsen *et al.*, 1992) have been identified in the ice core evidence – the most recent spanning the period from the warm conditions of GIS-1 to the cold minimum of GS-1 (Figure 2.5). D-O cycles describe the course of interstadials and are used interchangeably in the modern literature (e.g. NGRIP Members, 2004).

Neither the Vostok nor EDC cores are resolved enough to track the rapid climate change of the Upper Pleistocene D-O cycles. Identifying these patterns is further complicated by the differences that exist between the Greenland and Antarctic records. While the D-O cycles in the northern hemisphere are marked by sudden and high-amplitude temperature change, in the southern hemispheric ice record, similar transitions seem to be slower and temperature gradients less steep (1°–3°C). These records are, however, able to track the millennial scale variability of the longer D-O cycles. Direct Antarctic equivalents to these oscillations – with large millennial scale warming events, known as Antarctic Isotopic Maxima (AIMs) – have now been confirmed in the EDML core (EPICA Community Members, 2006).

An extensive study by Voelker *et al.* (2002), covering 183 sites from glacial, terrestrial and marine settings, has revealed many instances of millennial to decadal signals through MIS-3, indicating that while local characteristics may vary, the extent and tempo of these shifts appear to have been a global phenomenon (*see* Broecker, 1994). The exact cause of D-O cycles continues to be a matter of debate, it is generally thought to be tied to changes in oceanic circulation in the North Atlantic, though its impact clearly extends far beyond this region. Whether such oceanic changes cause these cycles to occur, or vice versa, is not yet clear. One thing is for certain: the longer D-O cycles are closely linked to episodes of iceberg calving (e.g. Huber *et al.*, 2006), so-called Heinrich events.

2.3.3 Heinrich Events

With the aim of elucidating the effects of the Earth's orbital geometry on iceberg calving in the North Atlantic, Heinrich (1988) examined coarse sedimentary layers from a series of cores collected from the Dreizack seamount and the surrounding deep plain of the eastern North Atlantic. He determined that these accumulations had been dropped by melting icebergs that had calved during six episodes between 14,000 and 70,000 years ago. These pulses had a periodicity of 11,000 ±1000 years that appeared consistent with changes in the Earth's precession cycle. Subsequent work confirmed what Heinrich had suspected: that the icebergs were the result of increased discharge from continental ice sheets but also that these events had a significant impact on prevailing conditions.

The sea floor of the North Atlantic is uniformly rich in ice-rafted debris (IRD) sediments (*see* Figure 2.6), but that which was dropped by these particular iceberg pulses carries a notable mineralogical signature. A gradient in the thickness of these layers also appeared from cores drilled in the Labrador Sea of several metres to 1–2 cm in cores from the eastern Atlantic, 3000 km away. These lines of evidence indicated that they were the result of massive iceberg discharges – *Heinrich events* – along the Laurentide ice sheet (Bond *et al.*, 1992; Broecker, 1994) and probably entered the Atlantic via the Hudson Straits. The long distances that these icebergs travelled, together with characteristically low levels of planktonic foraminifera, attested to the fact that during these periods, the North Atlantic waters were unusually cold – a picture confirmed by the greater geographic extent of the Arctic foraminifera species of left-coiling *Neogloboquadrina pachyderma* (Broecker, 1994).

Although dating of IRD layers indicates that Heinrich events are linked to the coldest phase of D-O cycles, the chronology also shows that they did not in themselves trigger the stadial portions of these cycles. It appears rather that they briefly intensified pre-existing stadial conditions ahead of abrupt warming back-up to interstadial conditions (Bond *et al.*, 1993). It has also become apparent that Heinrich events are only associated with the largest-scale stadial–interstadial temperature fluctuations. They appear to be superimposed over and

Figure 2.6. An ice floe from the Breiðamerkurjökull glacier, Jökulsárlón lagoon, southeast coast of Iceland. The dark streaks and staining are sediment (ice-rafted debris) that will settle on the bed of this coastal lagoon as the ice melts. (Photograph by F. Coward, reproduced with kind permission.)

are perhaps linked to a background of smaller-scale but more frequent (i.e. every 2000–3000 years) IRD events, themselves tied to the cool phase of intervening and similarly lower-magnitude D-O cycles (Bond and Lotti, 1995). These sequences of lower-magnitude IRD events and their associated D-O cycles occur over longer periods of time (*c.* 3000–8000 years) and are collectively termed *Bond cycles* (Broecker, 1994).

Although the ultimate cause of Heinrich events remains to be identified, progress is being made on identifying immediate causes. The calving appears probably to relate to ice sheet growth exceeding a critical height, causing it to become unstable. It has been calculated that in each case, as much as 10 per cent of the total ice sheet mass may have slid into the ocean – approximately 1 million km^3 – something that would have led to a short-term rise in sea level of 10–20 m (Bond and Lotti, 1995; Rahmstorf, 2002). The appearance of parallel episodes of discharge in other parts of the Canadian Arctic now suggests that Heinrich events may represent a pan–ice sheet phenomenon (Stokes *et al.*, 2005), whilst similar occurrences have also appeared in the North Sea, the result of calving from the Fennoscandian ice

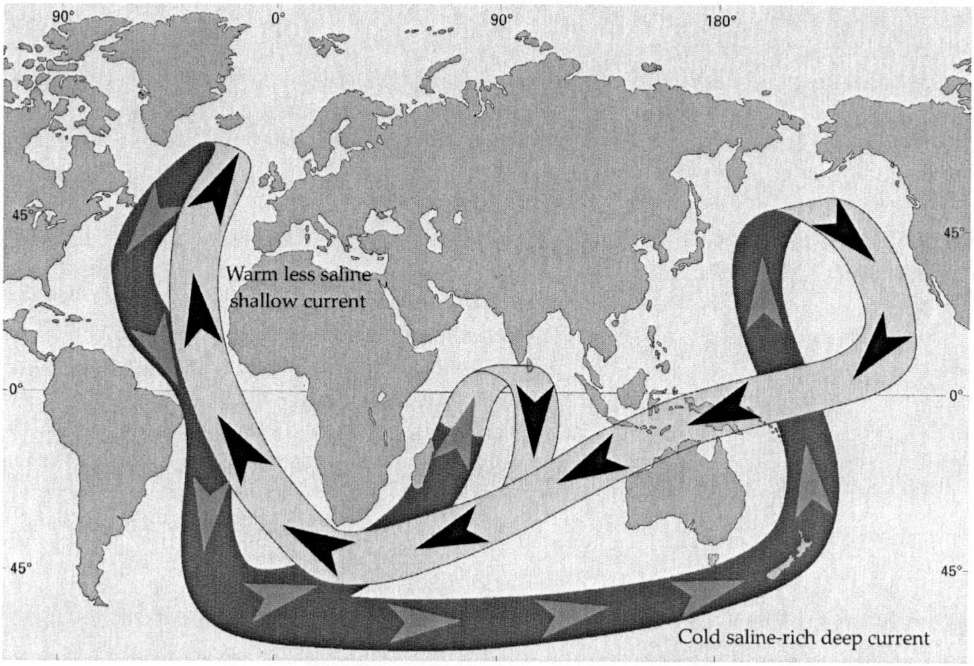

Figure 2.7. Schematic of the global thermohaline circulation.

sheet (Baumann *et al.*, 1995). Unlike D-O cycles, Heinrich events do not appear to have had a noticeable impact on temperature figures recorded in the Greenland cores, though commensurate changes do register in the Antarctic record (e.g. Vidal *et al.*, 1999). One of the major consequences of these iceberg discharges and the associated surge of freshwater into the North Atlantic is the extent to which this disrupted a critical climatic–warm water oceanic conveyor known as the Atlantic thermohaline circulation.

2.3.4 Atlantic Thermohaline Circulation

Ocean currents continually exchange water and heat throughout the oceans and from the surface to the depths, making them a key component in the Earth's climate cycle. The Atlantic thermohaline circulation (THC) constitutes part of this exchange system, wherein warm waters from the Caribbean flow north along the eastern seaboard of North America before heading northeast across the Atlantic towards Iceland (Figure 2.7). This current is a major contributor to the overall heat budget received by the North Atlantic region, currently allowing it to enjoy a milder climate than it should for its latitude. The warmth in this water is gradually exchanged with the atmosphere as it comes into contact with the cooler air of the north; it becomes cooler and saltier, making it progressively denser, until, in the Nordic Seas close to the modern sea ice margin, this layer of previously warm water has become denser than the water layers supporting it, causing it to sink through them. This mechanism forms the North Atlantic Deep Water (NADW) current, which then flows southwest, back across the floor of the Atlantic, perpetuating the northerly flow of warm water in a continuous cycle: the *North Atlantic Conveyor.*

Detailed modelling has shown the THC to be highly sensitive to changes in the North Atlantic freshwater budget; particularly so during the injection of the enormous amounts of freshwater during Heinrich events. By diluting the relative salinity of the northern waters, such injections would have impeded or may have even stalled this part of the conveyor, preventing it from becoming sufficiently dense to descend and circulate back across the Atlantic. The impact of freshwater pulses has been found to be strongly dependent on the nature of background climate conditions (i.e. glacial or interglacial). One of the prominent models to explain this was proposed by Ganopolski and Rahmstorf (2001). Using a simulation that portrayed a glacial, or cold, conveyor mode, they showed that prevailing conditions and the position of the sea ice margin would have meant that the NADW could form off the southern, rather than northern, coast of Iceland. Here, in the open ocean, their simulation suggested that the glacial North Atlantic Conveyor would have been more adaptive to surges in freshwater input than the modern conveyor – for which the formation of the NADW is 'geographically landlocked' within the more northerly Nordic Seas. The result is a cold conveyor system that is actually more stable than the modern warm one. This made the glacial D-O warming cycles short-term destabilizing episodes from which the system could rebound to its comparatively stable state. Under this scenario, D-O cycles occur when the conveyor briefly extends up into the Nordic Seas – characteristic of the modern (and less stable) warm conveyor mode. This temporary shift can be caused by a low-amplitude freshwater injection into the system such as would occur during low-magnitude IRD events. When the high-amplitude freshwater pulses of the Heinrich events occurred, they acted as a shock to the system: following a brief period of shut-down or near cessation (Rahmstorf, 2002), a more intense version of this sequence ensued. In each case, as the system gradually returned to its more stable (cold conveyor) state, the D-O cycle passed through its corresponding cooling-down phase. In essence, the more muted Holocene D-O-like Bond events (Bond and Lotti, 1995; not to be confused with Bond cycles) represent the same D-O cycle but obscured behind the dominance of the warm conveyor mode. Under interglacial conditions, the warm mode, with its sustained penetration of the Nordic Seas, represents one of two stable states that differ from the cold conveyor mode, the other state being 'off' (Ganopolski and Rahmstorf, 2001; Prange *et al.*, 2004) – and the reason why an increase in modern injections of freshwater from polar ice contraction and melting permafrost could have potentially far-reaching consequences.

2.4 Conclusion

Rather than simply putting life into a deep freeze, the Earth's glacial periods may be essential instruments in its creation and renewal. During the Pleistocene, the Earth was subject not only to a major series of long-term glacial–interglacial cycles but also to much shorter-term cycles of change, with transitions between colder and warmer climatic phases often being abrupt and precipitous. It was thought that the role of climate change in the emergence of early human behaviour was to have operated predominantly at a long-term evolutionary scale as a driver of adaptive responses to either deteriorating or improving conditions, but outside the more immediate perceptions of human groups. What has become clear as palaeoclimate records have become more finely tuned is that behavioural responses needed to be geared towards coping not only with prevailing glacial or interglacial conditions but also with changes that were sometimes occurring decade-on-decade and with a much more immediate impact to life. It is now apparent, in fact, that with the exception of the last *c.* 11,700 years, the greater part, if not all, of the Pleistocene history of *Homo sapiens* has taken place under a regime of increasingly marked climatic and ecological instability. What older models and interpretations of human behavioural development have potentially lacked is adequate attention to dealing with the pace of that change. The following chapters provide insight into this matter as they explore the relationship between such instability and regionalism in the trajectories of early human behaviour.

Chapter 3

Hominin Dispersal beyond Africa during the Lower and Middle Pleistocene

3.1 Introduction

The end of the Pliocene epoch (*c.* 5.3–2.58 Ma) marked the transition from a regime of high-frequency but low-amplitude climate oscillations to one of high-frequency and high-amplitude changes and the glacial–interglacial cycles of the Pleistocene. The extent to which changes in Plio-Pleistocene environments affected the emergence of our genus *Homo* and its expansion beyond Africa is still a source of considerable debate. At the present time, there is a broadening consensus that the selective pressures affecting these early hominin groups – such as inter-predator competition, faunal turnover events and adaptation to more arid and increasing open grassland conditions – do not yet sufficiently account for the causes of early dispersals, though each doubtless was implicated in one form or another. This chapter tracks those first expansions and explanations. While consideration is made of the European data, our particular concern is with the wider and earlier movement of hominins into Asia. Emphasis is placed on the timing, tempo, and nature of their adaptation to Asian landscapes. The strategies and mechanisms that enabled anatomically archaic humanity to colonize this great swathe of the Old World foreshadowed the diaspora of early *H. sapiens* and may have helped pave the way to our survival. Passage through the Upper Pleistocene would stretch hominin adaptability to its limit and in the process claim all species in the genus except our own.

3.2 Lower Pleistocene (2.58–0.775 Ma) Dispersals of Anatomically Archaic Hominins

Traditionally, the spread of early hominins beyond Africa has been most closely linked to one species, *Homo erectus*, though the extent of its dispersal through Africa and Asia, and evident differences in its morphology between regions, has courted uncertainty about whether two species are represented by this record. Since the 1980s, some researchers have argued that its earliest African representatives should be recognised by a different species name, *H. ergaster.* The identity of *H. ergaster* rests on a cluster of fossils from 1.8–1.5 Ma, sourced from the

Lake Turkana Basin of northern Kenya. Principal among these (Klein, 1999, table 5.1) are two skulls (KNM-ER 3733 and KNM-ER 3883), a partial skeleton (KNM-ER 1808) from Koobi Fora, East Turkana, and the skull and associated postcranial remains of KNM-WT 15000 ('Nariokotome Boy') from west Turkana (Walker and Shipman, 1996). Klein (1999) places *H. ergaster* as ancestral to all later species of *Homo*, including *H. erectus* fossils from within and beyond Africa (*see also* Dennell, 2009). Antón (2003) quotes an average value for the cranial capacity for the early African erectine adult specimens as 840 *cc*, compared to an average of 630 *cc* for its assumed ancestor, *H. habilis*. This moderate increase in brain size is presented as a possible explanation for the appearance of Acheulean (Mode 2) technology (*after* Clark, 1971) *c.* 1.7–1.6 Ma in Africa. The combination of enhanced encephalisation and this shift in technology had been seen as a possible aid to colonization beyond Africa after this period (Klein, 1999: 291).

That was the position a decade ago. Even at that time, however, complications to the model were arising in the form of claims that *H. erectus* had arrived in Java, Indonesia, as early as *c.* 1.81 ± 0.04 Ma based on the discovery of the partial cranium of a subadult individual from Mojokerto (Swisher *et al.*, 1994). This evidence seemed to place *H. erectus* almost simultaneously in two parts of the world removed by a straight-line distance of *c.* 8500 km, making it very unlikely that there had been a single exit from Africa in the same time period. The Mojokerto date has always been treated with some scepticism, though, given the complex geological circumstances of the fossil's deposition (see later). While the 1994 estimate of its age has now been credibly refuted (Huffman *et al.*, 2006), it is no longer the only uncomfortably early evidence from eastern Asia that has been attributed to archaic hominins. Associated dental and archaeological remains from the Yuamou Basin (Figure 3.1), in southern China, dated by palaeomagnetism to 1.7 Ma (Zhu *et al.*, 2008), continue to fire debate about colonization in the east Asia by this time. The diminutive *H. floresiensis* fossils (*see* Chapter 5) from the Indonesian island of Flores, in the Lesser Sundas chain (Morwood *et al.*, 2004) also continue to provoke taxonomic questions (e.g. Eckhardt and Henneberg, 2010). Although at the present time, they are recognised by the majority of researchers as a non–*H. sapiens* species (e.g. Lieberman, 2009), it is now thought more likely that the Flores population had a common but unknown ancestor with Asian *H. erectus* populations rather than having been derived from this species (Falk *et al.*, 2005).

Farther to the west, the current register of early hominin fossils from the central Asian country of Georgia (see later) is similarly dated to *c.* 1.77 Ma and is considered to be comparable to African *H. ergaster* samples. Meanwhile in Africa itself, two new fossil finds from the Koobi Fora Formation deepen questions about the identity of the earliest migrants. One of these is a very late-surviving habiline fossil (KNM-ER 42703, *c.* 1.44 Ma), indicating a much longer co-occurrence of these hominins with erectines than previously assumed. The other is a very small erectine fossil (KNM-ER 42700, *c.* 1.55 Ma) whose mosaic character parallels that seen in the Georgian fossils, suggesting against the latter being seen as in some way transitional as some have proposed (*see* Spoor *et al.*, 2007). Differences in morphological character between regional erectine populations also imply a degree of divergence that is out of step with the comparatively rapid expansion that the earliest data currently imply (Rightmire *et al.*, 2006). Though the significance of these differences is not agreed upon by all. For example, recent three-dimensional geometric morphometric analysis of overall variation in the shape of the cranial vault of *H. ergaster/H. erectus* (Baab, 2008) finds that the range of variability expressed is consistent with what one might expect within a single species.

There is still considerable uncertainty about just how many hominin species lived during the second half of the Quaternary, their movements, how they were related to one another, to what extent observed variation represents intra- rather than inter-species difference (*see* Kramer, 1993; Miller, 2000). For the purposes of the current discussion, which is primarily concerned with East Asian populations, *Homo erectus* will be the favoured nomenclature.

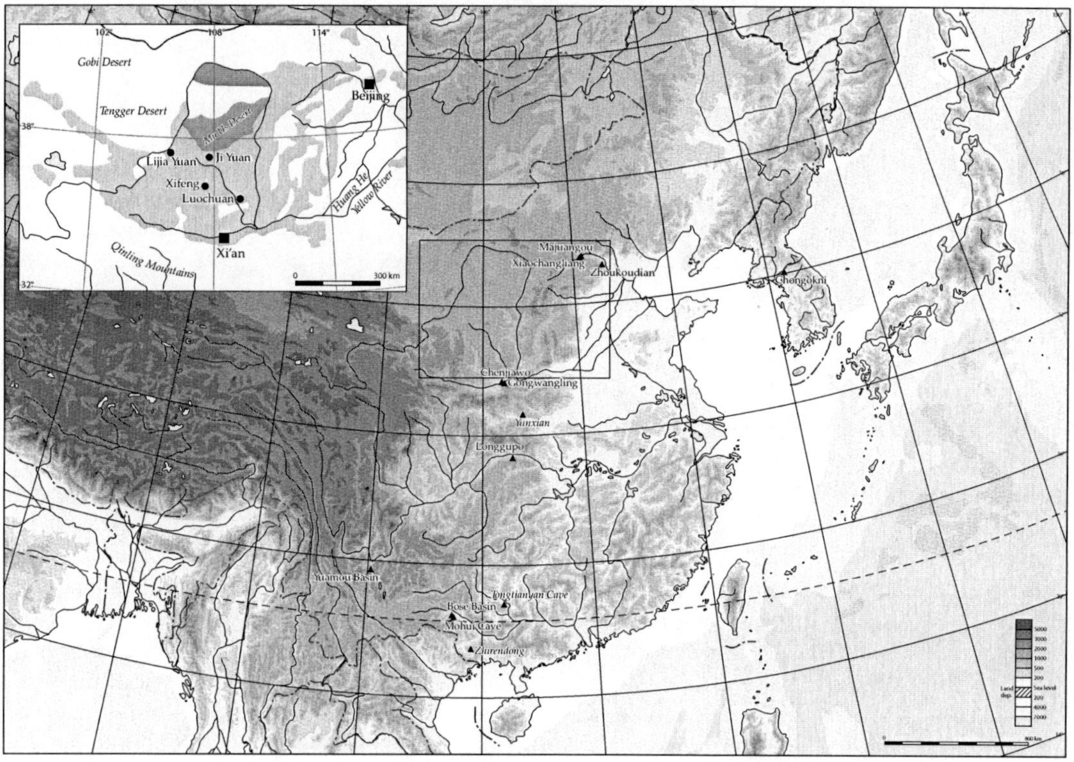

Figure 3.1. Map showing the East Asian Lower and Middle Pleistocene sites discussed in this chapter, together with the location of the Chinese Loess Plateau and key stratigraphic reference sections (inset); for sites in italics, see Chapter 5. (Maps drawn and prepared by D. Kemp.)

3.2.1 Western and Central Asia

On account of its geographic position, one of the most promising areas in western Asia for investigating Lower (and Middle) Pleistocene archaeology is the Saudi Peninsula (Petraglia and Rose, 2009). Work here is in its early stages and based primarily on surface scatters of artefacts, but already survey in areas such as the Wadi Fatimah, southeast of Jeddah and near the town of Dawādmi in central Saudi (Petraglia *et al.*, 2009) is revealing numerous Acheulean locales. The concentration of these suggests that hominin groups were probably exploiting river systems through the interior of the peninsula and possibly as far as the Persian Gulf. The positioning of sites in elevated locations, in particular, would have been advantageous to tracking game movements and for providing sight lines to other potential resource patches out in the landscape. This is thought to have been a transient settlement phase though, strongly controlled by oscillations between more humid, hospitable conditions and much more arid periods (Petraglia *et al.*, 2009: 114). The results of future work from here are awaited with interest.

One of the few sites with faunal assemblages dated to the Lower Pleistocene on the threshold to Africa, in the southern Levant, is that found in the 'Ubeidiya Formation (Tchernov, 1988). This lakeside site is situated on the western side of the River Jordan *c.* 3 km south of the present-day Sea of Galilee (*see* Figure 5.2). Lying as it does on a likely route out of Africa, the 'Ubeidiya locale has considerable importance for understanding the character of that initial expansion. The deeply folded nature of the site's geology means that it provides an accessible and long stratigraphic sequence of lake-shore sediments containing evidence of occupation (Mallol, 2006). Most of the archaeological material occurs within layers of the Fi member, one of four recognised depositional units (Li and Lu are predominantly lacustrine; the Fi and Fu members are predominantly fluvial) that collectively date 1.6–1.2 Ma. Feibel (2004) reports that the Fi member represents a palimpsest of deposits accumulated over a period of *c.* 40,000 years under a range of different environments but all under relatively dry climatic conditions. At the present time, palaeoenvironmental reconstruction has made it possible to show a hominin presence across a range of local microhabitats: from mudflats, marshes and palaeosols to wadi beds and pebbly shorelines. What had been less clear was whether these associations are an indication of actual activity centres and occupation or whether they might be reworked accumulations – a possibility attested by the variable state of preservation of both the lithic and faunal assemblages from the site. The soil micromorphological studies carried out by Mallol (2006) have been able to refine this picture, showing that the abraded portions of the assemblages were probably the result of prolonged exposure on the ground surface combined with localised, fluvial transportation. Her model suggests that hominins were possibly attracted to the lake-shore during the local dry season, both for the access it gave to freshwater and its potential hunting and scavenging opportunities. Under wet season conditions, she proposed, the accumulated anthropogenic deposits were then locally reworked and buried by low-energy fluvial processes.

The lithic evidence from the site is similar to contemporaneous assemblages from Upper Bed II at Olduvai Gorge – there classified as Developed Oldowan B (1.6–1.53 Ma). Comparable core-choppers, polyhedral and spheroids appear in the lowest layers at 'Ubeidiya, with the addition of bifacial elements at lesser or greater frequencies thereafter, leading it to be classified as either Developed Oldowan or early Acheulean, accordingly – a potential indication that different groups were coming to the site (Bar-Yosef, 1994). These data are taken by Bar-Yosef as evidence that hominins may have been exploiting the equitable conditions along the full length of the Levantine corridor, from its starting point in the Afar Rift region of East Africa northwards into the Near East. The nature of this occupation was examined by Gaudzinski (2004) through a study of well-preserved faunal assemblages from 17 archaeological horizons at 'Ubeidiya. Features such as the gnawing of bone shaft fragments and under-representation of epiphyses and taxonomic representation, together with the presence of coprolites and a comparatively high incidence of carnivore taxa, indicate that hyaenids, including the spotted hyaena (*Crocuta crocuta*), were major agents of accumulation in some of the assemblages (though some of the standard criteria used by Gaudzinski to define this have been questioned; Pickering, 2002); in others, these animals seem to have played a lesser role. Differences in the representation of larger-sized taxa certainly point to accumulators with different procurement strategies. In her characterisation of surface modification, Gaudzinski determined a low frequency of robustly identified cut-marks ($n = 16$) in both kinds of assemblage as demonstration of hominin involvement. What she was not able to show was evidence for bone marrow extraction in any of those she studied. This was of particular interest as it contrasted with expectations of the scavenging, marrow-focused subsistence strategy proposed by some for early African hominins (Gaudzinski, 2004: 71). Its significance would clearly depend on how activities were structured at the site and perhaps the season of its use. The alternative picture, which Gaudzinski favours, saw hominins obtaining protein from other sources. In addition to taking small mammals and gathering and breaking open nuts, an absence of marrow bone processing may simply have meant that hominins had greater direct and unchallenged access to carcass meat. The dominance of cervid remains raises the possibility that they were exploiting more wooded environments than has been supposed in Africa, but whether this signifies active hunting remains to seen.

Despite the early hominin presence in western Asia, evidence from its European peninsula is currently limited. What significance this has on the status of hominin occupation is unknown; nonetheless, at the very least, a periodic presence is to be found. One of the recent discoveries from the Sierra de Atapuerca, in Spain, at the site of Sima del Elefante, is a hominin mandible. This is attributed to *H. antecessor* – a possible descendant of *H. ergastor* – and has been dated by palaeomagnetism and cosmogenic nuclides to between 1.2 and 1.1 Ma, making it currently the oldest European hominin fossil (Carbonell *et al.*, 2008a). The find, though, remains chronologically isolated from the majority of fossil (and archaeological) evidence in Europe, which doesn't begin to appear until the Middle Pleistocene. Its existence has led Carbonell and colleagues to propose that the Sima del Elefante hominin, together with, potentially, the *c.* 0.8 Ma *H. antecessor* from the local site of Gran Dolina, may represent part of a separate, very early dispersal of this species of hominin into western Europe armed with a Mode 1, core and flake technology. Such a migration could have preceded the later dispersal of a *H. heidelbergensis* population (itself possibly related to *H. antecessor*) using Mode 2 (Acheulean) tools (Moncel, 2010). While this scenario (and indeed the taxonomic relationship between these hominins) is yet to be confirmed, additional evidence is beginning to surface that does bolster the possibility that hominins were already active in more than one European landscape before the beginning of the Middle Pleistocene. For example, the open-air sites of Barranco León and Fuente Nueva 3 in the Guadix-Baza Basin area of Orce, Spain, have now yielded well-stratified, largely *in situ* faunal and lithic assemblages (Barsky *et al.*, 2010; Martínez-Navarro *et al.*, 1997; Turq *et al.*, 1996). Rather than displaying any features attributed to the Acheulean, the comparable flake-based lithic assemblages from these sites are broadly Mode 1, and adapted to utilise the contrasting natures of local flint and limestone raw materials, introducing a degree of parochial variability. Dated by palaeomagnetism and a distinct faunal biochronology to around 1.3 Ma, these sites are located along the eastern margin of the Baza palaeolake. Hominins as well as other predators probably exploited the fact that this would have been a focal point in the landscape for grazing herbivores. The presence of numerous cut-marked and fractured elements provides corroboration of this likelihood (Barsky *et al.*, 2010). Further echoes of a fluctuating hominin presence in Europe's more northerly latitudes across the Lower to Middle Pleistocene transition have also been attested by discoveries, among other places, at Pakefield, in East Anglia, England and recently through the discovery of a small assemblage of cores, flakes and flaked tools (again with no trace of Acheulean components) from another East Anglian site, Happisburgh, at a date, based on palaeomagnetism and biostratigraphy, of 0.99–0.78 Ma (Parfitt *et al.*, 2010).

Limited but promising evidence has also been lately uncovered in the Taman Peninsula, on the southern coast of the Sea of Azov, Russia. Two sites here, Bogatyri–Sinyaya Balka and Rodniki, have produced lithic artefacts of a local Mode 1 techno-facies in association with faunal remains (Shchelinsky *et al.*, 2010). The culture-bearing deposits are reckoned to date from 1.6 to 1.2 Ma, based on geological situation, palaeomagnetism and biostratigraphy, and as with the north European evidence, they attest to the presence of hominins, even if here too the scale or success of that presence is hard to judge.

The first of the major Asian hominin fossil collections is undoubtedly that from Dmanisi, Georgia (*see* Figure 4.2). The site of a medieval village, which is itself located close to the confluence of two rivers, the locale lies *c.* 85 km southwest of Tbilisi at *c.* 41.31°N, 44.18°E. Systematic archaeological work carried out in the first half of the 1980s determined that house pits and storage cellars had been excavated into volcaniclastic alluvial deposits containing a rich faunal record, including hominin remains. The composition of this record was found to be consistent with what one would expect from around the time of the Plio-Pleistocene transition, paralleling the late Villafranchian fauna of western Europe. The palaeoenvironmental surrounding of the site is now interpreted, partially on the basis of the fauna, as having been a low-lying riparian woodland grading into more open savannah and grassland habitats outside of the valley (*see* Lordkipanidze *et al.*, 2007: supplementary material).

The great antiquity of this material has been mostly accepted by researchers (cf. Klein, 1999) and is secured by radiometric, palaeomagnetic and biostratigraphic dating. The burial context of the principal fossil-bearing deposits (Layer A) appears to be secondary, probably as a result of localised fluvial transport. This layer overlies a lava flow (the Masavera Basalt) that has been dated by $^{40}Ar/^{39}Ar$ to 1.85 ± 0.01 Ma. Given this age, the normal geomagnetic polarity of the Masavera Basalt and Layer A deposits indicates that they fall within the normal Olduvai subchron (*c.* 1.95–1.77 Ma). The high incidence of reversed polarity in samples from the stratigraphically higher Layer B is in line with a return to the prevailing Matuyama polarity (Gabunia *et al.*, 2000; Rightmire *et al.*, 2006).

The recovered technological suite comprises implements made on locally available fine-grained quartzite and basalt and is likened to pre-Developed Oldowan, on account of the lack of any bifacial elements (Vekua *et al.*, 2002). The frequency of tool-marks identified on the accompanying fauna is considered low (*c.* 1%) compared to that seen from European Middle Palaeolithic occupations (of *c.* 2%; Gaudzinski, 2004). It is not, however, far removed from that observed in many of the Southeast Asian Late Pleistocene *H. sapiens* assemblages (Chapter 7), leading one to refrain from equating cut-mark frequency somehow directly with hominin identity, involvement or ability. This assemblage is unusual, however, in that almost one-third of its *c.* 3000 bones are complete (some elements are even articulated), and this might be expected to improve the chances of tool-mark recovery. The presence of cut-marks (*see* Lordkipanidze *et al.*, 2007, whose supplementary figure S5 is a convincing cut-mark; figure S4 is clearly a scrape-mark, whose origin must be considered less certain) does imply that hominins were active participants in the accumulation of the assemblage. Marrow extraction and early access to carcasses are both inferred for some sections of the material but cannot yet be demonstrated with certainty. The presence of carnivore tooth-marks on a proportion of the bones ($n = 8$–9%), together with carnivore bones and carnivore coprolites, is testament to the attentions of other predators. Given the level of bone preservation at the site, this frequency is probably not sufficient to suggest it was, or was related to, otherwise more destructive carnivore denning activity (*see* Pickering, 2002). Although the sandy fossil and artefact-bearing deposits are alluvial in origin (Gabunia and Vegua, 1995), the presence of small bones (including hominin phalanges) suggests against significant fluvial sorting (*see* Dennell, 2005, for a discussion of this taphonomic process in another context). Given the uniformly minimal evidence of weathering or movement, deposition is interpreted as having occurred quickly in a low-energy environment (Vebua *et al.*, 2002), and the material possibly does not represent an extended palimpsest. This is most compatible with a lake-shore setting (e.g. *see* Mallol, 2006), but the actual circumstances of accumulation at Dmanisi remain to be fully unravelled and questions still exist about how this comparatively large, well-preserved assemblage was formed.

Turning to the hominin remains themselves, the inventory presently comprises three associated crania and mandibles (D2282/D211, D2700/D2735 and D3444/D3900), one cranium (D2280) and one mandible (D2600) as well as postcranial remains ($n = 32$ fragments) from a minimum number of four individuals. Although one of these is currently represented by a single metatarsal and is provenanced outside Layer A, two of the remaining three include D3444/D3900 and D2600, respectively; the third individual (D2700/D2735) is an adolescent. The near-complete facial structure of D2700 exhibits potentially significant resemblance to African *H. habilis* (*sensu stricto*; Rightmire *et al.*, 2006; Vekua *et al.*, 2002). Postcranial remains attributed to this skull carry a mosaic of features: some in keeping with selection for greater terrestrial locomotion; others are apparently retained archaic traits. This characterization leaves the sample possibly more comparable with *H. habilis* than *H. erectus/ergaster* though few postcranial remains of *H. habilis* exist anywhere for a thorough comparison. Following the discovery of the robust D2600 mandible in 2000, Gabunia *et al.* (2002) proposed that the

Dmanisi hominins formed a sufficiently divergent palaeodeme – or local breeding population (*see* Antón, 2003: 153) – from the African evidence that they should be assigned to a new and strongly sexually dimorphic species, *H. georgicus* – proposed as a forerunner even to *H. ergaster*.

Although the Georgian fossils are currently ascribed by most researchers to *H. erectus* (or occasionally *H. ergaster*), the differences that are evident in this sample indicate that the first hominins to expand out of Africa did not have the full derived suite of locomotive traits present in comparably dated African specimens or later *H. erectus* (Lordkipanidze *et al.*, 2007). The mixture of traits identified here, together with emerging evidence of sympatry between *H. habilis* and *H. ergaster* in Kenya, highlights both the dangers of overreliance on simple linear dispersal and evolutionary models and the probable complexity in the early Asian fossil record.

3.2.2 Northern and East Asia

Farther to the east, the central Asian site of Kul'dara at Khovaling, Tajikistan (*c.* 38.14°N, 69.84°E; *see* Figure 5.5), offers some of the earliest archaeological evidence for a hominin presence in this part of continent's interior, though the age of *c.* 850 kbp is comparatively late (Ranov, 1995; Ranov *et al.*, 1995). Kul'dara lies at the bottom of a gorge of the same name and contains only a small ($n = 40$) collection of verifiable artefacts. Mostly the assemblage comprises small flakes and flake tools, though two blades and part of a putative bifacial element are described by Ranov (1995). There is no indication of standardisation in form on any of these pieces. While pebble tools do not appear, the knapping technique does not involve core preparation and is still considered to most likely belong to a 'pebble tool' (or Mode 1) industry (Ranov, 1995).

Elsewhere in central Asia, putative bifacial (hand axe) industries have been found at sites in Kazakhstan and Turkmenistan, but evidence for substantial occupation of the region before the Middle Pleistocene is still very sparse and most assemblages are only surface collections. The technology is attributed to *H. erectus* (Shunkov, 2005), though there have been no hominin remains reported to demonstrate this. Beyond Dmanisi, then, the next and oldest Asian fossils are more than 5000 km to the east in China and Southeast Asia.

3.2.2.1 The Nihewan Basin (Majuangou and Xiaochangliang)

The corpus of palaeontological data from the Lower Pleistocene that has been accumulated by more than a century of research in China is sufficiently large that broad patterning and timing of faunal suites has been attempted. Drawing on evidence from sites with hominin or archaeological associations, Han and Xu (1985) reported a significant divergence between northern and southern faunas. Assemblages of a distinctly Palaearctic character appear in the north, while ones of characteristically Oriental fauna dominate at sites south of the Yangtze River (Aigner, 1981: 49; Figure 3.1). Between them lies a transitional zone (stretching from the Himalayas to the lower reaches of the Yangzi River) containing elements of both suites. The northern faunas are typically those from grassland and steppic landscapes and contain a mixture of taxa, some that had survived from the Tertiary, alongside more recent but nonetheless now mostly extinct species. A selection of key hominin and artefact-bearing sites, designated as type assemblages from each of these geographic zones of the Chinese Lower Pleistocene record, is discussed here.

One of the most intensively studied parts of the Lower Pleistocene sequence in China comes from the Nihewan Basin, in the northeast of the country (Shen *et al.*, 2010). The Nihewan Basin is estimated to cover an area of 150–200 km^2 from early survey work in the 1920s in the western part of Hebei Province and eastern Shanxi Province. This area forms part of a 1200-km series of active normal fault-controlled basins called the Shanxi Graben System, which snakes its way to the Qinling Mountains in the south (Li *et al.*, 1998). The geology of the basin's sediments indicates the expansion and subsequent shrinkage of a palaeolake, followed by incising of the landscape by the Sanggan River and eventual capping with (Malan) loess during the last glaciation – a sequence seen to parallel that of Olduvai Gorge (Wei, 2001). The complete history of deposition here spans the late Pliocene to Holocene epochs. In the eastern area of the basin, palaeomagnetic dating based on studies at two sections (the Hongya and Huabaogou) indicates that basal fluvio-lacustrine sediments were deposited just prior to the Gauss-Matuyama geomagnetic reversal, 2.58 Ma, and were underlain by red clays of a composition comparable to those observed on the Chinese Loess Plateau (CLP; Deng *et al.*, 2006; Figure 3.1). Known collectively as the Nihewan Beds, the lower portion of these was, for a long time, identified as the Nihewan Formation, the type sequence of the Lower Pleistocene in northern China. The term is now used to refer to the entire sequence beneath the upper glacial loess. Several sections have been excavated through the course of work in this area (*see* Schick *et al.*, 1991; Wei, 2001); discussion here will be confined to two of the principal sites – Majuangou and Xiaochangliang.

The Majuangou (MJG) section is a 96.6-m thick lacustrine and lake-margin sequence in the eastern portion of Nihewan. It contains four artefact-bearing units: from top to bottom, the Banshan, then MJG I, II and III. These were excavated (with breaks) between 1990 and 2002 by different teams, though initial discovery is credited to Wei Qi of the Institute of Vertebrate Paleontology and Paleoanthropology, Beijing. MJG III is the name adopted by some researchers for a site otherwise known as Goudi (Keates, 2010). All four units are located within a single magnetozone (R2), itself sandwiched between the termination of the Olduvai subchron and the onset of the Jaramillo (i.e. 1.95–0.95 Ma). On the basis of an averaged sedimentation rate, the four units are dated to approximately 1.32 Ma, 1.55 Ma, 1.64 Ma and 1.66 Ma, respectively, making the Goudi/MJG III material currently the oldest cultural material in eastern Asia (Zhu *et al.*, 2004).

The Majuangou lithic assemblages (Banshan $n = 95$, MJG I $n = 111$, MJG II $n = 226$, Goudi/MJG III $n = 443$, from varying sizes of excavated unit) are dominated by core fragments, which exhibit percussive platforms and multiple overlapping negative flake scars. While the range of pieces is deemed comparable to that found in African Plio-Pleistocene contexts, the choice of raw materials does not appear to follow the same trend for immediately local resources, pointing instead to the transportation of stone to this site from some distance away. Although hominin remains have not yet been recovered from the Nihewan Beds, the convincing lithic assemblages, coupled with impact scars and several large mammal shaft fragments within the Goudi/MJG III faunal component ($n = 1014$; Zhu *et al.*, 2004), are taken as dependable evidence of their presence in these lake-edge environments north of 40° latitude from *c.* 1.66 Ma.

The archaeological and apparently associated faunal remains from Xiaochangliang, *c.* 1.5 km southwest of Majuangou, come from a context that has been dated on the basis of palaeomagnetic evidence to *c.* 1.36 Ma (Zhu *et al.*, 2001). The lithic assemblage from this site *is* made on locally available raw materials, mostly of chert and volcanic rock (Dennell, 2009), perhaps suggesting differences in hominin mobility or landscape use to Majuangou. The recovered assemblage, including from less than systematic excavations in the late 1970s and more recent high-resolution work in 1998, comprises 3131 pieces (including 901 from the 1998 excavation). The greater part consists of flake débitage, unmodified flakes and cores; recognised tool types are infrequent (e.g. only seven have identifiable retouch in the 1998 sample) and include a predominance of side-scrapers and notches made using direct free-hand percussion (Shen *et al.*, 2010).

The faunal remains recovered from Xiaochangliang during 1998 comprised 3291 pieces. Although bone surfaces showed comparatively low levels of weathering – based on a

Table 3.1. Identified Mammalian Fauna from Goudi/Majuangou III and Xiaochangliang, Northeast China

Taxon	Common Name	Goudi/MJG III	Xiaochangliang
Rodentia			
Allophaiomys cf. *pliocaenicus*	Vole		X
Mimomys chinensis	–		X
Carnivora			
Felis sp.[a]	Cat		X
Pachycrocuta licenti[b]	Giant hyaena	X	X
Hyaena sp.[c]	Hyaena		X
Martes sp.	Marten		X
Viverra sp.[a]	Civet cat		X
Perissodactyla			
Coelodonta antiquitatis[a,b]	Woolly rhinoceros	X	X
Coelodonta sp.[b]	Rhinoceros		X
Equus sanmeniensis	Horse	X	X
Hipparion sp.[a,b]	Three-toed horse		X
Proboscidipparian sp.[b]	Three-toed horse		X
Proboscidipparian sinensis[b]	Three-toed horse	X	
Rhinoceros sp.[c]	Rhinoceros		X
Proboscidea			
Elephas sp.[c]	Elephant		X
Mammuthus trogontherii[b]	Steppe mammoth	X	
Palaeoloxodon sp.[a,b]	Straight-tusked elephant		X
Artiodactyla			
Bison sp.[a]	Bison		X
Camelus sp.	Camel		X
Cervus sp.	Deer	X	X
Gazella sp.[c]	Gazelle	X	X

[a] Identifications from the 1998 excavation at Xiaochangliang.
[b] Globally extinct.
[c] Locally extinct (*after* Louys *et al.*, 2007).
Sources: Peterson *et al.* (2003), Tang *et al.* (1981), Wei and Lister (2005), Zhu *et al.* (2004).

non-random small sub-sample (n = 123/3291, 3.7%) subjected to detailed taphonomic analysis (Peterson *et al.*, 2003) – 90 per cent of the material proved too fragmentary for confident element or taxonomic identification. The surviving 10 per cent provided six identifiable taxa (*see* Table 3.1). Whereas evidence for hominin involvement in the formation of the Goudi/MJG III assemblage is reasonably well attested, the same cannot be said for Xiaochangliang. Examination of the sub-sample of the 1998 fauna yielded no indication of cut-marks or percussive flaking. Although it was possible to positively identify some damage from carnivore action (gnawing and tooth-marking), the proportions (compared to African datasets) were not deemed high enough to make such action the likely primary accumulating source; rather, the presence within their sample of frequent evidence for sedimentary abrasion and a lack of conjoining fragments suggests that this was at least partly a hydrodynamically transported assemblage (Peterson *et al.*, 2003: 90) and, therefore, potentially of mixed origin.

The composition of the Nihewan fauna, as a whole (taking in evidence from several localities), is representative of a general landscape setting that juxtaposed forests with well-watered grasslands. The mammalian fauna from Goudi/MJG III was thought to be dominated by *Elaphas* sp. Revised identification of this material, though, has proven it to be the remains of a steppe mammoth (*Mammuthus trogontherii*), the immediate ancestor of the woolly mammoth (*Mammuthus primigenius*; Wei and Lister, 2005). The range of mammalian species from this site, combined with the composition from taxonomically parallel Xiaochangliang (Table 3.1), in many respects resembles that of north Eurasian steppic mammalian communities recently described for the late cool interstadial phase of MIS 3 (Markova *et al.*, 2010), perhaps suggesting broad comparability.

In addition to the fossil mammals, ostrich (*Struthio* sp.) has been recovered from Goudi/MJG III. Molluscan and aquatic fauna are also reported as part of the Nihewan Fauna. This includes six genera of freshwater molluscs – *Lamprotula, Cuneopsis, Corbicula, Pisidium, Limnea* and *Planorbis* (Tang *et al.*, 1981). Fish are represented by cyprinids and examples of the ninespine stickleback, *Pungitius* sp. – possibly *P. tymensis* (*see* Keivany and Nelson, 2000). Both of these have a wide ecological range but are cold-water adapted (Rawlinson and Bell, 1982) and are currently found in Arctic continental freshwater lakes and rivers. Tang *et al.* (1981) suggested that local mountainous margins held forests, while the hills, open grasslands and extensive river valleys typified the Canganhe Valley area of the Nihewan Basin.

Although we await confirmation through fossil evidence, the inferred presence of hominins in this region at this time appears to signal the comparative rapidity of their spread from western Asia (if they were *H. erectus*) and an ability to adapt to the cool environments and changing levels of aridity in northeast Asia that followed the onset of glaciation, as well as to the climatic and particularly monsoonal variability (as shown in the loess record) that begun in the late Pliocene (*c.* 2.6 Ma) and extended into the Pleistocene (An *et al.*, 2001; Zhu *et al.*, 2003). It may also show how the productivity of the vast grassland environments of Plio-Pleistocene Asia was a more defining factor than the colder conditions that accompanied movement into these higher latitudes (see later). The high incidence of trampling damage to the bones of the Goudi/MJG III locale may imply that hominins were again attracted to the lake-edge setting as a natural draw for game. It need not mean that groups were there year-round, but it does mean that these early hominins were living effectively in this northern landscape from an early stage.

3.2.2.2 Lantian (Chenjiawo and Gongwangling)

The Lantian locale lies on the southern edge of the CLP, at *c.* 34°N. Palaeomagnetic dating of the hominin fossil-bearing deposits at the sites of Chenjiawo and (10 km to the SE) Gongwangling here, and correlation with a period of palaeosol formation (S6) in the Luochuan loess sequence, suggest that remains at the former of the sites date to *c.* 0.65 Ma, the latter to *c.* 1.15 Ma (An and Ho, 1989; Hyodo *et al.*, 2002). The prevailing local conditions when the Chenjiawo *H. erectus* (represented by a mandible) was alive, were warmer and more humid than currently. By virtue of its extrapolated position in the Luochuan sequence (L15), the older Gongwangling *H. erectus* (represented by a partial cranium with nasal bones, maxilla and part of the upper partial dental row) appears to have lived during colder, more arid conditions – a further mark of the species' wide adaptive tolerance (illustrations of both specimens can be found in Wu and Dong, 1985: 81–2).

The lithic assemblage from Chenjiawo consists of four artefacts – a small scraper and three flakes. The scraper is described as having an edge formed by alternating retouch; all of the flakes are interpreted as having been produced through simple direct percussion (Zhang, 1985). The Gongwangling tool inventory is also small: 20 artefacts are reported in association with the cranium – 11 cores (all made on small cobbles), five flakes and four scrapers, with quartzite as the principal raw material. As at Chenjiawo, direct hard hammer percussion is identified as the main production technique. Ten of the 11 cores are unprepared and exhibit only a single striking platform; the eleventh is described as a polyhedral core. The four scrapers (two apparent side-scrapers and two end-scrapers) comprise the only formal tools (Zhang, 1985).

The fauna associated with the Gongwangling specimen includes forest-dwelling representatives of the southern faunal suite such as *Ailuropoda melanoleuca fovealis*, *Stegodon orientalis* and *Tapirus sinensis*. The faunal remains from Chenjiawo are mostly comparable to those from Gongwangling, with only a few taxa that are of more recent character, such as *Lepus wongi*, *Cuon alipinus* and *Elaphas* sp., as befits its younger age, but unlike Gongwangling, it contains no southern elements (Han and Xu, 1985). Why this distinction exists is not certain. An and Ho (1989) give two possibilities: one taphonomic, the other orographic. Either the fossiliferous beds at Gongwangling derived from a palaeosol-like deposit known to exist within L15 rather than from the loess itself, or uplift of the nearby Qingling Mountains subsequently prevented southern species from reaching the area by the time of the Chenjiawo occupation. Although the Qingling Mountains traditionally mark part of the modern boundary line between the northern Palearctic and the southern Oriental biogeographic regions of China, Tong (2006) has argued against these being an insurmountable obstacle to the passage of fauna. Northerly occurrences of southern fauna are generally restricted to lacustrine or fluvial deposits, and almost none is found preserved in typical loess deposits. It has been suggested that the particular composition of the Gongwangling fauna might stem from the fact that the site is atypically located close to the edge of the CLP (Tong, 2007).

There is a large discrepancy between the number of identified taxa at each site: $n = 41$ at Gongwangling and $n = 14$ at Chenjiawo. This is also not confined to just one or two families; Gongwangling shows greater species richness across all families of mammal except Proboscidea shared by the two sites (Dong *et al.*, 2000; Table 3.2). The pattern of species representation between Gongwangling and a key Lower Pleistocene locale from southwest China, Yuanmou (see later), also shows both a noticeably different species structure and that the latter contains none of the southern suite species found at Gongwangling (Table 3.2) – facts that argue against a simple extension of habitat northwards during the earlier occupation. Finally, recent faunal work by Wang *et al.* (2007) at four new cave excavations in the Bubing Basin area of southern China, and at the taxonomically similar site of Longgupo cave (see later), concluded that changes in species composition within the caves' Plio-Pleistocene to Holocene sequence draws into question the continued use of the traditional *Stegodon-Ailuropoda* fauna as a marker of Pleistocene fossil assemblages in southern China (*see also* Aigner, 1981). The distinctions observed between Gongwangling and Chenjiawo would seem to fit well with this observation.

3.2.2.3 The Yuanmou Basin

The Yuanmou Basin has been the subject of more than 60 years of palaeontological research. The principal hominin, artefactual and faunal finds were recovered from the Niujianbao section, member M4 (Layers 24–28) of the Yuanmou Formation during the 1960s and 1970s. In 1965, excavation of several isolated teeth of a horse (*Equus yunnanensis*) that lay semi-exposed on the surface revealed further teeth; among these, two were hominin: a left and right I^1. Both are described as having been found *in situ* – the one a semi-exposed crown, the other completely buried within the deposit (Qian and Zhou, 1991: 5–6). Both teeth were later described by Chengzhi Hu at the Geology Museum, Beijing, as being morphologically and metrically almost identical to *H. erectus* remains from the later Zhoukoudian fossils.

Table 3.2. Identified Mammalian Fauna from Chenjiawo and Gongwangling, Central China

Taxon	Common Name	Chenjiawo	Gongwangling
Insectivora			
Scaptochirus (= *Talpa*) *moschatus*	Mole		X
Primates			
Homo erectus[a]	Early hominin	X	X
Rhinopithecus lantianensis	Golden monkey		X
Lagomorpha			
Lepus wongi	Hare	X	
Ochotona cf. *thibetana*	Pika		X
Ochotonoides complicidens	–	X	X
Rodentia			
Apodemus sp.	Mouse		X
Apodemus cf. *sylvaticus*	Wood mouse	X	
Arvicola terrae-rubrae	Water vole		X
Bahomys hyposodonta	–	X	X
Cricetulus varians	Hamster		X
Cricetulus cf. *griseus*	Hamster		X
Cricetulus sp.	Hamster		X
Gerbillus sp.	Gerbil		X
Hystrix cf. *subcristata* (= *hodgsoni*)	Crestless Himalayan porcupine		X
Hystrix sp.	Porcupine	X	
Microtus epiratticeps	Vole		X
Myospalax fontanieri	Mole rat	X	X
Myospalax tingi	Mole rat	X	X
Myospalax sp.	Mole rat		X
Petaurista sp.	Flying squirrel		X
Carnivora			
Ailuropoda melanoleuca fovealis	Giant panda		X
Canis lupus variabilis[a]	Grey wolf subspecies		X
Cuon alpinus	Asian wild dog	X	
Cynailurus pleistocaenicus	Cheetah		X
Felis teilhardi	Cat		X
Megantereon lantianensis[a]	Sabre-toothed cat		X
Meles cf. *leucurus*	Badger	X	X
Pachycrocuta brevirostris[a]	Giant hyaena		X
Panthera cf. *pardus*	Leopard		X
Panthera tigris	Tiger	X	X
Ursus cf. *etruscus*	Primitive bear		X
Perissodactyla			
Dicerorhinus lantianensis	Two-horned rhinoceros		X
Dicerorhinus kirchbergensis	Two-horned rhinoceros		X
Megatapirus augustus[a]	Giant tapir		X
Nestoritherium sinense[a]	Clawed forest browser		X
Tapirus[c] *sinensis*[a]	Tapir		X
Equus sanmeniensis	Horse		X

Taxon	Common Name	Chenjiawo	Gongwangling
Proboscidea			
Elephas sp.[b]	Elephant	X	
Stegodon orientalis[a]	Stegodon		X
Artiodactyla			
Capricornis sumatraensis	Serow		X
Cervus (= *Pseudaxis*) *grayi*	Related to Sika deer	X	X
Elaphodus cephalophus	Tufted deer		X
Leptobos brevicornis	Large bovid		X
Leptobos sp.	Large bovid		X
Sinomegaceros konwanlinensis[a]	Giant deer		X
Megaloceros sp.[a]	Giant deer	X	
Sus lydekkeri	Pig	X	X

[a] Globally extinct.
[b] Locally extinct (*after* Louys *et al.*, 2007).
Sources: Dong *et al.* (2000), Aulagnier *et al.* (2009), Marlow (2007), Vislobokova (2009).

Some similarity to African *H. erectus*/*H. ergaster* (and even *H. habilis*) was also noted, though, rendering species-level identification as only tentatively *H. erectus* at the present time. Further excavations in 1973 yielded four stone artefacts – a unifacial scraper, a small bifacial core and two flakes – dispersed over a vertical distance that included the hominin find layer. These artefacts were made on quartz cobbles and are reported as bearing affinity to classic African Mode 1 industries (Zhu *et al.*, 2008).

Palaeomagnetic analysis of the deposits associated with the fossil teeth indicated that they lay within a tiny reversed polarity zone bracketed by periods of normal polarity. Different interpretations of this sequence have been offered. The deposition may have occurred during an excursion in the Gilsa (normal) subchron *c.* 1.64–1.79 Ma or *c.* 1.67 Ma (*after* Worm, 1997; *see* Qian and Zhou, 1991), or during an excursion within the early Brunhes geochron, making it almost a million years younger at *c.* 0.7 Ma (e.g. Hyodo *et al.*, 2002). The most recent assessment (Zhu *et al.*, 2008), based on samples taken in very close proximity to the site, places the finds within a reversal just above the upper boundary of the Olduvai subchron and at an age of *c.* 1.7 Ma. This is also the age most widely accepted by authorities in the field (Dong *et al.*, 2000).

The Yuanmou 'Hominid Fauna' includes taxa from both the M3 ($n = 9$ taxa) and M4 ($n = 32$ taxa) depositional units in the Yuanmou Basin (Pan and Zong, 1991). The M3 taxa were obtained from a different basin locality (Madahai) and comprised a mixture of predominantly forest-dwelling taxa and some open-ground elements (such as *Equus yunnanensis*), with a comparatively rich contingent of carnivores (*Canis yuanmouensis*, *Vulpes* cf. *chikushanensis*, *Pachycrocuta licenti*, *Panthera tigris*, *Panthera pardus* and *Cynailurus* sp.). The M4 taxa, by contrast, showed a clear dominance of Perissodactyla and Artiodactyla species, e.g. $n = 23/35$ (Table 3.3), indicating a shift from more closed to more open conditions during the Lower Pleistocene. The spectrum of mammalian taxa represented in M4 suggests that when hominins were active at the site, the surrounding landscape was made up of largely open grassland, with probably localised patches of forest – a picture that is echoed by the palynological evidence (Zhu *et al.*, 2008). The non-mammalian fauna (including the soft-shell turtle, *Amyda* sp.) indicate that this environment was well watered and probably contained silted freshwater sources such as sluggish-moving rivers or swampy lakes. It is

Table 3.3. Identified Mammalian Fauna from the M3 and M4 Components of the 'Yuanmou Hominid Fauna', Yuanmou Basin, North Central China

Taxon	Common Name	M3	M4
Primate			
Homo erectus yuanmouensis[a]	Early human		X
Lagomorpha			
Leporidae	Rabbit		X
Rodentia			
Arvicola sp.	Water vole		X
Hystrix subcristata (= *hodgsoni*)	Crestless Himalayan porcupine		X
Microtus sp.	Vole		X
Rhizomys sp.	Bamboo rat		X
Carnivora			
Canis yuanmouensis[a]	Wild dog	X	
Cynailurus sp.	Cheetah	X	
Felis sp.	Cat		X
Pachycrocuta licenti[a]	Giant hyaena	X	X
Megantereon nihowanensis[a]	Sabre-toothed cat		X
Panthera pardus	Leopard	X	X
Panthera tigris	Tiger	X	X
Viverricula malaccansis fossilis	Civet cat		X
Vulpes cf. *chikushanensis*[a]	Fox	X	
Perissodactyla			
Equus sanmeniensis	Horse	X	X
Nestoritherium (= *Hesperotherium*) sp.[a]	Clawed forest browser		X
Rhinoceros sinensis[a]	Rhinoceros		X
Rhinoceros sp.	Rhinoceros		X
Proboscidea			
Stegodon elephantoides[a]	Stegodon		X
Stegodon sp.[a]	Stegodon	X	X
Artiodactyla			
Axis cf. *rugosus*[a]	Deer		X
Axis shansius[a]	Deer		X
Axis sp.	Deer		X
Bibos sp.	Gaur		X
Bos sp.	Cattle		X
Cervocerus (= *Cervavitus*) *ultimus*[a]	Asian antlered deer		X
Eostyloceros longchuanensis[a]	Muntjak-related deer		X
Gazella sp.[b]	Gazelle		X
Metacervulus capreolinus[a]	Deer with three-tined antler		X
Muntiacus lacustris[a]	Muntjak		X
Paracervulus attenuates[a]	Deer		X
Procapreolus stenosis[a]	Deer		X
Rusa (= *Cervus*) *yunnanensis*	Eurasian (red) deer		X
Rusa (= *Cervus*) sp.	Eurasian deer		X
Sus scrofa	Wild boar	X	
Sus sp.	Pig		X

[a] Globally extinct.

[b] Locally extinct (*after* Groves and Grubb, 1987; Louys *et al.*, 2007).

Sources: Pan and Zong (1991), Zhu *et al.* (2008).

noteworthy that M3 does not include hominin remains and may be an indication of their avoidance of these more forested habitats.

3.2.2.4 Longgupo Cave

Excavations in the latter half of the 1980s at Longgupo Cave, in the Chongqing Province (formally part of Sichuan Province) of southern China, yielded a mandible fragment containing the crown and root of a P_4, M_1 *in situ* and an M_2 alveolus, and an isolated upper incisor. The primitive form of this dentition has raised speculation that this hominin may have been an *H. ergaster* or even *H. habilis* (Huang *et al.*, 1995; Wood and Turner, 1995). Though found in association with Lower Pleistocene fauna (representing 68 taxa), which again include woodland and open taxa, it is said that this sheds little light on the local environment (Wood and Turner, 1995). Two stone artefacts (a flake and hammer stone) were also recovered, though only one of these came from the same find context as the fossil hominin remains (Layers 7–8); the other came from Layer 5 and certainly does little to clarify the technological picture (Corvinus, 2004). In addition to fauna of Lower Pleistocene character, palaeomagnetic dates bracket the find locale to between 1.96 and 1.78 Ma, while an ESR date on a cervid molar from Layer 4 has produced an upper limit of 1.02 ± 0.12 Ma. Huang *et al.* (1995) proposed that the evidence from Longgupo may go some way to demonstrating that the first hominin to arrive in Asia was not *H. erectus* and that the latter may have evolved outside of Africa.

3.2.2.5 Mohui Cave

Associated with the Bose Basin, in the Guangxi Zhuang Autonomous Region of southern China is the smaller Buding Basin. This has recently produced putative evidence of an early hominin presence at Mohui Cave, one of four sites currently being investigated (Wang *et al.*, 2007). Mohui is located high in one of the karstic towers on the southwest of the basin and has yielded a faunal assemblage containing 28 animal taxa. Among these are two species of porcupine (*Hystricidae*) which, given the assemblage comprises mostly dental material with abundant evidence of rodent gnawing, is suggested as a likely vector of accumulation (Wang *et al.*, 2007: 376). The fauna is said to parallel in character that found at Longgupo, and on this basis a similar age has been proposed (Wang *et al.*, 2005). Included within the fauna are 33 hominoid teeth, two of which were initially claimed to be hominin: a right M_2 (MH001) and a broken left upper molar (M^1 or M^2, MH0018). Both were provisionally assigned to *H. erectus* on morphological grounds (Wei *et al.*, 2005: 2780). Their overall larger proportions, compared to specimens from other Chinese locales, were taken as an indication of greater antiquity. This, though, has been subsequently revised, and neither tooth is currently considered of erectine affinity. Sixteen of the 33 teeth are assigned to *Gigantopithecus blacki*, while the other 17 are now listed as of unclear affiliation but are considered comparable to those from Longgupo (Wang *et al.*, 2007). A small collection of eight stone artefacts was also recovered from different brecciated deposits at the site. These are considered to have technological similarities to those from sites in the adjoining Bose Basin but are not thought to be contemporary with the hominin fossils at Mohui.

In sum, the earliest fossil evidence for a hominin presence in China is heavily reliant on isolated cranio-dental fragments. A recurrent feature of interpretations of these fossils is that they are of insufficiently distinct character either metrically or morphologically to assign them categorically to *H. erectus*. This in itself raises certainly the intriguing possibility of pre-*erectus* dispersals into the Asian continent and of *in situ* evolutionary lines of descent. But much yet rests on the identity of these few pieces, and some researchers have begun to question if the fact that they tenaciously and repeatedly defy easy classification may not have another explanation (Dennell and Roebroeks, 2005; Schwartz and Tattersall, 1996). One of the original authors on the Longgupo discovery, Russell Ciochon, now sees reason to reassess his perspective on that and other related material such as the Mohui Cave teeth. Ciochon (2009) argued that the possibility must be at least entertained that these fossils belong not to hominins at all but to another extinct but as yet unclassified genus of ape that would have been contemporary with the giant Asian ape *Gigantopithicus*. Until this line of investigation can be ruled out, the Lower Pleistocene hominin fossil evidence from China is probably best treated cautiously though it is, nonetheless, supported by the presence of small collections of lithic artefacts. This may leave, for the present, the earliest confirmed evidence of *H. erectus* in eastern Asia as coming not from the mainland but from Island Southeast Asia (ISEA).

3.2.3 Southeast Asia

The large amount of fossil evidence for *H. erectus* in Asia comes from Java, commencing with the historic finds of the then *Pithecanthropus erectus* made by Eugene Dubois at Trinil (Dubois, 1896). The earliest colonization of Java had been posited at as early as *c.* 1.81 ± 0.04 Ma, from the dating of deposits thought to be associated with the Mojokerto cranium (Swisher *et al.*, 1994). Though widely quoted, this age has tended to be viewed with caution, not least because of the date's chronological proximity to the probable genesis of *H. erectus* in Africa. This fact, coupled with the ongoing uncertainty about just how robust the association was between skull and date, prompted a re-examination of the evidence. New fission-track dating of pumice deposits in the vicinity of the find spot do not support the 1.81 Ma age (Morwood *et al.*, 2003). Two separate horizons were redated: one, Pumice Horizon 6, located 15 m stratigraphically above the find area, 1.43 ± 0.1 Ma, is taken to represent a minimum age for the skull, though there is possible post-depositional mixing here, and the authors favoured the second of their dates as the most reliable. This was taken on a horizon 13 m stratigraphically below the find area (Pumice Horizon 5). These deposits show much less evidence of reworking and were interpreted as representing a new flush of volcanic material into the area. The date obtained on Pumice Horizon 5 was 1.49 ± 0.13 Ma, making this the new maximum likely age of the skull. After an exhaustive piece of detective work reconstructing the recovery site itself from original excavation records and photos, Huffman *et al.* (2006) successfully determined where the cranium was most likely found: at a location that turned out to be *c.* 20 m stratigraphically above the deposits dated by Swisher and colleagues and very close to the Pumice Horizon 6 dated location. While it is reasonably clear that the skull does not now date from *c.* 1.81 Ma, sufficient sources of ambiguity remain to make it likely that the new age range between *c.* 1.49 and 1.43 Ma will probably still be subject to some future revision (Huffman *et al.*, 2006).

A total of approximately 80 fossils, attributed by most to be those of *H. erectus*, have been found in the Sangrian Dome region of Java – 50 of these have known find spots. As with Mojokerto, however, obtaining secure dates from the complex deposits in this seismically active region continues to be fraught with difficulty. Currently, the accepted age of this material ranges from *c.* 1.66 to *c.* 1 Ma. This includes associated dates from five fossil find spots purportedly where the stratigraphy is less subject to deformation, and where dates fall between *c.* 1.5 and *c.* 1 Ma (Larick *et al.*, 2001).

An additional source of lively debate over the last few years has been the ancestry of the diminutive hominin remains from Flores (Aiello, 2010; *see* Chapter 5). Though exclusively Upper Pleistocene in age (*c.* 95–17 KBP), the type fossil of *H. floresiensis* (LB1) has been ascribed to a range of taxonomic positions (over and above microcephalic *H. sapiens*, which most now discount on postcranial as well as cranial grounds). The apparent combination of australopithecine and early *Homo* morphological features that this hominin exhibits currently suggests against it being a direct descendant of *H. erectus* (Argue *et al.*, 2006; Morwood *et al.*, 2005), and the possibility of relatedness to a more primitive hominin has been raised (Jungers *et al.*, 2009; Morwood and Jungers, 2009). This, together with genetic evidence now emerging of a shadowy archaic hominin population (the *Denisovans*) that potentially ranged from southern Siberia all the way to ISEA (Krause *et al.* 2010; Reich *et al.* 2010, 2011) serves to underscore just how little is known about pre–*H. sapiens* dispersals – not least that *H. erectus*/*H. ergaster* may not have been the only hominin migrating from Africa during the Lower Pleistocene. The recent announcement by Brumm *et al.* (2010) that lithic artefacts have been found at the site of Wolo Sege in central Flores beneath an ignimbrite, $^{40}Ar/^{39}Ar$ dated to 1.02 ± 0.02 Ma, will also continue to fuel speculation about the identity of the early hominin colonists of this island and how they got there.

3.2.4 Environmental and Climatic Conditions during the Lower Pleistocene

Addressing the mechanisms behind the Lower Pleistocene hominin dispersal(s) is essential to placing them in context. Species expansions are often linked to interglacial phases, where climate forcing and variability is considered a major contributing factor (e.g. deMenocal, 2011). A series of hominin speciation episodes in the early African record certainly shows correspondence with large-scale faunal turnover events – themselves likely the result of climate and habitat change. *H. erectus*, 'as a species', emerged during a period of comparatively high faunal and environmental variability in association with a major expansion of African grassland-adapted species *c.* 2 Ma (Bobe and Behrensmeyer, 2004). The environmental picture from Asia during this period is only now beginning to take shape; informed particularly by the long-term terrestrial records that have been recovered from the Central Loess Plateau in China, but also in the context of much more punctuated events.

3.2.4.1 Central Loess Plateau

Loess comprises wind-blown (aeolian), predominantly silt-sized particles (i.e. a grain size 2–64 μm in diameter; Walker, 2005). Quaternary loess deposits are formed where fine particles of wind-eroded glacial till or desert sands are lifted aloft and carried downwind from their source to where they form deep, loosely consolidated sequences. Large amounts of loess have accumulated across areas of North America, Europe and Asia throughout the Quaternary as well as in parts of Africa, South America and Australia. These regional sequences now provide a valuable source of palaeoenvironmental information – one that can be correlated with oceanic sediment cores (Anderson *et al.*, 2007). Within Asia, palaeoenvironmental analysis of the vast loess deposits of north central China are steadily accumulating in detail, making it possible to contextualise Plio-Pleistocene fossil and archaeological evidence with far greater clarity than was once possible.

The CLP covers 450,000 km^2 in the middle reaches of the Yellow River, northern China (Figure 3.1). Grading in particle grain size across the plateau suggests that the source areas of the loess deposited here over millennia are the inner Asian sand and stony deserts. Dust from these deserts is still captured and blown southeastwards on the northwesterly monsoon winds, with the finer grains settling out over the plateau (Huang *et al.*, 2000; Roe, 2009). The loess deposits, which are built up during arid glacial periods, are inter-stratified with palaeosols produced during more humid conditions of reduced dust transport and accumulation. Sections through the CLP show it to be *c.* 150 m thick but occasionally as deep as 300–500 m. Shifts in the palaeomagnetic properties of these deposits enable them to be tied to the Earth's known geomagnetic stratigraphy, giving a sedimentary record that stretches back to at least 8 Ma (Bai *et al.*, 2009) and potentially the last 25 Ma (Roe, 2009), making this the longest such sequence in the world. Most areas of the CLP are underlain by reddish silt-clay sediments (also of partly aeolian origin) that are of Pliocene age. The boundary between these red clays and the yellow loess has been established by palaeomagnetism to be 2.4–2.5 Ma (Liu and Ding, 1998), in keeping with the base of the Quaternary. In the last couple of decades, efforts have been stepped up using a range of proxy indicators (e.g. geophysical, geochemical and sedimentological) to exploit the CLP sequences to characterize long-term changes in the East Asian palaeo-monsoon.

The modern East Asian monsoon produces dry, cold conditions when winds are north-westerly (winter) and humid conditions when they are coming from the southeast (summer). Changes in the magnetic susceptibility of the deposits and changes in grain size have proven the best proxies for measuring the strength of the summer and winter monsoon, respectively (Lu *et al.*, 2000), creating a window in the long history of this hugely important component of the planetary climate system. The record that has emerged shows evidence of long-term fluctuations through the late Pliocene and the Pleistocene (Bai *et al.*, 2009). These are ultimately driven by variations in the Earth's orbit but indicate that the East Asian winter monsoon has been stronger under glacial conditions and weaker during interglacials. For the summer monsoon, the reverse is true: stronger during interglacials than during glacials. In addition to these larger cycles, secondary oscillations have also been identified within some of the more major soil and loess units. These shorter-term pulses of change, representing increases in the strength of the winter monsoon (through increases in grain particle size), have been found to closely match the sequence of Heinrich events in the North Atlantic, though fine detail, such as associated with Dansgaard–Oeschger cycles, remained elusive at first.

Two geological type sections located in the northern part of the CLP (in Lijia Yuan and Ji Yuan) were found to contain rates of deposition 2–4 times greater than that seen generally in the central part of the plateau (Liu and Ding, 1998). Covering the last glacial–interglacial cycle, high-resolution (sampling every 2 cm) magnetic susceptibility and grain-size studies produced sufficient resolution to reveal the presence of rapid, irregular oscillations at a millennial scale occurring within a longer-term trend in grain-size change and hence winter monsoon strength. This latter feature placed the monsoon system in a close relationship with evolving changes in global ice volume. With few exceptions, the rapid shifts identified in these sequences can be matched closely to the Upper Pleistocene D-O cycles evident in the Greenland cores (to an error margin of <1000 years), providing further evidence that these short-term climatic swings were a global rather than regional phenomenon (Leuschner and Sirocko, 2000; *see* Xia *et al.*, 2007, for similar monsoonal fluctuations in stalagmite data from Shanboa Cave, Shennongjia, central China). Millennial and sub-millennial fluctuations have now also been determined at other classic loess sequences, such as at the Luochuan and Xifeng sections in the central CLP, and for periods extending well beyond the last interglacial (Lu *et al.*, 1999, 2000) – for example, nine such oscillations have been recorded for the period between 865 and 840 KBP. This strongly suggests that early hominin populations in Asia would have been subject to long-term, climatically driven adaptive pressures (Dennell, 2009); but the potential for environmental selection would not have ended there.

3.2.4.2 Volcanic Eruptions

The possibility that instantaneous events in the Earth's recent geological history created population bottle-necks and, in their wake, colonization opportunities is a serious consideration. Two such major events occurred close to the transition between the Lower and the Middle Pleistocene: the Oldest Toba Tuff eruption and the Muong-Nong/Austalasian tektite strewn field event; we shall consider each with respect to its potential significance on hominin populations living in the region.

The Toba Caldera complex is located in northern Sumatra (Figure 3.2) and measures a colossal 100 km by 60 km, making it one of the largest Quaternary calderas on the planet

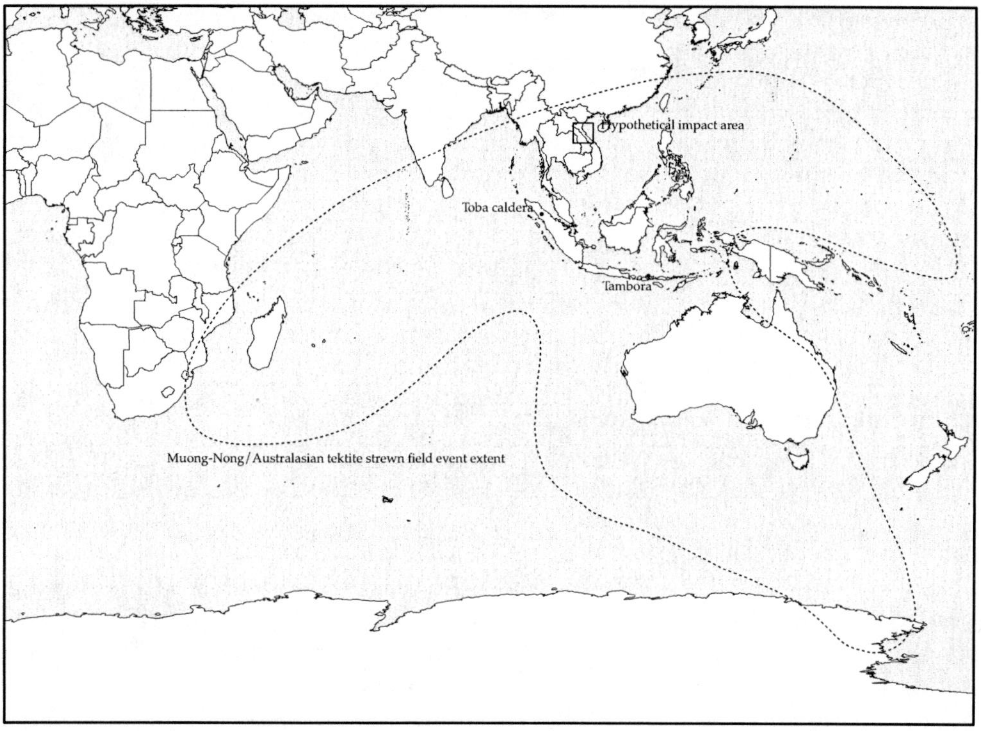

Figure 3.2. Map showing the locations of the Toba and Tambora volcanoes and the distribution of the Australasian tektite strewn field (dashed line); the square represents the hypothetical location of the source crater of the MN/AT aerial burst. (Redrawn *after* Folco *et al.*, 2010; map drawn and prepared by D. Kemp.)

(Chesner, 1998). Like other volcanoes dotted along the length of Sumatra, Toba owes its existence to the major tectonic forces that are subducting the Indian-Australian plate beneath the Sunda plate along the boundary known as the Sunda Trench. The caldera complex is the result of four separate pyroclastic eruptions that have taken place during the Quaternary: the Haranggoal Dacite Tuff (HDT), *c.* 1.2 Ma, the Oldest Toba Tuff (OTT), *c.* 840 KBP, the Middle Toba Tuff (MTT), *c.* 501 KBP and, most recently, the Youngest Toba Tuff (YTT), at *c.* 74 KBP. Both the HDT and MTT eruptions were comparatively small, ejecting *c.* 35 km^3 and *c.* 60 km^3 of dense-rock-equivalent (DRE), but were still in the same league as the 1815 Tambora eruption (Lesser Sunda Islands), the largest in recent history, which displaced *c.* 50 km^3 DRE (Chesner, 1998; Chesner *et al.*, 1991). The OTT and YTT were orders of magnitude more powerful, ejecting at least 500 km^3 and 2800 km^3 of DRE, respectively. This was the picture until recently.

The accepted date of *c.* 840 KBP for the OTT (Diehl *et al.*, 1987) has now come in for reappraisal on the grounds that the related tephra may not in fact have been dating this eruption. Correlating evidence from three oceanic cores in the South China Sea and one in the Indian Ocean, and using the distinct chronological markers of the Matuyama–Brunhes boundary and the microtektite layer of the Muong-Nong event (see later), Lee *et al.* (2004) relocated the OTT to just within the Matuyama geochron. In the Ocean Drilling Program (ODP) site 758 reference core (from the Indian Ocean), this links the eruption to tephra found in Layer E rather than Layer D, as previously thought. In which case, the OTT eruption falls during the deglaciation between MIS-20 (801–778 KBP) and MIS-19 (778–754 KBP) and, after astronomical calibration (based on Milankovich cycles), has been dated to between 788 ± 2.2 KBP (Lee *et al.*, 2004: 62) and another extrapolated age of 800 ± 5 KBP calculated by Glass and Koeberl (2006).

Lee *et al.* (2004) further proposed that the scale of the OTT event was larger and potentially more widespread in its effects than previously assumed (but still only around one-third that of the YTT eruption). The eastward fallout range, though, does appear to have significantly exceeded the distribution range of YTT. Evidence from ten marine sediment cores (Glass and Koeberl, 2006) in addition to those of Lee *et al.* (2004) confirms the presence of ash bearing the same chemical signature in the central Indian Ocean and *c.* 6900 km east of Toba in the western Pacific – a coverage area far greater even than Lee and colleagues had predicted. Without a clearer picture of prevailing atmospheric conditions between the two big Toba eruptions, though, these data remain suggestive rather than compelling.

The OTT revision has sparked debate within the geological community (e.g. Shane *et al.*, 2004; Chen *et al.*, 2004), but the consequences for early hominin movement and settlement in South and Southeast Asia, if it is correct, have yet to be fully appreciated within the archaeological community. One example of its potential impact (quite aside from survivorship of the eruption's immediate effects) is that it may have dating implications for archaeological and tephra associations in South Asia, which currently rely on there being only one Toba event recorded in the subcontinent: YTT (Lee *et al.*, 2004; cf. Westgate *et al.*, 1998). Also lying just within the Matuyama geochron and calculated as a mere 3.2 ± 1.5 KBP years before OTT (Glass and Koeberl, 2006) is the second event that could have impacted on early hominin dispersal and survival in eastern Asia at this time: the Muong-Nong–Australasian tektite strewn field (MN-AT) event.

Tektites are small, glassy objects created and ejected during extra-terrestrial impacts from upper crustal rocks or flash-fused surface sediments (Glass and Koeberl, 2006; Prasad *et al.*, 2007). Of the small number of tektite strewn fields known from around the world (n = 4), that resulting from the MN-AT, the youngest, has almost certainly the widest extent. Even before new discoveries of microtektites from this event in Transantarctic Mountains (Folco *et al.*, 2010) are factored in, the existing distribution field had been calculated to be approximately one-tenth of the Earth's surface (Prasad *et al.*, 2007) from Southeast Asia to Madagascar, from southern and central China – including the CLP (Wei *et al.*, 2003) – to Tasmania.

While there have been extended debates about whether MN-AT represented one or several chronologically and spatially separated impact events, it is now widely accepted to have been the result of a meteoritic strike in Southeast Asia, dated with broad consistency by $^{40}A/^{39}A$ to between 803 ± 3 KBP (Hou *et al.*, 2000) and 786 ± 12 KBP (Kunz *et al.*, 1995). The frequency of tektites in cored sediment samples and signatures in their chemical composition have been used to predict the approximate location of the impact crater as somewhere between *c.* 25°N, 102°E and 19°N, 106°E – from the area near the Vietnam–China border to land now submerged beneath the Gulf of Tonkin (Figure 3.2); however, as yet, no crater has been discovered. This has led some to suggest that the MN-AT may (like the much more recent, 1908, Tunguska event in Siberia) have been the result of an aerial burst rather than an impact meteor (Boslough and Crawford, 2008; Wasson, 2003). Either way, the immediate effects to the regional ecosystem would have been extensive, particularly so if from an aerial blast (Boslough and Crawford, 2008). To put this into perspective, a 'locally devastating impact' from a stony or metallic object *c.* 250 m in size would cause a crater on the order of 5 km in diameter and an area of incineration and destruction *c.* 10^4 km^2 (Chapman and Morrison, 1994). Estimates for the size of the crater needed to have caused the MN-AT distribution pattern vary widely due to levels of uncertainty in predictions from the available data, but even the most conservative estimate would require a crater seven and eight times this size (*see* Prasad *et al.*, 2007).

To date, no strong evidence has been uncovered to suggest that the MN-AT event had a long-term effect on global climate or ecology. There are, however, clues about its impact regionally. A study by Smit *et al.* (1991) looking at sediments from the ODP 758B marine core observed that the event did not result in the disappearance of any foraminifera but did affect the frequency of larger species. Haines *et al.* (2004) reported finding four MN-AT tektites within alluvial gravel and sand deposits exposed during quarrying activities at two locations in northern Thailand, 61 km apart. These deposits (Unit A) contained large quantities of preserved flood debris up to the scale of tree limbs and tree trunks, some >10 m in length. Fragmentary faunal remains were also recovered, including the extinct *Stegodon* sp. Two of the tektites were recovered from overlying deposits and were characteristically worn and abraded; the two recovered from Unit A were angular and sharp. Both of these latter were tested and found to have a geochemical signature that matched other material from the MN-AT event, whilst quartz grains from their associated sediment provided a compatible TL age of 0.75 ± 0.13 Ma. The authors proposed that these high-magnitude flood deposits would be consistent with the large increases in surface run-off one might expect following massive deforestation – the result of an impact event.

There is equally very little indication of the direct effect MN-AT had on resident hominin populations. The position of two tektites in hominin-related sediments at Sangrian, East Java, attributed to MN-AT and originally reported by von Koenigswald in 1957, has been assessed more than once (Hyodo *et al.*, 2002; Langbroek and Roebroeks, 2000). Although there is no doubt about the age of the tektites, such surety does not extend to their associations (Bettis *et al.*, 2009), leaving the relationship open to speculation.

The apparently *in situ* association between archaeological remains and MN-AT tektites in the Bose Basin, Guangxi Province, southern China (Hou *et al.*, 2000), holds greater promise, even if it is not itself immune to controversy. The Bose assemblage ($n = 991$ artefacts) comprises extensively flaked lithics made on cobbles from a range of locally available stone types (quartz, quartzite, sandstone and chert). They are considered to exhibit characteristic lithic reduction strategies and an appreciable degree of formal standardisation that technologically equates them quite closely with African examples of Mode 2 Acheulean technologies. Cultural transmission from African facies of the Acheulean, though, is considered a matter best treated carefully, as local independent invention cannot be ruled out (Dennell, 2009). The apparently isolated nature of the Bose assemblage is itself one indication for such divergence from the African record, which tends to boast a much longer stratigraphic succession. Here it is apparently a one-off, though the existence of reportedly comparable technology from the neighbouring Bubing Basin may hint otherwise (Wang *et al.*, 2007).

Like the artefacts, the tektites used to date them were recovered exclusively from one of seven terraces of the Youjiang River. The concurrent and highly localised occurrence of both is one of the main reasons why – despite the fact that MN-AT tektites have frequently been recovered from younger strata and are demonstrably prone to redeposition – in this instance, the associations are believed by the excavators to be valid (*see* Koeberl *et al.*, 2000). What this association does not demonstrate is whether hominins were occupying this terrace in the years, decades or centuries before or after the deposition of the tektites (Dennell, 2009). Citing abundant evidence of charcoal and silicified wood within the excavation, Hou *et al.* (2000) speculated that Bose could represent a recolonization of the basin by an hominin population in the wake of the MN-AT event; it could though just as easily be an indicator of what killed them. Newly released radiometric dates suggest that hominins were present in the Chinese northeast from as early as 770 ± 80 KBP at Zhoukoudian, Locality 1 (see later). If further evidence emerges to support this reassessment, it also certainly places *H. erectus* in northern China as well as ISEA within the same time window as the OTT and MN-AT events, with all that might entail.

3.3 Middle Pleistocene (775–127 KBP) Dispersals of Anatomically Archaic *Homo*

The Middle Pleistocene was marked by a series of glacial–interglacial cycles but on a scale that appears to have been less frequent than would be experienced in the Upper Pleistocene and in some instances also less severe. It is within this climatic context that we see evidence for an expansion in hominin settlement beyond previous enclaves, both in terms of the visibility and scale occupation, and in the range of different environments occupied. The scope of this development appears to be tied to a widening of the niches being exploited. Where previously open, dry and cool grasslands had been the dominant conditions to which hominins seem to have been adapted, over the course of the Middle Pleistocene sites began to appear in mountainous and coastal settings, as does evidence for deeper penetration into continental interiors and movement to higher latitudes; there is also continued evidence for island colonization. The mechanisms behind this phase of our genus' diaspora remain only slightly better understood than those of the Lower Pleistocene, with climate and associated shifts at the biological community scale still featuring strongly. Here we briefly consider the evidence from each region in turn.

3.3.1 Arabia and the Levant

During the Middle Pleistocene the Arabian Peninsula continued to occupy a potentially crucial geographic position for hominin dispersals out of Africa (Rose and Petraglia, 2009). With accurate Pleistocene site chronologies yet to be established here, and currently no hominin fossil remains, there is a heavy emphasis on documenting surface scatters, of which there are many. Two observations stand out from the archaeological surveys that have been conducted here over the past couple of decades pertaining to Middle Palaeolithic (*c.* 300–45 kbp) assemblages.

Firstly, the frequency of sites with these Mode 3 industries, such as have been revealed through investigations in the vicinity of the Wadi Hadramawt, Yemen and its tributaries (Crassard, 2009), is considerably higher than the number of (presumably earlier) Mode 2 locales and later Upper Palaeolithic ones (Petraglia and Alsharekh, 2002). This possibly indicates an expanding population and the establishment of a firmer foothold in this region, hypothetically starting 200 to 300,000 years ago. What is unclear is if there was occupational continuity here from groups making Mode 2 industries or if the appearance of Mode 3 assemblages marks the arrival of new incoming groups.

Secondly, sites containing Mode 3 artefacts have been recovered from a range of different ecological settings, now including inland basins and mountainous areas as well as coastal margins and river systems, seeming to indicate that whatever the scenario (*in situ* development or incoming communities), groups were ranging more widely across the landscape (Petraglia and Alsharekh, 2002). Without the benefit of a robust radiometric chronology for Arabia, placing any of this evidence in time is problematic. Crassard's (2009) analysis proposes some affiliation between the Hadramawt data and the Levantine Levallois Mousterian, tentatively pointing to a late date, probably of Upper Pleistocene age. Whether this is true of all the Middle Palaeolithic scatters remains to be seen. Petraglia and Alsharekh (2002) noted complementarity between Middle and Lower Palaeolithic sites, but the current state of the evidence does not permit any firm inference about the level of relatedness between these or the time frames of occupation.

Hominin presence regionally during the terminal Lower and early Middle Pleistocene is currently best represented at the northern Israeli site of Gesher Benot Ya'aqov (Figure 3.3), a location now on the River Jordan but which originally lay on the shore of the palaeo-Hula Lake. Dated (and recently redated) by palaeomagnetism, the largely *in situ* remains at this site

Figure 3.3. Excavations during the 1997 field season at the Gesher Benot Ya'aqov site on the River Jordan, Israel. (Photograph by N. Goren-Inbar, reproduced with kind permission.)

appear to have been deposited between 800 and 700 kbp (Goren-Inbar *et al.*, 2000). This correlates to a period between MIS-20 and MIS-18 (two glacials separated by an interglacial; Ashkenazi *et al.*, 2005). Given there are no further dates available for the occupation levels, situating the long (34 m) sequence into an accurate chronology continues to be something of a problem. Most of the excavated material comes from two layers (II-6 and V-5), 4 m and 13 m, respectively, above the Matuyama–Brunhes geochron boundary (Goren-Inbar *et al.*, 2000). The majority of vertebrate fauna comes from deposits situated in the upper part of the sequence and comprises a mix of African and Asian taxa – perhaps suggestive of interglacial conditions. The excellent level of organic preservation at the site has also allowed recovery of fossilised wood, fruit and seeds, with approximately half of these coming from wet habitat taxa, consistent with the palaeolake margin environment (*see* e.g. Goren-Inbar *et al.*, 2002). All lithic artefacts are attributed to an Acheulean facies (with bifaces, cores and core tools, flake tools and débitage; Goren-Inbar *et al.*, 2000). Evidence for the use of fire by early hominins (Goren-Inbar *et al.*, 2004) appears credible if perhaps not yet iron-clad (only eight of 23,454 fruit fragments are reported as burnt). The evidence from this site would seem to place it within a second expansion wave out of Africa and into western Eurasia, though whether the two primary periods of deposition represent different phases within that spread has yet to be determined.

The putative use of fire, forward-planning in tool production and indications of systematic hunting and butchery of medium-sized mammals (*Dama* sp.) have been suggested by Goren-Inbar *et al.* (2004) and Rabinovich *et al.* (2008) as revealing a level of behavioural complexity that is almost indistinguishable to that displayed by Upper Pleistocene anatomically modern humans. While this might be argued as a matter of degree rather than kind – the level of preservation here is rare elsewhere – the fact remains that these indicators do point to this expansion being undertaken by a hominin with both technical and intellectual resources at its disposal.

In the Levant, the Acheulean is followed by the Acheulo-Yabrudian (Garrod, 1956) – also termed the Mugharan Tradition (Bar-Yosef, 1994) – a mosaic package of tool elements, sporadically including tools made on blades but also thick side-scrapers and classic Acheulean hand axes. The Acheulo-Yabrudian marks the first distinctively Levantine industry and is thought to have emerged somewhat earlier than 350 kbp, possibly during the MIS-11 (428–352 KBP) interglacial (Barkai *et al.*, 2003). When in stratigraphic context (such as at Et-Tabūn Cave, Mt Carmel), the Mousterian overlies the Acheulo-Yabrudian, commencing *c.* 250 kbp (e.g. Tabūn D) and continuing until *c.* 45 kbp (within Tabūn B), and some suggestion that it emerged within the MIS-6 glaciation (Shea, 2007). The early Mousterian differs from the Acheulo-Yabrudian in its extensive use of Levallois core preparation, whilst typologically Upper Palaeolithic tool forms such as end-scrapers and burins are also common. Hominin fossil remains are few and too fragmentary to confirm the makers of this industry. It is also not yet clear if the early Mousterian represents an *in situ* development or signals a population replacement phase. The transition from the Acheulo-Yabrudian occurs close to the boundary of MIS-8/7 and a shift from comparatively cool and dry conditions to warmer more humid ones in the Levant; it also heralds a change in settlement foci from the northern and coastal Levant to a far more widely distributed occupation from the Negev and western Jordan to northern Syria (Shea, 2007).

3.3.2 Europe

In Europe, as already noted, fossil finds from the Sierra de Atapuerca, in northern Spain, have indicated the presence of *H. antecessor* in the terminal Lower Pleistocene (though the age of the TD6 *H. antecessor* from Gran Dolina does not yet appear to be firmly established, e.g. Carbonell *et al.* (2008a) quote it as *c.* 0.8 Ma, Blain *et al.* (2009) as *c.* 0.96 Ma) and the presence of *H. heidelbergensis* in the first half of the Middle Pleistocene by at least *c.* 530 kbp (Bischoff *et al.*, 2007). One or both of these could accord with the timing of the Gesher Benot Ya'aqov evidence. Palaeoclimate reconstruction carried out from Gran Dolina using amphibian and squamates remains (Blain *et al.*, 2009) indicated that the *H. antecessor* remains from level TD6 are associated with warm and humid conditions. For the later *H. heidelbergensis*, the environment accorded to level TD10 at Gran Dolina was more temperate. Blain and colleagues proposed that the absence of any evidence of burning from contexts associated with *H. antecessor* may be an indication that this species could not make and manipulate fire. By comparison, they noted the existence of hearths from *H. heidelbergensis* locales elsewhere in Europe (e.g. Terra Amata and Menez Dregan, France, and Vertessöllos, Hungary, each dated to *c.* 450 kbp), adding further weight to the possibility that by this period, hominins were able to exert a degree of buffering against inclement conditions as well as potentially control over their trophic position through this effective means of warding off scavengers and competitors. The type fossil of *H. heidelbergensis* comes from Mauer, Germany, and is dated to *c.* 700 kbp (Czarnetzki *et al.*, 2003), possibly also linking it to the second colonization pulse that Carbonell *et al.* (2008a) have envisaged. The age difference to the Spanish evidence for the same species, though, is considerable (*c.* 170,000 years), and grouping them as part of the same expansion would certainly suggest that this dispersal was more than a mere foray into the European heartlands. The recent discovery from Pakefield, eastern England, of a core and flake industry, also dated to *c.* 700 kbp (Parfitt *et al.*, 2005), underscores this possibility and also begins to raise the likelihood of sustained, if small-scale, occupation in landscapes north of the Alps and Pyrenees before 500 kbp (cf. Roebroeks, 2001). Whenever it started exactly, this phase of settlement does seem to have marked the establishment of a hominin population that was (or would become) capable of living in a range of different environmental settings over time in western Eurasia. Groups ranged from the Mediterranean northwards; sometimes during warm interglacials, such as at Boxgrove, 524–478 KBP (Roberts *et al.*, 1994), sometimes during more temperate ones, as seen at Sima de los Huesos *c.* 530 KBP, and even under boreal forest conditions with distinctly cool winters, as evidenced at Hoxne, *c.* 430–360 KBP during the MIS-11 interglacial (Ashton *et al.*, 2008).

All of the terminal Lower to Middle Pleistocene hominin fossil remains from Europe are interpreted as ancestral forms to *H. neanderthalensis* are distinct from *H. erectus* (Antón, 2003) and commenced a trajectory of localised, long-term adaptation and evolution once established. MtDNA evidence suggests that the last common ancestor between Neanderthals and modern humans probably lived in the first half of the Middle Pleistocene, and a divergence period of 569–465 kbp (95% confidence level) has been calculated, though it is itself currently dependent on the less secure period of divergence between hominins and chimpanzees (Green *et al.*, 2006). The typical morphological traits of *H. neanderthalensis* are thought to have emerged after *c.* 400 kbp (Stringer and Hublin, 1999), and it is believed likely that the specialist thermoregulatory adaptations that characterize this species of *Homo* were an evolutionary response to increasing climate stress and possibly in part a consequence of genetic drift, brought on by isolation from other populations (Roebroeks, 2001). These adaptations served *H. neanderthalensis* well, obtaining them an evolutionary window of existence that is still more than double that of *H. sapiens* and facilitated a high-latitude and wide geographic distribution (Finlayson and Carrión, 2007; Krause *et al.*, 2007). Beyond Europe, Neanderthal expansion may have included the whole of the Arabian Peninsula, to the east it appears to have extended into central and north Asia, and to the south potentially into the Indian subcontinent (Dennell and Roebroeks, 2005).

3.3.3 Central and Northern Asia

Moving eastwards, the incidence of occupation in central Asia also increases markedly during the Middle Pleistocene, with *c.* 70 sites recorded from river terraces, rock-shelters and open-air locations. Even with this larger sample, though, many collections comprise surface finds, and dating resolution is very patchy with chronologies still heavily reliant on typological seriation and geological formations (Vishnyatsky, 1999). The discovery of a series of sites in the loess palaeosols of southern Tajikistan that are stratigraphically relatable to one another and that can be linked into a geomagnetic chronology (Shackleton *et al.*, 1995) has provided a few temporal markers for material from this period. For example, the open-air site of Karatau I (*c.* 38.116°N, 69.117°E), situated 1700 m above sea level, is located within a pedocomplex attributed to the MIS-15 (*c.* 630–560 KBP) interglacial (Vishnyatsky, 1999). The knapping technique evident here, its components and absence of standardisation, is similar to what was seen in the likely earlier Kul'dara industry; both are classed as pebble tool based (Ranov, 1995). Another site also found in similar loess deposits is Lakhuti I (*c.* 80 km east of Karatau I; *c.* 38.21°N, 70.06°E). The lithic assemblage from here is regarded as a more developed form of that found at Karatau and is attributed to the MIS-13 (*c.* 512–480 KBP) interglacial (Ranov, 1995; Vishnyatsky, 1999).

Finally, Sel'-Ungur, in Kyrgyzstan, is one of the few sites radiometrically dated to the Middle–Upper Pleistocene boundary in this region. A U-series date on a travertine layer overlying the cultural deposits providing a minimum age of 126,000 ± 5000 BP (LU-936) is in accordance with both the palynological and faunal evidence recovered from the site. With the exception of one questionable biface, the assemblage is deemed overall to be akin to that found at Karatau and Lakhuti (Vishnyatsky, 1999). Thus almost the full span of Middle Pleistocene occupation of central Asia appears to be characterized by facies of Mode 1 pebble and flake industries, all made on locally obtained raw materials. Hominin remains are non-existent, leaving the identity to the fabricators of these pebble tool industries a matter for conjecture. Middle Palaeolithic industries begin to appear during the last millennia of the Middle Pleistocene. The fact that this happens after an hiatus of up to 60,000 years would seem to suggest the arrival of one or more new populations into central Asia. Middle Palaeolithic sites continue to appear in abundance thereafter into the first half of the Upper Pleistocene, developing into several apparently localised techno-facies (Ranov and Davis, 1979).

The Middle Pleistocene also sees the Siberian archaeological record beginning to emerge in earnest. The Altai Mountains are regarded as the most likely gateway through which the initial colonization of Siberia occurred (Derev'anko *et al.*, 1998) and traditionally thought to have taken place during interglacial periods as part of the process of warm-adapted biotic communities' northward expansion, possibly from Mongolia (Chlachula, 2001). The provisionally oldest evidence of archaeological occupation in the Altai comes from Karama (51.48°N, 84.52°E), a site discovered in 2001 in the Anyi Valley, but it is presently only dated by geological association to the very late Lower Pleistocene, its actual age to be determined. The pebble tool industry here is reputed to show indications of core preparation and secondary retouch alongside 'archaic' morphological features (Shunkov, 2005). Another potential candidate for early occupation in the Altai is Ulalinka (51.9°N, 85.9°E), also an open-air site and located on a tributary of the Maima River (Okladnikov and Pospelova, 1982). While there is little doubt that the split quartzite pebble artefacts recovered from the lower levels here are genuine, accurately dating the deposits in which they occur has proven a challenge (Chlachula, 2001). Shunkov (2005) attributes the upper end of a TL and palaeomagnetism date range, namely from 400 to 300 KBP, as its most likely age. This leaves the open-air site of Mokhovo I (*c.* 54.55°N, 86.35°E), in the Kuznetsk Basin, as currently the most securely dated of the first sites. A small collection of fewer than a dozen quartzite pebble flakes from here is associated with a faunal assemblage considered of Middle Pleistocene age. Both underlie a geological formation that has been TL dated to *c.* 430 KBP (Chlachula, 2001).

Fifteen kilometres upstream from Karama in the Anyi Valley, Denisova Cave (51.383°N, 84.666°E) contains one of the best-dated sequences of hominin activity through the second half of the Middle Pleistocene. The basal cultural deposits here (Layer 22.2) are TL dated to between 282 ± 56 KBP (RTL-548) and 223 ± 55 KBP (RTL-739; Derev'anko *et al.*, 2005). The associated lithic industry from Layer 22 is made on local materials and is described as containing evidence of incipient Levallois core preparation. The Mousterian, however, is more formally present at Denisova between Layers 20 and 12 (after which early Upper Palaeolithic elements begin to appear). Layer 20 is undated, though Layer 21 is TL dated to 155 ± 31 KBP (RTL-546); Layer 12 comes from a different part of the cave and is TL dated to 101 ± 25 KBP (RTL-612; Derev'anko *et al.*, 2005). The wide error margins mean that the Layer 22 material could be coming from as early as MIS-9 and as late as MIS-6. At present, it is unclear if the Mousterian emerged from earlier local traditions or if its appearance represents the work of an incoming population.

The record of hominin skeletal remains from Siberia is very sparse from all periods earlier than the LGM. Two teeth and one phalanx fragment have been discovered from Denisova (51.383°N, 84.666°E). The first, an M_2 molar, was recovered from Layer 22.1. This layer is TL dated to between 182 ± 45 KBP (RTL-738) and 171 ± 43 KBP (RTL-737). Large mammal faunal remains of both steppe and forest species indicate a mosaic environment, while pollen evidence tends to more particularly support cooler, drier and open steppe conditions for this horizon (Chlachula, 2001), leaving the matter equivocal. The second hominin tooth is an I^1, recovered from Layer 12, the later end of the Mousterian-dominated part of the sequence. The closest comparable remains come from the Neanderthal individual found at Teshik-Tash, Uzbekistan (which is undated; 38.336°N, 67.143°E). The mixture of traits in the Denisova specimen means that a positive identification to *H. neanderthalensis* cannot be ascribed with confidence (Shpakova and Derev'anko, 2000).

A recent genetic study has compared the Teshik-Tash fossil to Late Pleistocene hominin remains from another Altai site, Okladnikov Cave. The results of this study suggest that Neanderthals, whose previously confirmed Asian range stopped at Teshik-Tash, likely extended into Siberia (Krause *et al.*, 2007); though how early that expansion occurred is not known. A lack of evidence for a deep mtDNA divergence between the Siberian evidence and that from western Eurasian Neanderthals has been used to suggest that they were not present in most of central Asia and southern Siberia until comparatively recently. An exceptionally warm period, *c.* 125 kbp, may well have facilitated a northward spread (Krause *et al.*, 2007). In light of this timing and of repeated observations about similarities between the Siberian Altai, central Asia and Levantine sites (e.g. Brantingham *et al.*, 2001; Derev'anko *et al.*, 2000; Derev'anko *et al.*, 2005; Goebel, 2004; Rybin, 2005) the same period may even have seen an eastward spread rather than extinction of early Mousterian-using *H. sapiens* (Rabett, 2011). Intriguing new findings urge caution though. Indications are that the genetic history of late-surviving (i.e. post- 48 kbp) western Neanderthals may have differed from earlier populations there and from those living farther to the east (Dalen *et al.*, forthcoming); a salutary signal that we are still some way from grasping the complexities of Neanderthal dispersal.

Another recent genetic discovery, this time again from the sparing skeletal evidence at Denisova Cave, points to the existence in southern Siberia of an archaic hominin that is seemingly new to science. In 2008, a distal fragment of a fifth manual phalanx was recovered from Layer 11 here (Krause *et al.*, 2010). MtDNA was successfully extracted from this specimen, and since then also from a molar with the same on-site provenance (Reich *et al.*, 2010). These are believed to derive from two different individuals of the same population living 50,000–30,000 years ago, but who share their most recent common ancestor with *H. neanderthalensis c.* 640 kbp, and with *H. sapiens c.* 800 kbp (Reich *et al.*, 2010). This could mean that the timing and route of the radiation that brought the 'Denisovan' hominins to Siberia was different, both to the initial (earlier) dispersal of *H. erectus* into Asia and potentially from later settlement by *H. heidelbergensis*, *H. neanderthalensis* and *H. sapiens*. If nothing else, the findings imply Upper Pleistocene coexistence between different hominin species in southern Siberia. A situation once affirmed only for Europe appears now increasingly to have been a more common feature of late humanity.

3.3.4 South Asia

Evidence of early hominin dispersal within Pakistan and the Indian subcontinent continues to comprise mostly sparse examples of lithics from surface finds and comparatively little evidence from stratigraphically secure contexts (Chauhan, 2008; Dennell, 2003; Rendell and Dennell, 1985). Acheulean hand axes have been reported from Dina and Jalalpur, in northern Pakistan, and dated broadly by palaeomagnetism to 700–400 KBP (Rendell and Dennell, 1985). A selection of intentionally flaked stone artefacts from contexts in the Soan Valley at Riwat and Pabbi Hills also provisionally attest to a hominin presence there during the Lower Pleistocene (Dennell *et al.*, 1988; Dennell, 2003).

The fossil evidence of early hominin presence in the subcontinent continues to hinge on a single specimen, the Narmada cranium – and from possibly associated postcranial remains (Kotlia and Joshi, 2008; Sonakia and de Lumley, 2006) – discovered at Hathnora, Madhya Predesh, in the mid-1980s. Originally ascribed to *H. erectus*, the taxonomic identity of this fossil is now thought to be pre–*H. neanderthalensis*, possibly *H. heidelbergensis*, and conceivably a representative of a pre-Neanderthal migration from western Eurasia (Antón, 2003; Cameron *et al.*, 2004). The specimen is associated with Middle Pleistocene fauna, and a recent U-series date on a bovid scapula from this material has returned a supportive minimum date of *c.* 236 KBP. The extent of possible secondary reworking of the deposits at this locale, though, is unclear (a prevalent feature of many of the major fluvially deposited palaeontological sites in India), leaving this age estimate at best still a tentative one (Cameron *et al.*, 2004; *see also* Patnaik *et al.*, 2009; *see* Chapter 5). Chauhan (2008) regards Hathnora among a small number of other key Indian localities containing lithic evidence in association with faunal remains as of likely Middle Pleistocene age – the period to which most palaeontological sites in the subcontinent are currently thought to date. There is considerable promise in the rich Middle Son and Narmada Valley palaeontological deposits, currently receiving concerted archaeological and palaeoenvironmental reconstructive efforts (e.g. Chauhan, 2008; James and Petraglia, 2005; Kotlia and Joshi, 2008) for the retrieval of further hominin remains. At present, this crucial piece of the puzzle for early hominin dispersal continues to evade discovery, though it is probably only a matter of time before new material will surface here.

3.3.5 East Asia

The Middle Pleistocene record for Eastern and Southeast Asia carves a somewhat different path for hominin evolution to regions further to the west. *H. erectus* seems to disappear comparatively early there, seemingly to be replaced by *H. antecessor*, *H. heidelbergensis* and *H. neanderthalensis* during the Middle Pleistocene (a tally that may soon need to be revised in light of the Denisova results). The large amount of fossil evidence from the east, by contrast, describes persistent *H. erectus* population centres in both northern China (principally from the Zhoukoudian locale record) and ISEA, with the survival of the latter until at least the end of the Middle Pleistocene and perhaps beyond (Rightmire, 2001; Storm, 2001; Yokoyama *et al.*, 2008).

In 1919, a preliminary and productive survey was carried out by Andersson (a mining advisor to the Chinese government) of a fossiliferous deposit at Zhoukoudian, 50 km south-west of modern-day Beijing. This was followed in 1921 and 1923 by further survey and preliminary excavations by his associate Zdansky in another similarly appointed but richer set of deposits a short distance away – a site that would come to be known as Locality 1. One of several excavated locales in this vicinity to yield early hominin remains, the main 40 m sequence at Locality 1 lies in a vertical fissure cave on the north side of Longgushan hill, above the modern Zhoukou River. During the 1923 campaign here, Zdansky recognised a fossilised hominin molar and later, during his analysis of the faunal remains, a second (an unerupted lower premolar). The first announcement of these finds was made in Beijing in late 1926 (Black, 1934). Several further isolated teeth were recovered during more intensive work in 1927–8 before, at the tail end of the 1929 field season, the first *Sinanthropus* skeletal remains were uncovered: the adolescent (Locus E) skull, the fragmented (Locus D) adult skull and the adolescent (Locus F) mandible (Black, 1934). Through a series of further excavations before and after World War II, the site produced the remains of 51 hominin individuals (Boaz *et al.*, 2004).

The archaeological sequence is divided into 17 stratigraphic layers. Remains of *H. erectus* and more than 17,000 lithic artefacts have been found between Layers 10 and 3, with the majority split between a lower and upper concentration in Layers 8–9 and 3–4, respectively, and much lower frequencies through the intervening layers. There is, additionally, a large faunal assemblage appearing through the sequence to the depth of Layer 13 (Gaboardi *et al.*, 2005). This latter layer also contains two lithic artefacts as a very limited but potential marker of hominin presence (Dennell, 2009). Sediment analysis has linked the stratigraphy via palaeomagnetic and U-series dating to the regional climatic sequence (Zhou *et al.*, 2000). These authors reported that following the deposition of Layer 14 (correlating to MIS-19), sedimentation at Locality 1 appears to have been continuous, providing a reliable record of climatic variation at a millennial-scale resolution. From these data, part of the Upper Cultural Layer (Layer 4) was most likely deposited 500–470 KBP (correlating with the MIS-13 interglacial). Recently, this was independently corroborated by isotopic signals from herbivore tooth enamel (Gaboardi *et al.*, 2005). The Lower Cultural Layer (Layers 8–9), 660–590 KBP, is thought to straddle MIS-15 and 16. Both layers are breccias – indicative of cold climatic conditions – but Layer 8 is believed to record a warmer climate (Zhou *et al.*, 2000) leaving this association uncertain. The Bottom Cultural Layer (Layer 10) has been dated to *c.* 670–660 KBP, and is linked to the comparatively cool MIS-17 interglacial (*c.* 712–661 KBP). Six new $^{26}Al/^{10}Be$ dates (Shen *et al.*, 2009) on sediment quartz suggest that Layers 7–10 may fall closer to an average of 770 ± 80 KBP, though the full range of assays presented is very wide and appears to contain noticeable inversions: 2.78 ± 0.51 Ma (Layer 6) to 0.31 ± 0.74 Ma (Layer 13), leaving these results as promising but yet to be confirmed.

The nature of the hominin presence at Locality 1 has been a matter of considerable and long debate. Whilst the large lithic component from the site is the best indication that hominins were active here, previous conclusions about the site containing evidence for the early use of fire, beginning in Layer 10 (e.g. Liu, 1985), have now been shown to be reworked loessic silt or water-lain detritus deposits (Goldberg *et al.*, 2001). That said, evidence of charred bone does still hold up to scrutiny and indicates that although now drawing on a much smaller body of evidence, there does remain reason to believe that hominins were using fire periodically at the site (Boaz *et al.*, 2004; Goldberg *et al.*, 2001).

A number of lines of evidence point to hyaenids featuring prominently in the accumulations at Locality 1. A detailed taphonomic study (Boaz *et al.*, 2000, 2004) building on earlier observations has concluded that of 42 nondental hominin remains, 28 exhibit direct or indirect evidence of tooth-mark damage. The element representation of hominins at Locality 1 also is predominated by cranial and mandibular pieces. This is not an artefact of recovery strategies and is consistent with the taphonomic signature of a large-carnivore modified assemblage (Boaz *et al.*, 2004). The total assemblage of skeletal remains from the hyena *Pachycrocuta brevirostris* reported from the site is extensive: *c.* 2000 individuals representing all ages (Turner and Antón, 1996). The presence of juveniles and subadults within a hyaenid assemblage is taken as a strong indication of denning, as too is the presence of small carnivores (Pickering, 2002), which are also well represented in the overall assemblage at Locality I (*see* e.g. Dong *et al.*, 2000). Distinct stratigraphic differences in faunal assemblage composition (Watanabe, 1985), though, will clearly factor here in terms of temporal overlap with hominin remains – the details of which are not presently available in English-language sources. Although hominins were quite possibly visiting the site, the presence of their bones may largely, if not wholly, owe its existence to hyaena denning activities, whereby parts of hominin carcasses were periodically hauled back to the cave (Dennell, 2009).

The evidence from Zhoukoudian supports a scenario that a local *H. erectus* population existed in this vicinity (or repeatedly returned to it) and that it was sufficiently established in the environment to have done so over a long period of time. It is interesting that the early presence, now potentially dating from around the time of the MN-AT event, is followed by a period with a very low incidence of hominin remains, before higher frequencies return again in the upper layers – a consequence of this event?

Although there is evidence for internal evolution taking place within the *H. erectus* of this population through time (Wu and Dong, 1985) – notably in terms of increasing brain size – the Asian specimens of this species still group closer to each other than they do to

African representatives (Antón, 2003). This situation seems to support local continuity over a scenario of repeated long-distance migrations, but time will tell. Hominins occasioned to use the cave at Zhoukoudian, Locality 1, and had some probable mastery of fire, though this appears to be not as extensive as previously thought. An ability to use fire would go some way towards explaining the continuity of this population enclave across a series of environmental phases that, while predominantly interglacial in character, also included colder conditions as well. At the same time, this was evidently a population that was not itself immune to (or easily extirpated by) predation across the lifespan of its existence here.

3.3.6 Southeast Asia

The second and most enduring population centre of east Asian *H. erectus* during the Middle Pleistocene was in Java. The scale of research applied across the intervening landscape between Zhoukoudian and Java does not come close to matching that accorded within them, and this is almost certainly a major factor in the creation of these two islands of data. At present, though (*see* Marwick, 2009), there is no evidence to reliably indicate a sustained hominin presence at any location between north China and ISEA, despite the discovery and analysis of several sites yielding Middle Pleistocene fauna (e.g. Bacon *et al.*, 2006; Covert *et al.*, 2002; Vu The Long *et al.*, 1996; Tougard and Montuire, 2006). What survives are isolated collections of lithic artefacts dated, with varying degrees of confidence, to the Middle Pleistocene and periodic isolated dental finds attributed to *H. erectus* (e.g. Ciochon *et al.*, 1996; Ciochon and Olsen, 1991; Demeter *et al.*, 2004, 2005; Olsen and Ciochon, 1990). Among these materials, possibly the most securely dated are a small number of cores (or core tools) recovered from localities in the Central Lampang Province of Thailand. Three artefacts were retrieved from fluvial gravels near Ban Mae Tha. These deposits lay immediately beneath a palaeomagnetically reversed basalt flow, dated by K-Ar to between 0.8 ± 0.3 and 0.6 ± 0.2 Ma (Pope *et al.*, 1986). This material, together with a similarly small collection of artefacts from Ban Don Mun, and 'modified manuports' and brecciated fauna of Middle Pleistocene affinity from Kao Pha Nam, both from geologically equivalent contexts (Pope *et al.*, 1981:157–8), attests to early human activity and, hopefully, greater recovery potential of this part of the region.

One of the major lacunae in the reconstruction of this Middle-Upper Pleistocene faunal and hominin dispersal in Southeast Asia remains the Thai–Malay Peninsula. This landscape constitutes one of the world's key surviving but least well understood corridors of human global dispersal. The spread of vertebrate faunas, including hominins, likely passed from the Indochinese biogeographic province into the Sundaic Province (*see* Lekagul and McNeely, 1988) via this land route (Tougard, 2001). Work by archaeologists from Universiti Sains Malaysia (USM), Penang, to reconstruct the palaeogeography in the north central part of the peninsula has revealed several new sites with reportedly *in situ* lithic assemblages. Some of the associated deposits have been radiometrically dated from the latest Middle Pleistocene until the Holocene, indicating an extended hominin presence in this part of Mainland Southeast Asia and possibly either a third population centre, after China and ISEA, or at least the beginnings of that elusive infilling between them – though no fossils have yet been found. The Malaysian evidence and the record of existing finds from there are significant though rarely cited in the wider literature; therefore they are covered in some detail here.

3.3.6.1 Peninsular Malaysia

Three adjoining valley systems aligned diagonally from the western coast of the Malaysian side of the peninsula to the inland border with southern Thailand are at the centre of this discussion: the Lenggong, the Kinta, to the east and the Temangor to the west. During the first half of the twentieth century, anecdotal evidence of prehistoric remains in the Malay Peninsula were reported by officers in the British colonial service, prompting several to conduct archaeological excavations. However, no systematic Pleistocene zooarchaeological research and recovery effort was (or has ever been) mounted in these valleys.

Most early excavations yielded evidence that could be relative dated (and now radiocarbon dated) from the Late Pleistocene and Holocene. One locale in particular, though, Kota Tampan, yielded stone tools of a character comparable to ones being discovered in Java, at Pacitan. Initial conclusions about the antiquity of the Kota Tampan lithics, like those at Pacitan – that they were the result of Middle Pleistocene hominins (e.g. Movius, 1948: 403) – did not hold up under more recent scrutiny (*see* Bartstra, 1992). At Kota Tampan, overlying volcanic tuff attributed to Toba has always given it a potential chronological point of reference (Collings, 1938), and it is this association that underpins the current age estimate of the site to just before the YTT eruption (*see* Zuraina, 2003), confirmed by OSL dating of the culture-bearing sediments to *c.* 74 KBP (Hamid, 2007; *see* Chapter 5).

Research conducted by USM since the late 1980s in the Lenggong Valley followed up reanalysis of this site with the discovery of several other cave and open-air locations, the latter linked to palaeolakeshore occupations. The oldest of these is the probable Middle Pleistocene site of Bukit Jawa – its age estimation also based on the proximity of the cultural material beneath ash concentrations considered compatible to the Toba eruption (Mokhtar, 2000). Recent OSL dating of these culture-bearing deposits is pending but makes them almost certainly the product of archaic rather than modern hominins.

The predominantly lateritic soils in Lenggong have precluded the survival of much organic material at any of the early open-air sites so far discovered, while guano digging from the beginning of the nineteenth century has destroyed many local cave deposits, placing clear constraints on efforts to reconstruct the local environment and hominin subsistence within it. At least one well-preserved, brecciated faunal assemblage has been identified in the cave of Gua Badak (Ros Fatihah and Yeap, 2000). Speleothem samples taken from this site returned a Late Pleistocene U-series date range of *c.* 52–46 KBP but were later considered contaminated and unreliable (Ros Fatihah *et al.*, 2002). The potential for recovery, though, does clearly still exist.

Indeed a number of isolated macro-vertebrate fossils of Middle to Late Pleistocene character have also been reported through the course of the last century, from the adjacent Kinta valley as well as elsewhere on the peninsula. The list of finds includes individual specimens of teeth from what may be either the extinct *Elaphas namidicus* or *Elaphas maximus* (Peacock and Dunn, 1968) from the mountain of Bukit Besar (now Khao Khara Kiri, southern Thailand; Andrews, 1903) and also from tin deposits near Kuala Kangsar, Perak (Andrews, 1905). In 1967, a fossil tooth, likely of an *E. namidicus*, was found in a cave (out of context) near Kuala Bering, Kelantan, whilst two partially mineralised teeth, probably *E. maximus*, were recovered some years previously, again from an alluvial deposit at Lukut, western Negri Sembilan, as reported by Peacock and Dunn (1968). Richardson (1947) noted that teeth of *E. namidicus* and *E. antiquus* had been dredged from tin-bearing alluvium around Malim Nawar, south of Gopeng and from Chemor, north of Ipoh, Perak. Fragments of *Rhinoceros* sp., a deer (*Hydropotes* or *Muntiacus*) and a pig (*Sus* sp.) were collected from a mine near the town of Batu Gajah (Stauffer, 1973) – a recurrent feature as mining of the country's rich tin-bearing Quaternary alluvial sediments progressed. Although such finds invariably contain little or no contextual information, their presence does underline the fact that circumstances conducive to organic survival do exist despite prevailing surface conditions. The Kinta Valley as a whole (*c.* 101°10′E, 4°25′N) has the largest number of caves on the peninsula. In the 1920s, the mining of tin from calcified alluvium in a fissure (likely a fault line) lying east-northeast along the flanks of Gunung Temporung (Figure 3.4) yielded teeth of *Sus* sp. (Scrivenor, 1928). Despite the mining, commercially valueless deposits have survived here, including a 150-m column of undisturbed sediments in the large riverine cave of Gua Temporung (within Gunung Temporung) – a site already noted by speleologists as potentially a very rich repository for palaeoclimate and palaeoenvironmental information (Gebaurer and Price, 1995).

The alluvium fields of the Kinta Valley, especially the high alluvium of the Gopeng Beds, are known to comprise ancient deposits laid down by the Kinta River or a tributary thereof. The age of the Gopeng Beds was estimated to be from some time after the Miocene to as recent as the Late Pleistocene (Scrivenor, 1928, 1941). The largest and most oft-cited assemblage of cranial and postcranial fauna remains thus far discovered from these beds came in 1957 from beneath 30 feet of deposit in Gua Tambun, near the town of Ipoh in this valley (Hooijer, 1962; Lamb, 1959). It has been argued that the presence of *E. namadicus* within this assemblage lends support to the idea that the Isthmus of Kra (to the north) was the boundary between the Indochinese and Sundaic biographical regions, and until the Middle-Upper Pleistocene transition a division between archaic faunas to the south (into ISEA) and an increasingly modern fauna spreading from the north (Tougard, 2001). Thus far only one site from the area, Gua Naga Mas, has had relevant deposits radiometrically dated. Samples of flowstone taken for U-series dating, including that encasing an animal skeleton in the roof of this cave, indicate a wide age range of 228–119 KBP (Ros Fatihah *et al.*, 2002) or from approximately MIS-7d through to MIS-5e.

Given the apparent shoreline focus of early hominin groups in Lenggong, the presence of surviving raised beaches within the Kinta Valley and elsewhere in the northern part of the peninsula, for example, at the foot of Gunung Geriang, Kedah (Scrivenor, 1928), offer an additional line of investigation where deposits have survived the intense tin mining. Systematic borehole studies undertaken in 1977 of Quaternary deposits on the coastal plain from Port Weld to Taiping, Perak (northwest of the Kinta Valley), indicated the presence of concentrations of plant material and molluscs in the range −10 to −14 m below sea level at Port Weld (Bore Hole: P6). Farther inland, similar concentrations appeared at higher levels with notable occurrences in bores: Jd.1 (−2 to −13 m), Kd.4 (−4 to −13 m) and K3 (0–1 m). The molluscs are reported to be typical of recent Indo-West Pacific fauna; however, instances of species obtained from similar Plio-Pleistocene deposits elsewhere in Malaysia (Togopi Formation, eastern Sabah) were also found to be present (Suntharalingam *et al.*, 1977).

Finally, the Temangor Valley, now partially flooded by an artificial reservoir (*Tasek Temangur*), contains the largest single tract of 'undisturbed' rainforest surviving in the Malay Peninsula. Unlike the Kinta, this area was not exploited commercially during the last century, at least partly due to issues of politics and security. Tin-bearing deposits have been mined only along three rivers: Sungai Klian, Sungai Duablas and Sungai Jopai. These transect two isolated areas of valued mineral deposits in an area of otherwise barren granite. There is one elusive reference to the finding of a *Stegodon* sp. tooth from Temangor (Mokhtar, pers. comm.). In view of the evidence from the Lenggong and Kinta valleys, research here may yield significant undisturbed Pleistocene deposits. Continued systematic work in these three valleys and targeted examination of key identified sites such as Gua Badak and Gua Temporung should provide invaluable biogeographic information on the character and spread of vertebrate fauna from the Indochinese province into the Sundaic.

3.3.6.2 Island Southeast Asia

Using evidence from the detailed (if complex) prehistoric faunal chronology from Java, in conjunction with available faunal evidence from Mainland Southeast Asia and southern China, it is possible to trace a series of biotic dispersals into ISEA during the Middle

Figure 3.4. Gunung Temporung, Perak, West Malaysia. (Photograph by Ryan J. Rabett.)

Pleistocene (Tougard, 2001). The Javanese evidence suggests that these early dispersals, beginning around the Lower–Middle Pleistocene transition, drew most of their component species either exclusively from the Indian Siwaliks to the west, from Mainland Southeast Asia or more widely from mainland Asia (including China). Taking advantage of more prolonged continental connections through lowered sea levels that commenced *c.* 0.8 Ma, the suite of species identified as the Kedung Brubus (open woodland adapted) and the subsequent Ngandong fauna, including *H. erectus*, would have been able to enter Java repeatedly during this period (van den Bergh *et al.*, 2001) and were probably established as a population (*see* e.g. Hameau *et al.*, 2007; Kramer *et al.*, 2005; Sémah *et al.*, 2003). The small assemblage of 14 flaked stone artefacts from the site of Mata Menge, on Flores, dated by fission track to between 0.88 ± 0.07 Ma and 0.8 ± 0.07 Ma (Morwood *et al.*, 1998), suggests that an early hominin presence (assumed to be *H. erectus*) also still existed beyond Java under long-term insular settings (Marwick, 2009; van den Bergh *et al.*, 2001), something that also raises significant issues regarding the maritime abilities of archaic hominins.

The latest *H. erectus* surviving in Java are those from the Ngandong locality on the Solo River. The fossil record from here consists of skulls and skull fragments from 12 individuals as well as a small amount of postcranial material. It is possible that the paucity of the latter resulted from initial collection bias towards the more readily identifiable crania by the workmen excavating the site, though riverine sorting is another probable agent of selection given the context of the finds (Dennell, 2005). All of the material was excavated over two periods: 1931–2 and 1976–80. The brain size of the Ngandong hominins was found to be around 1100 *cc*, that is, larger than the *H. erectus* average (900 *cc*), and was initially thought most likely to represent an intermediate between *H. erectus* and *H. sapiens* (where brain size is around 1300 *cc*) – a position that accorded well with proponents of the Multiregional theory of human evolution (e.g. Lee and Wolpoff, 2003).

The alternate view, and the position reached though re-evaluation of the original data in the 1980s and since (*see* e.g. Storm, 1995), suggests that there are sufficient anatomical differences to modern humans to place the Solo specimens within the species *H. erectus*, as well as essential consistency within the East Asian evidence as a whole (Antón, 2003). Almost certainly both the hominin and faunal components of the Solo collection represent a natural accumulation of deposits along this stretch of the Solo River. If the association between these components is genuine, it suggests that the final Javanese erectines became extinct around the time of the last interglacial (in contrast to the prevailing model of the equitability of such conditions to hominin expansion) because they were unable to adapt to an increasingly closed rainforest environment (Storm, 2001). Uncertainty persists, however, about the nature of the depositional relationship between the hominin fossils and the fauna. In a detailed treatment of the geological situation at Ngandong, Barstra (1988) reported unexpectedly older U-series ages on bone samples obtained from Upper Pleistocene deposits at Ngandong compared to bone samples taken from the nearby site of Matar, which came from related deposits that should have been geologically younger. Barstra proposed that the cause of this discrepancy could have been related to methodological protocols in the dating technique. The possibility that the Matar bone could have been derived from older sediments upstream was not favoured but could not be ruled out.

Surprisingly late dates using ESR and U-series were obtained by Swisher *et al.* (1996) on isolated mammal teeth from the Ngandong fauna, suggesting an age range of 46 ± 3 to 27 ± 4 KBP (ESR) and 31 ± 0.2 KBP (U-series). Although these dates have been contested (van den Bergh *et al.*, 2001), recent gamma-ray spectrometric direct dates obtained from two of the Ngandong skulls (and one from the Sambungmacan locale, also on the banks of the Solo River) have produced an age range (dependent on how U-leaching is modelled) of *c.* 70–40 KBP (Yokoyama *et al.*, 2008). This might imply that at least a portion of the local *H. erectus* population persisted here much later than expected. On current understanding, survival into the Upper Pleistocene would place this population of *H. erectus* within the Punung rainforest unit (and potentially concurrent with *H. sapiens* in the region), but it would also place them out of sequence with the Ngandong, open-landscape-adapted fauna with which they are supposed to be associated – leaving real questions about the correspondence between the hominin and faunal components of the Ngandong assemblage. Despite this, ISEA currently has the distinction of being the only region of the world where there is good evidence that *H. erectus* persisted throughout at least the Middle Pleistocene, suggesting that the environments here may have played a unique role in the evolution of this species and that local adaptations over such a period should be expected (Antón, 2003; Kaifu *et al.*, 2008; Márquez *et al.*, 2001).

From a technological standpoint, it is hard to link the fossil evidence with the lithic. There is reason to equate, at least on geological grounds, an industry of small flakes and occasionally typologically discernible pieces (but without Acheulean elements) as possibly having been used by the Javanese *H. erectus* (Barstra *et al.*, 1988; Barstra, 1989, 1992). Unequivocal evidence of association, however, has yet to be found. Bifacial elements are said to occur in association with a similar small-flake facies and extinct fauna at the site of Ngebung 2, in the Sangrian Dome area of Java, in a layer radiometrically dated to 0.88–0.86 Ma (Simanjuntak *et al.*, 2010). However, the long-standing tradition among researchers in this region of attributing certain types of lithic industry to certain species of hominin has now been shown to rest on increasingly shaky foundations (*see* Moore and Brumm, 2007). Until this issue is resolved through a more robust understanding and dating of the geology of this island, and through a better grasp of the tropical mobility strategies of all early *Homo* – the principal features of which have been long assumed but are still only vaguely understood even for early modern humans in this region (Rabett and Barker, 2010) – technological associations must continue to be approached circumspectly.

Industries bearing Acheulean elements are to be found more widely if sporadically in Eastern and Southeast Asia than older accepted wisdom supposed. A greater level of tool

Figure 3.5. Two bifaces from Bukit Bunuh, West Malaysia. The implement in the upper image is made on suevite, that in the lower image is made on chert. Both pieces come from the *c.* 39,000 year cultural layer at this site (scale in 1 cm increments). (Photographs by Mokhtar Saidin, USM, reproduced with kind permission.)

diversity, extending to non-typical bifacially worked pieces, may also be obscured within this wider region's long-standing 'chopper-chopping tool' schema (Movius, 1948; cf. Yi and Clark, 1983). Such industries usually, though, contain bifacial elements only in low frequencies, and with some exceptions, such as the still debated age of stone artefacts from the Bose Basin (Dennell, 2004, 2009; Hou *et al.*, 2000), evidence seems to be appearing much later in the Asian archaeological record than elsewhere. Hand axe–like implements have been recovered from several surface and *in situ* locations in Korea, notably and probably earliest at Chongokni, where the lowermost artefact horizon has been recently estimated to date from 300 to 350 kbp (Bae, 2010). Although landscape survey in Korea has been intensive and excavations systematic, bifaces are only being recovered infrequently, and such radiometric dates as are available place many occurrences within the late Middle or Upper Pleistocene, or in association with laminar lithic industries (Bae, 2010; Lee, 2009). It might be sufficient to question the reliability of late dates on taphonomic grounds, were it not for the fact that this is not the only part of Asia where they occur. In Peninsular Malaysia, bifaces described from the open-air site of Bukit Bunuh, Lenggong, have also been recovered from deposits radiometrically dated to *c.* 39 KBP (Mokhtar, 2006a; Figure 3.5). A full description of the geological and archaeological sequences at this site is awaited with interest. Essentially, the weight of currently validated evidence suggests that while present in a number of Asian assemblages, biface numbers are few and appear to represent minor components of essentially the same pebble and flake industries (Corvinus, 2004). It is noteworthy that this is a situation also found in the Developed Oldowan A (1.65–1.53 Ma; Kimura, 2002), which should perhaps caution against equating their occurrence in Asia with sufficient reason to dismiss the 'Movius line' phenomenon completely at this stage (cf. Yi and Clark, 1983).

3.3.7 Environmental and Climatic Conditions during the Middle Pleistocene

Global climate records are more detailed for the Middle Pleistocene than they are for the Lower Pleistocene, but even so, they are still comparatively sparse for the beginning of this period. Carbon dioxide levels in the EPICA Dome C (EDC) core indicate that the MIS-18 glacial period (*c.* 754–712 KBP) was characterized by at least three millennial-scale AIM events, suggesting a deep antiquity to climatic cycles on the scale of the larger D-O cycles of the Upper Pleistocene (Lüthi *et al.*, 2008). The Levantine lacustrine occupation at Gesher Benot Ya'aqov (presumably of *H. erectus*) was established by or during this period; its exact timing in relation to the AIM events is unknown, but could be germane to understanding occupation here. New dates suggest that the Lower Cultural Layer at Zhoukoudian may have been formed during the early Middle Pleistocene (Shen *et al.*, 2009) – but the error margin produces a range from as early as the MIS-21 (*c.* 870-801 KBP) interglacial and as late as MIS-17 (*c.* 712–661 KBP) and, as mentioned, there are notable inversions. The actual timing of hominin appearance here, however, might have had more to do with the regional biogeography of Chinese hyaenids – their arrival and distribution in this part of the region – than it did with the initial entry of *H. erectus*, though clearly these events could be coeval. There are certainly corresponding indications for continuation of at least temporary colonization of the higher latitudes in northern Europe during this period (as indicated by the Mauer and Pakefield evidence) perhaps supporting its basic feasibility towards the later end of this interval in China.

Following this, two at least of the Middle Pleistocene glacials, MIS-16 (*c.* 661–630 KBP) and MIS-12 (*c.* 480–428 KBP), correspond with exceptionally severe conditions. There is some suggestion that these were of greater intensity than MIS-2, as evidenced by projections of *c.* 15 per cent greater ice volume (Shackleton, 1987); though, regional variation is also apparent. The EDC core suggests, for example, that MIS-12 was *c.* 1°C warmer than MIS-2 (Jouzel *et al.*, 2007). Overall, the current weight of evidence seems to support an age for the Lower Cultural Layer at Zhoukoudian either in the late phases of MIS-16 or the early part of the MIS-15 cool interglacial (*see* Jouzel *et al.*, 2007), with the MIS-13 interglacial is the likely period of the Upper Cultural Layer, though one acknowledges that these may prove to be under-estimations.

Near the end of the Lower Pleistocene, hominins had colonized deliberately, or by chance, the island of Flores; an act representing one of the earliest instances of maritime dispersal by hominins anywhere in the world. Isolated finds indicate a continued existence of *H. erectus* in ISEA throughout the Middle Pleistocene and perhaps beyond, though the error margins are too broad on available dates to assign clear glacial or interglacial timing for this evidence.

In terms of Middle Pleistocene climatic variability, during the period from MIS-15a to the end of MIS-11 (a combined period of *c.* 560–352 KBP), two of the glacial cycles in the Siberian depositional record appear to have had a relatively muted impact on the landscape, while fluctuations outside of these are much more evident. Analysis of the sediment core BDP-96-2 from Lake Baikal used the productivity of biogenic silica and diatom records as proxies for the amplitude of climate change at a resolution of *c.* 500 years (Prokopenko *et al.*, 2002). This evidence showed marked and dramatic deterioration in climate to glacial conditions during MIS-15b-d and MIS-10. For the intervening period, the evidence indicates comparatively lower amplitude climate swings. During the MIS-14 glacial, proxy production rates were almost unchanged from interglacial values. Although markedly reduced, and in contrast to other glacial phases, these rates also continued through almost the entirety of MIS-12. Diatom species diversity was also very low, taken as a further indication that environmental impacts through the period were less severe than during other glacial cycles. These data suggest that from MIS-15a to MIS-13 (*c.* 560–480 KBP), in particular, the landscape of central north Asia was both relatively stable and highly productive (Prokopenko *et al.*, 2002). It is perhaps no coincidence that during the interglacials of this time, settlement evidence begins to appear in central Asia. This period also contains both the probable *H. heidelbergensis* dispersal through Europe and the upper end of the projected emergence of *H. neanderthalensis*, implying that this branch of the genus was evolving during a period of comparatively low amplitude climatic oscillations compared to what was to follow after the watershed of MIS-11.

Considerable research attention has been directed towards the MIS-11 (*c.* 428–352 KBP) interglacial. In part this is because it appears to represent the closest analogue to the Holocene and may help track the nature and longevity of the present interglacial (e.g. Loutre and Berger, 2003). Equally important, though, it marked the frontier between two different kinds of climate system (Wang *et al.*, 2003). The Mid-Brunhes palaeomagnetic (MBE) event (Figure 2.1) corresponds roughly with the MIS-12–11 transition (*c.* 430 KBP) and signalled the beginning of high-amplitude changes that would typify the next four glacial–interglacial cycles leading to the present day. While the period before the MBE had been marked by less severe glacial maxima and cooler interglacials, the period after the MBE saw these swings between glacial and interglacial conditions become much more pronounced (EPICA Community Members, 2004; Petit *et al.*, 1999). Marine and ice core records confirm that MIS-11 commenced with a sharp warming *c.* 430 KBP from the glacial conditions of MIS-12. It was then characterized by a plateau of stable conditions for about 40,000 years, until *c.* 390 KBP. Thereafter, the rest of MIS-11 was marked by a series of warm–cold cycles that lasted for approximately a further 38,000 years before conditions plunged towards extreme cold in MIS-10 (*c.* 352 KBP; Ashton *et al.*, 2008).

In considering the potential points of access for hominins into the central provinces of Asia during the Middle Pleistocene, analysis of a recently obtained terrestrial record from the Azraq Basin in central Jordan, dated by geological association to 331 ± 9.7 KBP (Abed *et al.*, 2008), has shown this region to have been exposed to a wetter climatic regime than is currently experienced. This was possibly linked to deeper inland penetration of the Arabian–Indian palaeomonsoon, which during the long period of warming in MIS-9e (*c.* 338–322 KBP; *see* Roucoux *et al.*, 2006) resulted in localised lake formation, habitats that seem to have been favoured by early hominins.

Moving deeper into Asia, Middle Pleistocene hominin occupation of Siberia appears as early as MIS-12 or 11 (Mokhovo I), perhaps becoming more established between MIS-9 (338–280 KBP) and MIS-7 (242–190 KBP) or even MIS-6, based on the Denisova Cave material. Although chronologically much later, the new genetic evidence from this site hints that more than one species of archaic hominin had colonized Asia's higher latitudes at or before this time.

MIS-7, like MIS-9, was marked by a series of subphases, with interglacial maxima recorded for MIS-7a, c and e, the intervening resumption of glacial conditions during MIS-7d, and a marked cooling without return to full glaciations during MIS-7b, based on broadly concurrent records from the Austrian Alps, Siberia, China and the Antarctic (Prokopenko *et al.*, 2002; Spötl *et al.*, 2008). Analysis of the GRIP ice core also indicates that climatic instability became pronounced in the North Atlantic from 242 KBP onwards (Dansgaard *et al.*, 1993). The Narmada hominin provisionally appears during this period in South Asia, signalling a pre–*H. neanderthalensis* (but post–*H. erectus*) presence in (or expansion into) the subcontinent of unknown longevity. Archaeological evidence for lake-shore occupations yielding Mode 1 technology also looks to appear in Peninsular Malaysia at Bukit Jawa around this time.

The MIS-6 glacial record (*c.* 190–126 KBP) is well represented in a recent study from the East European Plain, where Molodkov and Bolikhovskaya (2009) have linked a marine mollusc-based chronological sequence with pollen-based vegetation signals from the European loess sequence. These authors identified three warming trends within MIS-6. The first, an early interstadial event *c.* 184 KBP, appears to have occurred very soon after the onset of glaciation. The second and third are dated by ESR to *c.* 172 KBP and *c.* 155 KBP, respectively. Each of these interstadial subphases is linked to a shift from tundra-steppe to the development of more open pine, and in the final event, birch-dominated woodlands. No error is given by Molodkov and Bolikhovskaya (2009) for their ESR dates, making cross-correlations with archaeological remains problematic; however, some correspondence might exist between the hominin fossil evidence from Layer 22.1 at Denisova and either the first or second of the interstadial proposed by Molodkov and Bolikhovskaya (2009); the final MIS-6 interstadial may correspond with the first appearance of Mousterian technology at Denisova Cave soon after 155 ± 31KBP, while the rapid warming at the very end of MIS-6 could be linked to the spread of the rainforest into Java and the establishment of the Punung fauna (Westaway *et al.*, 2007; *see* Chapter 5); modern humans were unlikely to be present on Java by this early time, but enduring pockets of the open-woodland Ngandong fauna may have facilitated the continued survival of *H. erectus* in parts of Java and possibly elsewhere in ISEA into the opening millennia of the Upper Pleistocene.

This period also marks the upper margin of (yet to be confirmed) claims for early *H. sapiens* in China (*see* Chapter 5). Hominin fossils from Mainland Southeast Asia suggest the presence of archaic hominin forms, but taxonomic uncertainty surrounds many of these specimens, all of which are highly fragmentary. While the Siberian evidence highlights the importance of considering dispersal events in connection with shorter interstadial events, the Southeast Asian material demonstrates that not all hominin (or faunal) expansions can be *a priori* linked to interglacials.

3.4 Dispersal Mechanisms: Faunal Turnovers, Carnivory and Habitat Specificity

3.4.1 Faunal Turnovers and Carnivory

Efforts to determine the mechanism(s) behind the Lower and Middle Pleistocene hominin dispersals discussed in this chapter are on-going. Linkage to an expanding Asian carnivore guild is one line of inquiry originally developed in reference to the European record (e.g. Stiner, 2002; Turner, 1984, 1992). Between *c.* 3.2 and *c.* 0.9 Ma, the Villafranchian fauna (defined from the fossiliferous beds at the type locality of Villafranca d'Asti, Italy) that had held sway across Europe underwent two major turnover events, firstly *c.* 2.5 Ma and then again towards the end of this period. Such major shifts, coupled with shorter-term changes in community composition, suggest that, depending upon when they first arrived in Asia, hominin groups would have encountered a different spectrum of prey (ungulate) species and carnivore competitors. Turner (1992) pointed out that while the diversity of 'carcass destroyers' here (chiefly hyaenas) declined *c.* 1.5–0.5 Ma to levels that paralleled those in East Africa, late persistence of large machairodont (sabre-toothed) cats may have left competitive openings for hominins and other omnivorous taxa to exploit. In this way, the patchy evidence for the earliest colonization of Europe may be a consequence of the severity of carnivore competition; forays may have been made but were unsustainable before *c.* 0.5 Ma. When the structure of the carnivore and herbivore guilds changed to favour hominins, the pattern of their occupation in Europe also changed significantly (Palombo and Mussi, 2006; Roebroeks, 2003).

The position of *H. erectus* within that carnivore guild could, in fact, also be of considerable importance to the late survival of archaic *Homo* populations in Southeast Asia. One feature may have particular significance: the incidence of other large carnivores. The Yuanmou 'hominid fauna' (*c.* 1.7 Ma) contains *Canis yuanmouensis*, *Cynailurus* sp., *Hyaena* (= *Pachycrocuta*) *licenti*, *Megantereon nihowanensis*, *Panthera pardus* and *Panthera tigris* among its large carnivores. The Nihewan fauna (*c.* 1.66–1.5 Ma) shows a more limited diversity but still contains two major scavengers, *Hyaena* (= *Pachycrocuta*) *licenti* and *Hyaena* sp. For the Zhoukoudian fauna from Locality I, collectively, the large carnivore record is extensive, though clearly not all of these were necessarily active in the vicinity of the site at any one time. The list includes *Canis lupus*, *Canis cyonoides*, *Canis lupus variabilis*, *Crocuta crocuta ultima*, *Cuon antiquus*, *Cynailulus* sp., *Homotherium ultimus*, *Hyaena sinensis*, *Hyaena* (= *Pachycrocuta*) *brevirostris*, *Megantereon inexpectatus*, *Nyctereutes sinensis*, *Panthera* cf. *pardus*, *Panthera* cf. *tigris*, *Panthera youngi Ursus arctos*, *Ursus* cf. *spelaeus* and *Ursus thibetanus* (Binford and Ho, 1985; Dong *et al.*, 2000). From the Javanese faunal chronology, the Trinil H.K. fauna (*c.* 1 Ma) associated with Dubois's *Pithecanthropus* (= *Homo*) *erectus* (Dubois, 1896) contained tiger, *Panthera tigris* and *Mececyon* (= *Cuon*) *trinilensis*. The subsequent Kebung Brubus (*c.* 0.8–0.7 Ma) fauna included *Panthera tigris* and *Hyaena brevirostris* (Storm, 2001; van den Berge *et al.*, 2001). The inference is that *H. erectus* populations in China may have been subject to generally higher levels of inter-specific competition from other large-bodied carnivores (and particularly scavengers) than *H. erectus* populations in ISEA. Under this scenario, one could posit that in mainland Asia, archaic hominin populations formed part of a large carnivore guild, one element of a highly productive faunal community with probably heavy inter-predator competition. Hominins possibly survived by filling a specialised daytime group hunting and scavenging niche bequeathed to them courtesy of their primate diurnal activity cycle and sociality (Rolland, 2002) or through marrow bone transport and extraction (Brantingham, 1998). Their numbers in any given area of the landscape were probably not large due to the confines of this trophic position. Their regional population, however, was widespread and successful enough in its niche to both absorb predation from other large carnivores, as likely demonstrated at Zhoukoudian and also, seemingly, to rebound from region-wide short-term ecological catastrophes such as the OTT and MN-AT events.

In terms of behaviour, mainland Asian hominins appear to have employed, possibly among other strategies, opportunistic scavenging (and hunting?) and targeted lake-edge environments that would have attracted animals in the landscape – both deeply embedded survival strategies that are also seen in Africa and the Levant (Mallol, 2006). While they may have processed game at these settings, as indicated by the Goudi/MJG III evidence, they probably did not camp there because it would have been too exposed to the elements and to nighttime predators. Control of fire is likely to have featured, even if not to the degree once claimed at Zhoukoudian; the way it was exploited (e.g. for defence or food preparation) is less clear. The notion that burnt bone was introduced to Locality I rather than having been burnt *in situ* raises the possibility that it could have been brought there from hominin camps by *Homo* or opportunistic scavengers.

In Mainland Southeast Asia, the earliest archaeological evidence comes in the form of small numbers of core tools, dating from the beginning of the Middle Pleistocene, with more extensive lithic assemblages appearing at later, as evidenced at Bukit Jawa, in the Thai-Malay Peninsula. Although the identity of the hominins fabricating these stone artefacts is unknown, they will almost certainly have been archaic forms. The occupational evidence also hints at persisting interest in lacustrine settings even in tropical environments. There is currently little fossil indication of the make-up of the carnivore guild here, though it will probably have included wild dogs (*Cuon* sp.), large cats (*Panthera tigris* and *P. pardus*) and probably hyaenids (*Crocuta* sp.; *see* e.g. Pope *et al.*, 1981) and crocodiles (*Crocodilus* sp.) – the latter being a predator that Dennell (2009) points out is often left out of interpretations when lake-shore assemblage formation is being considered.

In ISEA, *H. erectus* groups would have encountered a less diverse array of terrestrial carnivore competitors. *Hyaena brevirostris* appears to have been part of the immigrant taxa coming from Mainland Southeast Asia that formed the mixture of old and new forms in the Kedung Brubus fauna (Storm, 2001). Crocodila – *Crocodilus ossifragus* and *Garialis bengawanicus* – are also noted in both the Trinil H.K. and Kedung Brubus fauna. An innovative study by Hertler and Volmer (2008) examined prey competition between fossil carnivores in the Javanese sequence. They determined that *Panthera tigris* (the only large terrestrial carnivore consistently present throughout the Pleistocene sequence here) underwent a significant increase in body mass during the Ngandong faunal phase. They interpreted this as consistent with an increase in inter-specific competition within the carnivore guild during the preceding period of the Pleistocene, characterized by the Kedung Brubus fauna. Using detailed craniometric and phenetic analysis, Mazák and Groves (2006: 282) concluded that all *P. tigris* prior to and including the Ngandong specimens should be classed as *Panthera sondaica*, a distinct and separate species from that which would subsequently arrive here and give rise to the modern *P. t. soloensis*. The fate of *Hyaena brevirostris* – a long-time co-member of this community with *H. erectus* (Stiner, 2002) – was also to become extinct by the later Middle Pleistocene. Unlike the archaic tiger, it does not appear in the Ngandong fauna, though a subspecies of the leopard (*P. pardus*) may have arrived in Java at around this time – traced on genetic and craniometric grounds to South Asian populations (Meijaard, 2004).

Where these large carnivores failed, local populations of *H. erectus* may have endured. This does not necessarily imply a causal link between these trajectories; there are too many other factors unaccounted for. It would be safe to say, though, that whatever niche *H. erectus* had come to occupy in the Southeast Asian tropics, it appears to have been in a more resilient position than those held by the other large carnivores to inhabit Java. The possibility is there that it may have outcompeted them (*see* Klein (1999) for a parallel situation in Africa and Storm (2012), for similar thoughts on Java). It is equally possible that a combination of increased environmental selective pressures and reduced trophic ones may have enabled these hominin groups to expand their ecological niche (and population) in ways not previously possible, something also hinted at by changes in the fossil evidence (Antón, 2003). The most immediate, and early, archaeological illustration of this may well be the colonization of island environments, as suggested by the Wolo Sege and Mata Menge evidence on Flores.

3.4.2 Habitat Specificity

Models of hominin evolution based on habitat specificity are the most prominent in the modern literature (*see* e.g. reviews in deMenocal, 2004; Potts, 1998a). These portray adaptation as the process of responding to sets of challenges posed by particular environments. In Africa, for the emergence of early hominins, the emphasis has lain with how adaptation to open grassland savannah was a crucial element in the appearance of fundamental early hominin attributes such as bipedalism, tool making and meat eating. Using the extensive faunal record from the Omo Group of deposits from the Turkana Basin (northern Kenya and southern Ethiopia) as a proxy for habitat change, Bobe and Behrensmeyer (2004) identified major shifts in taxonomic abundance 4–1 Ma. They demonstrated that the *Homo* lineage first appeared in this region during a moderate-sized faunal turnover episode that also saw the appearance of grassland indicator species *c.* 2.4–2.2 Ma. This timing corresponds with the earliest evidence for lithic artefacts at Hadar and Omo in Ethiopia (*c.* 2.4–2.3 Ma), and is slightly younger than the current overall earliest evidence of lithic technology, which comes from surface and excavated localities along the Kada Gona River, Ethiopia, 2.6–2.5 Ma (Semaw, 2000). A second turnover and increase in grassland species occurred 2–1.8 Ma. This coincided with both the first appearance of *H. erectus* (Feibel *et al.*, 1989) and traditionally accepted dates for the Oldowan (Mode 1), *c.* 2–1.5 Ma. The nature of the linkage between the appearance of our genus and the expansion of grassland faunal communities is feasible but not yet proven, as there is as yet no clear evidence of its repeated co-occurrence with other particular grassland taxa. Equally, while the first appearance of stone tools may have been one of the ways hominins were adapting to these shifting conditions, the relationship is not precisely correlated and is probably more complex than current models allow (Bobe and Behrensmeyer, 2004).

One recent and novel development of this approach is the 'Savannahstan' model (Dennell, 2009, 2010; Dennell and Roebroeks, 2005). This proposes that the spread of Plio-Pleistocene hominins out of Africa had less to do with the attainment of an ability to migrate between continents and more to do with the ecological expansion of a successful species over contiguous grassland environments inferred to have stretched from Africa to East Asia during this period (Dowsett *et al.*, 1994), themselves making their first appearance in this epoch (Janis, 2003). One immediate advantage of this model is that it steps outside the modern geographic (and geo-political) boundaries between Old World regions and provides a new spatial dimension within which to consider the relationships between the many isolated points of evidence that exist within its confines.

The reach of habitat-specific hypotheses extends into technological studies and traditionally has had particular reference to eastern and Southeast Asia. Hallam Movius Jr reported that the most significant conclusion of the Joint American Expedition to Southeast Asia (1937–8) was the formal recognition that Palaeolithic technologies there appeared to have followed a developmental course that was distinct from that found in other areas of the Old World – 'in both Burma and Java we are dealing with a new, Far Eastern centre of the Lower Palaeolithic development to which the long-established European classification cannot be applied' (Movius, 1938: 412; *see also* Movius, 1944). Whereas the fact that pebble and flake (Mode 1) industries were transformed or superseded by what would be seen as progressively more technically complex forms in broadly the western portion of the Old World, in many parts of the east, Mode 1 industries appear to have persisted with much lower levels of technological change, and often alongside locally emergent forms, for hundreds of millennia afterwards, through the Upper Pleistocene and into the Holocene. When most instances of Mode 2 and higher modes of technology begin to appear here, incidents are patchy and frequencies are generally low. The oft-cited cause of this difference has been attributed to the limited availability of flakeable stone, a situation itself linked to the constraints imposed by the conditions of the Southeast Asian rainforests or the northern Chinese plains (e.g. Klein, 1999; Pope, 1988; Mellars, 2006a; Watanabe, 1985). While this was possibly a contributing factor to the character of the Asian lithics record, it is unlikely to have been a pivotal one (Yi and Clark, 1983) nor, as we shall see, is it particularly well supported by the archaeological evidence. In this instance, an alternative starting point for interpretation is that these regions were colonized by an ancestral population (or populations) that had left Africa *c.* 1.9–1.7 Ma, before the development, or at least rise to dominance, of bifacial industries (Foley and Lahr, 1997; *see also* Schick and Toth, 1993).

While exactly why this early regionalism in adaptation and behaviour occurred is still to be determined, what is apparent from the archaeological record is that from the late Middle

Pleistocene onwards, this nascent regional variation started to become more pronounced. In western Eurasia, the various techno-facies of the Mousterian mark the last of the long-term, widely distributed Pleistocene lithic industries; the Late Pleistocene here sees a period of diverse and comparatively short-lived technological suites coinciding with the period of extremely harsh glacial conditions that are traditionally seen as primary drivers across a range of adaptive spheres. It was also a period marked by highly unstable climatic conditions. By comparison, and in the face of comparable climatic upheaval, a different picture seems to have emerged in the central and northern parts of Asia. Here the Mousterian arose at a similar time, probably marking entry into the interior of Asia by a new, cold-adapted population (or populations) of hominins: likely *H. neanderthalensis*, though the presence of at least one other archaic form now is also indicated. From this point onwards, the technology shows considerable evidence of continuity into the Late Pleistocene, spanning the same time window when it is being supplanted in Europe. In Asia as whole, gradual technological changes are observed over most of the Upper Pleistocene, with growing emphasis on developments within previously minor components or occasionally the introduction of new ones, indicative of a process of internal transformation. There is little suggestion of widespread replacement (technological or demographic) until after the LGM,where replacement and recolonization by *H. sapiens* populations is especially clear in the Siberian record (e.g. Goebel, 2002).

3.5 Conclusion

Regional variations in early hominin adaptation and technology have been observed and deliberated over for more than 60 years. In the majority of cases, attempts by researchers to address these differences have been framed within grand inter-regional environmentally driven models, often typified by 'characteristic' faunal assemblages. There continues to be a vital role for these inclusive interpretations, but equally, and aided by increasing levels of data, there is a need to temper such models against the new reality that past environments were much more complex and unstable than we have assumed. Wide-scale and long-term changes continue to be apparent in the Earth's record, but within them we now know there to have been more localised and much higher-tempo changes, particularly through the Pleistocene, than traditional models are geared to deal with.

In this chapter, we have taken the Plio-Pleistocene Asian (and, to a lesser extent, European) record and, rather than incorporating the regional differences evident within it into a sweeping narrative of change, we have focused on them. The regional perspective being advocated remains habitat specific, but rather than using this as an excuse for why there were no uniform trajectories to hominin adaptation in different parts of the world, these differences can be seen not only as testament to our genus' remarkable adaptive plasticity but also as an essential part of its survival. Discussion in this chapter has also endeavoured to highlight the significance and effects that increasing climatic fluctuations may have had in these regional environments.

It is coincidental, but nonetheless perhaps important, that the onset of a trend towards even higher-frequency, higher-amplitude climate changes can be traced back to *c.* 242 KBP (MIS-7e-d; Dansgaard *et al.*, 1993) – a period closely tied to the calculated genetic appearance of *H. sapiens* (e.g. Relethford, 2008). MIS-5e would signal both an increasingly steep slide into the most marked phase of climatic instability and also the dispersal from Africa of the last Pleistocene species of *Homo*. The potential impact of this instability as a mechanism through which natural selection acts is now (perhaps not surprisingly, given modern concerns) being examined closely (e.g. Ash and Gallup, 2007; deMenocal, 2004; Potts, 1998a). With few exceptions, the high turnover in material culture forms and emphasis on representation seen in Late Pleistocene Europe is not repeated in any region of Asia or across the Sunda or Sahul palaeocontinents. Ultimately, only one species out of what appears to have been a diverse spectrum of early humanity would survive the quickening pace of global climate changes – but it would meet that challenge in different ways in different parts of the world.

Chapter 4

Regional Trajectories in Modern Human Behaviour

4.1 Introduction

For the last 30 years, there has been general, if not universal (e.g. Cabrera *et al.*, 2006; Clark, 1997a, 1997b), agreement that substantial differences exist between the Middle and Upper Palaeolithic cultural record of western Europe. This transition was marked by a pronounced shift in technology and is argued to have broadly coincided (over some thousands of years) with the demise of the indigenous hominin to this region: *H. neanderthalensis* and its colonization by *H. sapiens* (e.g. Mellars, 2005; Figure 4.1). Considerable uncertainty, however, still exists with respect to the process and timing of this replacement, as well as the kinds and levels of interaction between Neanderthals and modern humans (e.g. Bar-Yosef, 2002, 2007; Green *et al.*, 2010; Miracle, 1998; Vermeersch, 2005). The first part of this chapter considers the main spheres of cultural–economic change identified between Middle and Upper Palaeolithic contexts as they have been deduced from the western European evidence. In the second part of the chapter, these patterns are placed into the context of emerging and equally distinctive Pleistocene sequences from beyond Europe.

4.2 The 'Human Revolution' in Europe

From the early 1970s, when behavioural changes in the Late Pleistocene record of Europe were formally considered for the first time – based on data from the Périgord region of France (Mellars, 1973) – until the 1990s, the European sequence dominated discourse on the character of early humanity. It was here that the earliest examples of cultural practices that could both be accredited to *H. sapiens* and recognised as distinctively 'modern' were found. This was the time and the place where we first started to see behavioural reflections of our modern selves in the archaeological record. The Middle to Upper Palaeolithic transition appeared to mark a 'revolutionary' change (Mellars and Stringer, 1989) that would be articulated through Mellars' (1993) 'Population-Dispersal Hypothesis' of modern human origins in Europe, and characterized later by Henshilwood and Marean (2003) as the 'Later Upper Pleistocene Model' – an umbrella term proposed to cover the set of broadly comparable

Figure 4.1. A portrayal of early anatomically and behaviourally modern *Homo sapiens* during our species' Late Pleistocene dispersal through Europe. (© Wall to Wall Television Ltd., reproduced with kind permission.)

hypotheses seeking to characterize the emergence of modern human behaviour on the basis of this evidence. Following Mellars' (1973) original and, as he then noted himself, subjective selection of spheres where Upper Palaeolithic behaviour differed from the preceding Middle Palaeolithic, numerous trait lists of 'modernity' have since been compiled and/or critiqued from the European evidence (e.g. Bar-Yosef, 2002; Clark, 1997b; Gamble, 1993; Henshilwood and Marean, 2003; Klein, 1999; Mellars, 2005; Miracle, 1998; Mithen, 1996). This term tends to be used in a universalistic way but remains ill-defined. Following Mellars' original and subsequent publications on this subject (Mellars, 1973, 1989, 1993, 1994, 1996, 2005), the transition from the Middle to Upper Palaeolithic in Europe is characterized as involving the comparatively abrupt appearance of a range of technological and cultural–economic changes. Each of these traits is briefly discussed and assessed in the following sections.

4.2.1 Blade-Based Lithic Technology

The European Middle to Upper Palaeolithic transition appears to have been marked by a general shift from flake-based, Mode 3 to blade-based, Mode 4 lithic tool inventories. Although blade technology did feature in Middle Palaeolithic assemblages, the frequency of its occurrence was generally low: <10 per cent (Mellars, 1973). By contrast, most facies of the first Upper Palaeolithic industry, the Aurignacian, were dominated by blade and/or bladelet technologies, with evidence for the adoption of knapping techniques employing soft as well as hard-hammer reduction and indirect 'punch-struck' techniques. It is important to stress that this technological trend, though genuine, remains a generalisation of the evidence. The strength of applicability to all Early Upper Palaeolithic (EUP) industries in Europe is something that is still being wrestled with, and most agree that the transition was almost certainly complex and situational. This is well illustrated in the potential episodes of interaction and possible acculturation between modern humans and Neanderthals, such as proposed for the type site of another EUP industry, the Châtelperronian (Gravina *et al.*, 2005; Mellars *et al.*, 2007b), and with respect to the characteristics of industries and the identity of their likely makers from this period. For example, one of the richest EUP assemblages in Europe has been recovered at the Italian site of Riparo Mochi (Stratum G; Figure 4.2). Generally ascribed to a techno-facies called the 'Proto-Aurignacian', material from here is dated to 36–32 kbp (Kuhn and Stiner, 1998). The assemblage carries a predominance of bladelet forms (accounting for 43% of the lithic tool inventory) as well as bone tools, shell and stone ornamental pieces. Yet it also retains elements that bear affinity to the local Mousterian – the most dominant large tool forms are notches, denticulates and side-scrapers, all more commonly made on flakes than on blades. Diagnostic Aurignacian forms such as carinated and nosed end-scrapers and *busqué* burins are present but are quite rare. Such evidence could be used to argue that this represents a transitional industry, though equally, the high percentage of microlithic forms is not characteristic of the Aurigncian (or the European Middle Palaeolithic) but is more common to later Upper Palaeolithic industries in Europe such as the Gravettian. As another measure of the complexity to understanding even the European transition, the EUP assemblage excavated from the site of Lakonis I on the southern coast of the Greek Peloponnese has also been found to contain a retouched blade and bladelet component (*c.* 50% of the total assemblage) produced by the Levallois laminar method or on prismatic cores, and exhibits apparent continuity (rather than disjuncture) from the underlying Middle Palaeolithic material. It is also associated with a limited range of elements that can be accredited to the Aurignacian, but there is no bone tool component. A hominin molar of conceivably Neanderthal origin, but which displays a mosaic of features, has been recovered in direct association with this material, dated to between 44 and 38 kbp (Panagopoulou *et al.*, 2002–4). While the Mode 3 to Mode 4 distinction remains broadly valid, the position held until a decade ago that the Aurignacian was marked by continental-wide uniformity that contrasted with preceding localised facies of Middle Palaeolithic technologies is looking increasingly questionable.

4.2.2 New Lithic Tool Types

A second major departure observed in Upper Palaeolithic contexts is the appearance of new lithic tool types displaying clearly imposed form – including the distinctive carinated and nosed end-scrapers, burins (including multifaceted and *busqué* forms), retouched blades (such as 'strangulated' blades), retouched bladelet forms (notably Font Yves and Dufour forms) and tanged pieces, termed *Font Robert* points, which proliferated in Upper Palaeolithic industries (Mellars, 1996). There is recognition that some of the industries containing these forms, most particularly the Châtelperronian, were created by Neanderthals as they retain clearly traceable technological links to preceding Middle Palaeolithic industries (Mellars, 2004) – a position that may yet be extended to other EUP industries. Currently these are still normally accredited as being most likely the product of incoming modern human groups, though there is not yet unequivocal evidence that this was the case.

The recent cleansing of the earliest fossil record of modern humans in Europe – whereby *H. sapiens* from several key sites, including Vogelherd and Velika Pĕcina, have now been direct-dated to the Mid-Holocene – has led some to raise concerns about which species was fabricating the earliest Upper Palaeolithic industries (e.g. Conard *et al.*, 2004; Smith *et al.*, 2005). Others have argued that these revelations signal a need to revise the timing of modern human arrival in western Europe (Finlayson and Carrión, 2007). *H. sapiens* was certainly in Eastern Europe 34–32 kbp, based on direct radiocarbon dates taken on human remains from Mladeč, Czech Republic, $31{,}500^{+420/-400}$ (VERA-3076a) to 26,330 ± 170 bp

Figure 4.2. Map showing northwest Eurasian sites. For sites in italics, see Chapter 3. (Map drawn and prepared by D. Kemp.)

(VERA-2736; Wild *et al.*, 2005) and Pestera cu Oase 1, Romania, 34,290$^{+970/-870}$ bp (GrA-22810; Higham *et al.*, 2006). The Mladeč material is associated with a small number of artefacts (including bone points and awls and perforated animal teeth as well as lithics) that are considered to derive from the late Aurignacian – though clear association is still a matter of some dispute (Wild *et al.*, 2005). The Oase 1 remains have no archaeological materials associated with them and are notable for the degree to which they indicate modern humans from this period still exhibited a mosaic of modern and archaic features (Trinkaus *et al.*, 2003). The age of the Neanderthal remains from Level G_1 at Vindjia Cave, Croatia, was placed at *c.* 29–28 kbp, suggesting a late survival of this hominin here, similar to that proposed for parts of the Iberian Peninsula; however, subsequent redating placed the age of these remains in the vicinity of 33–32 kbp – earlier, but still within the transitional period (Higham *et al.*, 2006). In western Europe, the age of the Kent's Cavern 4 maxilla, previously thought to be 30,900 ± 900 bp (OxA-1621), has also been redated through reference to stratigraphically associated material, giving an age in the vicinity of 37–35 kbp. Although this specimen cannot yet be categorically assigned to *H. sapiens* (Jacobi *et al.*, 2006), this age is comparable to the recent redating of faunal material associated with the Fontéchevade human remains from Charente, France (Chase *et al.*, 2007). The state of the evidence demonstrates both the importance of chronological accuracy in discussions of behavioural change and the lack of clarity (and paucity of evidence) that still hampers our understanding of the earliest phases of the Late Pleistocene occupation of Europe.

4.2.3 Non-Lithic Technological and Representative Media

A growing emphasis on the imposition of form would appear not only in lithics during this period. The third sphere of innovation found in Upper Palaeolithic assemblages was much greater incorporation of durable non-lithic materials into cultural assemblages as

Figure 4.3. A portrayal of body ornamentation and pigment use (here, on the scalp) among Late Pleistocene *H. sapiens* around the time of their entry into Europe. (© Wall to Wall Television Ltd., reproduced with kind permission.)

technology and as vehicles for symbolic communication. This took four principal forms: the rise to prominence of media-specific techniques of manufacturing osseous materials, figurative pieces, parietal art and personal adornment (e.g. Figure 4.3). Although non-lithic materials do appear in Middle Palaeolithic assemblages, with the exception of a few minimally shaped awls, and occasional flaking of dense bone in simple *racloir* or denticulated forms mirroring lithics, there is very limited evidence for systematic and material-specific techniques of shaping outside of Upper Palaeolithic assemblages (Mellars, 1989; Verna and d'Errico, 2011).

This is not to say that more heavily modified artefacts were always apparent in Upper Palaeolithic contexts either. In the early phases of the Aurignacian (*c.* 43–40 kbp) in south-eastern Europe (Clark and Lindly, 1989), for example, and through much of the Levant (Marks, 1990; Newcomer, 1987), bone artefacts were present but only at low frequencies compared to what was seen in later industries such as the Magdalenian. Equally, there are indications of periodic localised symbolic expression in Middle Palaeolithic contexts, such as have been found in Iberia (e.g. Cabrera *et al.*, 2006; Zilhão *et al.*, 2010). Intensive material-specific working of media, such as bone, antler and ivory for tools (e.g. Gamble, 1993; Mellars, 2005) does, however, appear to have been an Upper Palaeolithic innovation. Some of the most ornate examples of this material date from comparatively early assemblages from sites in the Swabian Jura region of Germany such as Geißenklösterle, Hohle Fels, Hohlenstein-Stadel and Vogelherd. The arrival of Upper Palaeolithic material occurs abruptly here with the Aurignacian *c.* 35 kbp and is marked by the appearance of other novelties, including numerous examples of figurative art (Figure 4.4), musical instruments, abundant and diverse objects

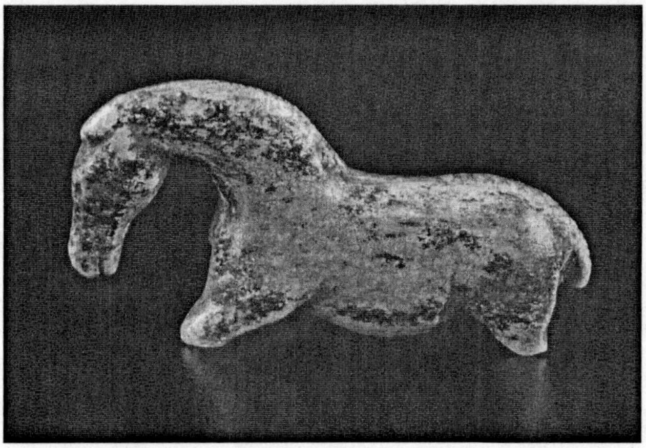

Figure 4.4. A miniature horse pendant (47 mm in length, 25 mm in height) sculpted from mammoth ivory (perforated between the front legs), dated to the Aurignacian, 30–36 kbp, Vogelherd (Conard, 2003), Germany. (Photograph by Hilde Jensen, University of Tübingen, reproduced with kind permission.)

of personal ornament and a characteristically high-frequency of blade-based technology – features entirely lacking from the area's Middle Palaeolithic (Conard, 2003; Conard and Bolus, 2003).

The early dates and wide diversity of forms appearing here have led to the proposition that this part of Europe was a significant crucible of cultural development, the products and the knowledge of which were then carried to other parts of the continent. The *Kulturpumpe* model (Conard and Bolus, 2003) offered three hypotheses. Firstly, significant change in symbolic expression and technological innovation was the result of competition between archaic and modern hominins following colonization by the latter along the Danube corridor *c.* 35–40 kbp. Secondly, dramatic and rapid change in climate may have strained socioeconomic systems during this period and hence been a driving force behind these developments. Thirdly, it considers that such innovations could have come about in connection with socio-cultural changes that were independent of the first and second propositions. The model stresses the importance of nuanced regional-based analysis to balance continental-scale narratives and provides testable hypotheses. At present, there is no clear evidence for co-existence between archaic and modern hominins in this area – indeed there are indications of quite a marked hiatus between their individual phases of occupation, leaving the first hypothesis doubtful. Human population density also appears never to have been very high in Germany as early site numbers are low compared to the later Magdalenian settlement record that began 13 kbp and following a hiatus of *c.* 10,000 years. This may make scalar stress, caused by larger than usual aggregations of people, less likely as a feature driving these innovations. Nonetheless, adaptations related to the effects of rapid climate change could still feature; equally so (and perhaps even related to this) changes in the creation and maintenance of social identities through durable media (*see* e.g. Farbstein, 2009; Gamble, 1982, 1999; Soffer *et al.*, 2000).

The new emphasis on creating varied forms of symbolic communication also extended to static engravings and representational paintings. These were either restricted to cave settings or have simply best survived in those contexts. While most classic examples of cave art, such as Lascaux, Altamira and Niaux, date to the Last Termination, earlier sites do also exist, such as La Grotte Cosquer, 27–18 kbp (Valladas *et al.*, 2001a, 2006), and Chauvet, which includes direct charcoal dates of 29.7 and 32.4 kbp (Valladas *et al.*, 2001b), though the accuracy of the Chauvet dates has been questioned (e.g. Pettitt *et al.*, 2009; Figure 4.5). On current evidence, an arc of artistic development appears to have commenced during the earliest millennia of the European Upper Palaeolithic. The impetus behind such an outpouring of creativity continues to be debated with ecological, demographic, social–ritual and cognitive approaches, all offering alternative takes on the data (e.g. Barton *et al.*, 1994; Conkey, 1997; Jochim, 1983; Lewis-Williams, 2002; Mellars, 2006b). The equally abrupt trend towards technological and representational simplification within the Early Holocene Azilian culture (Velasco, 1989) places some uncertainty on whether this should, however, be viewed as part of a fundamental change or, as may be argued for many of the traits under discussion here, part of a chronologically restricted and regionally specific set of cultural and environmental adaptations.

Also likely to be linked to social display is a trend towards durable body ornamentation (White, 1989, 2007). Although hints of this have appeared in some Neanderthal assemblages, such as in the Châtelperronian at Arcy-sur-Cure, and are now emerging from certain Middle Palaeolithic contexts as well, such as from the final Mousterian levels at the Italian site of Fumane Cave, dated to *c.* 45–42 KBP (Peresani *et al.*, 2011), the later evidence is sufficiently more substantial to suggest that body ornamentation was invested with a degree of formal significance within certain Upper Palaeolithic communities. This is especially evident, for example, in the ornate modern human burials from Sunghir, *c.* 200 km, northeast of Moscow. The main cultural horizon at this site has been dated to 28–27 kbp; the burials, though, have recently been direct-dated to two separate and subsequent episodes of activity: Sunghir I has an age of 22.9 kbp, whilst Sunghir II and III are dated to 24.1 and 23.9 kbp, respectively (Pettitt and Bader, 2000).

All three inhumations at Sunghir were found inside one of the dwelling structures uncovered at the site, and in each case, the bodies were laid on their backs with their hands folded across their pelvis. Sunghir I is that of a man, 55–65 years old. He was buried with a flint knife and dressed in clothing (including a head covering) that was likely decorated with *c.* 3000 beads and fox teeth. His forearms and biceps were decorated with 25 polished mammoth ivory bracelets with traces of black paint. He also wore a schist stone pendant around his neck, and lying with him in the grave was an additional horse pendant with drilled depressions and an ivory disk. Finally, overlying this burial was an ochre-stained adult female skull placed beside a stone slab.

The double grave contained two older children buried head to head (Formicola and Buzhilova, 2004). Pelvic morphology and bone DNA studies confirm that the burial contained a boy and a girl. The boy (Sunghir II) is the slightly older of the two based on dental development, at *c.* 12–13 years. The girl (Sunghir III) is estimated to have been *c.* 9–10 years. Both skeletons are essentially complete. The Sunghir II burial contained four shaft straighteners and 4903 beads, also a beaded cap with fox teeth. The boy had a belt around his waist decorated with 250 Arctic fox canines. Laid across his throat was an ivory pin, and at his shoulder was a mammoth sculpture. At his left side was the polished medullary cavity of a human femur packed with ochre, while at his right (and between him and the girl) was a long ivory lance (2.4 m), which researchers have suggested was too heavy to have been used. The Sunghir III burial also contained bone spears and 5274 beads (again, smaller than Sunghir I). The girl was also likely to have worn a beaded cap, but there are no fox teeth associated with the burial. Like the boy, she lacks a pendant around her neck, but does have a number of small ivory 'lances' – more proportionate to her smaller size.

The beads on all three bodies were arranged in what appear to have been strands, perhaps attached to clothing. Estimates suggest that each bead would have taken at least 45 minutes to 1 hour to finish. Just to put this into perspective, working a 10-hour day every day, the Sunghir II beadwork alone would have taken 367 working days. Soffer (1997) has suggested that the evident standardization of beads into a small number of forms could imply that craft

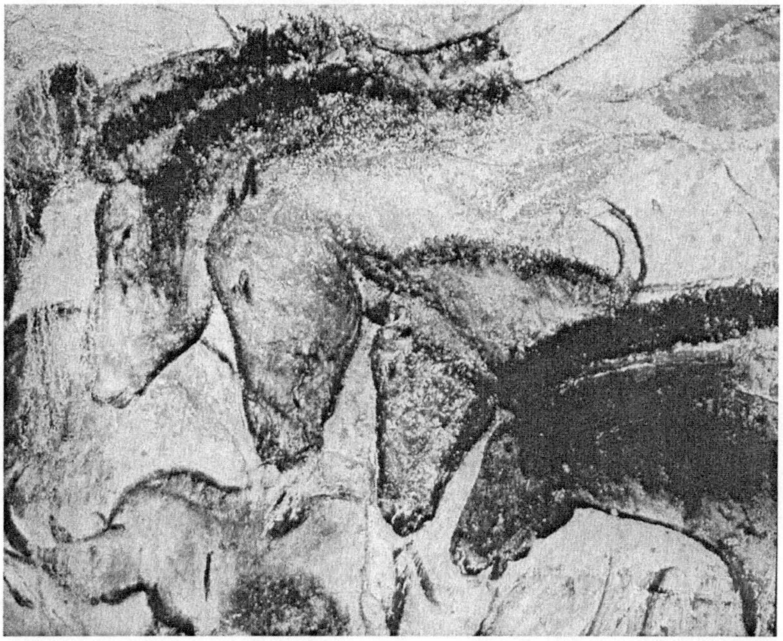

Figure 4.5. The panel of horses from the Grotte Chauvet, Ardèche, France. Interpretations of the depiction have suggested that it may represent either a particular herd of horses or a memorial of an individual animal depicted at different points during its life. (Photo no. 12, French Ministry of Culture and Communication, Regional Direction for Cultural Affairs – Rhône-Alpes Region, Regional Department of Archaeology.)

specialisation was beginning to emerge within the social structure. This potentially huge time investment in what were children's ornaments (greater, in fact, than had been invested in the adult male's decoration) also makes a good case that the status of the children was ascribed rather than achieved. In other words, the children were born into their status instead of having earned it through their own actions.

While ornamentation certainly seems not to have been confined to *H. sapiens*, it is, much more frequently represented in association with them. White (2007) laments the state of archaeological recovery strategies that in many instances have not been equipped or motivated to ensure that ornamental products are retained during excavation. The current pattern of occurrence is, therefore, probably a poor reflection of the distribution and frequency of such material during even the Upper Palaeolithic; it may equally raise questions about the thoroughness of recovery and genuine absence from preceding cultural contexts. However, multi-stage production and site-specific patterns of occurrence (some exhibiting more, others less evidence of manufacture) place the fabrication of personal ornaments within a wider, regional system of material movement by at least *c.* 36–30 KBP (*c.* 27-30 kbp; White, 2007). After this time, ornamentation continues to appear widely, with the most evident recovery occurring in burial contexts like Sunghir.

4.2.4 Mobility

Following the tendency towards intra-regional systems of communication, the fourth area of change is considered to lie in forager mobility. Middle Palaeolithic patterns of lithic raw material acquisition in western and central Europe show a predominant (60%–98%) use of local (0–5 km) materials. Only on rare instances had exotic raw materials moved more than

100 km (Féblot-Augustins, 2009). Lithic procurement strategies evident from Aurignacian assemblages suggest that there was a probable widening of foraging ranges among Upper Palaeolithic groups, with most material coming from within a 15 km range in western Europe (Blades, 1999; Rigaud and Simek, 1990) and a correspondingly larger range in central Europe, emphasising diachronic differences over synchronic ones (Féblot-Augustins, 2009).

The embedded nature of lithic procurement (Binford, 1979) within subsistence activities will also likely have exerted a profound effect on the proportion of exotics appearing in Upper Palaeolithic toolkits. Blades (1999) reports that during the early Aurignacian (*c.* 35 kbp), the proportion of tools from La Ferrassie, Abri Pataud and Le Facteur in the Périgord made on distant (25–45 km) raw materials was 15 to 25 per cent. During this period, faunal remains from Aurignacian sites tended to exhibit low species diversity, often they being dominated by reindeer (*Rangifer tarandus*). This contrasted with late Aurignacian (*c.* 27 kbp) assemblages at these same sites, which contained a wider spectrum of fauna believed to have been drawn from a reduced geographic area, either by choice or environmental constraints. Lithic inventories were found to contain a correspondingly reduced emphasis on distant lithic raw materials (Blades, 1999).

Conversely, evidence from early Eastern Gravettian Upper Palaeolithic sites in Moravia (30–27 kbp) shows intensive use of local lithic raw materials. Whether exploiting chert or quartzite, these sites were situated no more than a few kilometres from lithic sources. By the Evolved Pavlovian phase of Eastern Gravettian (27–25 kbp), these local outcrops, however, were being neglected in favour of remote source areas (Svoboda, 1994). The trend towards increasing levels of high-quality lithic imports here appears to have reached its apogee around this time, with 80 to 96 per cent of lithic raw materials were obtained from distant sources. Fifty to 90 per cent of silicites came from outcrops in the Silesian glacial sediments, *c.* 120 km away, or of even better quality from Jurassic origin in southern Poland. Ten to 50 per cent of radiolarites were coming from areas of the White Carpathians near the modern Slovakian–Moravian border (*c.* 100 km from sites in the Pavlovské hills of Moravia) or from the Austrian Danube Valley. After *c.* 20 kbp, procurement strategies in this area appear to have become more mixed, with some exotics, such as obsidian, but also greater exploitation of local cherts once again.

The emphasis on exotic materials is notably greater among Upper Palaeolithic sites than it is among Middle Palaeolithic ones. Blades (1999), though, has argued that the real change lay perhaps less with evolutionary differences in subsistence and lithic procurement, or mobility strategies, as these appear to fluctuate during both periods and may well be more closely tied to environmental parameters (*see also* Soffer, 1989); rather, the differences arose from how these basic features of mobility were integrated into maintaining the social structures of each population. In other words, the more general emphasis on exotic materials among Upper Palaeolithic groups, including non-lithic materials, such as marine (or fossil) shells (*see* e.g. Mellars, 1996; Vanhaeran *et al.*, 2004), is likely to have been linked with environmental changes but also the maintenance of and communication between more far-flung modern human socio-economic networks via direct or indirect (exchange) procurement.

4.2.5 Population Density

The fifth sphere of change linked to the Middle to Upper Palaeolithic transition is a rise in the population density of *H. sapiens*. Evidence to support such a rise in western Europe during the opening millennia of the Late Pleistocene comes from the increase in absolute site numbers and the intensity of occupation during this period (Mellars, 1973, 1989). Steadily accumulating genetic evidence now also indicates a major demographic expansion in *H. sapiens* during the second half of the Upper Pleistocene. On the basis of worldwide data, the average timing of this expansion is considered to be *c.* 50 kbp, with an emphasis on South and Southeast Asia; population expansion in Europe and the Middle East appears to have occurred somewhat later, *c.* 40 kbp (Atkinson *et al.*, 2008; Excoffier and Schneider, 1999; Harpending *et al.*, 1993; Relethford, 1995). While not an innovation in and of itself, population growth will have affected changes in many social and economic spheres of Upper Palaeolithic culture and inter-species competition for resources. Indeed, Shennan (2001) has observed that the broad correlation between reconstructed patterns of demographic change (specifically, population expansion) during the Upper Pleistocene and the timing of observed cultural innovations that begin to appear in quantity in Europe may be an indication of how essential a contribution this variable is to cultural change. The assumed overlap between archaic and modern human populations in some parts of western Eurasia may also have elevated the significance of this variable – harkening back to the *Kulturpumpe* model – though opinion continues to be divided about the scale or likelihood of interaction between them. The recent reconstruction of Neanderthal nuclear DNA suggests that a low level of inter-breeding may have occurred, though, if this happened, it was early and probably restricted to occupations in the Near East rather than in Europe (Green *et al.*, 2010; Yotova *et al.*, 2011).

4.2.6 Pace of Material Culture Innovation

If contact with modern humans in some way propagated or accelerated changes in material culture within existing local *H. neanderthalensis* groups (and/or vice versa), the pace of material change among moderns themselves seems to have been no less profound. New artefact forms appeared at an unprecedented rate of turnover for the Palaeolithic. It is this feature of Upper Palaeolithic culture that has provided archaeologists with patterned collections of material remains (principally lithic tool types) that are chronologically and geographically discrete enough to lend themselves to being identified as different 'cultures'. It remains a problematic assumption to see these entities as in any way parallel to ethnohistorical cultures (Clark, 1997a, 2006; *see* Chapter 8). However, taking the observed typological changes at face value, the evidence does indeed appear to suggest that the Late Pleistocene as a whole was marked by a quantum leap in the pace of material culture innovation, making this the sixth area of cited change in the western Eurasian sequence (Mellars, 2005).

4.2.7 Subsistence

Although the replacement in Europe of one hominin species with another has tended to focus commentary on differences in the respective subsistence strategies of *H. neanderthalensis* and *H. sapiens*, commentators acknowledge that within the context of local variations in habitat and climate change, the structure of the environment in Eurasia will have produced hunting economies that were at least broadly similar for both hominin species (Boyle, 2007; Chase, 1989; Clark, 1997b; Klein, 1999). Questions remain about how much of a departure was the evident reindeer specialization at many Upper Palaeolithic sites from practices at Middle Palaeolithic locales (*see* Chapter 8), while in the case of mammoth, even the extent to which they were deliberately hunted and their bones accumulated is also not known for certain (e.g. Svoboda *et al.*, 2005). Isotopic analysis of Neanderthal bones (e.g. Bocherens *et al.*, 2001, 2005) is adding a further dimension to this debate by providing support to the idea that the diet of *H. neanderthalensis* was predominantly composed of meat obtained from (probably) hunted herbivorous megafauna. These authors proposed that such a finding also raises questions about where such effective hunting of megafauna stands as a display of behavioural modernity (Bocherens *et al.*, 2005). This evidence portrays *H. neanderthalensis* as potentially the more specialized predator of the two hominin species, given long-term adaptations to the differently structured environments in Eurasia and Africa (Marean, 2007). If correct, the evidence for specialized hunting among Upper Palaeolithic communities may be the reflection of an acquired adaptation to the affordances of the Eurasian steppe rather than a qualitatively progressive development.

Whether the Middle and Upper Palaeolithic transition saw change or continuity in gathered resources is far less well documented (Klein, 1999), and whether any reliable distinctions existed in the gathering activities of late Neanderthals and early modern humans in this region is no better understood. Resources were certainly being collected from maritime margins around this time. For example, at Riparo Mochi, Italy, faunal evidence includes molluscan remains from the exploitation of local marine habitats during the Proto-Aurignacian (36–32 kbp). The larger species, such as limpets, were probably being exploited for food, whilst smaller ones, mostly tiny gastropods, appear to have been used exclusively for decorative purposes (Kuhn and Stiner, 1998). Marean's (2007) suggestion that the tropical heterogeneous and species-rich African habitats will have furnished *H. sapiens* with greater access to, knowledge of and reliance on plant resources compared to *H. neanderthalensis* across more northerly latitudes is definitely feasible and may have had importance for the colonization at other tropical landscapes. How modern humans learned to cope with settings they faced in the colonization of Eurasia is not well understood (*see though* Chapter 8). The drop in species diversity and more seasonally structured accessibility to resources may have been disadvantageous during initial movements into these environments. The often non-preserving nature of these materials further complicates our understanding. However, where taphonomic conditions have been favourable to preservation plant remains or even imprints of them have survived, the indications are that during the Late Pleistocene, these resources did feature prominently in economic activities. Evidence from two Greek sites, Klisoura Cave 1 and Franchthi Cave, offers rare insight into Late Pleistocene plant-gathering strategies in this region, while accumulating evidence from Moravia, the Levant and central Asia points to the wider importance of fibre-based technologies to Upper Palaeolithic communities.

Excavations at Klisoura Cave 1, on the northern edge of the Argive Plain, have revealed an occupational sequence, including several facies of the Aurignacian as well as Epigravettian and Middle Palaeolithic layers (Koumouzelis *et al.*, 2001). Six radiocarbon dates span the Aurignacian industries, producing an age range of *c.* 34–23.5 kbp. The upper part of the Aurignacian sequence included burnt fruit and seed remains, some of which could have been used as food or in the production of dyes. Another significant feature of the Aurignacian levels is the occurrence of a subset of hearths that were clay lined (Karkanas *et al.*, 2004; Koumouzelis *et al.*, 2001). This appears to be a chronologically discrete tradition restricted only to this part of the sequence, with no comparable evidence from the overlying Epigravettian or underlying Middle Palaeolithic levels. These basin-shaped features were consistently made to the same form and size (less than half a metre in diameter). Above each were found ash deposits, charcoal, burnt bone fragments and lithic artefacts. The majority of the associated bone fragments was scattered around individual features rather than inside of them. The body of each basin was made of reddish clay almost certainly derived from the red clay–rich alluvial soils of the nearby river's floodplain. Studies into the character of the basins indicate that they were not heated above *c.* 600°C – the typical kind of temperatures generated by campfires. The materials within them were apparently *in situ* and include burnt macroplant remains and grass phytoliths. On the basis of ethnographic sources, it has been suggested (Karkanas *et al.*, 2004) that the small size of these hearths may mean that they were used for specific cooking purposes, associated with incandescent fuels (e.g. hot coals) rather than fuels still at the flaming stage, and were perhaps satellite hearths around a central camp fire. The spatial arrangement of some of the hearths at the site begins to lend support to this hypothesis. Interestingly, at least one or two of the botanical taxa identified (e.g. the fruit of *Polygonum* sp. and seeds of *Chenopodium* sp.; Koumouzelis *et al.*, 2001: 532) may have

Figure 4.6. Franchthi Cave, southeastern Argolid, Greece. (Photograph by Ryan J. Rabett.)

required detoxification before eating (*see* Johns and Kubo, 1988), raising a further possible function for these features.

Lying to the east of Klisoura, Franchthi is a large cave overlooking a now submerged coastal plain off the Greek Argolid (Figure 4.6). Throughout its long occupational history – with the earliest dated contexts beginning *c.* 22.3 kbp; Middle Palaeolithic levels at the cave have not yet been excavated (Perlès, 2000) – the cave lay no further than *c.* 5 km from the sea. The site's hinterland comprises a range of hills, giving it a well-appointed ecotone position in the landscape (Perlès, 1999). Two major periods of Upper Palaeolithic occupation are recorded. Botanical remains from the earliest of these indicate cold, dry steppic vegetation in the vicinity of the cave. Their presence may represent human introduction, but given the low frequency, other non-anthropogenic vectors such as deposition by birds, rodents or wind are all equally plausible. After *c.* 21 kbp, there appears to have been a significant depositional hiatus of *c.* 6000 years, before a more intensive period of occupation *c.* 13–10 kbp, when botanical remains, including carbonised seeds, nuts and grasses, reflect a warmer, wetter environment of open woodland and appear at a frequency that makes it much more likely that they were brought to the cave by humans. The flowering periods on all of the identified plant species fall between April and October, suggesting a summer occupation, though all of the edible species represented could in theory have been stored and utilised during other times of the year; however, no evidence of storage facilities has so far been found (Hansen, 1991).

Evidence for the use of plant fibre–based technology also begins to appear in the European archaeological record *c.* 27–22 kbp, in association with modern humans. At sites such as Kostenki I and Zaraisk on the Don and Osyotr rivers, respectively, in Russia, and at Pavlov I, in the Czech Republic, impressions of various forms of cordage have started to be recovered in fragments of fired clay (Adovasio *et al.*, 1996; Soffer *et al.*, 2000). These impressions, together with the fortuitous recovery of extensive organic remains from Late Pleistocene waterlogged deposits elsewhere in the wider region – for example, at the *c.* 19.4 kbp open-air site of Ohalo II, Israel (Nadel *et al.*, 1994, 1995, 2004; Weiss *et al.*, 2004), and Dzudzuana Cave (*c.* 30 kbp), Georgia (Kvavadze *et al.*, 2009) – give a sense of the range of objects that perishable materials were used to make, including rope, netting, basketry and bedding. They also indicate how plant materials probably constituted a much larger proportion of cultural inventories than has been previously considered. What is less clear is what portion of these technologies represent Late Pleistocene innovations, as opposed to their appearing in the archaeological record at this point simply because of taphonomic (and, to a degree, recovery) constraints. Rare finds of plants from earlier dated sites such as Kebara and Douara Caves in the Levant (Bar-Yosef, 2000) may be taken as an indication that the preservation bias is probably considerable.

Thus far we have briefly considered seven behavioural spheres that broadly encapsulate the trends that have been identified within the European Upper Palaeolithic: the dominance of blade technology; the introduction of new and distinct formal tool types; the appearance of new media, ornamentation and symbolic representation; long-distance procurement networks; demographic expansion; an increased pace of change and shifts in subsistence. The current state of evidence, though, demonstrates that our picture of Late Pleistocene socio-cultural change in this region continues to evolve, and while genuinely profound, it is also becoming increasingly evident that the reasons behind these changes were many and varied rather than being progressive in the traditional sense. This has come into sharp focus over the last decade in particular, when there has been a growing realization of the Europe's specifically regional character, as Upper Pleistocene records from elsewhere have begun to exert distinctive independent trajectories of their own (Haidle and Pawlik, 2010).

4.3 Modern Human Behaviour Beyond Europe

Since the turn of the millennium, there has been a widening ground-swell of opinion that the Upper Pleistocene of Africa and, increasingly, the Asian and Australasian records raise questions about the evolutionary significance of the Middle to Upper Palaeolithic transition in Europe and whether changes seen there represent an accurate and universally applicable model of our species' emerging character during the Upper Pleistocene. To an extent, though, this has always been accepted implicitly, even when the greatest emphasis lay on the European evidence. While discoveries from here dominates the cultural traits highlighted in the Population Dispersal Hypothesis, as its very name presupposes, the origins of that movement were presumed to lay outside western Europe.

4.3.1 The Near East and North Africa

The eastern Mediterranean and north Africa both contain archaeological sequences that include a technological (lithic) shift from Middle to Upper Palaeolithic character earlier than that seen in Europe, and seemingly continuity between characteristically Middle and Upper Palaeolithic tool industries – though such continuity in the Levant is now questioned (Shea, 2008). Radiometric dates from sites in the Levant indicate that while Mousterian Middle Palaeolithic industries were still dominant 48–46 kbp, by 45–42 kbp, assemblages were of predominantly Upper Palaeolithic character (Bar-Yosef, 1994, 2000). Other traits of the dispersal model, however, are less well represented in this sequence. For example, in the so-called transitional industry of the Emiran, worked bone and antler artefacts are very rare, and while end-scrapers are comparatively common in the Ahmarian, burins are scarce (Bar-Yosef, 2000). Representational art is also extremely sparing before *c.* 13 kbp in the Levant (Bar-Yosef, 1994). Body ornamentation in the form of pierced marine shells – such

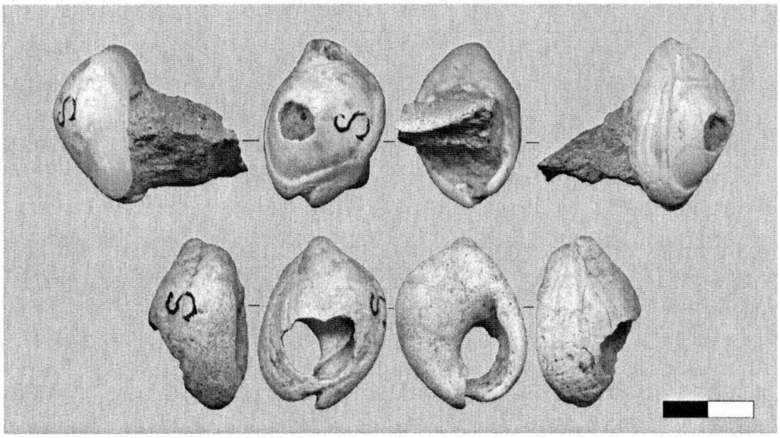

Figure 4.7. Two *N. gibbosulus* shell beads from Es-Skhūl, Israel. (© Vanhaeren/d'Errico/The Natural History Museum London, reproduced with kind permission of M. Vanhaeren; scale in 5 mm increments.)

as at Ksar 'Akil, Lebanon, Layer XVII–XIV – was likewise a rarity but seems to have begun appearing in the Mediterranean region tens of millennia before other manifestations associated with the Middle to Upper Palaeolithic transition. Archived material from Es-Skhūl (Layer B) in Israel ($n = 2$) and at the site of Oued Djebbana, Algeria ($n = 1$), was dated by association to 135–100 KBP (ESR and U-series) and 90–35 KBP (Vanhaeren *et al.*, 2006; Figure 4.7); whilst at Taforalt, Morocco ($n = 13$) perforated *Nassarius gibbosulus* shell beads are dated to *c.* 82 KBP (Bouzouggara *et al.*, 2007). These finds indicate an exceptionally early and widespread incidence of such material and, with Oued Djebbana reported not to have ever been closer than 190 km from the coast during the Upper Pleistocene, they carry suggestion of early long-distance procurement. In addition to this evidence, there is clear indication that early *H. sapiens* were in the Levant long before they entered Europe and, like (though not necessarily co-resident with) local *H. neanderthalensis* populations, there, they were employing Middle Palaeolithic technology. While there is little doubt in many researchers' minds that developments in the western Eurasian Upper Palaeolithic were presaged by developments here, in the Levant, the time may be coming when it becomes prudent to consider the European and Levantine sequences more as adaptively distinct entities.

4.3.2 Sub-Saharan Africa

Africa is widely accepted as producing the genetic progenitor of *H. sapiens*. Its late Middle and Upper Pleistocene archaeological record has also yielded some of the earliest, if sporadic, instances of the behaviours that were used to define the European transition (McBrearty and Brooks, 2000). Most well-known, until recently, has been the Howieson's Poort backed crescent blade–bladelet and bone tool industry, which appeared at a number of sites in southern Africa. This material is thought to have been part of a composite technology and had been broadly dated to 80–50 KBP (based on ESR, OSL and U-series; Feathers, 2002) and chronologically associated with early *H. sapiens*. OSL reassessment of classic Middle Stone Age (MSA) sites – such as Klasies River and a series of lesser known, but increasingly significant sites such as Meilkane, in the Lesotho Highlands (*see* Stewart *et al.*, *forthcoming*) – has significantly and reliably narrowed the time window of the Howieson's Poort to *c.* 5.3 ky (64.8–59.5 KBP; Jacobs *et al.*, 2008).

Another, more controversial discovery from Zaire, the barbed and plain bone points from Katanda, is now dated with reasonable certainty to *c.* $89^{+22/-15}$ KBP (based on TL, U-series and ESR; Brooks *et al.*, 1995) and 95.6 ± 13.9–59.1 ± 10.4 KBP (OSL; Feathers and Migliorini, 2001). Subsequently, finds, especially from southern Africa but also elsewhere in the continent (Ambrose, 1998b; McBrearty and Brooks, 2000), are confirming behavioural forms parallel to those seen during the European Upper Palaeolithic but which were appearing tens of millennia before their European counterparts.

Key material from another MSA complex, the Still Bay Industry, has been excavated at Blombos Cave, on the Southern Cape shoreline of the Indian Ocean (Henshilwood *et al.*, 2001, 2002). This industry and this site have figured prominently in the reappraisal of human behavioural evolution. Compelling evidence from here includes bone points from both the M1 (*n* = 10) and M2 (*n* = 20) phases of the MSA (Figure 4.8), dated to (M1) 75.6 ± 3.4 KBP (OSL) and *c.* 77 ± 6 KBP (TL) and to (M2) 84.6 ± 5.8 KBP–76.8 ± 3.1 KBP (OSL) and 105 ± 9 KBP–76 ± 7 KBP (TL; d'Errico and Henshilwood, 2007; Henshilwood and Sealy, 1997; Henshilwood *et al.*, 2002; Jacobs *et al.*, 2006); pierced *Nassarius kraussianus* (an estuarine gastropod) shells from M1 (*n* = 39) and M2 (*n* = 2; d'Errico *et al.*, 2005); inscribed ochre from M1 (Henshilwood, 2007; Henshilwood *et al.*, 2002) and a large assemblage (*n* = 400) of bifacially worked lanceolate lithic points, the type fossil of the Still Bay Industry from this same phase. The new radiometric dating of the Howieson's Poort has been matched by a similar assessment of Still Bay from four sites. The results are less robust than those for the Howieson's Poort but suggest not only that the Still Bay Industry also came from a very narrow time window of potentially less than a thousand years (71.9–71.0 KBP) but also that there appears to have been a gap of *c.* 6700 years between these two entities rather than continuity (Jacobs *et al.*, 2008). Such evidence also suggests that there were abrupt technological changes taking place within the MSA (Stewart, 2009), adding further texture to the African developmental sequence.

These discoveries and a growing appreciation for an African-centred behavioural sequence for pre–Late Pleistocene *H. sapiens* encapsulated in McBrearty and Brooks's (2000) seminal paper have resulted in the proposal of a new set of models. These have focused on the development of human modernity within or even before the MSA and have foregrounded 'gradualist' and more 'punctuated' viewpoints on the trajectory that development took (e.g. Foley and Lahr, 1997; McBrearty, 2007; McBrearty and Brooks, 2000; Mellars, 2006a). Several commentators have wisely raised concerns about the validity of retaining markers of behavioural modernity that were devised for the European sequence in non-European contexts. Henshilwood and Marean (2003) have, for example, proposed that one alternative approach might be to focus on analysing the ways that early symbolism was used to organise behaviour rather than simply the capacity for it. Many of these approaches, though, do still tend to portray modern human behaviour as an obtained *faculty*; an interpretative position that is questioned herein.

4.3.3 Asia

The current trend towards more region-specific records constitutes a revisiting of older assertions about regionalism in the development of early human technology and culture within the discipline of Archaeology. The possibility that there should have been behavioural differences between Afro-European and Asian early hominins has been one of the most enduring and perplexing points of discussion in Palaeolithic archaeology – even though in most cases interest has been focused on Lower and Middle Pleistocene data. One of the leading proponents to argue that there had been more than one cultural trajectory to Old World human evolution was the Russian archaeologist A. P. Okladnikov (Davis, 1987). Though in broad agreement, in this respect, with earlier scholars in this field (notably

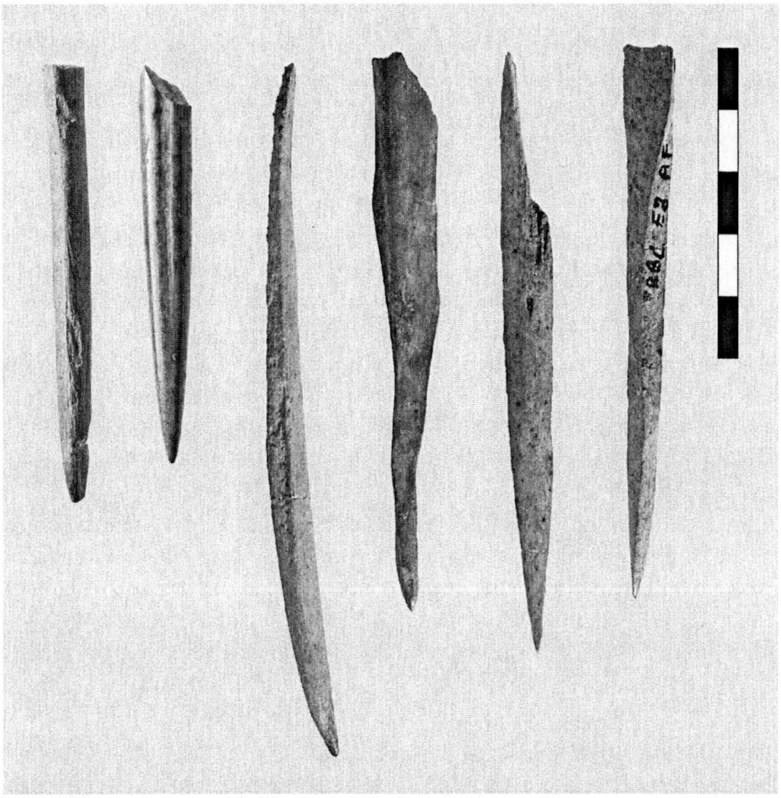

Figure 4.8. A selection of MSA bone points from Blombos Cave, South Africa (scale in 1 cm increments). (Reprinted in modified arrangement after Henshilwood *et al.*, 2001, Blombos Cave, Southern Cape, South Africa: preliminary report on the 1992–1999 excavations of the Middle Stone Age levels, *Journal of Archaeological Science* 28: 421–48; reproduced with permission from Elsevier.)

H. Movius and Z. N. Zamiatnin), Okladnikov proposed that the similarities that set the Asian Palaeolithic apart should be tempered by internal variability of its own, that Asia should not be treated as a 'single Siberian-Chinese Palaeolithic province' (Okladnikov, 1978: 324). It appears, indeed, to have been the case that a different picture emerged in the central and northern part of Asia, compared not only to Europe and Africa (Brantingham *et al.*, 2004) but also to Southeast Asia and Australasia. In central and northern Asia, the Mousterian (Mode 3) arose at a similar time to that seen in Europe – a development that, while not proven, probably marks the entry into these regions by new cold-adapted hominin groups, likely (though perhaps not confined to) *H. neanderthalensis*. Unlike Europe, however, from this point onwards, technological inventories exhibited marked continuity though into the Late Pleistocene, a span including the arrival of *H. sapiens*.

Southeast Asian archaeology, like African, is enjoying a resurgence that will, it is hoped, bring greater clarity to what has been the most enigmatic of sequences. The region has seen the longest survival of *H. erectus* anywhere in the Old World contains good evidence for local biological evolution in that species and, in all likelihood, related species that would ultimately give rise to *H. floresiensis*, and even possibly other hominin forms (e.g. *see* Mijares *et al.*, 2010). In this region, stone tool technology seems outwardly to have been extremely conservative and enduringly so, though not without localised developments and innovations (*see* Chapter 6). Predominately Mode 1 industries linked to the earliest phases of colonization continued to support hominin adaptation through the Middle and Upper Pleistocene, and not just early species but *H. sapiens* also. In this context, some of the components traditionally associated with modernity appear within MIS 3 (*c.* 59–29 KBP), but occurrences are rare until after the LGM. Some degree of similarity in terms of this sporadic and seemingly delayed occurrence is seen elsewhere in the South Asian evidence (James and Petraglia, 2005) and in evidence from the Late to post-Pleistocene record from Sahul (modern Papua New Guinea and Australia; Brumm and Moore, 2005; O'Connell and Allen, 2007). Habgood and Franklin (2008) go so far as to suggest that there is no evidence that a 'package' of traits for modern behaviour was exported from Africa to other regions of the world. Although we have yet to tackle comprehensively with the impact of taphonomic issues (*see* e.g. Langley *et al.*, 2011), the data that are now emerging from well-excavated Upper to post-Pleistocene sequences outside of western Europe are starting to provide convincing evidence that early humans were temporally and inter-regionally heterogeneous in their behaviour (*see* Shea, 2011).

4.4 Conclusion

In this atmosphere of reappraisal, the current volume looks beyond the trait list approach to early human behaviour. To do this, it integrates macro-scale evolutionary trends in hominin dispersal and adaptation with high-resolution, short-term adaptive responses expressed through archaeological evidence from a single region, Southeast Asia, in an effort to begin to better understand the mechanisms behind inter-regional behavioural variability. This part of the world has featured in debates about human cultural development for nearly 70 years (Movius, 1944), but its uniqueness has still to be integrated satisfactorily into a wider narrative. In one of the more recent attempts to do this, Mellars (2006b) sought to explain the apparent lack of Upper Palaeolithic technologies east of the Indian subcontinent in relation to three factors (aside from undisclosed environmental causes). The first of these was a scarcity of high-quality, fine-grained stone for tool production. The second supposed that if colonization of Southeast Asia and Australasia were primarily undertaken via a coastal route, this would have centred on the collection of marine resources, with less emphasis placed on hunting larger animals and, with it, less need for the kinds of technologies that appear elsewhere. The third factor, he argued, may have been 'cultural drift', whereby there may have been a progressive loss in the complexity and diversity of material culture and technology as colonizing groups became more remote from their demographic points of origin. While there is considerable evidence to challenge these particular assertions (e.g. *see* Habgood and Franklin, 2008; *see also* Chapters 5-7), Mellars' paper takes this debate forwards through its acknowledgement of and efforts to articulate these observed regional differences in the behaviour of *H. sapiens*. The remaining chapters of this volume chart the initial spread of modern humans into Asia and, especially, Southeast Asia with the same aim.

Chapter 5

The Initial Upper Pleistocene Dispersal of *Homo sapiens* Out of Africa

5.1 Introduction

For much of the last 30 years, significant changes observed in the character of the European Late Pleistocene record between techno-facies of the Mousterian and ascendant Upper Palaeolithic industries have been tied to the arrival of modern humans and, with them, a major shift in behavioural capacities: the 'Human Revolution'. Only in the last decade has invigorated research in Africa and Asia, including some sensational discoveries, begun to reveal that the European evidence is only one part of a much wider story. A new generation of models is emerging that aspires to address the changing shape of global Palaeolithic research; with many now seeking closer integration between archaeological, anthropological, genetic and palaeoclimatic work (e.g. *see* Mellars, 2006a; Verpoorte, 2009). Since the late 1980s, of these, genetics has probably had the greatest direct impact on the study of Pleistocene humanity. For many, the coding within mitochondrial (mtDNA) and Y-chromosome DNA has proven to be an extremely powerful tool in tracing the origins and spread of *H. sapiens* during the Upper Pleistocene via the genetic heritage of modern populations. Nonetheless, methodological issues and refinements mean that this cannot be treated as a silver bullet approach. Not all interpretations have been met with universal acceptance, and the evidence continues to provoke debate about source populations, the timing and routes of hominin dispersal within Africa and to all points beyond (e.g. Oppenheimer, 2004, 2008; Templeton, 2007). This chapter examines the primary *H. sapiens* radiation starting specifically from the earliest documented expansion of our species out of Africa during MIS-5e. While there is no doubting the importance of the European cultural sequence to the story of our past, since it does not now appear to have formed a prominent part of this initial colonization process, emphasis is placed on the state of evidence through Asia and especially along the perimeter of southern oceans, which is arguably the most prominent of the current dispersal models: the 'Southern Route' hypothesis.

5.2 MIS-5e (126–116 KBP)

MIS-5e is broadly, but not exactly, in temporal concordance with the Eemian interglacial of the northwest European system (Shackleton *et al.*, 2002). For convenience, the two are taken as interchangeable. Here we examine the climatic conditions and excavated evidence of early modern humans, first in the African and Levantine provinces of the Mediterranean during MIS-5e, then farther eastwards into Asia.

5.2.1 Conditions During the Last Interglacial

The chronologically refined Greenland ice cores indicate that there was pronounced climatic changeability throughout MIS-5 (126–74 KBP; Figure 5.1) – a pattern that is now supported by palaeontological, geomorphological and sedimentological records at least within Europe (Bardijí *et al.*, 2009). For short periods of time, multiple climatic downturns created conditions similar to those experienced during the preceding MIS-6 glacial. Only the first part of MIS-5e shows interglacial conditions at the NGRIP and GRIP cores, and even during this period, there is fine structure to the curve suggestive of transient climatic features, during which conditions appear to have turned cooler (more like interstadials; GRIP Members, 1993: 203; Rousseau *et al.*, 2006). Climatic shifts appear to have been most prevalent between the early and middle stages of MIS-5e but occurred throughout the interglacial. Some of these subcycles appear to have been very short-lived, completing in as little as one to two decades; others took between 70 and 5000 years to complete. Two such subcycles have been studied in detail (GRIP Members, 1993). The first of these, 'Event 1' occurred near the culmination of the interglacial and is estimated to have lasted for *c.* 70 years. During this interval, oxygen isotope levels plunged to mid-glacial values. 'Event 2' represents one of a series of 'massive and sustained oscillations' (GRIP Members, 1993: 206) during the first *c.* 8000 years of MIS-5e. During this interval, isotope values suggest that temperatures similar to those experienced during the GS-1 (the Younger Dryas) were sustained over a period of *c.* 750 years. Through the final 2000 years of MIS-5e, conditions attained a level of stability more comparable to that of the Holocene. This may have been followed by a rapid transition at the end of the interglacial (GRIP Members, 1993). The fine-grained NGRIP core points towards a more gradual overall deterioration in conditions but with abrupt cooling *c.* 119 KBP (Figure 5.1). This stadial has been defined as the Late Eemian Aridity Pulse (LEAP), which lasted for 468 years, based on terrestrial (varve and lacustrine) records in northern Europe, starting and ending abruptly over periods of less than 20 years (Sirocko *et al.*, 2005). A wide-scale hiatus in coral reef growth has also been noted from around this time (122–121 KBP; Stirling *et al.*, 1998), suggesting that its effects were felt globally. After a brief respite, this was closely followed by the deeper GS-26 (*c.* 118–116 KBP), which marked the end of MIS-5e (Shackleton *et al.*, 2002). Conditions, though, continued to fluctuate into MIS-5d (in Europe, the onset of the Weichselian glaciation; NGRIP Members, 2004).

The causes of the MIS-5e climatic shifts are not yet clearly understood, though marine sediment evidence from the Norwegian, Iceland and Greenland seas indicates correspondingly rapid changes in ocean circulation and heat transfer. This evidence does not, however, match that associated with the D-O events during the course of the Weichselian glacial (Fronval and Jansen, 1996; cf. Bond *et al.*, 1997), suggesting that other climatic mechanisms were likely to have been responsible. Indeed, there are also indications for other kinds of localised variance. For example, pollen evidence from northern Europe suggests that differences to true glacial conditions during the transient cold snaps may have existed at a seasonal level. More extreme fluctuations may have been limited to winter, whilst summer temperatures remained relatively warm even during these spells of pronounced climatic deterioration (Field *et al.*, 1994).

In the face of these differences across a range of spatial scales, attributing *H. sapiens* expansion out of Africa to the benefits presented by an apparent amelioration of conditions during MIS-5e needs to be weighed carefully and possibly even suspended until greater chronological resolution of the archaeology is available. High-resolution climatic data are

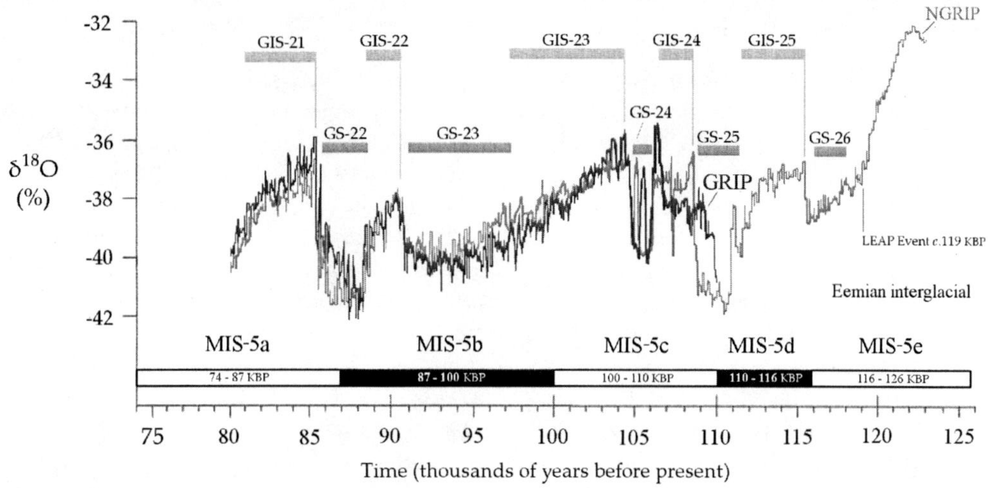

Figure 5.1. MIS-5 (74–126 KBP) oxygen isotope curve (*c.* 80–123 KBP represented); GIS are Greenland Interstadials, GS are Greenland stadials; data are from the GRIP and NGRIP ice cores. (Figure redrawn from NGRIP Members, 2004, and Rousseau *et al.*, 2006, and annotated after Sirocko *et al.*, 2005.)

making it increasingly apparent that traditional ebb-and-flow models of hominin range expansion and contraction are too simplistic. Different elements of a biological community respond to environmental change at differing rates (*see* Section 5.3). This means not only that conditions will exact greater selective pressure on some community elements than others at any given time, but also that increasing the frequency of climatic realignments has the potential to greatly escalate the complex nature of this adaptive process, reinitializing change even before communities have stabilized in response to preceding fluctuations. Although it had been a feature of global climate since MIS-11, the increasing level of climatic volatility appears have reached a crescendo during the period from the MIS-6 to the end of MIS-2 (Grootes *et al.*, 1993), including the period of the last interglacial.

5.2.2 North and Northwest Africa

In north Africa, new research combining space-borne radar imaging of extinct river systems with isotopic evidence for palaeo-freshwater discharges into the southern Mediterranean has revealed an alternate route for dispersal out of sub-Saharan Africa during the latter millennia of MIS-5e (Osborne *et al.*, 2008; Figure 5.2). This work has identified an extensive river system, draining northwards into the Mediterranean that would have provided a viable and sustained means of egress for early modern human populations during this period. It chimes too with new zoogeographic work, which also has identified this network of waterways as a viable corridor across the Sahara during humid phases including the last interglacial (Drake *et al.*, 2011). To the west, recent direct dating by U-series of the Jebel Irhoud juvenile *H. sapiens* (Irhoud 3) from Morocco to 160 ± 16 KBP (Smith *et al.*, 2007) does, however, imply that early humans may not only have been in northern Africa during humid episodes but, potentially, also during more arid periods. This may even lead us to wonder if the Levantine evidence, rather than indicating dispersal, could simply represent the edge of an enclave of north African-based humans that was periodically isolated from the continental interior. On-going work in this part of Africa (*see* e.g. Barker *et al.*, 2010) will hopefully bring further clarity to this situation.

5.2.3 Northeast Africa

Walter *et al.* (2000) discuss the discovery of an assemblage of lithics, including flakes, blades and bifaces, from contexts within an exposed MIS-5e interglacial reef terrace on the Red Sea coast of Eritrea at Abdur (Figure 5.2). The composition of the assemblage is likened to that found at other African MSA sites, though the inclusion of bifacially flaked elements is not normally an associated component of MSA industries and consequently a transitional (or 'Final Acheulean') status was tentatively accorded to Abdur material. The discovery since, however, of another mixed MSA-Acheulean assemblage associated with the *H. sapiens idaltu* (*c.* 160 KBP) from Herto, Ethiopia, suggests that technology of this nature was perhaps part of a more enduring (i.e. not necessarily transitional) and complex adaptive situation (Clark *et al.*, 2003; Stringer, 2003; White *et al.*, 2003). Thirteen stratigraphic sections of the Abdur terrace were measured by Walter and colleagues and divided broadly into three components – a basal cobble lag deposit, a lower shell zone and an upper coral zone. Tools were found in abundance within the basal cobble zone and the lower part of the lower shell zone. Coral samples were taken for U-series dating from the upper coral zone. Accepted assays (six accepted, one rejected) yielded an average age of 125 ± 7 KBP and the likely antiquity for the archaeology. While there is no fossil evidence to demonstrate that the cultural remains in this terrace were deposited by *H. sapiens*, it is thought to be the case based on finds from around this time in neighbouring regions of Africa (Stringer, 2000). Walter *et al.* (2000) proposed that these findings may mark part of an emergent pattern of early modern human behaviour: namely, adapting to exploit marine resources in the face of growing climatic fluctuations and environmental stress. They postulated that human dispersal out of Africa may have been linked to increasing competition for marine resources during periods of worsening climate and hyperaridity in Africa. If so, part of that dispersal may have involved movements of people along the Red Sea coast (towards the Levant) and perhaps across the sea into Arabia during the MIS-6 sea level glacial low-stand.

Recently published research (Armitage *et al.*, 2011) has identified MSA lithic technology of northeastern African affinity at the east Arabian rock-shelter site of Fay-Ne 1 at Jebel Faya. Although of a similar age to the Mousterian assemblages of the Levant, no direct relationship is seen to exist technologically. OSL dates from associated contexts give this material an age of between 127 ± 16 KBP (OSL Tr19-03) and 94.8 ± 13 KBP (OSL Tr19-04; Armitage *et al.*, 2011: supplementary material). Although attributing lithic industries to particular species of hominin is inadvisable in Asia, the contention by the authors that this industry is probably the work of *H. sapiens* may mean that the sphere of human activity outside Africa this early was not confined to movement towards the Levant, as generally supposed. It may also indicate that those early migrants came from more than one source region.

5.2.4 Southwest Asia: The Levant

It was, nonetheless, in the Levant that the first clear evidence for the expansion of modern humans out of Africa was uncovered. The person who initially drove forward this work was Dorothy Garrod. A pupil of the renowned French Palaeolithic prehistorian the Abbé Breuil, Garrod's fascination with the origin of modern humans and the demise of the Neanderthals was likely inspired by her mentor's own great interest in these topics (Smith, 2009: 73); these were questions that for her, like him, required global answers. Garrod looked to western Asia as the source of population movements into Europe (Clark, 1999: 401). This search would take her to the Kirkuk region of northern Iraq, where her excavations uncovered the first evidence for the Mousterian in that area, and then to northern Palestine, where she began her seminal work in the Wadi-el-Mughara (Valley of Caves), principally at three caves on

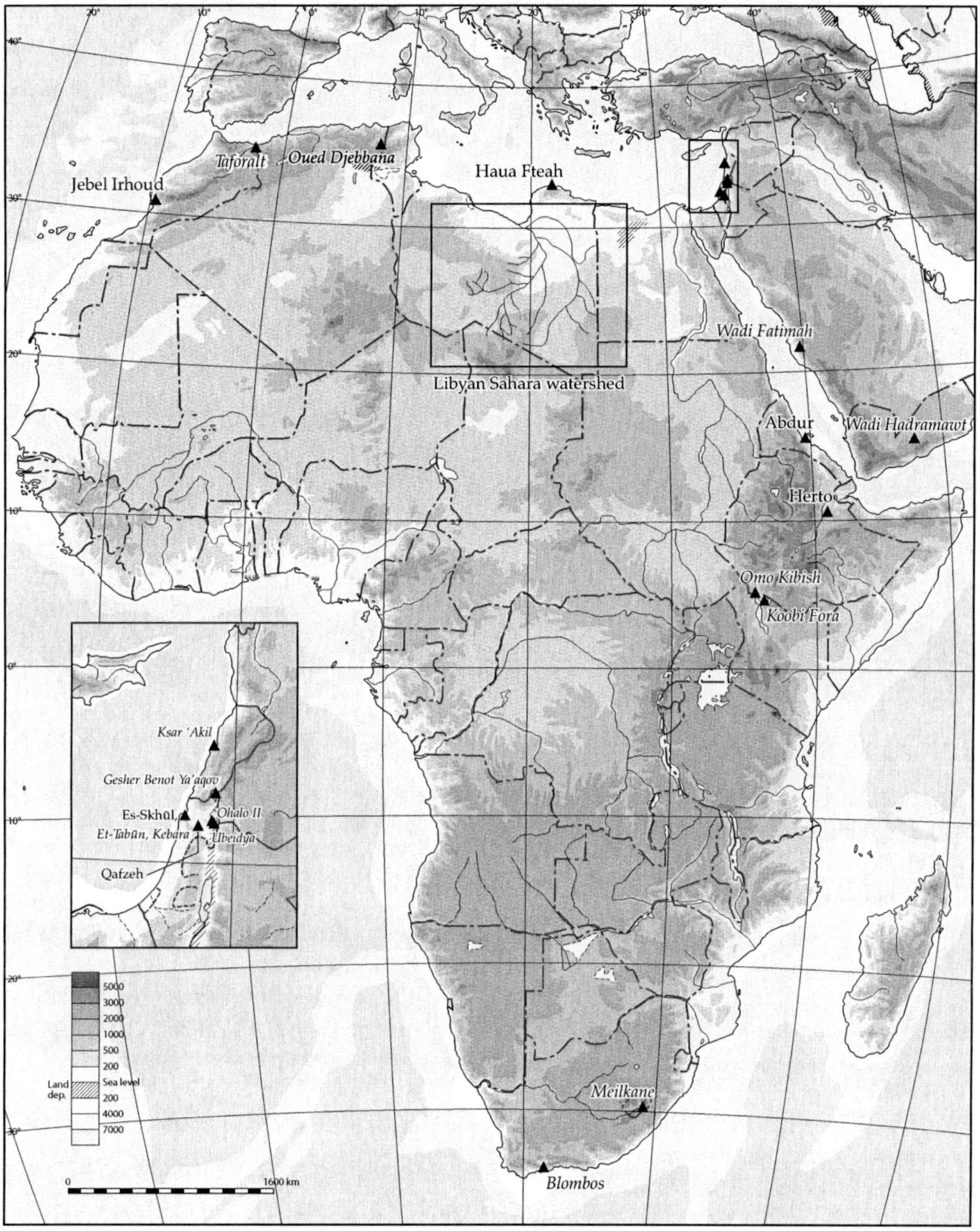

Figure 5.2. Map showing the trans-Sahara watershed (redrawn after Osborne *et al.*, 2008), together with key African and Middle Eastern sites discussed in this chapter (inset: Levantine sites); for sites in italics, see Chapters 3–5. (Maps drawn and prepared by D. Kemp.)

the northwestern slopes of Mt Carmel – et-Tabūn and Mugharet el-Wad (Figure 5.3) and Mugharet es-Skhūl.

Between 1929 and 1934, directing an Anglo-American-funded project, Garrod carried out what would be one of the most productive archaeological investigations into Pleistocene humanity. Although isolated finds, such as an enigmatic fragmented skull from Zuttiyeh Cave in the Wadi Amud, had come to light previously (1927) in this area, Garrod's excavations

Figure 5.3. The Mount Carmel caves viewed from the northwest. Et-Tabūn is just visible on the right-hand side of the frame at the top of a large talus slope; El-Wad is the third cave entrance from the right; Es-Skhūl lies out of view, around the edge of the massif (left of frame). (Photograph by Ryan J. Rabett.)

would uncover one of the largest groups of early human remains from anywhere in the world: at least ten men, women and children emerged from calcified deposits in the small cave of Mugharet es-Skhūl (Cave of the Kids; Figure 5.4; McCowan, 1937).

The Skhūl remains would all prove to be those of early modern humans and included evidence of probable grave goods – most notably in the case of Skhūl V, a male lying on his back with his legs drawn up under him and a wild boar mandible held within his arms.

Garrod's simultaneous work in El-Wad would uncover a total of eight cultural deposits from the Mousterian and Natufian through to the Bronze Age. As significant as the El-Wad finds would be for studies into the emergence of agriculture and sedentism in the Levant, it would be the relationship between the Skhūl and Tabūn remains that would stir interest into the Pleistocene antiquity, early radiation out of Africa and behaviour of *H. sapiens*.

Through Garrod's pioneering work, supplemented and expanded upon by later researchers, neighbouring Tabūn Cave would produce the longest sequence of cultural deposits in this part of the world (*c.* 25 m), stretching back 600,000 years. Hominin remains here lay at a much greater stratigraphic depth than the Skhūl individuals and comprised a smaller sample: the most notable specimens among which were uncovered in late 1932 on the western part of the terrace. These finds, the partial skeleton of a Neanderthoid woman ('Tabūn C1'), with apparent vestiges of a very young infant found lying close to her left upper arm, were recorded as excavated from near the surface of Layer C, but so close to surface of that layer that it may have come from the one above it, Layer B (Smith, 2009: 85).

The apparent contemporaneity of the robust Skhūl modern humans and the Tabūn Neanderthals gave rise to interpretations either that both were part of a transitional population (*Palaeoanthropus palestinensis*) containing Neanderthal and near-modern forms (Smith, 2009:

Figure 5.4. The small north-facing Es-Skhūl cave and terrace, Mount Carmel. (Photograph by Ryan J. Rabett.)

85-6), or that they were the result of local hybridization (*see* Simmons *et al.*, 1991). Both interpretations, though, relied on that assumption that the populations were overlapping; as work continued, evidence began to emerge that this was probably not the case. Analysis of the macrovertebrate fauna from each cave suggested the Tabūn hominins to be older than those at Skhūl, whilst perceived evolution in the lithic industries also suggested that Skhūl was best fitted at the end of the Tabūn sequence (Stringer *et al.*, 1989). This now placed the age of the Tabūn individuals as *c.* 50–60 kbp and Skhūl at *c.* 45–40 kbp.

Between 1989 and 1993, ESR, TL and U-series dating of materials associated with the Skhūl fossils provided a range of dates of unexpectedly early antiquity. ESR on bovid teeth from the hominin burial level (Layer B) produced (early uptake) dates of between 81 ± 15 KBP and 101 ± 12 KBP (Stringer *et al.*, 1989); a TL age of 119 ± 18 KBP was obtained from burnt flints also within Layer B at Skhūl (Mercier *et al.*, 1993), and McDermott *et al.* (1993) reported five U-series (early uptake) dates between 46 ± 05 KBP and 68 ± 05 KBP, and a single date of 88 ± 13 KBP. This range of dates was taken to reflect greater complexity in the Layer B stratigraphy than previously thought and left as unresolved the actual depositional associations of the hominin remains: they were located within Layer B but could have been cut into these strata from a higher level.

Although an age of *c.* 34 KBP was obtained by U-series direct dating of the C1 mandible, a much younger date of *c.* 19 KBP was obtained on the C1 femur. This discrepancy was put down to differences in uranium uptake by the respective elements, and a late uptake age of *c.* 33 KBP for the femur was more comparable to the mandible (Schwarcz *et al.*, 1998). The stratigraphic provenance of C1, though, was still unresolved, and separate ESR and TL chronologies of the site sequence indicated that Layers B–D were all older than these Late Pleistocene U-series dates indicated. The accuracy of the U-series assays was itself also questioned (Millard and Pike, 1999). Reconsideration of Garrod's notes suggested that C1 was probably deposited in Layer B, and a direct ESR date, this time on a fragment of left M_1, gave an age estimate, based on a series of projected depositional circumstances, that suggests C1's age to be between 112 ± 29 KBP (early uptake) and 143 ± 37 KBP (late uptake; Grün and Stringer, 2000). Direct dating of the Skhūl V skeleton (Grün *et al.*, 2005) showed that this individual and the associated pig mandible to have approximately the same age at $93^{+21}/_{-12}$ KBP and $98^{+19}/_{-10}$ KBP, respectively – a result that now, ironically, even raises the possibility of broad contemporaneity between the Skhūl and Tabūn occupations, though the dating is still not yet ironclad (Shea, 2007).

In addition to the work at Mt Carmel, the site of Qafzeh in lower Galilee would also figure prominently in the evidence of an *H. sapiens* occupation of the Levant during MIS-5e. Another large sample of 16 early modern humans was uncovered at Qafzeh through the work of Neuville (1932–5) and Vanderneersch (1965–79). The levels where these finds were made (XVI–XXI) have returned U-series dates (early uptake) of 103 ± 19 KBP and 105 ± 02 KBP (Layer XIX; McDermott *et al.*, 1993), while TL assays on 20 burnt flints from Qafzeh – layers, with flint samples in parentheses, were: XVII (6), XVIII (3), XIX (4), XXI (3), XXII (3) and XXIII (1) – produced a date range of 109.9 ± 9.9 KBP (Layer XXI) to 82.4 ± 7.7 KBP (Layer XIX; Valladas *et al.*, 1988: table 1). Although not without reversals, these results do clearly indicate the presence of early *H. sapiens* here during MIS-5. Depending on the provenance of the Tabūn C1 burial, the data from Qafzeh could represent further indication that *H. sapiens* and *H. neanderthalensis* were in the same region during a broadly similar time frame (Stringer, 2007). Although both populations employed facies of Mousterian technology, they appear to have been utilising it in different subsistence strategies (Lieberman and Shea, 1994), which may have moderated contact opportunities in any case. At present, interaction between these two populations can be neither proved nor discounted on the dating evidence, though the recent genetic findings of Green *et al.* (2010) do now suggest that at least some early inter-breeding between them may have occurred in the Levant at around this time (Green *et al.*, 2010: 721), or at least that Neanderthals contributed to the genetic make-up of modern humans outside Africa (Yotova *et al.*, 2011).

As evidence of an early pulse of modern human movement beyond Africa, the hominin discoveries from Skhūl and Qafzeh have been generally thought to represent a population taking advantage of amelioration in environmental conditions associated with MIS-5e, and that they entered the Levant from an as yet indeterminate source as part of a northward expansion of Afro-Arabian community biota (Tchernov, 1998) . The archaeological record from the Nile Valley during this period is inconsistent with that of the Levant (Marks, 1990), and with the possible exception of the Station One site in the Sudan (Rose, 2004), there is little evidence for a northward expansion of the African MSA (Mode 3) technologies up the Nile, leaving the probability of this corridor being the primary exit route out the continent at best uncertain. For the present time, and if we are dealing with a genuine dispersal, then the trans-Saharan route looks to be the best bet, though the emerging evidence from Arabia may point to the existence of more than one exit route. It should be remembered that none of the published dates for the Levantine occupation is sufficiently accurate to demonstrate an interglacial occupation – even if we do not discount the possibility of human presence there during MIS-5c (*c.* 110–100 KBP) or MIS-5a (*c.* 87–82 KBP), also both more humid phases. Until there is higher chronometric resolution to early *H. sapiens* occurrences in the Levant, linkage between the interglacial of MIS-5e specifically and the supposed advantages of its climatic 'equitability' to dispersal remains provisional. By the same token, while it tends to be accepted that this early expansion of *H. sapiens* was an abortive one, it does not necessarily follow that the absence of substantive fossil evidence for modern humans in Europe before *c.* 36–34 kbp (*see* Trinkaus, 2005) means that there were no modern humans outside of Africa during the intervening period.

Recently obtained speleothem data from five caves in the central and southern Negev Desert of Israel reveal a cluster of wet climatic episodes that likely tempered the southern desert barrier here to a species radiation between 140 and 110 KBP (i.e. the end of the MIS-6 glaciation until MIS-5d). These data, presented in Vaks *et al.* (2007: figure 2), show peaks around the end of MIS-6, possibly corresponding with the period of rapid increase in temperature at the end of this glacial (Brewer *et al.*, 2008) and at the start of MIS-5e, *c.* 126 KBP. Secondary peaks are noted *c.* 124 KBP (elevated warm period), *c.* 117 KBP (during GS-26) and *c.* 112–113 kbp (within GIS-25 during the downward trend towards GS-25; Figure 5.1; NGRIP Members, 2004). Palaeoclimatic data from southern Arabia (Parker, 2009) also indicate that predominantly arid conditions in MIS-6 were periodically interrupted by pulses of monsoon-related precipitation, while the onset of MIS-5e brought a significant increase in rainfall across the peninsula until *c.* 120 KBP, followed by a second peak between 82 and 74 KBP. These changes had the potential to have periodically reduced desert barriers, drawing people into the Arabian interior.

Technological similarities between Levantine, central Asian and Siberian sites (Brantingham *et al.*, 2001; Derev'anko *et al.*, 2000, 2005; Goebel, 2004; Rybin, 2005) may even hint at an eastward movement. In essence, the Levant need not have been a cul-de-sac for early human settlement. Small groups of humans could have travelled south-east through the Negev to reach the Red Sea coast. The Fay-Ne 1 data certainly raise the possibility that modern humans had already reached the Persian Gulf during MIS-5 (Armitage *et al.*, 2011). Although still tentative, lines of evidence from the eastern coast of Africa, Arabia and from the Asian interior could be interpreted as indicating that the earliest *H. sapiens* populations beyond Africa were more successful than has been generally assumed, even if most (if not all) enclaves did not survive long-term.

5.2.5 South Asia

The earliest fossil evidence of *H. sapiens* from the Indian subcontinent does not appear in the archaeological record until early MIS-2 at Fa Hien-lena (38–30 KBP) in Sri Lanka (6.648°N, 80.215°E; Perera, 2010). This fact, coupled with the scarcity of accurately provenanced finds and limited resolution in the delimiting of lithic industries, makes it difficult for a detailed assessment of early modern human activities in this region (James and Petraglia, 2005). Hominin activity of an unknown scale and species is indicated during the period leading up to, during and after MIS-5e. For example, the 19 m of dated stratigraphy of the 16R Dune from the Thar Desert of central India contains cultural material (interpreted as Middle Palaeolithic) in association with a date of *c.* 150 kbp. A scraper-based industry from Patpara, in the Middle Son Valley, indicates a hominin presence *c.* 103 KBP (James and Petraglia, 2005). There is also reason to suspect that populations were in the subcontinent at the time of the most recent Toba supervolcanic (YTT) eruption (see later) *c.* 74 KBP. Based on recently excavated finds, Petraglia *et al.* (2007) have proposed that the hominins living here (whoever they were) were capable of behaviours that gave them sufficient flexibility to survive this super-eruption. These groups used a lithic technology that bore greater similarity to that of the MSA than the Middle Palaeolithic, providing a potential link back to Africa. Dates obtained from cultural layers at the site of Jwalapuram in the south central part of the subcontinent (Figure 5.5) indicate occupation between 77 ± 6 KBP and 74 ± 7 KBP (OSL). Although the date ranges overlap, they were obtained for layers below and above YTT ash – the grounds for this assertion. Predictions of an early presence of *H. sapiens* in this region on genetic grounds (Atkinson *et al.*, 2008; Oppenheimer, 2009) does also lend support to the possibility that the cultural layers at Jwalapuram represent occupations by modern humans. Substantive proof, though, will have to await the results of future research.

5.2.6 Northern and Central Asia

Moving north, in central Asia, Sel'-Ungur Cave (*c.* 39.94°N, 71.34°E) on the Sokh River, Kyrgyzstan, is one of the few sites in the region radiometrically dated to the Middle–Upper Pleistocene boundary. A U-series date on a travertine layer immediately above the cultural deposits has provided a minimum age for the archaeology of 126 ± 5 KBP (LU-936), which is also reported to be consistent with the character of both palynological and faunal evidence from the site (Vishnyatsky, 1999). The lithic industry (made primarily on jasper and slate pebbles) comprises a range of choppers, side-scrapers, notches and denticulates, and retouched flakes of Mode 1 character, marking the enduring persistence of such technology in this region (*see* Rabett, 2011). Six hominin teeth have been recovered from Sel'-Ungur, but they exhibit a mixture of dental features placing them somewhere between *H. erectus* and *H. neanderthalensis* (e.g. Vishnyatsky, 1999). Verifiable evidence of *H. sapiens* from the Upper Pleistocene is very scarce and not helped by the current paucity of sites of suitable or dateable antiquity. The only other notable material evidence is fragments of two mandibles and teeth found in 1962 and 1966, respectively, during excavations at the site of Samarkandskaya, Uzbekistan (*c.* 39.65°N, 66.94°E), but their exact provenance is also uncertain (Vishnyatsky, 2004).

Hominin presence continues to be recorded in the Altai region of Siberia during this period, but the evidence about which species are represented remains equivocal, and there are large error margins on the available radiometric dates. The current position is that groups are more likely to have been *H. neanderthalensis* than *H. sapiens*, though the evidence of Krause *et al.* (2010) does open the door to the possibility that another hominin species was also present (*see* Chapter 3). At Denisova Cave, tools of European Upper Palaeolithic character, such as burins, end-scrapers and blades, reportedly appear throughout the technological sequence but rarely exceed 10 per cent of the total lithic assemblage by layer in the lower layers, which are otherwise dominated by Mousterian elements. This ratio begins to reverse around Layer 13. Layer 12 is dated to 101 ± 25 KBP (RTL-612) by TL, where the difference is 30.7 to 69.3 per cent Mousterian. Thereafter, Upper Palaeolithic elements at first balance and then by 46 ± 2.3 KBP (GX-17062) – that is, Layer 9 – account for more than 90 per cent of tools, though 'archaic' elements are still present (Derev'anko and Markin, 1998).

Three kilometres to the southwest of Denisova Cave, near the confluence of the Karakol and Anui rivers, is the site of Ust'-Karakol (51.383°N, 84.683°E). This open-air site was excavated in 1986 and then again 1993–7 and contains a series of cultural layers inter-digitated between sterile lithological strata (Derev'anko and Markin, 1998). The site's stratigraphy is also subject to the effects of complicated post-depositional processes. The oldest cultural layer here (Layer 19A) is dated by TL to 133 ± 33 KBP (RTL-661; Derev'anko *et al.*, 2005) but contains artefacts that are described as 'poorly diagnostic' (Chlachula, 2001), and attributing a particular hominin species to their fabrication is not possible on current evidence.

5.2.7 East Asia

In northern China, a mandible from Tianyuan Cave (near Zhoukoudian) is considered likely of early modern human status, though no date is available at present to establish its antiquity (Trinkaus, 2005). From southern China, a skull and some postcranial remains belonging to *H. sapiens* were discovered in the late 1950s at Tongtianyan Cave (24.183056°N, 109.432222°E) in the Guangxi Zhuang Autonomous Region. The stratigraphy at this site is complex, and the skeletal remains referred to as the 'Liujiang hominin', though excavated carefully, were not removed by professional archaeologists, making their exact provenance open to question. Reconstruction of their most likely depositional context and U-series

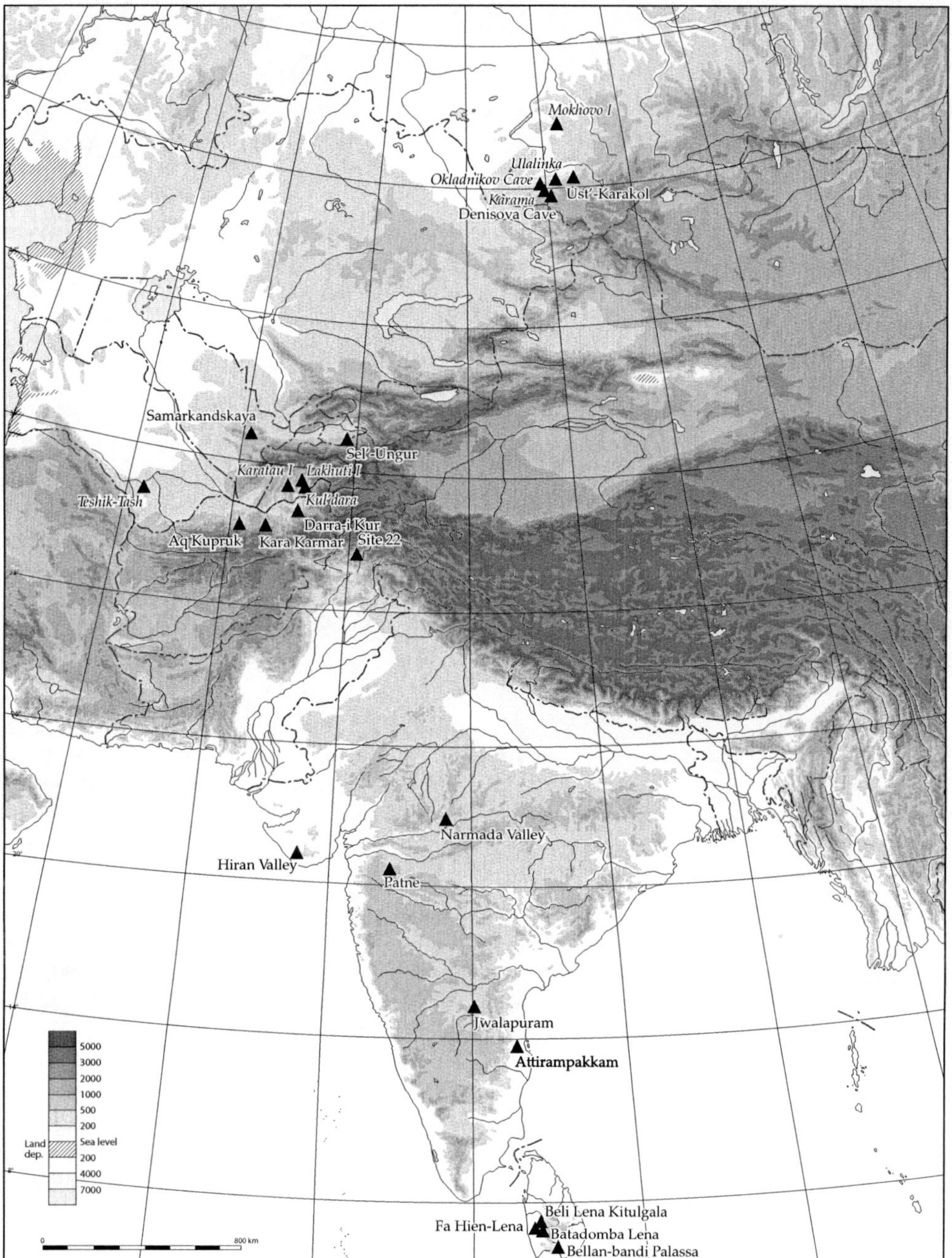

Figure 5.5. Map of South, Central and northern Asia, showing the locations of sites discussed in the text (sites in italics are discussed in Chapter 3). (Map drawn and prepared by D. Kemp.)

dating of accompanying sediments suggests a likely date of *c.* 139–111 KBP (Shen *et al.*, 2002). This antiquity would place it broadly within MIS-5e (and potentially as early as the end of MIS-6 and as late as MIS-5d). Early human remains have now also come to light at Zhirendong, in the Guangxi Zhuang Autonomous Region (22.287111°N, 107.512528°E; Liu *et al.*, 2010). This material, comprising two molars and the anterior portion of a robust *H. sapiens* mandible, was found together with a small faunal assemblage of late Middle to early Upper Pleistocene character. The fossil-bearing deposits were uncovered beneath a sequence of flowstone units, the basal unit of which is dated by U-series to 106 ± 6.7 KBP. Further work is clearly needed here, including, ideally, the direct dating of these fossils, before this evidence can be considered robust. The existence of Middle Pleistocene hominin fossils from China bearing a combination of *H. erectus* and *H. sapiens* features, such as evidenced in the crania from Yunxian (Tianyuan and Etler, 1992), and the possibility of other unknown ape (*see* e.g. Ciochon, 2009) and hominin species in the region, complicates identification. It signals a need for caution and the anticipation of further surprises within the evolutionary trajectory of the Asian record. These caveats accepted, the Liujiang and Zhirendong fossils do begin to present credible candidates for a very early *H. sapiens* presence in eastern Asia during the early part of the Upper Pleistocene, with all the consequences to the Out of Africa dispersal model that would entail (e.g. Wu, 2004).

5.2.8 Southeast Asia

In Island Southeast Asia, the penultimate of the Javanese biostratigraphic units, the Punung fauna, has now been dated by TL, providing a degree of confidence that it became established no earlier than 128 ± 15 KBP and no later than 118 ± 3 KBP (probably early MIS-5e; Westaway *et al.*, 2007). The loss of five teeth, putatively identified as human, that were associated with faunal remains from the Punung locales is a major blow; however, the recovery and analysis of a single specimen (P^3; PU-198) raises the possibility that *H. sapiens* had established some kind of presence in Java contemporaneous with the establishment of the Punung fauna (Storm *et al.*, 2005). This connection can be considered as no more than tentative at present, however.

The Punung represents the first appearance from Mainland Southeast Asia of tropical species such as *Pongo pygmaeus* and *Hylobates* sp., which require closed-canopy tropical forests. This presents, though, something of a problem. The humid conditions needed for closed-canopy rainforest to develop are generally thought to require an interglacial climate (*though see* Cannon *et al.*, 2009, for a different perspective). Rising sea levels, however, would have minimised the existence of effective land bridges and, consequently, would have inhibited rainforest spread into Island Southeast Asia. Fauna dispersal does not require the existence of a land bridge, but for these arboreal taxa to succeed in establishing themselves, it would be necessary. This in turn requires colder, glacial conditions to be prevailing at higher latitudes in the northern hemisphere in order to lower sea levels.

From *c.* 130–122 KBP, the seas in Island Southeast Asian (based on data from the Huon Peninsula in eastern Papua New Guinea) matched or exceeded modern values by up to 6 m (Chappell and Shackleton, 1986; Lambeck and Chappell, 2001: figure 3; Lambeck *et al.*, 2002a: figure 2a). Sea levels then dropped dramatically (*c.* 122–112 KBP) with current values attained *c.* 118 KBP, a low-stand of *c.* –62 m by *c.* 112 KBP and the recovery to *c.* –19 m by 106 KBP (Chappell and Shackleton, 1986; Lambeck and Chappell, 2001). If one takes the values presented in Westaway *et al.* (2007) without error margins, their entire span falls within a period when Java would have been an island, making the spread of strictly arboreal fauna an unlikely proposition (*see* van den Bergh *et al.*, 2001: 391–2). Within the error margins of the more recent of the Westaway *et al.* (2007) dates, there is a window of opportunity. This would have to be after 118 KBP, coming forwards in time, and before *c.* 112 KBP, when sea levels were below current values and specifically during the period when sea levels were at a minimum of –30 to –40 m below modern; any higher than this, and there would no longer be a contiguous land bridge between Java and Sumatra–Peninsular Malaysia (Sathiamurthy and Voris, 2006; Voris, 2000).

Going backwards in time, the next available period for rainforest colonization occurred during the rapid warming at the end of MIS-6 (Brewer *et al.*, 2008). Between *c.* 140 KBP and *c.* 130 KBP, sea levels rose from a low-stand of lower than −135 m to 0 m (*after* Lambeck and Chappell, 2001: figure 3; Lambeck *et al.*, 2002a: figure 2a). Interestingly, climatic amelioration may have reached interglacial temperatures up to 6000 years before the sea attained its modern value at *c.* 130 KBP (Lambeck *et al.*, 2002a). This is perhaps the more likely time when one might expect to see a forest expansion episode into Java. If the Holocene is a reasonable point of comparison (and this is by no means certain), in the tropics, this earlier transition will have coincided with increasing humidity and elevation of temperatures climbing towards interglacial values. It also falls within a known period of major regional vegetation change, *c.* 135–125 KBP (Hope *et al.*, 2004; van den Kaars and Dam, 1995). A modern human presence in Java at this time, while not impossible, on current evidence seems rather unlikely, but raises the possibility that human movements into Java may have been out of step with the initial colonization of the island by rainforest.

In summary, however it was that our early antecedents arrived in the Mediterranean provinces north of the Sahara, this early dispersal of *H. sapiens* is generally considered to have been ultimately an abortive one. Either groups retreated into Africa because of worsening environmental conditions *c.* 75 kbp or they died out. Oppenheimer (2009: 7) stated that there are no mtDNA or Y chromosome lineages outside of Africa that come close to the kind of time depth of the early population of *H. sapiens* known to have been living in the Levant at around this time (see later). A lack of evidence, he argues, indicates that these people did not contribute to the modern gene pool. That said, the appearance of haplogroup L3, the African clade that gave rise to the two basal non-African clades haplogroups M and N, is quoted by Macaulay *et al.* (2005: 1035) as dating to *c.* 84 kbp – the molecular clock calculation given by Oppenheimer (2008) is 83,500 ± 8400 bp. On this basis, one could just as easily argue that the dates are in fact considerably closer to the time period covered by the new Skhūl V U-series dates. This and the intriguing data emerging out of the rest of Asia suggest that the legacy of this nascent population expansion should probably not be written off just yet.

5.3 MIS-5d to the Last Termination (116-22 KBP)

As we come forward in time into the last glacial, the footprints of modern human activity beyond Africa become more discernible and manifold. In this section, we follow that trail through the countries bordering and within the southern oceans.

5.3.1 Climate Conditions During the Last Glacial Period

Although generally thought to be less of a primary driver of climate change during the Quaternary, the uplift of the Tibetan Plateau affected prevailing climates at middle to high latitudes across Asia, Europe and even North America (Anderson *et al.*, 2007). The Asian monsoon is intimately tied to the effects of summer heating of the lower atmosphere caused by the vast expanse (*c.* 2.5 million km^2) and altitude of the plateau (Harris, 2006). Currently, extreme low pressure develops over Asia during the northern summer. The pressure gradient that is created between the continental land and sea surface temperatures draws in moisture-laden air from the ocean. As this air reaches the Himalayas and is forced to rise, it gives up its moisture and rain is deposited across southern Asia before it reaches the continental interior: this is the rainy summer monsoon. By contrast, the intense cooling of the Asian continent during the winter reverses this, pushing cold, dry air out over the southern fringes of the continent towards the ocean: the dry winter monsoon (Kellman and Tackaberry, 1997). Large parts of the European and Asiatic landscapes were already prone to being comparatively open and quite arid environments and this situation would have been exacerbated by the Pleistocene glaciations.

The presence of the northern ice sheets and colder sea surface temperatures (particularly during Heinrich events) would have arrested the summer thaw across the Tibetan Plateau. Extensive and enduring winter snow cover over this landscape, together with the cooler seas, would have depressed the important thermal differential between the land and ocean, weakening the strength of the rainy summer monsoon and increasing the strength of the dry winter monsoon (Cosford *et al.*, 2008). In Europe, the expansion of the Fennoscandian ice sheet and corresponding extension of this winter Asiatic high-pressure zone blocked incoming warm air circulating across the Atlantic and displaced southwards the westerly jet stream, which is also associated with precipitation-bearing cyclones (Fernández, 2006). The result would have been the intensification of existing aridity and the creation of a steppic landscape without modern analogue that would have been underlain by permafrost and inhabited by predominantly cold and arid-adapted flora and fauna (Guthrie and van Kolfschoten, 2000; Klein, 1999).

Palaeoecological understanding of the way that species respond to climatic shifts has undergone considerable revision since the 1970s. Where previously it had been assumed that adaptation occurred predominantly at the level of a biological community, with a spectrum of presently inter-related taxa oscillating as a single unit in response to climate, it became apparent that in fact a great deal of that response occurs at an individual rather than community level (Huntley, 1991; Huntley and Webb, 1989; Roy *et al.*, 1996). The models changed from bands of broadly stable, enduring habitats oscillating back and forth in sync with glacial cycles to ones in which the palaeobotanical and palaeofaunal evidence was indicating species mixtures for which there are no immediate modern analogues: novel faunal associations, for example, included species found at high latitudes, with those today found in continental or temperate landscapes.

Although trees invaded as far north as Siberia during MIS-3, in Europe the landscape remained largely open during this comparatively warm phase of the last glacial. Forests tended to be confined to riparian environments and scattered refugia in the three southern peninsulas – the Iberian, Italian and Balkan – and possibly also parts of Greece, Turkey, and the Caucasus mountains (Anderson *et al.*, 2007; Médail and Diadema, 2009). These areas contain a higher genetic and taxonomic species diversity compared to more northern parts of Europe (Hewitt, 2003); this may have been a factor in explaining why they provided favourable conditions for human settlement. The key to the landscape's composition, though, lay in the sheer frequency of glacial climate fluctuations. In Europe, the climate was oscillating too quickly for slower sylvan taxa to recolonize the land during periods of amelioration before cold snaps cut them back again. The frequency of climate change, therefore, applied strong selective pressure against slower-adapting taxa, favouring those that were either fast colonizers or had wide adaptive ranges. In essence, swathes of the western European and Asian steppic environments were populated by species that were the best adapted to cope with climatic instability. This biological community certainly included bands of *H. sapiens*. Their presence suggests that the climatic characteristics of Upper Pleistocene Eurasia may not only have encouraged colonization of this landscape by mobile human groups but was probably also a force in structuring the subsistence strategies they employed to establish that presence.

5.3.2 Modelling the Eastward Spread of Modern Humans Out of Africa

The Out of Africa II genetic model, which marks the radiation of *H. sapiens* and was originally proposed by Cann *et al.* (1987), has marshalled considerable support over the last 20 years, with studies repeatedly placing the deepest ancestry for *H. sapiens* clearly in sub-Saharan Africa (e.g. Cavalli-Sforza *et al.*, 1988; Forster and Matsumura, 2005; Harpending *et al.*, 1993; Goldstein *et al.*, 1995; Harpending and Relethford, 1997; Lahr and Foley, 1998; Relethford, 1995, 2008). Despite this, reflection on the subject does continue. For example, Relethford (2008) explains that while a range of genetic analyses supports an African origin for *H. sapiens*, the degree of observed genetic variability is not necessarily an indication that there has been greater time depth for the accumulation of these genetic mutations; it can also be explained by a larger long-term effective population. Templeton (2007) also argues that as mtDNA and Y-DNA have the shallowest coalescence times of all genetic regions found in humans, and as such, they do not have sufficient time depth to accurately account for the distribution of human populations at a chronological horizon before this recent phase of expansion. Striking a similar tone, Balloux (2009) cautions against the use of mtDNA markers as a reliable way of tracking demographic change through time, speaking out against the crucial assumption that these markers are evolving neutrally – and so are not subject to natural selection; some studies suggest this may not be so. While acknowledging the contribution of mtDNA studies, Balloux proposes that it is time to end our overreliance on these problematic portions of the human genome. Although they offer fascinating new windows into the possible journeys of our past, we cannot yet be certain that what we are seeing through them is actually an accurate portrayal of what lies on the other side of the temporal glass. What they give us are possibilities, and the following models are probably best viewed from that standpoint.

Presented in Harpending *et al.* (1993) as the best fit for the mtDNA evidence that was available at that time, the Weak Garden of Eden hypothesis was one of those proposing that the modern global population of humans emerged from an enclave of sub-Saharan *H. sapiens*. Harpending and colleagues suggested that the data were consistent with either an early expansion, *c.* 100 kbp, and subsequent bottleneck event (a significant reduction in population size) or early modest growth and then slow expansion. The former of these two positions was later adopted formally. The colonizing daughter populations appeared then to remain relatively isolated from one another until after *c.* 50 kbp. During the second half of the Upper Pleistocene, a more rapid expansion then occurred independently from within each of these daughter populations (Harpending *et al.*, 1993: 495). The model did not call for population replacement (archaic with modern human), though the final rapid expansion that resulted was argued to have been facilitated by a technological change from Mode 3 to Mode 4.

While there remains some uncertainty about exactly when morphologically *H. sapiens* appeared in Africa (*see* e.g. Bräuer *et al.*, 1997), the molecular clock continues to indicate a Middle Pleistocene age for the oldest branch of the surviving human genetic tree (Oppenheimer, 2009). The proposed coalescence date of Harpending *et al.* (1993) has been revised backwards to *c.* 180,000 kbp based on a number of studies into mtDNA and Y chromosome microsatellite loci (Goldstein *et al.*, 1995; Kivisild, 2007; Lahr and Foley, 1998). The relative contemporaneity of these findings is consistent with a loss of genetic diversity around this period – supporting the likelihood of a bottleneck that was potentially tied to climatic events within the MIS-6 glaciation, perhaps culminating during its final glacial maximum *c.* 140 kbp (Lahr and Foley, 1998). This was followed by the expansion episode northwards, perhaps across the Sahara and certainly into at least the Levant, and southwards into southern Africa, before supposed population contraction rendered some of these groups extinct and left the remaining groups on what would be relatively isolated trajectories until the next and much larger radiation *c.* 50–40 kbp (*following* Harpending *et al.*, 1993).

There are potential issues with using glaciation as the cause of this root population contraction though. Firstly, while severe, MIS-6 was not without interstadial events that were elsewhere possibly even linked with archaic hominin expansions as far north as southern

Siberia (see earlier). There is nothing that particularly marks the MIS-6 glaciation as inherently worse than any other; leaving us to question why contraction occurred then.

Secondly, if the size of the human gene pool were cut back during this period, one would expect that this should appear in the ledger of fossil discoveries as a noticeable lacuna by virtue of simple probability (one is less likely to find a smaller population); however, this does not appear to be the case: many of the early *H. sapiens* discovered in recent years come from this time window and from a wide geographic spread, from northwestern to eastern Africa. For example, indirect dating of the Omo I and II hominins from Kibish, Ethiopia, by $^{40}Ar/^{39}Ar$ (McDougall *et al.*, 2005) has produced a weighted mean age of 196 ± 2 KBP, very close to the boundary between MIS-7 and MIS-6. Still in Ethiopia, the Upper Herto fossiliferous deposits from the Middle Awash region are effectively sandwiched between two dated contexts, placing them within MIS-6. The underlying sediments are dated using, principally, the $^{40}Ar/^{39}Ar$ method to and age of *c.* 160 ± 2 KBP. The tuff which caps the fossil-bearing layers could not be dated by $^{40}Ar/^{39}Ar$; however, a second tuff layer 6 m above it has been ascribed to a known second tuff that is dated to 154 ± 7 KBP (Clark *et al.*, 2003; White *et al.*, 2003). In the northwest, there is the U-series date of 160 ± 16 KBP on the Jebel Irhoud juvenile (Smith *et al.*, 2007). The oldest genetic expressions of expansion do seem to lie in the east (Excoffier and Schneider, 1999; Lahr and Foley, 1998), but whatever the cause(s) of contraction here beforehand, the African *H. sapiens* fossil record does not seem to have become correspondingly depleted. In other words, whatever caused contraction in the *H. sapiens* enclave that would later give rise to the world's current population, we cannot yet be certain it affected all parts of the wider early human population in Africa in the way that a glaciation might.

The detailed proposition by Ambrose (1998a), suggesting that the YTT super-eruption may have been an alternative cause of the bottleneck during the early Upper Pleistocene, warrants consideration. Potentially commensurate with the beginning of the early human expansion, the Toba eruption could have inhibited movement for several millennia. While Ambrose cogently presents an argument for the severity of Toba, there remains uncertainty about the true impact this event had on human history. 74.5 KBP is considered to be the end of GIS-20 (Shackleton *et al.*, 2004), a transition that accords closely with the best current estimate for the Toba eruption of 74 ± 2 KBP (Oppenheimer, 2001). Clive Oppenheimer argued that several unresolved issues about the YTT eruption's duration, intensity and height make it problematic to determine its effect on global climate. If the recent redating of the OTT becomes orthodoxy, this earlier (albeit smaller) eruption of the super-volcano occurred coincidentally with the transition from glacial (MIS-20) to interglacial (MIS-19) conditions, casting any necessary linkage between such events and climatic deterioration into some doubt (Lee *et al.*, 2004).

As noted, evidence has now emerged for continuity of hominin occupation across the YTT tuff horizon in India (Petraglia *et al.*, 2007). In Peninsular Malaysia, immediately adjacent to northern Sumatra, numerous ash deposits ascribed to Toba are reported as occurring at several locations in Perak, including the northern half of the Lenggong Valley (Chung and Yin, 1977; Walker and Sieveking, 1960). Similar deposits are reported from four other states (Selangor, Negeri Sembilan, Trengganu and Kelantan) besides Perak (Zuraina, 1990). Archaeological material from excavations by University Sains Malaysia (USM) in the vicinity of Kota Tampan – USM re-excavations are labelled herein as KT1987 and KT2005 – was recovered interspersed within and below an ash layer identified as YTT (Zuraina, 1990). This ash was first considered likely to be from Toba by Collings (1938: 576) and was locally dated within the peninsula by fission track to 68 ± 7 kbp (Chesner *et al.*, 1991). Recently, a new OSL date has also been obtained from the main cultural layer uncovered at KT2005. Details about the error margin were not available at time of writing, but the date is reported as *c.* 70 KBP (Hamid, 2007; Mokhtar, 2007b). The refined age will be of considerable importance in placing hominin occupation at this site relative to the eruption, as will further investigations in this archaeologically rich valley. At the present time, while there are older sites locally, a hiatus does appear to exist between KT2005 and the next oldest radiometrically dated site, Bukit Bunuh, which has Upper Pleistocene levels dated by OSL to 39 KBP (Mokhtar, 2007a). Nonetheless, reliable radiocarbon dates from post LGM sites in Lenggong, such as Gua Gunung Runtuh (5.1175°N, 100.9675°E; Zuraina, 1994), testify to a long history of occupation – one that may or may not have been interrupted around the time of the Toba event. For the moment, it is this issue of tighter chronological convergence that raises the largest question mark over Ambrose's thesis. In particular, while the date of the Toba eruption itself is comparatively robust, the same cannot be said for the degree of resolution accorded to the bottle-neck event ahead of the rapid *H. sapiens* expansion (Oppenheimer, 2001). Although a plausible cause of population constriction, the true impact of the YTT eruption on early human populations remains to be determined.

Turning to the diaspora itself, during the early part of the last glacial, two routes have been proposed for what is argued to be the first sustained migration beyond the African continent: a northern and a southern route. The northern route represents the traditional view that humans colonized the Levantine corridor before entering Europe, Asia and ultimately the Americas. A northern exit required the negotiation of hyperarid desert zones. Although windows of opportunity have been identified within the kind of time frame required for the early colonization of the Levant (as indicated earlier), it remains fair to say that climatic conditions are likely to have played a significant role in dictating where and when the desert (be it through the Sinai Peninsula or the western Sahara) could have been breached, or re-entered. This model has difficulty accounting for the evidence of modern humans in Australia by *c.* 50 KBP or the earliest secure date on a *H. sapiens* fossil in Southeast Asia: 35–37 KBP – a U-series direct date taken on the Deep Skull from the West Mouth of the Niah Caves (Barker *et al.*, 2007) – coming as they do from as early if not earlier than the most likely first entry of *H. sapiens* into Europe (Trinkaus, 2005). The genetic evidence also suggests a chronologically deep inter-regional split between populations from African, Asian and European (and Oceanic) groups to the tune of 50,000–75,000 years – assuming a generational turnover rate of 25 years and a constant global population size model of *c.* 10,000 individuals (Mountain and Cavalli-Sforza, 1997). The contemporaneity of modern human appearance in the archaeological records of such geographically removed regions as well as the existence of considerable genetic distance between them has increased attention towards identifying an alternative southern exit route out of sub-Saharan Africa.

5.3.3 The Southern Route Hypothesis

The Southern Route hypothesis (Lahr and Foley, 1994; Stringer, 2000) proposes that early modern humans left Africa during the first half of the Upper Pleistocene, following a path of dispersal along the coastlines of the Indian Ocean into Southeast Asia and Australia. Interpretation of the increasingly central genetic evidence for this Upper Pleistocene spread has fallen into two camps: one that favours a single dispersal via this route with daughter populations budding off it, the other favouring multiple migration events from an African source.

Advocates of the single dispersal model argue that a subset of the population following the coasts of the southern oceans struck north, probably from or through South Asia, into Eurasia. Under this scenario, the ancestral genetic enclave in question is calculated to have arisen in the vicinity of the Indus Valley *c.* 50 kbp before heading northwest via the Zagros Mountains and the Tigris and Euphrates valleys towards Europe and the Levant (rather than from Africa via the Levant). A second movement *c.* 33 kbp is calculated also to have arisen in South Asia and tracked first northwards into central Asia and southern Siberia before also turning westwards to colonise Europe (Oppenheimer, 2004, 2009). The genetic evidence, though, remains equivocal on the matter, and the possibility exists that the early southern dispersal happened independently of the spread into Eurasia. Recent work by Rasmussen *et al.* (2011), for example, sequencing genetic data from historical Aboriginal hair finds good reason to support this contention.

The significance of the Southern Route hypothesis is strongly indicated in the fact that of the two daughter mtDNA haplogroups of L3 (M and N), which populate the world outside of Africa, both are present (and to a high level of diversity) in South and Southeast Asia – one of two areas suggested to have an ancient history of being 'population generators' (the other being China; Atkinson *et al.*, 2008; Webb, 2006). M and N also appear with considerable time depth and regional specificity (Kivisild *et al.*, 2003). By comparison, only representatives of haplogroup N are present in western Eurasia (including the Levant), where they are also less diverse and hence interpreted as younger (Oppenheimer, 2009). Further support had also come from mtDNA analyses of modern-day populations in the Andaman and Nicobar Islands in the Indian Ocean (Endicott *et al.*, 2003; Thangaraj *et al.*, 2005). These data suggested that the Nicobarese show a close genetic relationship to other Southeast Asian populations and hence probably colonized the islands from the east at some point since the LGM. The Onge and Great Andamanese sections of the population, however, exhibit two ancient maternal genetic lineages (M31 and M32) that evolved as a result of long-term isolation after the islands were colonized and were at that time not known from any modern South or Southeast Asian populations. This suggested them to be vestiges of a much older population and a candidate for ancient population dispersal into the region (Thangaraj *et al.*, 2005). Recent reassessment of the M31 and M32 haplogroups and refinements in protocol, however, now finds evidence of M31 lineages in northeast India and considers this, rather than Mainland Southeast Asia, as the probable source region of Andaman populations (Endicott *et al.*, 2006; Wang *et al.*, 2011). Significantly, the recovery of these lineages in India means that the current estimate for the initial colonization of these islands may have been later, sometime between 30 kbp and the LGM.

Analysis of mtDNA motifs among the Orang Asli indigenous groups in the Malay Peninsula currently shows many to be unique, implying a time depth to the last recent common ancestor of between 44 and 63 kbp (Hill *et al.*, 2006; Macaulay *et al.*, 2005). Although the reliability of the ρ (rho) statistic, which is commonly used in dating estimates of the mtDNA haplogroups, has been called into question (Cox, 2008), an early divergence of the Malaysian groups still finds strong support (Stoneking and Delfin, 2010). Macaulay *et al.* (2005) proposed that the antiquity of L3, with a calculated coalescence time of 83,500 ± 8400 bp, makes it likely that there was a single early exit from Africa and that this probably progressed via the Red Sea and coastal margin around the Indian Ocean to Southeast Asia. They concur that the colonization of Europe could have appeared as an offshoot of this single dispersal, which may have been halted by climatic conditions, and that the ancestors of modern populations here and in the Levant entered via Asia rather than separately from Africa. The possibility that the flagship culture of *H. sapiens*' arrival into Europe, the Aurignacian, may have had its origins in the Zagros Mts, south of the Caspian Sea in Kurdistan, along this route, does have its proponents in archaeological and genetic fields of research (e.g. Finlayson and Carrión, 2007; Oppenheimer, 2003: 133; Otte, 2007; Otte *et al.*, 2007), but it is not yet widely accepted.

The genetic picture from the Asian interior suggests that a significant proportion of that region's present population derives from the second half of the Upper Pleistocene. Work by Comas *et al.* (2004) on mtDNA from a sample of 12 modern central Asian populations concluded that most of its present genetic distinctiveness derives from groups that appear to have come from eastern as well as western Asia, with elements of the East Asian haplogroups G and D calculated to have appeared here *c.* 30–25 kbp. Y chromosome analysis carried out on modern central Asian populations by Wells *et al.* (2001) suggested that a very small fraction of the gene pool may date to the populating of central Asia by anatomically modern humans as early as 50–40 kbp.

The clarity of mtDNA evidence for East Asian and Siberian populations is still poorly resolved. Most detailed studies have not focused on patterns within Asia but have rather been more concerned with the spread of humans into the Americas (e.g. Fagundes *et al.*, 2008; Schurr and Sherry, 2004). Derenko *et al.* (2003) have identified that modern south Siberian populations show considerable mtDNA diversity that could link them to an early phase of Asian colonization. The highest incidence of west Eurasian lineages, marking a clear Late Pleistocene immigration event, appears in the Altai region. In all, 17 per cent of the total mtDNA gene pool (480 individuals sampled) were found to comprise contributions from the west, as compared to 81 per cent of Asian origin. This was taken as an indication that most *H. sapiens* migration into southern Siberia came from central and/or eastern Asia and not the west, though their appearance in the region is calculated to a similar period based on coalescence dates. For example, haplogroup C has a calculated date of 38,400 ± 9900 bp, while there is a date of 37,500 ± 6700 bp for haplogroup D. Rather than colonization from the west, the Asian population may well have been a major contributor to west Eurasian populations. Oppenheimer (2003: 150–1) proposes how mtDNA and Y chromosome evidence for a post-Aurignacian second movement of *H. sapiens* into Europe (attributed to the Gravettian cultural complex) could come from a genetic population ultimately arising from South Asia that moved east of the Caspian Sea and into southern Siberia before moving westwards into Europe.

Although the genetic evidence is both fascinating and compelling in many respects, it is important to reiterate that aspects of the process of constructing descent trees continue to be inexact and liable to significant shortcomings (*see* Endicott *et al.*, 2009, for a detailed discussion). In addition to the controversy now surrounding the accuracy of coalescence dates (Cox, 2008), the dates themselves remain relative rather than absolute indicators of population antiquity (*see* Goldstein *et al.*, 1995, for efforts to establish absolute genetic dates). These chronological estimates are dependent upon a range of assumptions such as the age of the split between the line leading to hominins, on one hand, and to the other great apes, on the other. It also relies on estimates of early human population size, which may be questioned. Webb (2006) has argued that Upper Pleistocene populations may have been considerably larger than is generally believed. Furthermore, the rate of generational turnover used in genetic studies varies considerably: for example, 20 years (Takahata *et al.*, 2001), 25 years (Mountain and Cavalli-Sforza, 1997) or 27 years (Goldstein *et al.*, 1995), with amplified consequences over the kind of time periods in question. Analysis by Smith *et al.* (2007) used dental development to demonstrate that the *c.* 160 kbp Jebel Irhoud juvenile *H. sapiens* had a similar developmental history to contemporary humans. This at least increases the probability that early *H. sapiens* generational turnover rates were broadly comparable to modern ones for modelling purposes, though many other environmental, nutritional and disease-load factors will still have affected turnover rates through time as well as geographically for different human enclaves. The genetic record is also only that of surviving populations. It has little to contribute to any understanding of the that elusive component of early *H. sapiens* diaspora: dispersals or enclaves of people that did not survive to contribute genetic material to the current global population. This may well have been the majority. It is also the case that comparatively minor expansion events may have been rendered genetically invisible by much larger scale demographic dispersals (Lahr and Foley, 1998). Finally, it is also the case that a single dispersal event is itself an unusual and somewhat 'counter-intuitive' scenario that is not seen in any other continent-to-continent movement of modern humans (Oppenheimer, 2009). As Webb (2006) has argued, if Africa was the original generator of hominin populations, what event or combination of circumstances conspired to halt such a long-established production process during the first half of the Upper Pleistocene? And if something did stop further emigration, why is there no evidence that Africa's internal population continued to grow in the way that South Asia's and China's have done? Such questions could be partially answered by the multiple dispersal hypothesis, including later movement of African people via the northern route into the Levant *c.* 45 kbp. Indeed, there is some genetic support for a Late Pleistocene dispersal from eastern Africa into north Africa and the Levant based on evidence from particular Y chromosome lines such as E-M78δ; however, this migration is calculated at 14.7 ± 2.7 kbp (Cruciani *et al.*, 2004) and therefore much later than the primary L3 (M and N) dispersal. Another possibility, more promising for multiple dispersal advocates, is that whereas the M haplogroup is linked to an east African dispersal – and a calculated migration to South Asia *c.* 64.8 ± 15 kbp – the N haplogroup is more widely dispersed within Africa. It could have followed the same southern route to South Asia or, potentially, could have arrived there via a northern Levantine migration *c.* 50 kbp (Barnabas *et al.*, 2005).

Genetic evidence is opening up a vital new perspective on the Upper Pleistocene dispersal of *H. sapiens*, but there is some way yet to go in honing its reliability. More optimistically, the recent sequencing of a complete *H. neanderthalensis* genome (Green *et al.*, 2008, 2010) raises the realistic possibility of tracing genetic signals from dead and extinct populations. We can expect that future developments in ancient DNA research will provide equally important new, controversial and probably complicating perspectives on the understanding of our Pleistocene history.

5.3.4 South Asia

If we assume that the genetic picture of human dispersal is essentially correct, and that the southern route played a key role in early moderns' dispersal, one of the main centres for investigation into the archaeological evidence has to be South Asia (Atkinson *et al.*, 2008). Environments across the subcontinent during the Upper Pleistocene appear, broadly, to have been characterized by cold and arid conditions with widespread thinning of vegetation cover to grassland and forest-patch habitats (Misra, 2001). Ice core evidence from both the Qinghai–Tibetan Plateau in the east (Thompson *et al.*, 1997) and sediment cores off the coast of Pakistan in the west (Schulz *et al.*, 1998) indicates that the punctuated climate of more northerly latitudes was also experienced here between *c.* 110 and 10 KBP. The stadial–interstadial oscillations revealed in the Greenland ice cores are paralleled by shifts in the intensity of the summer monsoon across the subcontinent. This would have had the effect of temporarily opening up and then reclosing habitats to occupation by human groups, as is suggested, for example, by palaeoenvironmental and archaeological evidence for this period in the Thar Desert, in the northwestern part of the subcontinent (Kar *et al.*, 2001).

A number of sites in Afghanistan and Pakistan show signs of occupation from close to the base of the Late Pleistocene. These include, Darra-i Kur, Aq Kupruk, Kara Karmar and Site 55. Excavated in the 1960s (Dupree *et al.*, 1972), deposits in the cave of Darra-i Kur in northwest Afghanistan (36.733333°N, 69.983333°E) contained a stone tool industry interpreted as Mousterian, comprising *c.* 800 primary, utilised and worked flakes, as well as Levallois flakes and points, hand-axe forms, flake-blade forms, débitage, flake cores, various forms of scraper (including side and oblique-scrapers) and one possible dihedral burin. A fragment of hominin skull was also found in association in these Mousterian levels (from Cut LCII: 200). Comparison against modern human and Neanderthal remains was carried out by J. L. Angel of the Smithsonian. Angel proposed that although the specimen carried a mosaic of traits, it was closer to *H. sapiens* than *H. neanderthalensis*, drawing parallels to the Skhūl and Qafzeh fossils as possible contemporaries. Only one date is available from the Mousterian levels at this site: 30,000 + 1900/−1200 bp (GX-1122). Geochron Laboratories, who ran the original sample, advised considerable caution over its accuracy (Dupree *et al.*, 1972: 13). Although still treated as contemporaneous with the Sri Lankan fossil remains (e.g. Sonakia and de Lumley, 2006), the antiquity of this specimen is highly questionable. Its associated cultural material could imply an older or a younger age just as readily, given that it is a feature of the South Asian sequence that there is a late survival of Middle (and even Lower) Palaeolithic technologies (e.g. Baskaran *et al.*, 1986; Misra, 2001). Petraglia (2007) has stated that in all likelihood, modern human groups entering the region were probably doing so bearing similar Middle Palaeolithic technologies to resident hominins – something that may have reduced any competitive advantage the former might have had but which is also going to conflate our understanding of who made what industries. The Darra-i Kur fossil may yet provide an important part of the jig-saw towards reconstructing the route modern humans took into the subcontinent.

To the west of Darra-i Kur, but still in Afghanistan, more than half a dozen rock shelters and open-air sites have been identified on the terraces of the Dalkh River, close to the town of Aq Kupruk (36.083056°N, 66.834444°E). Four of these localities were excavated by Dupree and colleagues in 1962 and 1965. Aq Kupruk (AK) II and III were found to contain Palaeolithic facies, termed the 'Kuprukian'. The overwhelming majority of struck, but unmodified, core products were primary flakes; something taken to suggest that most of the AK tools were probably made on site. Scrapers, often carinated, steeply retouched end-scrapers made on flakes or blades, constituted a prominent part of the formal tool assemblage. Burins also were in evidence, and the limited variability of forms (most being of the dihedral type) is considered a feature in parallel with other Upper Palaeolithic technology in the Middle East and elsewhere. Retouched flakes outnumbered retouched blade forms, which reached up to 10 cm long and did not conform to any clear techno-typology. A limited number of points were recovered – most being classified as 'pseudo Levallois points'. Significantly, there was also a strong micro-blade component to the industry, including tiny, expertly struck, micro-cores that rarely exceeded 50 mm long. These micro-blades formed the largest non-débitage type in the assemblages. In the European sequence, microlithic forms only start to be found in quantity within post-Pleistocene, 'Mesolithic' cultures. In South Asia, there is no such separation between micro and macro-blade industries, with the former appearing in association with otherwise Upper Palaeolithic toolkits. The level of tool standardization beyond these micro-blade forms is also not comparable with that seen in the Upper Palaeolithic industries of Europe. As such, they are viewed as an internal regional development rather than evidence of an incoming population (James and Petraglia, 2005). AK III remains undated. AK II, an open-air location, has two Upper Palaeolithic components separated by water-lain deposits. Though considered similar enough to be lumped together (Dupree *et al.*, 1972: 14), only the upper of these (B) has been dated, giving an age of 16,615 ± 215 bp (Hv-1358). If calibrated, this would be close to the end of the LGM *c.* 19.7 cal. KBP. The AK II (A) assemblage may in fact be slightly older (Kolb, 1980: 130; cf. Davis and Ranov, 1980). Both, though, are younger than Kara Karmar, a cave site to the northeast, near the town of Samangan (36.265°N, 68.016389°E).

Excavated in 1954, C. S. Coon reported four cultural levels at Kara Karmar, apparently alternating between amorphous flake industries (I and III) and an Upper Palaeolithic blade industry (II) – biased towards scrapers and some micro-blades but with no evidence of burins – and a micro-blade industry (IV). Early radiocarbon dates suggested an age of *c.* 34 kbp for component II (Dupree *et al.*, 1972: 74). Currently, the earliest industry classified as Upper Palaeolithic (i.e. flake and blade; including some micro-blade forms) comes from Pakistan, at Site 22 (*c.* 33.52°N, 73.164167°E) near Riwat (Dennell *et al.*, 1992), where loess deposits overlying the cultural remains have been dated by TL to *c.* 45 kbp. With no evident organic preservation at the site, there remains little to distinguish who made it.

Field *et al.* (2007) sought to compare the Middle Palaeolithic record of this region against a mathematical model of early *H. sapiens* dispersal and later population growth. Using geographic information systems (GIS) and 'least-cost' route modelling, they concluded that a western coastal entry is more likely to have been favoured than one via the Hindu Kush Mountains and the Asian interior. Although groups may have followed maritime and interior riparian corridors, residual archaeological evidence of this route remains to be found, though as they say, coastal inundation is likely to factor here. Some near-coastal Palaeolithic sites do exist in India. For example, in the northeast, in the Saurashtra Peninsula, assemblages dated by $^{230}Th/^{234}U$ have been recovered from a number of locales in and around the Hiran Valley. These assemblages contain Lower or Middle Palaeolithic industries but, again, have comparatively recent dates of 190–60 kbp and *c.* 58 kbp, respectively (Baskaran *et al.*, 1986). Investigation at another location, Attirampakkam (13.230556°N, 79.888889°E), lying on the floodplain of the Kortallayar River *c.* 40 km from the modern southwest coast of India, has produced a deep but undated occupational sequence containing typologically Lower, Middle and Upper Palaeolithic assemblages (Pappu *et al.*, 2003) and seemingly long-term interest in this part of the landscape.

A further proposal by Field and her colleagues – that the rapidity of any early modern human dispersal may also have contributed to the paucity of evidence – relies on the 'Wave of Advance' population model. In its original form, this predicted that population growth and migratory activities create a diffusionary process whereby a population expands at a steady rate (Ammerman and Cavalli-Sforza, 1984: 67). Variants of the model have had a long and often contentious history, especially in relation to the interpretation of the initial colonization of the Americas, starting with that proposed by Martin (1973). The appropriateness of its application to South Asia remains to be seen. Nonetheless, as an exercise in the predictive modelling of early population dispersal along the edge of the southern oceans, the work of Field and colleagues provides an important benchmark.

The wide distribution of Lower and Middle Palaeolithic sites across most parts of India – excepting the Ganga plains, Western Ghats, the northeast and extreme south of the country (Misra, 2001) – is a reasonable indicator that it has sustained a stable and long-term human population, but there are problems. As already indicated, in India, populations were supposed to have been among the highest in the Upper Pleistocene and the source region of major expansions into central, northern and western Eurasia *c.* 50 kbp, yet there are fewer Upper Palaeolithic assemblages than Middle Palaeolithic ones. This may be attributable to a lack of survey coverage across what is a vast country, though there has been more than a century and a half of archaeological work here (Field *et al.*, 2007; Newbold, 1844). Of course, there is the possibility that, given the region's technological character, modern humans may have been responsible for making some Middle as well as Upper Palaeolithic industries.

The extreme paucity of Pleistocene fossil hominin remains certainly complicates this issue and is particularly curious given that considerable quantities of faunal remains have been recovered, if often in secondary contexts, from many of the country's river systems (Chauhan, 2008; Jones and Pal, 2009). At present, there is only the Narmada specimen. Though currently dated (if provisionally) to *c.* 236 KBP (Cameron *et al.*, 2004) and interpreted as *H. heidelbergensis* (Kotlia and Joshi, 2008), the results of new fieldwork re-examining the Quaternary deposits of the Narmada Valley may call for yet another revision of its place in the country's Palaeolithic record. Patnaik *et al.* (2009) have shown that the gravel beds at Hathnora that yielded the fossil are considerably reworked. This could potentially reposition the date of the fossil itself to as late as *c.* 48 kbp. Given its archaic physiology, this would exacerbate efforts to sift early modern human settlement from that of other resident archaic hominins. Darra-i Kur aside, the oldest *H. sapiens* fossils currently recognised from the subcontinent have come from Fa Hien-lena Cave and Batadomba-lena, both in Sri Lanka (Perera, 2010).

Although the archaeological potential of Fa Hien-lena had been known for many years, systematic excavation here, in one of the largest caves in Sri Lanka, was not undertaken until the second half of the 1980s. Two trenches were opened, and in the second of these (Area B), the excavators found rich habitation deposits with human remains and a non-geometric microlithic industry. The mixing of skeletal elements from different bodies hampered concrete MNI (minimum numbers of individuals) estimates per level, but at least 12 are noted. The repeated occurrence down through the archaeological sequence also shows that the site was a focus of human interment throughout the Late Pleistocene and on through until the Mid-Holocene. Fragmentary human remains from Levels 2–5 are associated with ^{14}C dates, the oldest of which (from Level 5) is 38,465 ± 430 cal. BP (Beta-33294; specimen YF-86B-M6 6). The Pleistocene remains that have received the closest attention though, owing to their greater completeness, are those of a 5- to 6-year-old child. Potentially a secondary burial, these are also mingled with the bones from two other infants, a juvenile and a young adult female, in what is thought to be a purposeful placement (Kennedy and Zahorsky, 1997). The child (specimen YF-86B-N7-3, or Fa Hien 2) is dated by associated charcoal to 30,060 ± 380 bp or 35,460 ± 386 cal. BP (Beta-33299).

Approximately 40 km to the southeast, the skeletal evidence from Batadomba-lena Cave (6.766667°N, 80.2°E) comes from its basal deposits (Stratum 7C). The ten pieces of bone are fragmentary, comprising teeth, phalanges, cranial fragments and the body of a thoracic vertebra. They do, however, occur in the same layer as a small number ($n = 17$) of microliths and associated charcoal (Kennedy and Deraniyagala, 1989). The ^{14}C date retrieved from 7C is 28,510 + 2150/−1710 bp (PRL-857) – or *c.* 33,703 ± 4120 cal. BP. Interestingly, Stratum 6B, 18,867 ± 1538 cal. BP (PRL-858), at this site, together with slightly younger fossils (14,228 ± 1029 cal. BP (PRL-861)) from the cave of Beli lena kitulgala, *c.* 35 km to the northwest (6.998333°N, 80.410833°E), have between them produced additional remains of approximately 38 individuals (Kennedy *et al.*, 1987). It is unclear why the frequency of recovery should suddenly improve during the Last Termination, but Kennedy's morphometric analysis comparing this latter sample with remains from a Mid-Holocene site to the southeast, Bellan-bandi Palassa (*see* Simpson *et al.*, 2008), indicated close affinity between the two populations and likely long-term continuity in this region.

On technological grounds, Mellars (2006a) has drawn parallels between the geometric micro-blade character in techno-facies of the African Howieson's Poort and micro-blade evidence from Patne, western India, and Batadomba-lena, Sri Lanka. Although demonstrably associated with *H. sapiens* on both continents, they are chronologically very remote from one another: 64.8–59.5 KBP (OSL; Jacobs *et al.*, 2008) and *c.* 33.7 cal. KBP (James and Petraglia, 2005). Whether these similarities amount to anything more than convergent evolution, therefore, is open to question. James and Petraglia (2005: 14) further note that while some South Asian industries display transitional tool elements, there is little indication of any distinct cultural break that could be equated with the arrival of an incoming population during the time span when late Middle Palaeolithic flake and prepared core industries transitioned to more characteristically Upper Palaeolithic industries (*c.* 45–28.5 kbp). Despite these trends toward the inclusion of more blade and micro-blade elements, the overall level of tool diversity also does not readily separate these from the region's Middle

Palaeolithic industries, which are also described as 'exceptionally diverse' (Field *et al.*, 2007: 89).

If the lithics are ambiguous on the matter of the arrival of modern humans in South Asia and the site distribution evidence for dispersal is tentative only, an alternate avenue might be the appearance of transformations in behaviour. If one follows such markers as are used in the European Upper Palaeolithic (Mellars, 2005), it is evident that they do not appear in the South Asian record until the middle to later MIS-3. For example, perforated marine shell beads appear from the earliest level at Batadomba-lena (Layer 7c), dated to 35,977 ± 409 cal. BP (Wk-19963), and from Fa Hien-lena at *c.* 38 cal. KBP (Perera, 2010: 184–5; Perera *et al.*, 2011), while ostrich eggshell beads have also been reported from Patne (from levels dated to 25 ± 2 kbp), including putative decorated fragments (James and Petraglia, 2005). In short, although the genetic evidence points to the early arrival of modern humans into South Asia and the subsequent importance of this region demographically, there is little in the way of concrete evidence yet in the material culture, despite its richness, to support the arrival of a new population during the early part of the Upper Pleistocene. This is not a situation confined to South Asia, it is one that applies to many parts of Asia: the probable initial colonization by *H. sapiens* did not seem to lead to wholesale changes in technology or, apparently, in (at least preserved) forms of complex or symbolic behaviour. There *are* examples of the traditional markers of symbolic representation associated with *H. sapiens*, but these often appear either later or in quite different contexts to those in Africa or Europe. Where authors have previously considered this feature of the Asian record reducible to factors such as limited access to 'quality' lithic raw material or a progressive 'loss of complexity' with increasing distance from a demographic source, in light of the growing body of evidence for genuine inter-regional difference, significant reassessment is now needed. There is currently little archaeologically to indicate that the Upper Pleistocene population of the subcontinent was growing through this period, yet equally, there also remain few indications for what other motivation(s) might have caused the emigration of people out of this region in either of the two northern pulses *c.* 50–45 kbp and *c.* 33 kbp described by Oppenheimer (2003) or prompted dispersal eastwards towards neighbouring Southeast Asia, our next port of call.

5.3.5 Sundaland

In comparison to the severity of Pleistocene conditions at higher latitudes, environmental changes occurring in the tropics were once thought to have been rather muted (*after* Flenley, 1997). In the 1970s and 1980s, research efforts, such as the CLIMAP programme, prompted this picture to be revised. Oxygen isotope data from planktonic foraminifera were now indicating that tropical oceans had been 1°–2°C cooler during the LGM than they are presently, and increased aridity was thought to have restricted lowland rainforest to refugia in an otherwise more open, savannah-like landscape (Broecker, 1996; Kellman and Tackaberry, 1997). Pollen data from mountain cores in Papua New Guinea and Taiwan have consistently indicated a drop of 6°–10°C with a concomitant depression of the snowline by up to 1000 m (Verstappen, 1975; Bellwood, 2007; Maloney and McCormac, 1995). The gathering of oxygen isotope data from tropical corals also now pushes estimates of lowland cooling at the LGM to 5°–6°C below modern values (Broecker, 1996; Flenley, 1997; Gagan *et al.*, 1998). The effects on moisture-laden systems travelling over the continental landmass of Sundaland, coupled with resultant changes in the positioning of the atmospheric Inter-tropical Convergence Zone (ITCZ – a comparatively narrow region of the atmosphere, confined to lower latitudes, where the rising air of the northern and southern hemispheric Hadley circulation cells meet) and this drop in air and sea temperature, combined to exert great influence over Southeast Asian climate. Temperature depression on this magnitude would have meant that tropical typhoons (which require critically high sea surface temperatures to form) would have been less frequent during the Pleistocene, though the increased pressure gradient may have increased the velocity of monsoon winds (Petersen, 1969). Cooler seas would also have reduced convectional precipitation – another of the chief components of tropical climate (Kellman and Tackaberry, 1997). The glacial conditions in northern Eurasia would have tended to cause an increase in the pressure gradient between the Asiatic and Australian high-pressure cells (anticyclones) and the ITCZ. Bellwood (2007) suggested that this may have moved the ITCZ away from the Indo-Malaysian region for longer periods than it does currently, resulting in more pronounced dry seasons here. Even on the equator, rainfall maxima would have been smaller and for a briefer period in the year. It has been estimated that annual rainfall during the Pleistocene may have been of the order of 30 per cent down on contemporary values (Urushibara-Yoshino and Yoshino, 1997).

The current suite of genetic studies suggests that the first concerted spread of modern humans into Mainland Southeast Asia probably occurred mid-way through MIS-4 (*c.* 63 kbp), that these people were probably the ancestors of modern-day Australo-Melanesian aboriginal groups such as the Orang Asli of the Malay Peninsula, and that some degree of maritime technology was already in use. At the present time, though, archaeological contexts with well-dated fossil evidence of *H. sapiens* do not reach back this far in time. The oldest fossil evidence firmly attributed to modern humans from Mainland Southeast Asia comes from the site of Moh Khiew in southern Thailand (Thai–Malay Peninsula) and is an inhumation dated to 30,989 ± 707 cal. BP (TK-933 Pr; 30,989 ± 707 cal. BP; Matsumura and Pookajorn, 2005, see later). Additional, isolated fragments have been reported from three sites in northern Vietnam – Hang Hum I and II and Keo Leng – with projected ages based on faunal associations of sometime between 180 and 40 kbp at the Hang Hum locales and *c.* 30 kbp at Keo Leng. Material from another site, Tham Om, may be *H. sapiens* or may be *H. erectus* (Olsen and Ciochon, 1990). The mixture of archaic and modern characteristics that have been noted on this and other recent finds of predominantly isolated teeth (e.g. Demeter *et al.*, 2004, 2005) precludes certain identification and continues to highlight the multifarious nature of the region's fossil record. Such mixed-affinity fossils, dated broadly to the late Middle and early Upper Pleistocene, leave open the possibility that there may have been more than one physically distinct population of hominins present on the mainland. Even so, with the exception of the Kota Tampan evidence at 74–70 KBP, no other archaeological sites from the southern part of Mainland Southeast Asia are dated to a time significantly before the beginning of the Late Pleistocene (as defined herein), though the infinite date from the basal unit at Lang Rongrien (see later) may qualify. After 40 kbp, the evidence becomes comparatively more extensive and diverse, but the reasons for this horizon in the record are unknown and, given the range of dating techniques now available, is probably not an artefact of chronometry. Despite more than a century of regional study, most analysis has been piecemeal, and survey coverage could be an important factor given that until the 1980s, most of this work was confined to cave investigations, and many of these had already been mined of their deposits because of their value as agricultural fertiliser.

5.3.5.1 Thailand

One site from the Thai side of the Thai–Malay Peninsula that has received considerable attention over the last couple of decades but, it must be noted, only contains human burial evidence (two primary and two probable secondary interments) from its uppermost, Mid-Holocene unit (Unit 1) is Lang Rongrien rock-shelter, Krabi Province (8.177375°N, 98.879361°E; Anderson, 1988, 1990, 1997; Mudar and Anderson, 2007). Ten stratigraphic units are identified here through 3.5 m of deposits (to bedrock; Figure 5.6). Units 10–8 contain cultural remains that predate the LGM. Unit 10 has an infinite date > 43 kbp; the same unit also yielded a ^{14}C date of 37,265 ± 1000 bp (PITT-1248) – or 42,358 ± 885 cal. BP – together with one intrusive Holocene date. This places occupation around GIS-11 (Figure 2.5). The small faunal assemblage from this unit comprised remains that were predominantly from mammals (*c.* 74% n = 248/336). Of these, only a fraction (n = 25) could be identified to taxa, including elements from bovids, cervids and a single canid maxillary incisor. The remaining fauna was chelonian (23% n = 77/336), two bird and two fish fragments. Two chipped bone fragments were identified, but intentional shaping could not be confirmed (Anderson, 1990: 59); lithics comprised 14 pieces (*see* Chapter 6).

Units 9 and 8 have dates spanning about 10,000 years, from 42,108 ± 1589 cal. BP (SI-6819) to 32,421 ± 673 cal. BP (SI-6816). Unit 7 is ambiguous. Excepting for a single pocket of charcoal, fossilised bone, five chert flakes and a groove and snap fragment of antler, the entire 1.5 m of this unit comprised only friable limestone rubble. An infinite radiocarbon date was obtained from the charcoal deposit, suggesting it to be older than the material in the underlying units. Anderson (1990: 53) surmises that if this dating is correct, the material may have come from a small cave or fissure above the main shelter, which subsequently collapsed as part of the Unit 7 accumulation. This could point to an older presence at the site than even the basal units propose. The processes responsible for the Unit 7 formation are unclear, though, given the chronological gap, the sequence may just as readily have been related to changes in environmental conditions during the LGM.

The faunal pattern of Unit 10 is broadly repeated in the higher units, with the addition of *Elephas* sp. tusk fragments and two amphibian limb elements from Unit 8 and three reported antler artefacts from Units 9 (n = 1) and 10 (n = 2; Anderson, 1988). A total of six unidentified molluscan fragments were identified from Units 9 and 8; none was identified from Unit 10. By studying the habitat preferences of these taxa, together with available palaeoenvironmental evidence, Mudar and Anderson (2007) conclude that occupation was taking place at a time when the area around the cave was made up of savannah-like conditions and open-canopy forest. It is unclear if such habitats reflect general conditions during the Upper Pleistocene or pertain to occupation specifically during stadial phases.

The site is located close to the transition between the (now flooded) coastal plain and the interior lowlands. From this, Mudar and Anderson (2007) suggest that the site may have formed part of a seasonal settlement system that incorporated both habitats. The distance of the site from the coast for Units 10-8 is predicted to have been 85 and 110 km, respectively (*after* Lambeck and Chappell, 2001). The absence of arboreal fauna from Lang Rongrien is accorded to a small sample size and to the brevity of occupational intervals. Finally, although some of the lithics from these three units were made on locally occurring quartz stone, the majority of the assemblage (n = 45 pieces, dominated by flakes) was made on materials, such as siliceous limestone, chalcedony and chert, which were not of local origin – a situation which contrasted to that seen during the post-LGM occupation of the site (Anderson, 1988, 1997).

Approximately 5 km from Lang Rongrien, the rock-shelter site of Moh Khiew is situated in an isolated limestone massif in the northern part of the southern Thai province of Krabi (*c.* 8.162725°N, 98.92345°E). A focus of attention by local farmers for the last 30 years excavating the soil as a fertiliser, there has been considerable disturbance of the surface layers. A 2 × 5 m trench was excavated at the eastern end of the shelter by Silpakorn University to a depth of 4.1 m. Following the natural stratigraphy, five cultural levels were identified and dated. All but Levels 1 and 2 date to the Holocene. The single date of 30,989 ± 707 cal. BP (TK-933 Pr) was obtained from the transition between these two levels (Pookajorn, 1996).

Unifacial and bifacial core tools, flake and waste flake pieces appeared throughout the sequence at Moh Khiew, with fragments of polished 'adzes' (including blanks) and ceramics

Figure 5.6. Lang Rongrien rock-shelter under excavation 1984–5. (*After* Anderson, 1990, reproduced with kind permission of the author.)

beginning to appear in Level 4 and increasing in number through into Level 5. On the basis of microwear analysis, these tools are thought to have been used in hunting and manufacturing activities (such as woodworking and the making of traps; Pookajorn *et al.*, 1994). A considerable quantity of waste flakes was recovered (notably, in Level 4) consistent with on-site stone tool manufacture. There is evidence for changes in lithic raw materials over time, though irrespective of whether chert or quartzite was used, the same kinds of core tools were being made, suggesting that stone quality did not exert a strong influence over production decisions. The nearest chert and jasper deposits are within 2 km of the cave.

In contrast to the picture from Lang Rongrien, a wide range of fauna was exploited at Moh Khiew, including bovids, suids, cervids, cercopithecids, sciurids, rodentia, carnivora and fish. Larger animals were dominant in the first level – including pigs – as well as arboreal taxa. Marine molluscs were not present but increased in significance higher up in the sequence when the site was within a couple of kilometres of the coast (Pookajorn *et al.*, 1994). At *c.* 31 cal. KBP, it would have been *c.* 100–120 km from the coast, following the line of a series of palaeorivers (*after* Mudar and Anderson, 2007).

5.3.5.2 Malaysia

On the Malaysian side of this peninsula, the only reported evidence dating to this period comes from the recently discovered (2003) open-air site of Bukit Bunuh (Figure 5.7), a reportedly palaeo-island setting within the same inland lake along whose shores Kota Tampan was situated, 1.5 km to the southeast (Mokhtar, 2006a, 2006b, 2007a). Characterized as a lithic workshop through the occurrence of anvils, cores, hammer stones, pebble and flake tools and débitage, it is considered distinct from other similarly featured sites in the Lenggong valley on account of the raw materials used and the presence of putative hand axes ($n = 19/347$ classified tools from the site). Whilst quartz and quartzite (commonly used raw materials for tool making at other Pleistocene sites in Lenggong) were also employed at Bukit Bunuh, chert and an impact breccia (suevite) were also used here. The hand axe component of this Late Pleistocene assemblage (*see* Figure 3.5) reportedly comprises bifacially and unifacially trimmed pebbles with convergent laterals and no specific raw material selection preference (Mokhtar, 2006a). Two sediment samples were taken from this site for TL dating, using the infrared stimulated luminescence method (IRSL): one from the cultural layer and one from the overlying deposits (BB2 and BB1, respectively). The TL date for BB2 returned as 39,000 ± 2600 BP; that for BB1 was much more recent at 13,000 ± 800 BP (Roberts *et al.*, 2005). The error margin on the cultural layer remains too large to realistically assign occupation of the site to a particular climatic period within MIS-3. The lack of organic preservation at this site leaves open what species of hominin was responsible for its creation.

The oldest evidence of modern human activity currently available from the Thai–Malay Peninsula does not seem to be consistent with the kind of strand-loping settlement system suggested for initial colonization along the Southern Route (cf. Mellars, 2006a). It should, however, be noted that the location of the Thai sites, close to the heads of palaeoriver systems (Mudar and Anderson, 2007: figure 2), the appearance of sites adjacent to palaeolakes in the valley systems of northern Peninsular Malaysia and an emphasis on non-local lithic raw materials at Lang Rongrien may indicate comparatively mobile groups who were possibly using waterways to aid movement. Any settlement on the coastal plain has now been lost or buried, and perhaps these inland sites should be viewed as part of a later 'filling in' of the landscape, as suggested for South Asia by Field *et al.* (2007). This, though, remains speculative and does not accord with the Kota Tampan evidence either, which is both early and inland. At *c.* 70 KBP, the palaeocoast would have been at approximately the −50 m mark (Lambeck and Chappell, 2001); Kota Tampan is currently 54 km inland (in a straight line). At −50 m, there would have been a significant channel between the northern part of Peninsular Malaysia and northern Sumatra, but the site would still have been *c.* 94 km away from the coast (Sathiamurthy and Voris, 2006; Voris, 2000). The issue that continues to hover over these early occupations, though, is that none can yet be confidently attributed to *H. sapiens*. The evidence we are seeing could be part of the settlement system of a different hominin altogether.

5.3.5.3 Vietnam

East of the Thai–Malay Peninsula, there is also a long history of archaeological research in Vietnam. This country has provided a major corpus of data on Late Pleistocene and particularly Early Holocene human settlement. In 1927, the results of geological surveys in this area by the Ecole Française d'Etreme Orient, Hanoi, demonstrated the presence of a generalised pebble and flake tool industry that would come to be called the 'Hoabinhian' (*see* Chapter 6), after the province of Hoa Binh in the northern part of the country, where it was first discovered. Most Hoabinhian sites in Vietnam are largely confined to the uplands south of the Song Hong River and tend to date to the period *c.* 18–7 kbp (Nguyen Viet, 2000). There is not yet consensus about the early evolution of this industry, however, or its relationship to the 'Sonvian', its purported (but overlapping) predecessor, thought to

Figure 5.7. Photograph of the immediate environs of the Bukit Bunuh island site from a palaeolake, Lenggong Valley, West Malaysia. (Photograph by Mokhtar Saidin, reproduced with kind permission.)

span the period *c.* 23–13 kbp (Nguyen Gia Doi, 2005; Reynolds, 1990). The vast majority of the *c.* 140 Sonvian sites from northern Vietnam are open-air locales discovered along elevated river terraces above the Bac Bo plain (Ha Van Tan, 1997). Although preservation conditions for organic remains have been poor and there has been little scope for subsistence reconstruction, their position does further mark the early importance of river systems in the region. The other pre-LGM industry recognised in Vietnam is the Nguòmian. All three of these industries are believed to be represented at the Nguòmian type site: the Nguòm rock-shelter in Bac Thai Province.

At Nguòm, the Nguòmian appears in the basal cultural unit and is currently undated. It is distinguished on account of a high frequency of small retouched flake tools, points, scrapers and knives and possible blades and blade tools; bipolar as well as multiplatform cores; and a small but confirmed pebble tool component (Anisyutkin and Timofeyev, 2006; Nguyen Khac Su, 2008; Reynolds, 1993). This predominance of flake tools is taken to refute the long-held notion that a core-tool-based technology extended in unbroken line deep into the Pleistocene (Ha Van Tan, 1999). Details of the accompanying faunal remains are limited, but identified taxa include a mixture of terrestrial and arboreal taxa, including *Sus scrofa, Cervus unicolor, Macaca* sp. and *Pongo pygmaeus*, and the faunal suite is seen as part of the *Ailuropoda–Stegodon* community (Hoáng Xuân Chinh, 1991). No molluscs were recovered at this level.

The middle sedimentary unit at Nguòm is assigned to the Sonvian and appears to show technological continuity from the lower unit but with the addition of a higher frequency of pebble tools – some of which bear a resemblance to later Hoabinhian lithics, though there has been disagreement over the likely extent of that affiliation (Ha Van Tan, 1997; Reynolds, 1993). Changes in the character of the cave sediments have been taken as indication that the middle cave infill may have accumulated under different climatic conditions. This unit contains two published radiocarbon dates (on shell and seemingly uncorrected) that are almost identical: 23,000 ± 200 bp (Bln 2692/I) and 23,000 ± 300 bp (*c.* 27 cal. BP; Hoáng Xuân Chinh, 1991; Reynolds, 1990), providing an indication both of the age of the Sonvian and the presumed greater antiquity of the Nguòmian. Land snails (*Cyclophorus dodrans* and *Cyclophorous speciosus*) begin to appear at this point in the sequence, but quantified numbers are not reported, and there is a considerable drop in frequency of two of the previous vertebrate marker species: *Ailuropoda* and *Pongo pygmaeus* (though isolated teeth of the latter are still reported). There is also a continued presence of other previously occurring vertebrate taxa, such as *Cervus unicolor* and *Macaca* sp., though no details are available about the structure of this prey population. Given the predominantly open-air nature of Sonvian sites, there is no way to determine if the spectrum of taxa represented at Nguòm is typical of this complex.

Faunal remains from the upper cave sediment unit show continued importance of cyclophorid molluscs (*Cyclophorus dodrans*, *C. siamensis*, *C. fulganatus* and *C. specious*), though with increasing frequency of freshwater molluscs such as *Brotia* (= *Antimelania*) *siamensis*, *Brotia* (= *Antimelania*) *costula* and *Angulyagra* sp. By this time, *Ailuropoda* and *Pongo pygmaeus* no longer occur; leaving all identified vertebrate remains as extant mainland species. A radiocarbon date on shell of 18,600 ± 200 bp (Bln-2691/2; *c.* 22 cal. BP) was obtained from the base of the upper sedimentary unit and suggests that this phase of occupation probably ended during the LGM. Notable elements of the lithic assemblage are unifacial core tools and edge-ground tools, consistent with the Hoabinhian (Hoáng Xuân Chinh, 1991).

In sum, the current state of fossil evidence from Mainland Southeast Asia does not show a visible modern human presence here until the Late Pleistocene – at least 20,000 years after the arrival time predicted by the genetics (*see* Endicott *et al.*, 2009). Some indication of whether occupations occurred under warmer or cooler conditions is retrievable through site faunal remains, and inter-site differences may indicate variability in subsistence strategies, but the picture is far from clear. Absolute dating resolution is currently such that it is also hard to allocate occupations to specific climatic episodes (stadials or interstadials). It is all the more interesting, therefore, that all of the oldest confirmed *H. sapiens* fossil evidence in this region come from locales in ISEA.

5.3.5.4 The Philippines

The earliest evidence for modern humans in ISEA appears in the east, in the Philippine archipelago and Borneo. The cave site of Tabon (9.275497°N, 117.979975°E), on the north coast of the Philippine island of Palawan, was originally excavated in the 1960s by Robert Fox, and although fragmentary human remains were recovered from Pleistocene layers, none were commented upon until a report appeared in the late 1970s. Macintosh (1978) discussed a mandible fragment that appeared in neither the original site reports nor the final monograph. Although a cast of this specimen still exists, the whereabouts of the actual piece and its significance are unknown. Besides this specimen, other human material that was reported from Fox's excavation (and recovered in 1962) included two further mandibular fragments and a front bone from disturbed deposits within the cave (Fox, 1970). He proposed that the fossils were associated with deposits dated by ^{14}C to *c.* 25–22 kbp. Direct U-series dating of the frontal bone fragment (Dizon *et al.*, 2002) confirmed a Late Pleistocene age, but one somewhat younger than Fox's estimate, at 16.5 ± 2 kbp (early uptake). During renewed excavation in 2000, an additional 12 human skeletal fragments were recovered from Pleistocene levels (Detroit *et al.*, 2004). Further direct U-series dates were attempted on another right mandible fragment (PXIII–T436 Sg19) from the original excavations and a fragment of right tibia (IV-2000-T-197) from the new campaign. These returned dates of $31^{+8}/_{-7}$ KBP and $47^{+11}/_{-10}$ KBP, respectively. Although the authors regard the association of fossil fragments of variable ages as something that is consistent with the disturbed nature of the deposits, they view the early date with caution until it can be verified. What is notable is that even within this small total corpus (old and new material) of human remains from Tabon Cave, the evidence suggests the presence of potentially two morphologically distinct groups of humans. The majority of fragments represents a slightly built population; a single unlabelled and (unfortunately) undated mandibular specimen appears to be from a very much more robust *H. sapiens* (Détroit *et al.*, 2004; Dizon *et al.*, 2002). This possibly echoes the position from the mainland and further cautions that the record of modern human occupation in this region may contain complexities that we are only just beginning to realise.

Two additional aspects of the Tabon archaeological evidence are significant to the current discussion. Firstly, the associated lithic assemblage comprises mostly flake débitage, made on locally available chert. Large unifacially trimmed basalt choppers, cores and quartz and basalt hammer stones were also excavated from the Late Pleistocene contexts (*c.* 30–23 kbp). Only a comparative few (<10%) retouched tools were identified (Fox, 1978). Establishing a formal typology for these pieces proved problematic, though Fox (1970) was able to identify a range of different scraper types. The overall impression from this assemblage is that finished tools were mainly being removed from the site for use elsewhere in the landscape, perhaps signifying that the use of Tabon was related to its proximity to this lithic source.

Secondly, despite careful excavation and sieving strategies, no marine shells were excavated from the Pleistocene or Early Holocene levels. Fox (1970: 38) attributes this to the distance from Tabon to the sea during these occupational phases, suggesting it to have been 30–35 km. While such a distance at the LGM has now been confirmed (Piper *et al.*, 2011; E. Charles, pers. comm.; *see* Charles *et al.*, forthcoming, referenced with kind permission), the absence from Early Holocene levels, when the sea would have been much closer and on occasion exceeding modern values (Maeda *et al.*, 2004), is harder to explain if the determining factor for marine resource use is simply a matter of proximity.

Still, in the Philippines, note needs to be made of archaeological remains from two further sites. First is the discovery in 2001 at Arubo 1, in central Luzon Island, of a small ($n = 18$) collection of reputable lithics, including cores, core tools (with one bifacially worked core tool, described by Pawlik (2004) as of 'proto–hand axe' character) and flakes. All material was made on a variety of locally available chert and remains undated. Secondly, Mijares (2005, 2008) has reported on the discovery of flakes with blade-like proportions (made predominantly on chert, but also on quartz, using simple percussion) from the lower levels at the site of Callao Cave (Layer 8, spits 21–22), one of the Peñablanca caves, of northern Luzon. Charcoal from this layer is dated by ^{14}C to 31,195 ± 430 cal. BP (Wk-14881). Callao is currently 36 km from the Pacific coast. Given the very steep drop-off along the northeastern coast of Luzon, even at the height of the last glaciation, this distance was not significantly different (*see* Sathiamurthy and Voris, 2006). The recent publication (Mijares *et al.*, 2010) of fragmentary fossil hominin evidence (a third metatarsal) from lower levels at this site, and direct U-series dated to 66.7 ± 1 KBP, is set to further excite discussion about early colonization of the region given the island setting. This is particularly so as the morphometric analysis of this small gracile piece places it within the ranges not only of *H. sapiens* but also of *H. floresiensis* and even *H. habilis*.

5.3.5.5 Borneo

A similar situation to that a Tabon arises at the northern Sabah island site of Gua Balambangan, located on the western coast of Pulau Balambangan (*c.* 7.200725°N, 116.8531°E; Jeffrey, 2005; Zuraina *et al.*, 1998b). The current basal cultural levels at this cave have been dated to 19,954 ± 244 cal. BP (Beta-105172). They are also associated with a locally available chert source (<1 km) and show evidence of a flake tool industry (including cores and waste flakes). Despite its coastal location, even during the final Pleistocene, there is little indication that marine resources featured in the diet of people using this site (deer and wild cattle

Figure 5.8. Looking out through the West Mouth entrance to the Niah Caves. The main archaeological excavations took place in an area to the right of the picture. (Photograph by Ryan J. Rabett.)

predominated) until the Early Holocene – between 11,453 ± 319 cal. BP (Beta-109141) and 10,007 ± 227 cal. BP (Beta-109140), when estuarine and freshwater molluscs begin to appear. This suggests that although the coast is seen as a major route for early human migration, proximity to this resource-rich environment did not automatically mean that people immediately became reliant upon it (*see* Chapter 7).

Given the current uncertainty surrounding the Tabon evidence and the intriguing but still slightly ambiguous evidence from Callao, the *H. sapiens* fossil evidence from the Niah Caves, Borneo, currently represents the most securely dated evidence of early human occupation in ISEA (Barker *et al.*, 2007). During the Sarawak Museum's second (1957) field season of excavation at the West Mouth of this large system of caverns (Figure 5.8), efforts were concentrated on expanding a large test pit dug in 1954. Long trenches were sunk to the same depth as the original sounding (i.e. eight feet) along its southern and northern flanks. Two weeks into the season, a trial sondage was dug into dark sediments that had been exposed in the base of one of these trenches, E/B1. While work continued apace in other parts of the West Mouth the following year, excavation in this new area (referred to as 'Hell') progressed far more slowly and painstakingly. These detailed efforts, however, were rewarded with the discovery, just four days into the 1958 season, of fragmentary modern human remains: the so-called 'Deep Skull' (Figure 5.9). Detailed analysis of this partial cranium and maxilla was published soon afterwards (Brothwell, 1960), indicating the remains to be those of a late adolescent of *c.* 17 years or slightly older, of Australo-Melanesian affiliation and of unknown sex. Subsequent study revised the age to that of a young adult, 20–30 years, and sexed the finds to very likely those of a female (Birdsell, 1979).

Figure 5.9. (left) Looking north across the 'Hell' trench system as preserved today towards the rock overhang, just visible in the background. The Deep Skull was recovered from contexts, currently suspended in mid-air close to the centre right of the frame. (right) The Deep Skull still *in situ*. (Photographs by Ryan J. Rabett and Sarawak Museum Harrisson Archive no. 363, reproduced with kind permission.)

The Deep Skull was originally dated by association to samples taken from a charcoal-rich layer (collected in 1957) in an immediately overlying deposit. Two uncalibrated dates were published for the skull: 39,820 ± 1012 bp (GrN-1339C; Vogel and Waterbolk, 1963) and 41,720 ± 1012 bp (GrN-1338C; de Vries and Waterbolk, 1958), though the latter date was later considered to have been a mix of samples and deemed potentially invalid (Krigbaum, 2001). Despite the fact that this find emerged from the most fine-grained part of the Niah excavation, both the skull's age and the lack of published supporting documentation made the find a controversial one. The presence nearby in the same cave mouth of a large Neolithic burial area raised the chance that the skull was intrusive from these later levels (e.g. Bellwood, 1997; Bulbeck, 1982; Kennedy, 1979; Solheim, 1983; Storm, 2001).

Reinvestigation of the West Mouth between 2000 and 2005 as part of the Niah Caves Project (NCP) resolved these uncertainties by unravelling the site's complicated stratigraphy, including the depositional context and original find depth of the skull. This was aided by an suite of 30 new AMS ^{14}C dates (using both ABOX and A-B-A treatment protocols) from these deep deposits and by the direct U-series dates on fragments of the skull itself (Barker *et al.*, 2002a, 2007). This reanalysis largely confirmed Harrisson's original conclusions. Two archived charcoal samples were taken, one from immediate association with the skull and one from less than 1 m away from the find spot at the same level. These returned dates of 40,984 ± 305 cal. BP (OxA-V-2076-16) and 40,344 ± 415 BP (OxA-15126), respectively. The samples subjected to U-series dating returned consistent ages of $37^{+5}/_{-4.7}$ KBP and

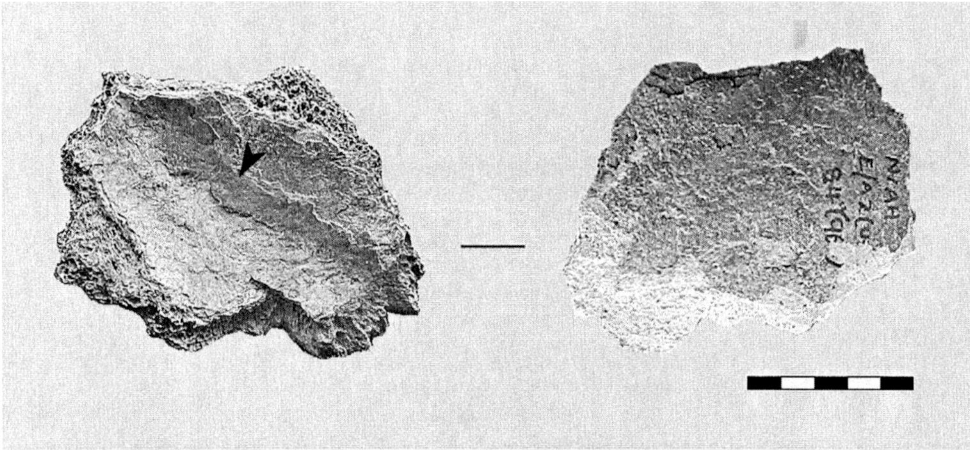

Figure 5.10. A fragment of hominin cranium with pigment adhering to the interior surface. Provenance: E/A2 84″–96″, excavated 01/04/1958 (scale in 5-mm increments). (Photograph by Ryan J. Rabett.)

$34.5^{+3}/_{-3}$ KBP. The methods combined provided an accurate calibrated range for the skull of 35–42 KBP.

A less well-known feature of this excavation is that the Deep Skull did not constitute the entirety of human remains from these early levels. Nearby were also found a near complete left femur and a right proximal tibia fragment (Krigbaum and Datan, 1999, 2005). These were almost certainly of the same individual on the basis of their size and fusion stages, though direct association with the Deep Skull can only be surmised. A human talus was also found and sent as part of the original sample of fauna to Leiden for analysis (Hooijer, 1963) but could not be relocated. During the course of analysing the archived fauna from this site (by the author and P. Piper), 27 additional human cranial fragments were identified from deposits corresponding taphonomically and, as far as can be ascertained, stratigraphically to those which yielded the Deep Skull but *c.* 10 m to the south (Rabett and Barker, 2007). Neither the fragments nor the sediments that they came from have yet been dated to confirm their suspected antiquity, leaving something of a new Niah skull enigma. Added to the potential significance of these finds is the presence of a red 'wash' to the inside of three of these fragments (Figure 5.10). Whether this represents the deliberate application of colour and the associated ritual practices that might pertain to is hard to determine; the skull could have served purely as a receptacle for the mixing of pigment. Analysis of the coating from one of the fragments using a variety of techniques, including microscopy, energy dispersive microwave analysis and infrared spectrophotometry, suggests that rather than being an iron oxide (ochre), this coating is more likely to be a plant-based substance, possibly a tree resin (Pyatt *et al.*, 2010) reminiscent of other pigments studied from later periods in the same cave complex (Pyatt *et al.*, 2005).

Until the new cranial material can be directly dated it can be considered only with due caution; however, a further bone fragment, this time a piece of Geoemydidae plastron (Figure 5.11), was also identified during reanalysis of the faunal remains from a trench (Y/3) excavated by Harrisson beneath the rock overhang adjacent to the Hell trench system – designated as 'Area A' by the NCP. The basal deposits in this area of the cave are now known to date from the same period as the Deep Skull (and in some instances, earlier; Barker *et al.*, forthcoming). This particular fragment was originally recovered at a depth of 66–72″ below ground surface. AMS ^{14}C dates for spits 60–66″ and 72–78″ of 41,164 ± 281 cal. BP (OxA-15163) and 41,675 ± 278 cal. BP (OxA-15164), respectively, from adjacent trenches Y/E3

Figure 5.11. Pigmented Geoemydidae plastron from Area A of the West Mouth, Niah Caves, dated to *c.* 41 cal. KBP (scale in 5-mm increments). (Photograph by Ryan J. Rabett.)

and W/E2 indicate that its probable antiquity is similar to that of the Deep Skull. Like the new skull fragments, this piece may have been a receptacle or mixing palate, but in this instance, there is a very clear and straight dividing line between the coloured and uncoloured portions of the plastron, which may imply that this was part of a more deliberate application, perhaps of a design – full details of the chelonian remains from the Pleistocene deposits at the West Mouth of the Niah Caves are presented in Pritchard *et al.* (2009). To the best of the author's knowledge, this fragment currently represents the earliest evidence for the use of pigmentation almost certainly attributable to *H. sapiens* in Southeast Asia.

A second feature of these deposits that was never made apparent in the published literature is that the Deep Skull was an element of a much larger assemblage comprising mostly animal remains (*c.* 10,000 of which were macrovertebrate fragments). Some of these remains were still partially articulated and probably represent deliberate dumping episodes, likewise patches with high concentrations of burnt bone. A moderate-sized collection of butchered bone fragments (n = 30) from all Hell and Hell-related trench contexts (though excluding similarly aged deposits beneath the rock overhang), in addition to a small collection of bone and lithic artefacts, offer further strong evidence of human use of the site (Barker *et al.*, forthcoming; Rabett *et al.*, 2006; Rabett and Barker, 2007).

Dates from these and other deposits indicate that humans had been coming to Niah, probably intermittently, for several thousands of years prior to the deposition of the Deep Skull. The oldest calibrated date from the site is currently 48,732 ± 818 BP (OxA-V-2076-15) from context 3.1 under the rock overhang. There are three further charcoal samples with calibrated dates of 48,247 ± 620 BP (for both OxA-V-2057-29 and OxA-V-2057-30; contexts 3140 and 3143) and 48,370 ± 570 BP (OxA-V-2057-27; context 3131) from Hell; one further Hell date remains outside the current available calibration ceiling of 50 KBP using the Fairbanks_0107 calibration curve.

Lithics from the Upper Pleistocene deposits at Niah are predominately flakes (Figure 5.12) – some of which may have been curated and have developed use polish – though certain classic tool forms can be discerned, such as burins and scrapers (Reynolds, 2007). Residue analysis of 80 Late Pleistocene lithics suggests that 25 per cent of the sample was used against hard or siliceous material (such as palm wood) and 12 per cent was used against softer materials (Barker *et al.*, 2007). The main lithic raw material used appears to have been metamorphic sandstone, with rare instances of limestone, chert, jasper and quartz. Some of these types of stone are available in the vicinity of Niah, though a petrographic comparison suggests that the nearest source area for jasper was 48 km away (Zuraina, 1982).

Figure 5.12. A selection of struck flakes from the Hell and lower E/B1 trenches, Niah Caves (scale in 5-mm increments). (Photograph by Ryan J. Rabett.)

The earliest instance of bone technology at Niah is a small ($n = 8$) assemblage of pieces from deposits in the West Mouth entrance. The oldest specimen, a newly discovered artefact (Rabett *et al.*, 2006), is dated by association (i.e. from the same depth as a closely adjacent trench) to an AMS ^{14}C date of 44,941 ± 329 cal. BP (OxA-15630). This was a point-form with modification characteristics most consistent with use as a piercing implement (Figure 5.13). All but one of these early tools is characterized by minimal production intervention and alteration of pre-existing fractured bone fragments. All pieces appear to have experienced considerable, perhaps protracted use. None is likely to have been used as an armature of any sort.

To this may be added two artefacts from the earliest phase of occupation (Phase I) identified in the deposits beneath the rock overhang (Area A): a worked Suidae tusk fragment and a bone edge-form tool (traditionally, variously referred to in the literature as bone 'spatulae' or 'axes' but carrying a clearly utilised working edge). These data begin to suggest that from an early stage, although tool frequencies were very low, a range of different osseous raw materials were being used to create tools that probably served different functions – a feature that would still be present when bone tool frequencies increased substantially here and elsewhere during the Last Termination (*see* Chapter 6).

Botanical and geochemical evidence from the part of Upper Pleistocene occupation at Niah have provided sufficient resolution to link this record approximately to the global ice core chronology (*following* Johnsen *et al.*, 2001). On the basis of two calibrated ^{14}C charcoal dates obtained just above the top of an analysed sediment column sample from Hell (Hunt *et al.*, 2007), evidence within the column likely coincides with a major marine transgression and appears to record oscillating temperatures and vegetation structure near the cave across a number of interstadial and stadial conditions commencing *c.* 52 KBP during GIS-14 (Hunt *et al.*, 2012; *see though* Stimpson, forthcoming). The faunal record from this early period does include a comparatively high frequency of *Pongo pygmaeus* remains (Rabett *et al.*, 2006) that points towards the existence of closed-canopy lowland forest in the area around the cave during at least some spells of occupation.

Outside of the interstadial phases recorded in the pollen sequence, there are indicators, such as, high levels of herbaceous taxa and limited evidence for mangrove pollen, suggestive of a drier, more open environment. Hunt *et al.* (2007, 2012) conclude that the character of the landscape around Niah was cyclic, alternating between lowland and mangrove forest and these more open habitats. Evidence of localised burning that is potentially anthropogenic occurs within the column sample from as low as the penultimate pollen zone (H-2). Frequent levels of *Justicia* in the overlying and provisionally younger pollen zone (H-3) are taken as a

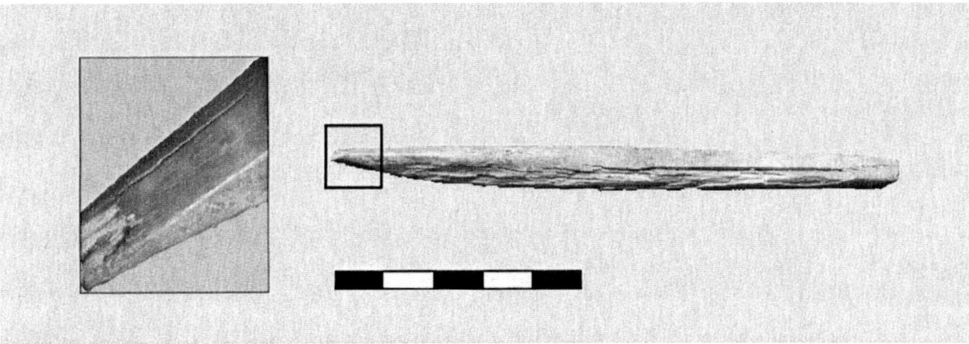

Figure 5.13. Photographs of the earliest bone tool from the Hell trench, Niah Caves (scale in 5-mm increments). (Photographs by Ryan J. Rabett.)

clear indication of fire in the landscape and appear in association with episodes of dense forest. The current position is that deliberate burning was instigated to maintain forest clearings around Niah from the earliest times (Hunt *et al.*, 2012: 77).

Hunting strategies have been extrapolated from the macrovertebrate fauna of this period at Niah (Barker *et al.*, 2007; Barker *et al.*, forthcoming; Rabett *et al.*, 2006; Rabett and Barker, 2007). The greatest proportion of the represented taxa comes from Suidae (almost certainly *Sus barbatus*; *see* Cucchi *et al.*, 2009), closely followed by chelonians and Cercopithecidae. The age structure of the represented pig population does not indicate the targeting of any particular age category – contra to modern hunting practices (e.g. *see* Cranbrook and Labang, 2003) – suggesting a non-selective procurement technology, with some kind of trapping a good candidate. Given the mass migratory and reproductive cycles of *Sus barbatus* (e.g. Curran and Leighton, 2000) – linked closely to periods of cyclic fruit abundance in the rainforest – there are hints, based on the appearance of juvenile pigs, that some hunting and trapping activity may have been timed to coincide with these periods of pig abundance. The possibility certainly exists that the proximity to a major pig migration route was an attraction that repeatedly drew foragers to the Niah area.

Addressing the extent, timing and nature of rainforest exploitation was one of the main objectives of the renewed work at the Niah Caves, including the role and availability of carbohydrate food sources. This aspect of the NCP was undertaken as a follow-up to debate into the accessibility of these critical resources in closed-canopy rainforest and hence the viability of preagricultural rainforest occupation (e.g. Bahuchet *et al.*, 1991; Bailey *et al.*, 1989; Brosius, 1991; Colinvaux and Bush, 1991; Headland, 1987, *see* Chapter 7). The long history of human presence at Niah in conjunction with excellent organic preservation made this an ideal location to test the hypotheses put forward by both sides of this debate. Examination of the area immediately behind the Hell trench system, on the cave interior side, in a now fragile standing section (NCP 2.1), revealed a series of inter-cutting pit-like features (Figure 5.14). These have been AMS dated (on charcoal recovered from within the features) to between 39,167 ± 356 cal. BP (OxA-11302) and 34,483 ± 264 cal. BP (OxA-11303); analysis of the pit fills is on-going, but it is thought that they may have been used to provide a means of nut detoxification.

Evidence of the deep-rooted yam (*Dioscorea* sp.) and of the sago palm (*Caryota mitis* or *Eugeissona* sp.) has also been obtained from contexts excavated and sampled during the NCP (Barton, 2005). The latter taxon was the most frequently identified and was also found across the most contexts in Hell, including from two of the earliest (3143) and (3131) with dates of 48,247 ± 620 cal. BP (Ox-V-2057-30) and 48,370 ± 570 cal. BP (OxA-V-2057-27). Its presence is viewed as almost certainly the result of human foraging activities. Similarly, several *Pangium edule* seeds were recovered during an earlier brief re-excavation of the site (Zuraina, 1982). Consumption of *P. edule* first requires the removal of toxins (Johns and Kubo, 1988) – further indication that plant resources were being processed from an early period to make them edible. Contexts within 0.5 m of two finds of *P. edule* and at the same level have been dated by AMS-ABOX to 39,431 ± 450 cal. BP (OxA-15622) and to between 39,044 ± 535 cal. BP (OxA-15624) and 44,097 ± 436 cal. BP (OxA-15623; Barker *et al.*, 2007).

In sum, following early indicators of fire in the local landscape, clear evidence of hominin (presumably *H. sapiens*) occupation in the West Mouth of the Niah Caves appears by at least *c.* 48 cal. KBP and probably somewhat earlier. At that point, visiting groups were capable of successfully locating, exploiting and processing forest carbohydrates. Evidence over subsequent millennia shows the use of lithic and non-lithic technologies (both at low frequencies) – hunting strategies that targeted terrestrial game especially and probably linked to periods of game abundance but which also included (unidentified) methods for taking arboreal game. Use of botanical forest resources is also indicated in the creation of indirect products for non-subsistence activities such as the mixing of plant-based pigment. These spells of occupation may correlate to phases of forested, if changeable, interstadial conditions during MIS-3, though exact correlations with particular climatic stages remain to be determined with certainty.

Three further sites in Sabah, Malaysian Borneo, offer considerable interest for the future in terms of early human colonization in this region. Gua Samang Buat is a cave with a cultural sequence currently 4 m deep and OSL dates of 29 KBP at *c.* 2.5 m and 46 KBP at the present base of excavation. The second location is Mansuli, an open-air site with limited organic preservation but lithic cultural material reported to date from *c.* 100 KBP (OSL; Mokhtar, 2007b). Finally, two open-air sites, known as Tingkayu locales I and II (*c.* 4.727256°N, 118.068372°E). These were found to contain unexceptional large pebble tools, cores and flakes, all of which are made on brown chert pebbles or grey tabular chert. A subset of these assemblages, however, consisted of bifacial pieces and lanceolate points (Bellwood, 1984, 1988) of a form that has not appeared anywhere else in Southeast Asia. The sites would have been located on the margin of a large palaeolake (comparable to the situation seen in the Lenggong Valley) that was originally formed when a lava flow from the nearby Mt Mostyn caused the damming of the Tingkayu River. The approximate age for the lake's formation was determined by a ^{14}C date on charcoal sealed beneath part of this lava flow of *c.* 28 kbp (Bellwood, 1997). The sites are deemed to be no older than this and probably no younger than *c.* 18 kbp, when evidence from nearby caves suggests the lake had drained. Bellwood has recently voiced uncertainty about the stated age estimate, on the grounds that in the 27 years since its discovery there continues to be nothing like this industry appearing in any of the new well-dated archaeological sites in the region with Pleistocene records (Bellwood, 2007). He speculates that the Tingkayu material might be of Holocene age, though, it has to be said there is little to compare with it during the Holocene either. In 2007, Universiti Sains Malaysia reported finding five new open-air sites in local river terraces to Tingkayu, so further developments in this area can be anticipated.

5.3.5.6 *Java*

Moving south from Borneo, recent work on various islands of Indonesia has revealed considerable antiquity and new ramifications for the dispersal of hominins through Upper Pleistocene ISEA. In East Java, renewed excavation (in 2004) at the cave site of Song Gupuh (8.161111°S, 110.979939°E) yielded a 12.5 m cultural sequence stretching back to *c.* 50 kbp – inferred on the basis of available ^{14}C and luminescence dates. The earliest faunal deposits below this (at 14.7 m) show evidence of percussive fracturing that might infer hominin agency at a date of *c.* 70 kbp (Morwood *et al.*, 2008). This early fauna is characterized by

Figure 5.14. Inter-cutting pit features in the Pleistocene deposits of the West Mouth (left, centre frame) during the 1959 excavations of this part of the cave. (right) The section (NCP 2.1) sampled by the Niah Caves Project. (Photographs by (left) Sarawak Museum Harrison Archive no. 625 and (right) G. Barker, both images reproduced with kind permission.)

a limited range of cervid, bovid and suid remains, taken as indicative of a relatively open landscape, though the presence of tapir (*Tapirus* sp.) and small numbers of macaque (*Macaca* spp.) suggests at least periodic forest cover. The character of the stone tool assemblage appears to have been comparatively consistent from the earliest clear phase of occupation onwards (until the Neolithic): a predominance of apparently expediently flaked pieces, including a small number of flake 'choppers' made from minimally reduced, multiplatform cores.

If one accepts that the presence of fully modern fauna, as exhibited at Song Gupuh, provides grounds for arguing against the presence of *H. erectus* (e.g. Storm, 2001), the Pleistocene portion of this cultural sequence might be the product of *H. sapiens*. However, this remains to be demonstrated. The earliest fossil evidence of *H. sapiens* currently available at the site, including a very robust juvenile skull, date from the beginning of the Holocene 11,410 ± 180 cal. BP (Wk-14650) – calibrated using IntCal04 (Reimer *et al.*, 2004).

Similar uncertainty surrounds the nearby (<10 km) site of Song Terus (*c.* 8.160003°S, 111.013525°E). Faunal remains from the basal, 'Terus' levels at this site have been dated recently by coupled ESR/U-series assays to between 341 ± 51 KBP and 254 ± 38 KBP (Hameau *et al.*, 2007). The middle layers are termed the 'Tabuhan' and are inferred to date to *c.* 45 kbp. These contain comparatively small quantities of lithics – including chert and limestone flakes and cores – faunal remains and putative evidence of hearths (Sémah *et al.*, 2003, 2004). ESR/U-series dating is pending on these levels, but it is complicated taphomically by the presence of volcanic ashes stratified within them (Hameau *et al.*, 2007). There is no indication as to which species of hominin was responsible for producing the Tabuhan assemblage. Modern human evidence occurs in overlying ('Keplek') levels, and one assumes an archaic hominin species from the Terus levels below, leaving the middle layers open to speculation, and potentially, to be the product of either. The discovery that has probably reignited interest in this region's Palaeolithic archaeology more than any other in recent decades, though, arose to the east of Java in the Lesser Sunda Islands.

5.3.6 Wallacea

With the aim of documenting more about early hominin colonization of Flores, following the discoveries at Mata Menge, in the Soa Basin locale, a joint Australian–Indonesian team began excavating in July 2001 at the large cave of Liang Bua Approximately, 31 km from the

Figure 5.15. (left) Indonesian–Australian excavations at Liang Bua, Flores. (right) Deep excavations in Sectors VII and XI. (Photographs by M. Morwood, University of Wollongong, reproduced with kind permission.)

north and south coasts (Figure 5.15), this site had received episodic attention from Dutch and Indonesian archaeologists in the past, but only yielded Holocene-aged evidence to depths of up to 4.25 m. In the third season of the new work, diagnostically hominin finds began to appear at much deeper levels and of much greater antiquity; firstly, in the form of an isolated left P_3 with broadly human characteristics but, seemingly, not belonging to *H. sapiens*. This find came from a sector of excavation that was about 20 cm above a horizon (4.5 m below ground in a different part of the cave) that had yielded a coupled ESR/U-series date (on a *Stegodon florensis insularis* molar) of $74^{+14}/_{-12}$ KBP (LB-JR-8a). Further fragmentary hominin remains were found in this sector (Sector IV) back to a depth of 7.5 m and a maximum sediment age (by TL) of 95 ± 13 KBP (LBS4-32a), including an right ulna and left radius, both ascribed to the period dated by (LB-JR-8a). The full description of these and other elements from an MNI of six individuals is presented in Larson *et al.* (2009; based on upper limb bones), and an MNI of nine individuals, the current site maximum, recovered from Sectors IV, VII and IX, is presented in Jungers *et al.* (2009; based on lower limb bones).

Within a week of finding the P_3 (in September 2003), the team had uncovered a fairly complete skeleton (LB1) comprising skull, mandible, pelvis and leg bones (minus arms, which were recovered later from the adjoining sector) at a depth of 5.9 m (Figure 5.16). LB1 would be presented as the *H. floresiensis* holotype (Brown *et al.*, 2004; Morwood *et al.*, 2004) and was dated by close association to two AMS ^{14}C samples: (5.88 m) 18.7–17.9 cal. KBP (ANUA-27116) and (5.90 m) 18.2–17.4 cal. KBP (ANUA-27117). Calibration was carried out using the INTCAL98 dataset and CALIB4.3 (Morwood *et al.*, 2004: supplementary material). The remains were also bracketed between two TL sediment dates of (quartz) 35 ± 4 KBP (LBS7-42b) at 5.93–5.98 m and (feldspar) 14 ± 2 KBP (LBS7-40a) at 4.85–4.90 m, though this latter is deemed to be an underestimate of the true age for this sample (Morwood *et al.*, 2004: supplementary material).

No stratigraphic or artefactual evidence from the recovery location suggest that this was a deliberate burial. The individual, likely a female, appears to have died at the edge of a pond and may have sunk quickly into the fine sediments (Morwood *et al.*, 2005). Although few lithic artefacts were found at the same level as LB1 ($n = 32$) in Sector VII, another trench, Sector IV, produced a dense concentration (up to 5500 per m^3) at this depth. Small flakes predominate but in addition and struck bifacially from small radial chert and volcanic stone cores, there is a formal component including points and perforators, blades and microblades. This latter component was found only in association with *Stegodon florensis insularis* remains (van den Bergh *et al.*, 2009), and appeared throughout the sequence from *c.* 74 KBP to immediately below six water-lain layers of white tuffaceous silts from volcanic eruptions that coincide with the extinction of both *H. floresiensis* and the stegodon, now thought

Figure 5.16. Excavating LB1 in Sector VII. (Photograph by Wayhy Saptomo, Indonesian Research Centre for Archaeology, reproduced with kind permission.)

to be *c.* 17 KBP (calibration unknown; van den Bergh *et al.*, 2009). This date appears at the opening of the most intense period (17–6000 KBP) of volcanic activity globally for the last 110,000 years, as revealed in the GISP2 core (Zielinski *et al.*, 1996). Morwood *et al.* (2004) suggest that the juxtaposition of fauna and lithics implies selective hunting of stegodon by *H. floresiensis*. Van den Bergh *et al.* (2009) add to the discussion on the subsistence practices of this hominin by asserting that the predominance of juvenile stegodon, with only a few older individuals, indicates that *H. floresiensis* was either unable or unwilling to tackle adult *Stegodon florensis insularis*. Further taxa taken (hunted or scavenged) are thought to have included komodo dragons, endemic giant rats and terrestrial birds, giving a broad sense of the exploitative niche the hominin occupied, seemingly for the full sequence of its occupation at Liang Bua. There is no evidence for the exploitation of marine fauna such as mollusc collection or for the use of shell artefacts or adornments, though evidence of systematic butchery and the probable use of fire is indicated (Morwood *et al.*, 2005; van den Bergh *et al.*, 2009).

Finally, it is intriguing to note that despite some differences, there is significant overall technological similarity (particularly in the form of bifacial and radial reduction strategy) between the Liang Bua and Soa Basin lithics, despite their removal from one another by 50 km and 700,000 years. This has led to the suggestion of possible relatedness between the two hominin lineages producing these widely separated industries and that *H. floresiensis* was more than capable of producing stone artefacts of recognisable sophistication – a fact which may seem at odds with the small brain size of this new species (Brumm *et al.*, 2006). Until further evidence is amassed, however, this association between populations based on technological similarities must remain a tentative one.

Since its discovery in 2004, controversy about the taxonomic status of *H. floresiensis* has raged between those who consider the LB1 specimen representative of a new species of hominin and those who consider it to be modern human afflicted by developmental abnormalities (e.g. Aiello, 2010; Argue *et al.*, 2006; Ecknardt and Henneberg, 2010; Falk *et al.*, 2005; Jacob *et al.*, 2006). In addition to anatomical issues there is observed similarity (if with some variation in technique and raw material sourcing) in the lithic reduction sequences attributed to *H. floresiensis* and modern human inhabitations of Flores (Moore *et al.*, 2009). For the present, though, those advocating the new species position for *H. floresiensis* appear to have the stronger case. This is most particularly so as the evidence in favour rests not on a single specimen but on the, albeit incomplete, remains of multiple individuals, all of similar stature (Jungers *et al.*, 2009; Larson *et al.*, 2009; Morwood *et al.*, 2005). There are also distinct changes in faunal composition and exploitation strategies between the periods when *H. floresiensis* is thought to have lived on the island and when fossil evidence of *H. sapiens* first appears (van den Bergh *et al.*, 2009) late in the sequence (*see* Chapter 6). In Sector IV, median calibrated dates of 11.1–10.7/10.6 KBP (ANU-19209 and ANU-19211) have been obtained on charcoal 3.97–4.00 m below the ground surface (Morwood *et al.*, 2004: supplementary material; Morwood *et al.*, 2005). This suggests that there is currently no evidence of a direct population overlap on Flores, though clearly at a regional level, such an overlap would have existed.

The colonization of Flores, insular for most of its history, remains an enigma, both in terms of the timing and the means of arrival. While water crossings by modern humans do rely heavily upon cognitive and creative attributes, it does not follow that these are required for successful island colonization. Ancestral species of elephant and species of giant tortoise were capable of traversing the kinds of distances by sea that would have been necessary to establish viable breeding populations on Sulawesi (*c.* 50 km), which had been isolated from the Sunda Shelf since at least the Palaeogene (van den Bergh, 1999) – a feat achieved without the problem solving, scheduling and innovative abilities of modern or archaic humans. Colonization of Flores by *H. floresiensis* or an earlier hominin, either of which could have been responsible for the finds at Wolo Sege and Mata Menge, remains contentious. Although the earliest island occupation has often been attributed to *H. erectus* there is as yet no fossil evidence to support this.

Arrival may have been accidental. Morwood and Jungers (2009) have argued for the plausibility of colonization by groups clinging to natural rafts of vegetation after something like a tsunami event. Queiroz (2005), one of the authors cited in support of this model, outlines a change in research opinion about the possible mechanisms of biotic dispersal. While vicariance-inducing processes, such as those caused by the movement of plate tectonics, are thought to explain many occurrences where genetically related taxa appear in widely separated contexts, there is a growing realisation that oceanic dispersal (by rafting, swimming or floating) may be an alternative where it can be shown that genetic divergence was more recent than could have occurred as a result of geological forces alone. This may be a mechanism of dispersal that is better suited to explain the translocation of plants rather than animal taxa, though the jury is still out on this point.

Tsunamis are known to impact Flores (*see* e.g. Tsuji *et al.*, 1995, for detailed treatment of such an event in 1992) and have inflicted severe damage to areas of coastal habitation. They can trigger landslides and deposit sand sheets sometimes kilometres inland. One example of sediment coring of sand sheets along the northwestern coast of Sumatra, following the 2004 tsunami there, produced a record of such events throughout the last millennium (Monecke *et al.*, 2005). In short, although other factors must be taken into account, tsunamis leave tangible geomorphological signatures. The discovery and robust dating of such discontinuities in coastal Pleistocene deposits on Flores, given its long insular history, should be achievable and would be an important step towards testing the credibility of this model. Monecke *et al.* (2005) do note, however, on the basis of the Sumatran evidence that while repetitive, tsunami events over the last thousand years have not occurred all that frequently, having generational or larger gaps between them. To establish a viable founding population on an island by this mechanism, there would need to be repeated arrivals within a short period. As things stand, using a tsunami to account for the successful colonization of Flores seems uncertain. Even if multiple events did occur, one unintentional landing could easily be followed by others that miss the island altogether. Such events are also just as likely (if not more so) to extinguish communities as to establish them. Tsuji *et al.* (1995) talk at length about the severe damage and death toll wrought along the coast of Flores. There is no mention of incoming floating debris during the 1992 tsunami.

If, conversely, greater intentionality or navigational skills are accorded to *H. floresiensis* or its ancestors, there is a danger of undermining the very isolationism necessary to induce insular evolution such as dwarfism among the hominin population. Although such abilities can be lost by established populations (as is well documented among Tasmanian aboriginals), as a hypothesis to explain the case of Flores, it, too, awaits tangible evidence. At the moment, there are certainly wisps of possibilities about how this island could have been colonized by archaic hominins, but there is no direct support yet for either of the proffered scenarios – a fact which for some may not help affirmations about the authenticity of the occupation record in general here. For the moment, we must await, with considerable interest, the outcome of further work being carried out by Morwood and his colleagues on Sulawesi, the nearby island they believe to have been the most likely point of embarkation for emigrants to Flores (Morwood and Jungers, 2009).

5.3.6.1 East Timor

Travelling southeast from Flores, the island of Timor has long been considered a good prospect for finding evidence of early *H. sapiens*' necessary movement across the waters of Wallacea between the palaeo-continents of Sunda and Sahul (New Guinea and Australia). While the dates for the initial colonization of Sahul are presently thought to be sometime between 65 and 45 KBP (see later), there has been little indication of modern human occupation consistent with 'island hopping' through Wallacea to get there; nothing recovered so far has dates that are this early. Work undertaken by O'Connor and colleagues (beginning in 2000) in East Timor drew upon these issues and the promising initial studies by Glover (1966–7; *see* Chapter 6), which, despite extensive survey and test-pitting programmes devised with the aim of tracking migration to Australia, had obtained no evidence older than *c.* 13 kbp. This site of Lene Hara on the east coast had already been identified and excavated in the early 1960s by the Portuguese archaeologist, Almeida, but was only partially described. The site contained a locally distinct lithic assemblage and was undated, making it a good candidate for renewed study (O'Connor *et al.*, 2002).

Two occupation phases are recorded at the site: a Neolithic occupation in the upper 10 cm of the 84 cm re-excavated sequence and a series of Upper Pleistocene occupations beneath this. Seven ^{14}C dates on shell through most of the remainder of the sequence (12–75 cm) indicated it to be of MIS-3 age, with dates falling quite tightly between 30,110 ± 320 bp (ANU-11398) and 34,650 ± 630 bp (ANU-11418). Some internal reversals are noted. Coupled with the fact that sherds of pottery do appear within the top 25 cm, and that two shell artefacts recovered at depths of 24–28 cm and 36–40 cm, respectively, have been direct-dated to the Mid-Holocene (O'Connor *et al.*, 2002; O'Connor and Veth, 2005), the deposits have experienced a degree of reworking.

The character of the lithic assemblage shows continuity through the entire sequence, with an emphasis on flakes with associated cores and, reportedly, bearing similarity to early industries from northern Australia (O'Connor *et al.*, 2002). The small size of both was taken as indication that the red chert on which they were made was not locally occurring, though further analysis needs to demonstrate the validity of this. In light of depositional disturbance and the large time gap between the Neolithic and Palaeolithic occupations, the apparent technological uniformity may also need to be reassessed.

Small quantities of reptile and animal bone, marine fish and shell were found throughout the sequence with little apparent change in composition. *Strombus luhuanus*, *Trochus niloticus*, *Lambis lambis*, *Turbo* sp. and *Nerita* sp. – species of rocky platform marine mollusc – are conspicuous components. Similar to the situation in northern Luzon, Lene Hara's proximity to the sea is unlikely to have changed significantly between the Upper Pleistocene and the present, owing to a steeply shelving marine topography. This in itself means that both these areas have the potential to provide valuable evidence of the extent of early human occupation along and exploitation of coastal margins in this region that is unaffected by the perennial issue of inundation. No *H. sapiens* remains have yet been recovered from the Pleistocene levels at Lene Hara, but whoever was here, the faunal assemblage provides good evidence that they were using marine resources during the Late Pleistocene. Indeed, the productivity of these resources is attested by the continuing emphasis on fishing into the Early Holocene, as indicated through the documentation of a shell fish hook from this site (O'Connor and Veth, 2005; *see* Chapter 6).

A kilometre away, excavations at the Jerimalai rock-shelter carried out in 2005 yielded a series of rich ceramic horizons dating to the Mid-Holocene. Beneath these, the test pitting (two 1 × 1 m squares) produced Late Pleistocene deposits. The earliest of these provided calibrated dates on marine shell of 42,696 ± 435 BP and 42,140 ± 260 BP (O'Connor, 2007). The lithic assemblage thus far recovered from the site comprises mostly cores, small flakes and some flakes carrying rudimentary retouch. The lack of any standardized forms or recognisable 'types' makes it difficult to characterize, and all the more so by the fact that there is, again, no discernible shift, however slight, in the make-up of the toolkit between the Pleistocene and Holocene site components. The comparatively abundant and well-preserved faunal assemblage is dominated by marine resources. The rapid coastal drop-off will have meant that at no point in its occupation history was Jerimalai ever more than *c.* 5 km from the sea – though the shoreline itself would have been further removed in a vertical sense during glaciation. Fauna from the Pleistocene levels is dominated by turtles and fish. Judging from element size, the turtle remains are probably those of the green turtle (*Chelonia mydas*). If this is the case, and given that this species inhabits very different environments at different life stages (Bonin *et al.*, 2006), further analysis of this material may well be informative in reconstructing local marine habitats. In addition to inshore reef species, the fish remains include fragments from large pelagic (open-sea) species such as tuna (Scombridae); finds indicating that groups using the site had the technological means (fishing kits and water craft) to exploit offshore fisheries (even if these were in truth still comparatively close by due to the rapid drop-off in the local sea-bed). No human remains have yet been recovered here either.

5.3.6.2 Sulawesi

To the north of Flores lies Sulawesi, the largest island in Wallacea (*c.* 174,600 km^2). Pleistocene occupation here is presently known from two sites located on the southwest peninsula. The first of these, Leang Burung 2 (5.003611°S, 119.661389°E), was identified in 1950 by van Heekeren, but the principal excavation carried out here took place in 1975 (Glover, 1981). The cave is cut back into the base of high limestone cliffs, *c.* 20 km inland of the modern coast. Intended as only a supplementary effort to other more detailed work being undertaken at nearby Ulu Leang 1 and Leang Burung 1 (*see* Olsen and Glover, 2004; *see* Chapter 6), only a comparatively small total area of 15 m^2 was excavated. The upper layers were marked by considerable modern disturbance, but beneath this emerged a complex (tilted) shell midden stratigraphy. Although ash and burnt bone were comparatively common, charcoal was rarely encountered – a situation also noted at certain other places in the region, for example, at Lobang Hangus (Niah Caves) and at Jerimalai, East Timor. Radiocarbon assays were obtained on freshwater shells of the gastropod *Brotia perfectora*. Efforts were made to account for the uptake of ancient carbon into shell formation through provision of a calculated correction of *c.* 1350 years based on comparison to modern samples (Burleigh, 1981; Mook, 1981). When applied to four Late Pleistocene ^{14}C shell dates from the excavation, this produced an occupation range between 33,198 ± 330 bp (GrN-8649; or 31,848 ± 330 bp, corrected) and 20,755 ± 250 bp (BM-1492; or 19,405 ± 250 bp, corrected; *after* Glover, 1981: 16). The presence of brecciated bone and stone fragments underlying this stratum suggests that this may not represent the oldest phase of occupation.

The technological assemblage consisted of a small fraction of retouched pieces (n = 52/5485, <1%), utilized flakes (n = 124), flakes with edge gloss (n = 31), large flakes (n = 26) and small scalar cores (n = 53) but was dominated by débitage (n = 5198, 95%). Approximately 94 per cent of the assemblage appeared to have been made on three varieties of chert that are presently available in a river bed *c.* 15 km away. Among the small retouched tool component were seven Levallois-like points recovered from three different levels. Although no classic Levallois cores were recovered, bifacial cores (n = 3) capable of producing such points were found in related contexts. Glover (1981: 29) considered that this minor but continuing element was probably a local, independent invention, one that remains unique to this site. The capability to produce different flaking techniques clearly did exist among some early populations in this region, but it was neither a widespread nor persistent characteristic of stone technologies here.

Approximately 20 km to the northwest of Leang Burung 2, a single 2 × 2 test pit was excavated at Leang Sakapao 1 (4.816667°S, 119.6°E), a cave situated in a massif outlier *c.* 70 m above the surrounding lowland landscape (Bulbeck *et al.*, 2004). The upper layers (down to spit 3) showed some signs of post-depositional disturbance, though ^{14}C obtained on freshwater and marine shells (again, in the absence of charcoal) produced consistently pre-LGM ages for all spits (1–5). Uncorrected and uncalibrated ages ranged from 25,120 ± 260 bp (Wk-4261) to 31,280 ± 570 bp (Wk-3821); leading to the suggestion that abandonment of the site may have been tied to alterations in the landscape that coincided with the onset of the LGM.

The technological sample did not (with a single exception) bear any resemblance to local 'Toalean' tool types from the first half of the Holocene (*see* Bulbeck, 2006; *see* Chapter 6) and was also treated as being of Late Pleistocene origin. The assemblage (n = 821 pieces) included cores, retouched and utilized flakes (among them, flakes displaying silica gloss) and a predominance of débitage pieces (n = 654). A form of white chert was the principal raw material for tool production. The inclusion of 39 unworked nodules of this material suggests that it was acquired locally and brought to the site for processing. Only small amounts of bone, including pig teeth, were recovered, but no detailed information is available about the spectrum of fauna found. The presence of a small amount of saline-tolerant molluscs is in notable contrast to the situation at Leang Burung 2, where only freshwater molluscs were found. By the onset of the LGM, the nearest coast to Leang Sakapao 1 would have been 50–70 km to the west. The inclusion of brackish species of mollusc at the site is possibly explained if collection occurred in an estuary setting, which would have reached farther inland. In terms of early human colonization, both Leang Burung 2 and Leang Sakapao 1 have been interpreted representing occupations that saw the incorporation of inland areas into previously coastal-based settlement (Bulbeck *et al.*, 2004: 112).

5.3.6.3 North Malaku Islands

Farther eastward again, one comes to the North Malaku archipelago. Several sites here have been explored, but to date, only Gebe Island, between Halmahera and islands off the western tip of Irian Jaya, has provided securely dated Pleistocene occupation levels (Bellwood *et al.*,

1998). Fauna and technological evidence from the site in question, Golo Cave, have revealed valuable insights into the exploitation of marine resources as components of technological as well as subsistence strategies by the hominins (assumed to be *H. sapiens*) who were coming here (Szabó *et al.*, 2007). Four principal occupation phases are noted through the 2.4 m sequence examined; only the earliest, Phase I (*c.* 190–240 cm), concerns us here. Uncalibrated ^{14}C dates (on shell) indicate that this period of cave use took place between 32,210 ± 320 bp (Wk-4629) and 21,780 ± 160 bp (ANU-11007). Four of the six dates, though, cluster in the period *c.* 29–32 kbp, marking this as likely the main period of use; the sixth is an Early Holocene date whose out-of-place occurrence at this level is linked to some vertical movement of material. Shell artefacts are observed throughout the sequence, with noted shifts in the raw materials utilized and the approaches taken to modification. The Phase I artefacts comprise a collection of pieces knapped on fragments from the operculum of a large species of sub-tidal zone gastropod (*Turbo marmoratus*). Consistency in the knapping procedure led the authors to propose that these artefacts embody a degree of imposed form in the way they were produced. Having assessed the taphonomic situation of the flaked *T. marmoratus* artefacts and discounted alternate potential raw material sources, Szabó *et al.* (2007) proposed that direct collection from the species' reef and lagoon habitats is the most likely explanation. This strategy contrasts with the *c.* 88 per cent of molluscan remains from the shell midden that would have come from rocky inter-tidal and 'splash' zones. The distinction is taken to indicate quite deliberate and targeted collection strategies that were different for subsistence and artefact manufacture. The Pleistocene shell artefacts at Golo appear alongside a small collection of lithics (n = 54, though this total combines Phase I with the lowermost part of Phase II) made on metavolcanic rock. The authors contend that this suggests knapping probably came from brief episodes of free-hand flake removals using direct percussion. Flakes predominate, and there is no indication of intentional retouch or significant curation, though incompatible comparison of the flakes and core flake scars and a lack of débitage (after recovery and taphonomic processes are discounted) implied that the flakes were struck elsewhere and then brought to the cave. The lithic assemblage has the markings of being expediently produced, with little attention to standardization.

Like East Timor, Gebe was noted by Birdsell (1977) as a likely stopping point on the migration route to Australia. Szabó *et al.* (2007) point out that a land bridge has not existed between Gebe and New Guinea at any point during the Pleistocene and that any colonization of the island would have involved navigation of a 40 km stretch of ocean. How people arrived here and how often they came is not known. The clustering of the Phase I dates may intimate that the visits were brief and purposeful. While furthering the range of capabilities of Late Pleistocene hominins in this region, this evidence, as at so many other sites, does not seem quite old enough to represent part of its initial colonization. In terms of hominin behaviour, the authors rightly argue that the contrasting picture between the shell and lithic artefacts indicates that the people coming here during the Upper Pleistocene were not only targeting resources but were also employing material-specific reduction strategies, demonstrating that they were knowledgeable about what they could find here and how it could be used to meet their requirements. Importantly, work here chips away at one of the traditional pillars of European lithic analysis: that stone tools are a reliable indicator of behavioural capacity; clearly this is not always the case.

The Upper Pleistocene record from Southeast Asia is not without parallels to the kinds of behavioural components used to define human modernity deriving from the African and, ultimately, interpretation of the European records. The invasive flaking of the Tingkayu lithics produced a small number of pieces that could be readily lost in a European Upper Palaeolithic assemblage. The use of pigmentation and bone technology both appear early in Southeast Asia, with the latter easily matching the timing for the first instances of this technology in Europe and the Levant, but these isolated, often localised, instances seem to remain just that until the Holocene – a pattern of occurrence that also appears to have existed in Sahul (Habgood and Franklin, 2008; O'Connell and Allen, 2007).

5.3.7 Sahul (Greater Australia) and Western Melanesia

Considered part of the 'New World', Sahul comprises the modern landmasses of Australia, New Guinea and Tasmania, together with sundry other islands lying on the Sahul continental shelf and outliers to it in western Melanesia (such as New Britain, New Ireland and the Solomon Islands). A detailed picture of environmental conditions across this vast area during the Upper Pleistocene is still being established and records prior to *c.* 60 KBP continue to be patchy, though a broad picture is available. By early MIS-3, conditions appear to have favoured lake development in what are today arid and semi-arid parts of Australia (e.g. Chappell, 1991; Cohen *et al.*, 2011), including the Willandra Lake system in New South Wales. Now all that remains of these latter lakes is a chain of 13 dry basins, varying in area from 6 to 400 km^2; 55–50 kbp, these held *c.* 1000 km^2 of freshwater up to 10 m deep. High water levels prevailed for about 15,000 years, during which time modern humans began to camp along their shores, before a change in prevailing climate brought increasing aridity. The lakes then began to dry up, and dunefields began to spread southwards through the area, until by *c.* 15 kbp, almost all were completely dry (Mulvaney and Kamminga, 1999). Broadly, this shift from well-watered lacustrine environments to more arid ones during the second half of the Upper Pleistocene characterizes the continental Australian record.

In western Victoria, another system of palaeolakes – Lake Bolac and Lake Turangmoroke – have recently been subjected to a detailed ecological study by Cook (2009). This work has revealed a discontinuous record of lake levels and local environmental conditions dating back to *c.* 90 KBP (during MIS-5b). One of the few sequences dating back this far, and one of the few available from the vast arid–semi-arid continental zone, this record opens with dry-land vegetation and freshwater input into the lakes. Between this time and the onset of MIS-3 (59–29 KBP), the lakes were surrounded by an open woodland habitat and, based on represented taxa, the possibility of a rainfall budget that was more equally delivered between the seasons. Within MIS-4 (74–59 KBP), there are indications that the environment had experienced a temporary (though of unknown duration) shift back to drier conditions and reduction in lake levels. Then, about midway through MIS-3 (*c.* 47 KBP by OSL), the area appears to have experienced a period of considerable environmental change. Initially conditions were comparable to those seen previously, though with greater evidence for periods of lake drying. This is followed by a shift towards open grassland and steppe conditions, with more saline, shallow and stagnant lake environments, which persisted until towards the end of MIS-2.

To the north, in the Lake Eyre Basin, Johnson *et al.* (1999) examined emu eggshells to track the relative occurrence of C^3 and C^4 plants using the different $\delta^{13}C$ values each of these two photosynthetic pathways produce. Their analysis also points to the existence of different vegetation regimes 65–45 KBP and 28–15 KBP. The Lake Eyre basin extends across approximately one-sixth of Australia's currently arid and semi-arid environments, making it a good marker of general trends across a large part of the continent. Since the geographic distribution of C^3 and C^4 plant communities in Australia is linked to rainfall received through summer or winter monsoons, changes in $\delta^{13}C$ provide a proxy indicator of the predominant season of rainfall in the palaeo-record. These data for the 65–45 KBP period show highly variable $\delta^{13}C$ values. This would be consistent with one of two scenarios, both involving a variable monsoon. Either climatic oscillations were operating at a periodicity less than the dating resolution allowed at this period (±8000 years), or the palaeoenvironment itself contained a characteristically more mixed combination of C^3 and C^4 plants, making it potentially more comparable to African savannah. On the basis of the Lake Bolac and Turangmoroke evidence, the latter may be the more likely.

Pollen diagrams from Tasmania, South Australia and Queensland all indicate that temperatures around the time of the LGM were 5°–8°C lower than today with depressed levels of precipitation, while glacial advances in the Snowy Mountains and Tasmania also testify to the deterioration of conditions (Chappell, 1991). Lower and less variable $\delta^{13}C$ values in the Lake Eyre Basin are interpreted as reflecting a near-total absence of plants following a C^4 pathway, possibly as a result of reduced summer monsoon precipitation (Johnson *et al.*, 1999). Arid conditions appear to have prevailed across much of Sahul during this period, and not only did they persist through the LGM but they also continued into the Last Termination, without immediate recovery. Indeed the greatest local landscape aridity in the Bolac and Turangmoroke area peaked with the last glacial transition rather than with the LGM (Cook, 2009). Modern humans first appeared in the Sahul archaeological record during the continent's more watered, lacustrine phase, and by 35–25 kbp, almost all of the major environmental zones of Greater Australia, from the equatorial areas of the north to Tasmania in the south, had been colonized (Cosgrove *et al.*, 2007; Habgood and Franklin, 2008).

It has long been recognised that to establish a reproductively viable population on this continent, it would have been necessary to cross (and probably cross repeatedly) wide stretches of open sea. Using a minimal array of assumptions – combining variables of a target's distance, its size and visibility – Birdsell (1977) predicted there would have likely been two basic routes for crossing Wallacea for Sahul, each with optional paths contained within it (Figure 5.17). Whichever route colonizers opted for, however, would still have required at least one crossing of >70 km. The first hypothesised route (the southern) embarked from Java, proceeded along the Lesser Sunda Islands chain (though, seemingly missing out Flores), through Timor to northern Australia; the second route (the northern) commenced in Borneo and threaded a path via Sulawesi to either Halmahera and Irian Jaya or via the islands of Buru and Seram, before landing on the northern Sahul Shelf between Australia and New Guinea. Without full consideration of other key environmental variables, which would have impacted on the viability of these routes (including crucial data on the effects of prevailing winds, ocean currents, local tides and tectonic movement that were largely unavailable), Birdsell acknowledged that the model, though informative, remained incomplete (Birdsell, 1977: 133). So it has remained for 30 years. A detailed reassessment and simulation using detailed GIS-based oceanic and geographic data would be a worthy follow-up to this seminal and oft-referenced piece of work.

The current spread of dates at sites like Lena Hara and Jerimali makes the southern passage the more likely, but even so, their antiquity (particularly Jerimali) is comparable to the Australian evidence rather than being appreciably older. As Bellwood (1998) has argued, it is possible that the available dates from sites in Wallacea are not the product of a single wave of migration. Rather, they could be the product of more than one radiation event and only of 'in-filling' of the island seascape much later.

Periods of significantly lowered sea levels during the Pleistocene would have created opportunities for travel from Sunda to Sahul. Even so, all such voyages would still have had to cross permanent deep-water channels between continental shelves beyond sight of land. Speculation about the motivating forces varies from single factors, such as the sighting of large smoke plumes rising from natural bush fires in Australia and New Guinea (Webb, 2006) to gradual undirected but step-wise colonization between favoured resource patches within Wallacea (Allen and O'Connell, 2008). Most agree, though, that given the scale of the distances involved and the need to establish a stable population, it is unlikely that Sahul

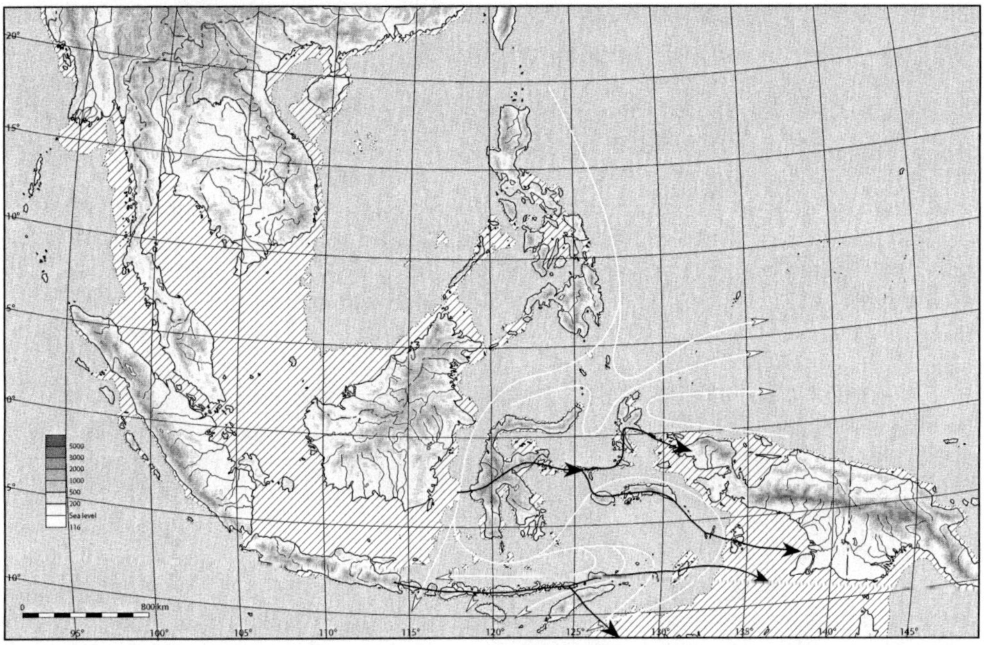

Figure 5.17. Map of Pleistocene Southeast Asia, showing the exposed Sunda Shelf to −116 m, prevailing ocean currents and possible human migration routes through Wallacea. (*After* Dennell, 2009, and Birdsell, 1977; map drawn and prepared by D. Kemp.)

was colonized by accident: a degree of intentionality is probable, as is a pre-existing capacity for maritime travel. These factors gave Latinis (2000) grounds for proposing that such a traverse is unlikely to have been conceived of by inland tropical forager communities, and for his suggestion that a divergence of subsistence systems had occurred in Sundaland – a proportion of whose population must have occupied its coastal habitats, where they would have developed the means and likely had the desire to develop offshore activities (*see also* Bowdler, 1990). Davidson and Nobel (1992) argued that the colonization of Sahul demanded complex language structures to conceptualise, construct and navigate ocean-going boats. Their central contention that this episode of the Pleistocene diaspora provided the first clear expression of 'behaviour that is distinctly human' may be questioned in light of recent discoveries in Africa and re-pondered in light of discoveries in Flores. Nonetheless, it remains a significant and demonstrable achievement, and very likely borne out of an accumulated pool of local maritime knowledge, traditions and skills held by people already well acquainted with the ocean.

The date of the first arrival of modern humans in Sahul from Southeast Asia continues to be a source of lively discussion, with opinion divided between 'long' and 'short' chronologies. The best evidence for a pre-45 KBP arrival has come from five sites. Four of these are on the northern fringe of Sahul: Bobongara, in modern-day New Guinea (Groube *et al.*, 1986), and the sites of Nauwalabila and Malakunanja II in Australia's Northern Territory (Bird *et al.*, 2002; Roberts *et al.*, 1990, 1994). The remaining two are in the southern half of the continent. The Devil's Lair (Western Australia; Turney *et al.*, 2001) and Lake Mungo (New South Wales; e.g. Bowler *et al.*, 2003), between them, provide broadly contemporary evidence to that uncovered in the north, though only at Lake Mungo have human remains been recovered from this period.

The certainty for such antiquity has been questioned by other researchers on the grounds of low frequencies of material remains and/or uncertainties about the depositional relationships between the artefacts and dates (e.g. O'Connell and Allen, 2004). These authors favour a more conservative estimate of 45–42 KBP by focusing on what they consider to be the most securely dated portions of these site sequences (O'Connell and Allen, 2004, 2007; Allen and O'Connell, 2008). New work in the Ivene Valley of New Guinea (Summerhayes *et al.*, 2010), which also gave rise to the famous Late Pleistocene site of Kosipe (see later), is adding to this chronological debate. Occupational evidence has recently been uncovered here at seven additional sites. The dates from the earliest of these, Vilakuav, lie between 46,180–42,900 cal. BP (Wk 27073) using the Calib 6.0.1. Incal09 curve and 48,690–42,970 cal. BP (Wk 27072; Summerhayes *et al.*, 2010: supplementary material).

Whether the long or short chronology proves correct, the colonization and in-filling of habitable zones across this vast continent with its humid and semi-arid and cool temperate habitat zones followed thereafter through the remainder of the Late Pleistocene. This would bring people to the remote inland fastness of the Bird's Head Peninsula, in Irian Jaya, by *c.* 20 kbp (Pasveer *et al.*, 2002; *see* Chapter 6) and persistence in highlands of Papua New Guinea; to the Sahul tropical outliers such as the Bismarck archipelago – commencing as early as 39 kbp (uncorrected for marine reservoir effect; Leavesley *et al.*, 2002) – and the Solomon Island archipelago, from *c.* 28 kbp (Wickler and Spriggs, 1988); to the continental interior such as at Cuddie Springs, *c.* 33–14 kbp (Fullagar and Field, 1997; *though see* Grün *et al.*, 2010); and to the glaciated highlands of Tasmania (*c.* 34 kbp; Pike-Tay and Cosgrove, 2006). The way in which colonizers successfully incorporated the resource bases of such varied environments itself is argued to mark a major adaptive divergence for a source population more acquainted with tropical seas and rainforests (Bowdler, 1990).

O'Connell and Allen (2007) have proposed that apparent parity between the earliest robust dates from northern and southern sites in Sahul and those from the neighbouring Sunda Shelf to 40–45 KBP may mean that there was a rapid primary dispersal across both regions, irrespective of the number of sea crossings that might have been involved. Whilst this may well have been true (certainly pre-LGM hunter-gatherer group mobility estimates indicate that it may have been higher in both regions than it would be after the LGM – see later), one does need to place 'rapidity' into context. While future refinements in dating could tighten the short chronology the same refinements could just as easily expand it by as much as 10,000 years; either way, it is still a far cry from the swiftness of the spread that continues to be attributed to the dispersal of Clovis-bearing groups across North America – a mere *c.* 200 calendar years (Waters and Stafford, 2007).

If 40–45 KBP is a genuine event horizon across Sunda and Sahul (*see* Endicott *et al.*, 2009), it might alternatively indicate a point at which a demographic threshold was crossed and archaeological visibility increased rather than a period of primary migration. The significant rise in site numbers during the post-LGM period (Morwood and Hobbs, 1995; O'Connell and Allen, 2007) certainly supports the significance of demographic change in later millennia, but as we shall see later (*see* Chapter 8), demographics, too, may well have factored earlier on. The second possibility is one of recovery. Recent genetic studies into the ancient population structure of Sahul and northern Melanesia propose an initial settlement of the latter 70,000–40,000 years ago (i.e. hedging their bets), either by a single and genetically diverse population or else through multiple but smaller diverse migratory events. The weight of opinion currently lies with the former of these hypotheses (Hudjashov *et al.*, 2007). Indications are, though, that population expansion within Sahul did not begin until towards the end of this time (Ingman and Gyllensten, 2003). People appear to have arrived in New Guinea by at least *c.* 50,000 years ago, something that seems to accord fairly well with the archaeological picture. This was followed by a population expansion perhaps 10,000 years later into the Bismarck and Solomon island archipelagos. Previously, it was thought that the Q branch of macro-haplogroup M had been confined to New Guinea and Melanesia. New evidence now has haplogroup Q identified in northeast Australian Aboriginal populations

(Hudjashov *et al.*, 2007). This appears to represent an isolated secondary expansion into the continent from New Guinea before the end of the last glacial. Such evidence aside, the existence of geographically localised haplotype patterns and deep coalescence dates suggests that there is unlikely to have been significant genetic mixing along female or male lines (mtDNA or Y chromosome) from outside groups or between the main colonizing groups prior to the Holocene (Friedlaender *et al.*, 2005; Hudjashov *et al.*, 2007; Merriwether *et al.*, 2005). Population expansion also appears to have been restricted to within comparatively immediate areas, arguing against the presence of Pleistocene human enclaves across Southeast Asia and Sahul being regarded as a loosely unified 'Australo-Melanesian' type (Merriwether *et al.*, 2005). At the same time, while there exists a long-known mixture of robust and gracile phenotypes among the earliest Sahul colonists (e.g. Storm, 2001), no-one would argue that these early populations were anything other than modern humans. MtDNA taken from a sample ten individuals (both gracile and robust) representing both Late Pleistocene and post-Pleistocene ancient 'Australians' (Adcock *et al.*, 2001) concluded that there is no genetic evidence that either morphological variant reflects the retention of specifically 'archaic' lineages. Genetically, these physically different populations cannot be distinguished: both are clearly *H. sapiens* (*see also* Hudjashov *et al.*, 2007). These same data, however, sound a note of caution against demographic models that are reliant entirely upon the genetic relatedness of modern-day populations to map Pleistocene population dispersals. MtDNA recovered from the Lake Mungo III – a male articulated burial dating to *c.* 40 kbp (Bowler *et al.*, 2003) – places this individual in a genetic lineage that may have diverged *before* the most common recent ancestor of all current human populations (Adcock *et al.*, 2001). In other words, it begins to suggest that some people living in Pleistocene Australia may not have come from the ancestral palaeodeme that ultimately gave rise to the present worldwide population. The proposition, though, remains controversial, with some questioning the authenticity of the DNA recovered (Smith *et al.*, 2003).

Lithic raw material sourcing in Southeast Asia suggests that a greater frequency of exotic stone was entering sites before the LGM, declining in the post-LGM period in favour of local sources – a situation that parallels changes reported from some sites in Southeast Asia (e.g. Anderson, 1990; Shoocondgej, 1996; Zuraina, 1982). In Sahul, the early movement of stone and other materials (including marine shell and ochre) reached in excess of 100 km (Habgood and Franklin, 2008: table 1). For example, evidence from the New Ireland site of Matenbek points to a small but persistent appearance of obsidian, sourced to New Britain – a straight-line distance of *c.* 350 km – as early as 20–19 kbp (Allen *et al.* 1989). Obsidian formed part of a comparatively wide range of raw materials that were exploited on New Ireland from the earliest occupation here – *c.* 32 kbp at Matenkupkum. Over time, that diversity (which included fine-grained siliceous pieces) is reported to gradually narrow in favour of metamorphic rock. Allen *et al.* (1989) argued that this picture is consistent with a strategy of exploiting local river pebbles where favoured stone becomes steadily harder to find. This interpretation, however, seems slightly unsatisfactory, for with obsidian clearly available and exploited from an early date, access to high-quality stone was not tied exclusively to local sources anyway. Furthermore, river systems will likely replenish stone as they move material downstream. Access appears to have become more restricted, but this may be linked to changes in group mobility rather than stone availability. This shift in lithic raw material use from sites across Sahul is now also being tied to the model of a region-wide rise in population density during the Last Termination, prompting reduced mobility and a focus on local resources (O'Connell and Allen, 2007).

Late Pleistocene stone tool technology itself shows some parallels to that from Southeast Asia its likely source region (Mulvaney and Kamminga, 1999). There is a predominance of flaked tools, with many falling into the broad categories of scrapers and cores, which, despite the presence of prepared cores, give it a strong Mode 1 character (Foley and Lahr, 1997).

Although there is very little evidence for formal types, some temporal and intra-regional variation has been noted. Often this has been attributed to raw material access (e.g. Allen *et al.*, 1989), but in some instances, localized differences may be linked to a degree of technological patterning. For example, the southwest Tasmanian site group, including the deep sequence at Parmerpar Meethaner, are typologically linked through the gradual development of relatively standardized thumbnail scrapers during the Late Pleistocene (Cosgrove, 1995).

Early tool assemblages (some of them pre-LGM) also contain a precociously early ground stone component. Ground stone and waisted stone 'hatchets' (without evident formal standardization) have been excavated from dated contexts at a number of sites in the north of the region, though occurrences, while less common, do extend beyond there (Mulvaney and Kamminga, 1999). The northern sites include Malangangerr, Nawamoyn, Nauwalabila I and Malakunanja II in Arnhem Land (Habgood and Franklin, 2008) and Bobongara, Huon Peninsula (Groube *et al.*, 1986) and the Ivane Valley (Summerhayes *et al.*, 2010; White *et al.*, 1970) in the Papua New Guinea highlands. The Arnhem Land examples ($n = 14$) had all been bifacially flaked and then had their cutting edges ground; various hafting mechanisms appear to have been employed based on placement of lateral grooving or presence for a tang. The three stone hatchets recovered at Bobongara were associated with TL-dated tephra beds; however, uncertainties about the nature of the contexts dated have led to the potentially pre-45 KBP age of these finds being treated with caution. The large excavation at Kosipe, in Ivane Valley, yielded five flaked 'axe-adzes' (as well as one definitely ground piece probably dating from the Early Holocene), eight waisted blades (flaked) and other flake pieces concentrated within the Pleistocene part of the sequence (particularly around Level 5). Two charcoal dates (from Levels 4 and 5, respectively) give estimates of this period of site use as between 31,699 ± 985cal. BP (ANU-190) and 32,165 ± 650 cal. BP (ANU-191; White *et al.*, 1970). Additional specimens have been recovered from South Kov, Airstrip Mound, Joe's Garden and Vilakuav sites here, all dating from before (and then disappearing after) the LGM (Summerhayes *et al.*, 2010: 79). The likelihood that these implements were hafted is also not questioned. What may be inferred about their likely function (or range of functions) from this and their distinctive character is less certain, though use in forest clearance activities is assumed; something that would have been one way to encourage new growth and attract game (Groube, 1989; Jones, 1989; Mulvaney and Kamminga, 1999; White *et al.*, 1970). Residue analysis of lithics from Ivane Valley (Summerhayes *et al.*, 2010) and previously from Kilu Cave, in the Solomon Islands, certainly demonstrates the early role of stone tools in plant exploitation in this region. Here, the recovery of starch residue from taro (*Alocasia* sp.) indicates that the long-term exploitation of this tuber commenced as early as *c.* 28 kbp (Loy *et al.*, 1992).

While this kind of stone tool is less common outside of northern Sahul, early instances of grinding and milling implements have been identified at a number of Australian sites of Pleistocene age. Occurrences are sporadic, though this recovery picture may be at least partly a product of the generally low frequency of these artefacts and the small size of most Pleistocene site assemblages (Gorecki *et al.*, 1997; *see also* Langely *et al.*, 2011). At Cuddie Springs, New South Wales, worn and polished grinding stone fragments may begin to appear from *c.* 33 kbp and continue to the top of the sequence (Fullagar and Field, 1997). While the range of purposes for which these implements were utilised has also not yet been established with certainty, plant processing is indicated. The appearance of this technology is linked here to increasing local aridity and associated resource stress (Fullagar and Field, 1997: 306). The association between archaeological remains and megafauna at Cuddie Springs, long advocated for earlier levels at this site, has recently come into question on taphonomic grounds (e.g. Gillespie and Brook, 2006; Grün *et al.*, 2010), leaving the robusticity of other components here in need of reassessment as well.

The Late Pleistocene period saw the extinction of many large terrestrial mammalian reptilian and avian species. This occurred globally – including in Southeast Asia (e.g. Piper *et al.*, 2007a, 2011) – but rates of loss were highest in North and South America and in Australia (72–88%; Price *et al.*, 2009). The broad timing of these extinctions appears to coincide with the expansion of modern human populations into these regions, with the consequence that many researchers link the demise of megafauna to the impact of human hunting pressure. Although the 'overkill' hypothesis, or at least strong human culpability, has remained popular for more than 30 years (*see* Alroy, 2001; Brook *et al.*, 2007; Martin, 1973), others continue to argue that this hypothesis requires much stronger empirical evidence for megafaunal hunting than actually exists (Grayson and Meltzer, 2003; Mulvaney and Kamminga, 1999; Wroe and Field, 2006). Recent work in South America by Politis and Messineo (2008) reports clear evidence of human exploitation of megafaunal species, but they also demonstrate that survival of three genera of megafauna into the Holocene is not consistent with an overkill scenario. While the scale of climatic shifts at the end of the Pleistocene is often proposed as the most likely alternative selection pressure, just as with assigning human agency, the often coarse nature of the dating and issues over sampling strategies continue to make it unclear exactly when extinctions occurred. In Sahul, the loss of these taxa towards the end of the last glacial may have stretched across millennia (Wroe and Field, 2006), though extirpation at the hands of humans in some areas does remain plausible and favoured (e.g. Rule *et al.*, 2012).

In light of this, recent work to direct-date bones of late glacial megafauna in Tasmania (Turney *et al.*, 2008) offers welcome reappraisal at a local level. The Tasmanian remains examined were taken from four sites (approximately half of the known number of sites on the island containing megafauna). The dated samples cover a time span from MIS-5e to MIS-3. They indicate that the Tasmanian megafauna – which are all also represented on the mainland – survived longer here than they did elsewhere. On the mainland, almost all are thought to have become extinct by *c.* 48.9–43.6. cal. KBP. The Tasmanian evidence at the most recent end of this timescale – from Titans Shelter and Mount Cripps caves – has provided dates of *c.* 44.9–40.9 cal. BP (Turney *et al.* 2008: supplementary data). The younger age postdates the establishment of a land bridge between Tasmania and the mainland *c.* 41 cal. KBP and is close to the time of the first occupation of Tasmania by humans, at the site of Warreen *c.* 40.1 cal. KBP. Given that there appears to be no evidence for major swings in environmental conditions around this time, Turney *et al.* (2008) opt for human predation as the most likely cause of megafaunal extinction. An earlier study by Roberts *et al.* (2001), using optimal dating of sediments and $^{230}Th/^{234}U$ dating of associated speleothem deposits from 28 sites across mainland Australia, determined that while the arid conditions of the LGM could be discounted, as the dates were around the onset of the Late Pleistocene, other climatic and/or human causes of extinction could not be ruled out. The fact that there remains considerable uncertainty about whether Late Pleistocene groups were actively and routinely hunting megafauna in North America or even Europe, makes the weight of empirical evidence not yet sufficient to call it one way or the other.

A detailed survey of faunal remains from 216 sites, dated 27–5 kbp, was undertaken by Balme (1995) in the lower Darling River region of New South Wales – of this total, 32 open-air middens were dated either directly or by inference to 27–22 kbp. Balme's analysis revealed site middens to be dominated by one of three prey species: freshwater mussels (*Velesunio ambiguus*), freshwater crayfish (*Cherax destructor*) or golden perch (*Macquaria ambigua*). She drew parallels between the thin horizons that make up the Pleistocene sites and similarly structured ethnographic 'dinnertime' camps (e.g. Meehan, 1982), where the bulk of food collected by comparatively large groups of people, may have been consumed before the left-overs were transported to a residential camp. A very similar picture is recorded from the ephemeral pre-LGM midden sites in the Willandra Lakes system. With few exceptions, the same three species dominated assemblages. Although they do contain a range of other taxa, including small mammals and reptiles, representation outside of these three species occurs at very low MNI levels (Allen, 1998: table 3). Pleistocene encampments within the Darling River area appear to have been predominantly of a short-term nature and are interpreted as sporadic but large-scale foraging events – either representing aggregations or comparatively large residential groups – coming together to take a single resource in quantity (Allen, 1998; Balme, 1995). The focused nature of this exploitation strategy is further underlined by evidence of what was available but *not* hunted in the vicinity of the Willandra sites. For example, although remains of the hairy-nosed wombat (*Wombatus lasiorhinus*) are abundant from the Upper Mungo Unit – OSL dated to *c.* 40–30 KBP (Bowler *et al.*, 2003) – but despite it being easy prey and broadly contemporaneous with human occupation, there are no archaeological indications that this animal was ever targeted; most occurrences appear to be of animals which had apparently died in their burrows (Bowler, 1998).

This kind of subsistence specialization does not seem to have been confined to the watered interior of Sahul either; it also emerged as a common feature among southwest Tasmanian sites dated 35–11 kbp. There assemblages are frequently characterized by the dominance of a single prey species, *Macropus rufogiseus* (Bennett's wallaby), accounting for an overall occurrence of 70–80 per cent of identified taxa (Pike-Tay and Cosgrove, 2006) – on occasions, for example, Kutikini Cave, this can rise to >90 per cent (Garvey, 2007). A seasonal component is also now emerging in the early exploitation strategy towards this animal. Using growth-increment analysis, a pilot study on *Macropus rufogiseus* by Cosgrove and Pike-Tay (2004) points to a late fall or early spring hunting period at the site of Warreen, visited between 40,141 ± 518 cal. BP (Beta-42122B/ETH-7665B) and 19,114 ± 289 cal. BP (Beta-42993). The logic behind entering upland areas during the winter, particularly during LGM winters, the densest occupation period at Warreen, is that fat levels in the meat would have been higher and the fur thicker and that during this time of year, the populations of *Macropus rufogiseus* would have been more sedentary. Occupation at these sites seems to have cycled according to changes in the local development of the grassland habitat favoured by this species, while other hunting opportunities were largely ignored (Holdaway and Porch, 1995; O'Connell and Allen, 2007). This may relate to the particular cold-season nature of this exploitation strategy, which potentially limited the availability of other resources; alternatively, if there is a seasonal component to this economy, the dominance of this one species at excavated sites might suggest that only part of that system has yet been identified. Either way, the age profiles compiled of *Macropus rufogiseus* for Warreen, Bone Cave (29–13.7 KBP) and the later (19–14 KBP) Kutikina Cave suggest that predation rates were comparatively low (Pike-Tay and Cosgrove, 2006). A low hunting pressure, coupled with what appear to have been ephemeral occupations and non-exploitation of all available habitats and possible resources, is seen as consistent with a model of comparatively low human population density for the pre-LGM in Sahul (O'Connell and Allen, 2007).

The low population density argument is further developed by the notion that both the lacustrine fauna of the Darling and Willandra River areas and the upland wallaby populations of southwest Tasmania may have represented 'ecologically tethered animal resources' (Balme, 1995; Pike-Tay and Cosgrove, 2006). People were organising their procurement strategies around resources which were dependable under the increasingly variable conditions of the Late Pleistocene (e.g. Bowler, 1998; Allen and O'Connell, 1995; *see* Chapter 7). As shall be discussed in Chapter 8, this additional scenario is also consistent with the kind of strategies one might expect among colonizing communities.

The Late Pleistocene colonization of Sahul appears to have been marked by comparatively few (surviving) technological innovations prior to the LGM, though the impact of taphonomy on this is only now being seriously considered. Those innovations that do appear signal the opening up of new resource opportunities. Ground stone and waisted implements can be linked to intervention in the creation of forest gap dynamics as a means of encouraging new growth and attracting forest-edge game, whilst grinding and milling stones suggest diversification towards alternate, *possibly* lower-ranked resources. Otherwise, a generally minimal investment flaked lithic toolkit appears repeatedly in conjunction with specialized subsistence strategies focusing on a narrow selection of available resources. The existence of this relationship must disassociate any necessary linkage of the latter to the formal and diverse typologies that are evident in the European Late Pleistocene sequence.

5.4 Conclusion

This chapter has discussed the initial dispersal of anatomically modern humans out of Africa and has placed particular emphasis on the state of genetic, environmental and archaeological evidence with respect to the Southern Route hypothesis. Three key points emerge.

Firstly, within the corpus of genetic evidence, whatever its shortcomings, the picture of modern humans radiating from Africa (be it via a single or multiple events) remains dominant after 20 years of research. The purported antiquity of the Chinese evidence might signal that some revision is on the horizon, though this has yet to convince. Indications are also that the Southern Route hypothesis, while still light on archaeological evidence, looks to have been either the primary line of that diaspora or at least a major component of it. The European sequence, as important as it is during the Late Pleistocene, probably did not form part of this initial spread. As a result, it is no longer justifiable for that sequence to be held up as the standard against which the rest of Pleistocene humanity should be judged. Although there remains something of a mismatch between the genetic and archaeological streams of evidence for the African exit, the orthodox position that the MIS-5 expansion of modern humans into the Levant was a dead end may also require reassessment.

While the world's current population can trace its ancestry to a single African lineage, this may not have been the only lineage of *H. sapiens* to have left Africa – something that we are perhaps glimpsing in the Mungo III evidence and which we may anticipate will become more mainstream in the future. Conversely, there is comparatively little evidence of human presence for the 'arrival' period predicted by the genetics. Major inundation of the wide continental shelves of South and Southeast Asia has inevitably severely truncated the available record, though places do exist (e.g. northern Luzon and East Timor) where ancient coastlines have survived. At least one of these two areas is predicted to lie on one of the most likely migration routes into Australia, yet even under these prime conditions, the evidence does not immediately reach back to the *c.* 60 kbp horizon that it needs to for the genetic model – potentially, this will be something that the new evidence from the Philippines is set to change, if the hominin there is *H. sapiens*. It may also be time to begin reconsidering the models we have been using to accommodate the archaeological data we do have. The occupation and exploitation of coastal resources features prominently in the Southern Route hypothesis, but the significance of their role on the ground turns out to be more equivocal. What archaeological evidence that does exist for this time window indicates that hominins were capable of occupying areas up to 100 km inland just as easily as along the coast – a fact that also has important consequences for modelling the speed of the dispersal process. The problem is that we do not yet know which species was responsible for most of these pre-40 kbp sites (or even some of the more recent ones). More problematic is archaeological evidence that the exploitation of coastal resources was not uniform. Even when sites were located close to the coast, maritime resources do not always appear to have been exploited consistently – an issue to which we shall return.

Secondly, growing evidence from the Pleistocene human record supports the need for more regionalised assessment and less grand narrative. It is abundantly clear that although *H. sapiens* as a species are very unified (and allowing for the holes in this picture that extinct genetic lines have left), they were not entering into every region of the world and impressing a single behavioural suite onto the environments and materials they found there. There is strong evidence that most of the Asian record was marked by pronounced continuity between Middle Palaeolithic material culture and Upper Palaeolithic – something that may even apply in the Levant. The arrival of modern humans in Asia does not appear to have precipitated major changes or sudden technological turnovers, and identifying hominin species through tool typology cannot be taken as read, as Moore and Brumm (2007) have excellently demonstrated for ISEA. Only in Europe can it be argued that inter-regional arrival coincided with a marked break in technological and economic practice, and recourse to raw material and cultural retardation arguments to explain differences to this standard elsewhere cannot be supported. It does remain intriguing that changes were not precipitated on the archaeological record with the arrival of modern humans, but this is on the pretense of our expectations and models of modern human behaviour in the Upper Pleistocene, which are almost entirely derived from the European experience.

There is almost certainly no simple answer to the questions posed by the differences we see, but it is time to set aside models that seek to explain this variability through purely degenerative and restrictive criteria. Hunting and gathering economies persisted through the length of the Upper Pleistocene in no small part because these were the most suited to the conditions people encountered. The much-vaulted emergence of agriculture in the Holocene simply could never have developed under the environmental and climatic instability that typified the human experience in the Pleistocene, finally rendering as obsolete the view that hunting and gathering is no more than a 'basic cultural-economic stage'. One might even go as far as to say that if the future holds a return to more volatile climate conditions, with all the consequences that this will have on biological communities, updated elements of these prehistoric foraging subsistence strategies may provide an essential formula for long-term economic adaptation and survival. The spread of our species across the Pleistocene planet needs to be re-examined not only because the archaeology demonstrates the existence of genuinely distinct localised adaptations but because the apparent pan-species unity of humanity's hunter-gatherer past is itself a relic of a unilinear and progressive model of our evolution.

The third point to emerge from this chapter is that the mosaic character of the meagre Asian hominin record needs to be taken seriously. The history of hominin occupation of this vast continent is not even close to being understood; the Dmanisi, Denisova, Ngandong, Narmada, Liang Bua and Callao evidence all demonstrate emphatically that there is much here that we do not yet have a handle on. Even the modern human evidence from East and Southeast Asia and Sahul raises the possibility of early physiological diversity in *H. sapiens*. The recent extension of the *H. neanderthalensis* range into southern Siberia, the new late dates of *H. erectus* in Java and the survival of *H. floresiensis* until the LGM mean that overall hominin diversity lasted for much longer in Asia than in Africa. This in itself complicates interpretation of an already problematic archaeological record in determining who created what industries. A further feature is that large amounts of data in Southeast Asia and Sahul only seem to begin appearing after *c.* 40–45 KBP. We are left to consider whether this marks the beginning of population expansion in the region and the creation of higher archaeological visibility rather than the first appearance of people here and what impact taphonomy has on this picture.

Although there is good reason to believe that the climatic and environmental instability of the Late Pleistocene are strongly linked to regional developments in hominin behavioural adaptations, there is not yet sufficient chronological resolution to confidently match archaeological trends to environmental ones. Fortunately, these two lines of evidence do converge during deglaciation, where some of the most sudden and far-reaching climatic and environmental changes are matched by archaeological contexts with consistently sub-centennial error margins in calibrated dates. In the following two chapters, we shall consider evidence for the range of human technology settlement and subsistence strategies from the Southeast margin of the Asian record during the Last Termination of the Pleistocene and into the Early Holocene.

Chapter 6

Climate, Dispersal and Technological Change during the Last Termination and Early Holocene in Southeast Asia

6.1 Introduction

Over the last decade, a new vibrancy in the archaeological investigation of Southeast Asia's prehistory has been overturning many traditional views about the timing, extent and nature of hominin presence in this region. Long-held perceptions that saw cultural and technological restrictions as explanations of inter-regional differences, particularly to Europe, are being challenged, and staid portrayals of prehistoric tropical foragers have been found to mask complex, subtle and archaeologically accessible evidence of behavioural variability. Such work is still in its early stages, but already, meaningful progress has been made in reconsidering the way prehistoric Southeast Asians adapted to the particular demands and opportunities that the environments of this region presented.

During the last glacial, Southeast Asia was less prone to the extreme temperature shifts experienced at higher latitudes, though variation was still observed. The region was particularly subject to short-term oscillations in forest cover, changes in rainfall pattern and profound geographical remodelling with post-glacial rising global sea levels. New genetic evidence suggests that from the final millennia of the Pleistocene onwards there was significant intra-regional population dispersal, possibly following a bottleneck event. The archaeological evidence from the early post-glacial is marked by notable innovations and economic realignment; this was a time of substantial transformation for people and their way of life. In this first of two chapters examining in detail how humans adapted to Late Pleistocene and Early Holocene conditions in Southeast Asia, we review the late to post-glacial environmental history of the Sunda region, and the genetic and technological changes that were occurring during this period; in Chapter 7, new evidence is presented about the impact environmental changes had on local forager economies. All radiocarbon dates from archaeological sites mentioned in both chapters are presented in the Appendix.

6.2 Regional Climate and Conditions across the Last Termination (22–11.7 KBP)

The Last Termination of the Pleistocene is, for the moment, the most suited for studying the relationship between changing climatic and environmental conditions and human adaptive mechanisms. Climatic shifts across this period were pronounced, with adrupt temperature changes and incremental inundation of coastal margins punctuated by episodes of rapid sea level rise. The archaeological evidence is reasonably abundant from this period, and the error margins on calibrated ^{14}C dates are more consistently at the sub-centennial scale necessary to examine correlations between human and environmental change with reasonable accuracy.

Though geographically far removed from Southeast Asia, the Greenland ice core record still provides the global benchmark for the chronology of climatic events during the Last Termination. Severe cold gripped middle and high latitudes during the LGM, tempered only by subdued warming during the first part of deglaciation (ahead of GS-2; Blockley *et al.*, 2006; Lowe *et al.*, 2008; Rasmussen *et al.*, 2006; *see* Figure 2.5). The first significant amelioration in the climate *c.* 14.6 KBP marked the beginning of an approximately 2000-year period of predominantly warm conditions (though with at least two, possibly four, very brief cold episodes) lasting until *c.* 12.8 KBP: this was GIS-1 (in northwest Europe, the Bølling–Allerød interstadial). The severe cold of GS-1 (Younger Dryas) brought a return to new full-glacial conditions between *c.* 12.8 and 11.7 KBP, before the onset of rapid and sustained warming from *c.* 11.7 KBP – now formally recognised as the global stratotype section and point (GSSP) for the base of the Holocene (Walker *et al.*, 2009).

The most far-reaching and direct consequence of deglaciation in both hemispheres was the worldwide transgression of sea levels from their minimum of *c.* −116 m at the LGM. There is no simple relationship between the decay (or growth) of ice sheets and sea level change (Lambeck *et al.*, 2002a): changes in ice volume, complex gravitational and orbital variables between the Earth, Sun and Moon, tectonic activity and isostatic recovery all factor. Evidence collated from seven regions around the world – Barbados, the western Atlantic (Florida), Tahiti, the Huon Peninsula, Papua New Guinea, northwest Australia (Christ Church), New Zealand, and the Sunda Shelf (Lambeck *et al.*, 2002b) – reveals a general picture that the melting of the ice sheets began about 19 cal. KBP. For the first 500 years, the average global rise in sea level was comparatively rapid at *c.* 30 mm/yr (3 m/100 yr). It then slowed to *c.* 3.3 mm/yr (0.33 m/100 yr) before again accelerating *c.* 16–12.5 cal. KBP at an average rate of about 16.7 mm/yr (1.67 m/100 yr; Lambeck *et al.*, 2002b: 358). It had been thought that the cause of this increased rate was a response to the first major post-LGM meltwater pulse from the Laurentide ice sheet (equivalent to two-to-three Greenland-sized ice sheets); however, there is no conclusive evidence that this was the case (Clark *et al.*, 1996). More recently, it has been shown that there was a major retreat of the East Antarctic ice sheet *c.* 15.3 and 13.4 cal. KBP. The huge injection of water this caused, referred to as meltwater pulse IA (mwp-IA), may have been the root cause of the rapid rise (Verleyen *et al.*, 2005). Data from marine sediment cores in the South China Sea (site/cores 17940, 17954 and 17961, 18252-3 and 18287-3) give a commencement date of *c.* 15 cal. KBP and a termination 14.4–14.2 cal. KBP. They also show this to be synchronous with an abrupt 1°C rise in sea surface temperature, commencing 14,700 ± 300 cal. BP (Kienast *et al.*, 2001, 2003) – the beginning of GIS-1. Using high-resolution coring (core 18287-3 at 5.65°N, 110.65°E) with sampling at intervals equivalent to 100–300 years, indications are that this temperature rise also happened over a very short period of less than 200 years (14.7–14.5 cal. KBP; Steinke *et al.*, 2001). Close study of the Southeast Asian data (Hanebuth *et al.*, 2000) has shown that *c.* 19–14.6 cal. KBP, sea levels rose in this region on average 4.1 mm/yr (0.41 m/100 yr). For a 300-year period between 14.6 and 14.3 cal. KBP, they then rose sharply at 5.3 m/100 yr, synchronous with the beginning of GIS-1. Although through the rest of this interstadial, sea levels rose at a slightly slower rate – for example, from 14.3–13.1 cal. KBP increasing *c.* 1.3 m/100 yr – the fact and pace of inundation can be expected to have had far-reaching consequences for early tropical hunter-gatherer populations living at the time.

Of particular importance to changes in Southeast Asia was the exposure of the Sunda Shelf (Figure 6.1). At the height of the LGM, this incorporated the present islands of Borneo,

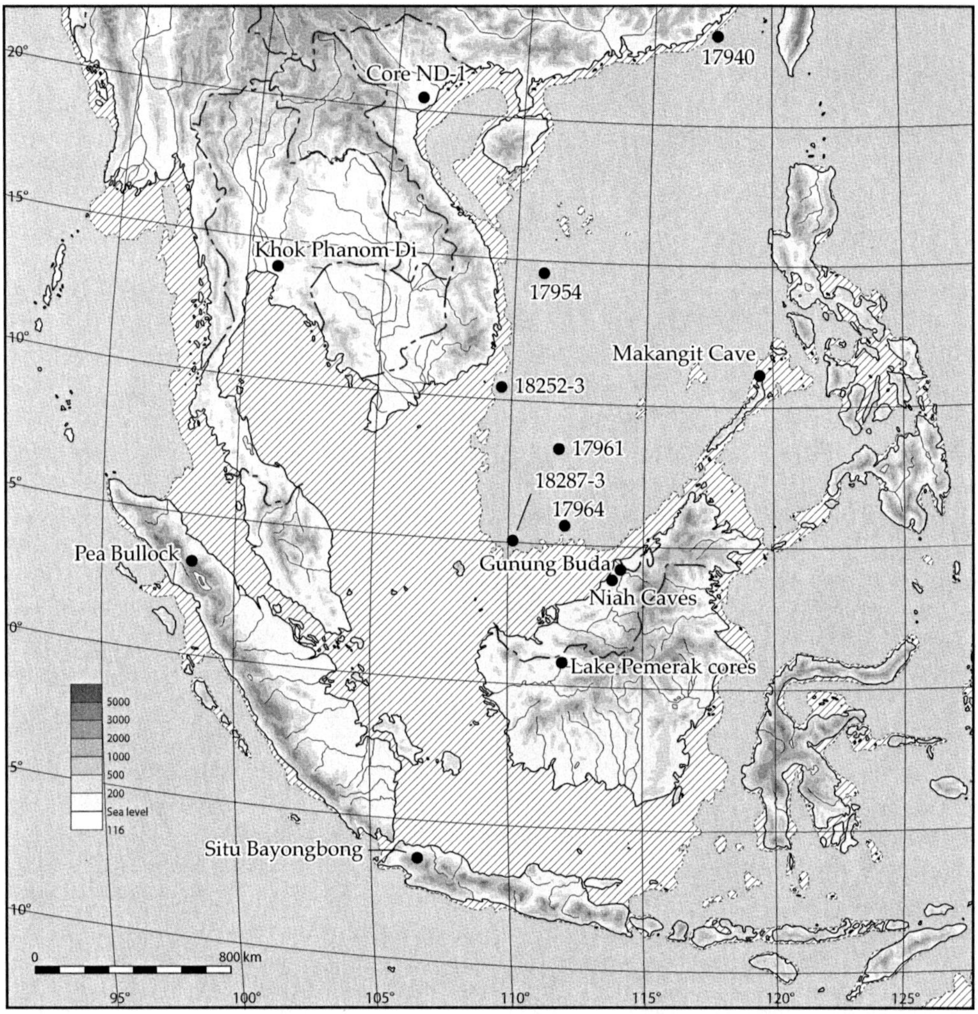

Figure 6.1. Map of Pleistocene Southeast Asia, showing the locations of terrestrial and ocean sediment core records referred to in the text. (Map drawn and prepared by D. Kemp.)

Sumatra, Bali and Java, and it had a total breadth near the equator of *c.* 2200 km, creating a landmass *c.* 75 per cent larger than exists today (Verstappen, 1975). The greater land surface would have had impacted on regional climate in a number of ways. It would have affected the local albedo, and computer models suggest that this would, in turn, have contributed to making prevailing conditions cooler and drier than currently (Bush and Fairbanks, 2003), as indicated by the development of local glaciers in Borneo and Sumatra. The corresponding reduction in local sea surface area would also have resulted in lower levels of evaporation for the monsoon winds that bring rain to many parts of the region (Heaney, 1991). The net result would have been an extension of more seasonal forests and savannah-like conditions from the Asian mainland through much of the centre of the continent (Bird *et al.*, 2005; Heaney, 1991; Wurster *et al.*, 2010).

In terms of vegetation changes through the Last Termination, several palynological sequences now exist, though coverage is still comparatively patchy. In the highlands of western Java, an 8-m core was obtained from a small swamp at Situ Bayongbong at 1300 m (Figure 6.2; Stuijts, 1993). The basal pollen zone here was between 19,952 ± 360 cal. BP

Figure 6.2. Photograph of the Situ Bayongbong location during coring. (*After* Stuijts, 1993, *Modern Quaternary Research in Southeast Asia* 12, front plate, reproduced with kind permission.)

(GrN-10522) and 14,246 ± 310 cal. BP (GrN-11791). Globally, this covered the period of gradual warming from the end of the LGM to the beginning of GIS-1. In the highlands of Java, the palynology indicates a montane forest with a noticeable coniferous element (*Dacrycapus imbricatus*), currently found at an altitude of *c.* 1800 m, and 500 m higher than Situ Bayongbong. This lowering of vegetation zones can be correlated with a mean annual temperature *c.* 3°C lower than today's *c.* 18.7°C for this altitude. The lower temperature does not, however, seem to have been accompanied by any significant change in rainfall pattern: the tree taxa represented all favour ever-wet conditions.

Between *c.* 14.2 cal. KBP and 12,770 ± 64 cal. BP (GrN-11521), the incidence of conifers declined in this core, consistent with a rise in temperature. After *c.* 12.7 cal. KBP, local forest began to show a dominance of the family Fagaceae (oaks and beeches), characteristic elements of modern submontane forests at this altitude and an indication that forest composition was approaching that of today rather than dipping back into a colder phase.

On the northeastern edge of the Sunda Shelf, Sun and Li (1999) and Sun *et al.* (2000) reported findings from ocean core 17490 (23.116667°N, 117.383333°E). These showed that between 21,550 and 19,010 BP (the late LGM), a biotic community contained high but very variable levels of herbaceous taxa (particularly *Artemisia*). Three rapid cycles were observed containing either high frequencies of *Artemisia* or montane conifer taxa. Tropical and subtropical broad-leaf species show a moderate increase from lower in the core, and mangrove taxa begin to appear in very low quantities around this time. Between 17,660 cal. BP and 13,280 cal. BP, the core contained a high frequency of tropical and subtropical broad-leafs, significant levels of *Artemisia* but low frequencies of montane rainforest and conifer taxa. This phase is also noted by a peak in mangrove pollen, extrapolated as occurring at *c.* 14 KBP, and a sharp drop in *Artemisia*, a change the authors equated with the establishment of GIS-1.

In West Kalimantan, Borneo, a series of lowland cores have been obtained from the Upper Kapuas River Basin (Anshari *et al.*, 2001, 2004). The study area (0.733333°N, 112.1°E) is 35–50 m above sea level within a catchment basin that is enclosed on three sides (north, south and east) by mountains up to 2000 m. The cores provide the first evidence of lowland climate change from the island's interior. Core HN3 was obtained from the margin of Lake Pemerak and has a basal date of 34,181 ± 172 cal. BP (Wk-5779). Despite the value from its antiquity, there are major discontinuities through the sequence, obscuring changes during most of the LGM and the Pleistocene–Holocene transition. The period of the early deglacial is covered, though. Biozone B4 is dated between approximately the end of the LGM to the beginning of GIS-1 – 20,758–19,393 BP (Wk-6277) and 16,190–14,652 BP (OZE-135) – calibrated by the authors using CALIB 4.2 (Stuiver and Reimer, 1993). Palynologically, the high frequencies of *Gluta renghas* during this zone were taken as indicators of closed riparian forest near to the coring site, whilst *Palaquium* indicated that there was also peat forest in the vicinity. High percentages of fern spores and montane species also indicated that the forest community was diverse and not analogous to its modern swamp forest counterpart. The presence of charcoal throughout the core sequence is taken as evidence that fire was an active agent of these lowland forests back to at least *c.* 30 KBP; its cause, though, is less clear.

Three further cores were obtained from around and within the same lake (Anshari *et al.*, 2001, 2004). Two produced dates relevant to the current discussion. The first (Core A) contains one ^{14}C date 14,353 ± 148 cal. BP (OZE-137). This seems to straddle a change in local conditions from a mixed peat–montane forest to a forest in which the montane components appear only at very low levels. The second (Core C) contains one ^{14}C date of 17,722 ± 201 cal. BP (OZE-147) sandwiched between one of *c.* 46 cal. KBP and a Mid-Holocene date. The fact that all appear within a 0.82 m sequence indicates that deposition was almost certainly disrupted. That said, the zone containing the Last Termination date is characterized by a higher frequency of herbaceous taxa compared to that lower in the profile. Taken together, the evidence from these four cores suggests that the landscape around the Lake Pemerak area was comparatively open *c.* 17.7 cal. KBP, with a subsequent trend towards mixed lowland peat and montane forest. This forest structure persisted until *c.* 15.2 cal. KBP, but the montane components had largely dropped out by *c.* 14.3 cal. KBP during GIS-1.

Looking north to the Philippines, the first long-term terrestrial record of climate and environmental change has been recovered from Makangit Cave (11.199481°N, 119.500617°E), on Palawan (Bird *et al.*, 2007). The position of Palawan Island, towards the northern latitudinal limit of Holocene rainforest, means that it should be particularly sensitive to shifts in the prevailing climatic regime. Interpretation of carbon-isotope (δ^{13}C) values obtained from the lower and largely undisturbed 0.7 m portion of a 1.8 m profile indicated that grassland or sparsely treed savannah-like conditions existed in the vicinity during the LGM; a picture consistent with regionally reduced rainfall and taken as evidence that Sundaland contained stretches of drier, more open habitats at least periodically during the Late Pleistocene (Bird *et al.*, 2007: 68). While the sequence is anchored by two radiocarbon dates, these are at either end of the profile. The portion of the sequence attributed to the

Pleistocene–Holocene transition is very narrow and precludes any certainty over whether this local transformation was comparatively rapid or more protracted, and whether it was interrupted by a GS-1 event.

In the North Atlantic, GS-1 had been marked by an abrupt thermal depression and rebound. Using the GISP2 data, Mayewski *et al.* (1993) reported that the onset and termination of this stadial occurred within 10–20 years and that its duration (1300 ± 70 years) was characterised by massive, frequent and short-lived (decadal or less) changes in atmospheric composition. Steffensen *et al.* (2008), using data from the NGRIP core, found similar evidence for decadal to centennial-scale change; all of which would have been immediately apparent to hunter-gatherer groups on generational timescales.

The timing and particularly the effects of GS-1 in Southeast and East Asia were varied. Bordering this region, high-resolution pollen data from a varve sediment core in Lake Suigetsu, Japan, spanning 15,701–10,217 BP (Nakagawa *et al.*, 2005) certainly contains a late cold event; however, it lagged behind that seen in the North Atlantic records by 250–400 years and was of much reduced amplitude. The mean annual temperature depression was 4° ± 2°C – only one-third of the drop experienced in the North Atlantic region, where deterioration almost matched that experienced between full interglacial and glacial conditions (*c.* 13°C).

On the northern edge of the Sunda Shelf, the palaeogeographical evolution of the Song Hong deltaic region of northern Vietnam is one part of the region that is now reasonably well understood through a combination of onshore and nearshore studies (Hanebuth *et al.*, 2006; Hori *et al.*, 2005; Shimanski and Stattegger, 2005; Tanabe *et al.*, 2003a and b, 2006). Seven recently drilled onshore cores have provided a composite of sedimentary sequences from the LGM to the present, with a single core (HV) extending back to 40–46 KBP.

The sedimentary facies within one of the key Vietnamese cores (ND-1, 20.372778°N, 106.146667°E) has been divided into three units (Tanabe *et al.*, 2006), the upper two of which concern us here. Unit 1 dates from as early as 14,950 ± 580 cal. BP (Beta-142418) and 14,840 ± 580 cal. BP (Beta-142417), until sometime prior to *c.* 11.4 cal. KBP, and the beginning of the overlying Unit 2. As such, Unit 1 appears to have spanned both the GIS-1 and GS-1 substages. Over this period, the area's hydrology shifted from that of a meandering river to floodplain sediments. Regionally, during the GS-1 period, the rate of sea level rise did slow to an average of *c.* 1.14 m/100 yr (*after* Hanebuth *et al.*, 2000), probably changes in the river system were in relation to this.

Sediments in ND-1 ascribed to Unit 2 date from between 11,430 ± 200 cal. BP (Beta-142414) and 9210 ± 180 cal. BP (Beta-164867) to the beginning of the Holocene and appear to have been deposited under conditions influenced by tidal action. Tanabe *et al.* (2006) interpret the presence of wood and plant fragments as well as molluscs (*Potamocorbula* sp. and *Corbicula* sp.) as indicating that this phase was characterized by brackish water habitats and an estuarine environment, both of which should have offered new subsistence opportunities to local hunter-gatherer communities.

On the eastern edge of the Sunda Shelf, data from marine sediment core 17964 (Sun *et al.*, 2000; 6.158333°N, 112.213333°E) have shown that for the period including the LGM, 25,400–19,700 cal. BP (calibration curve unspecified), there appears to have been frequent cyclic alteration between herb-dominant and montane conifer-dominant phases. Oscillating conditions between montane rainforest and lowland rainforest taxa are also apparent 19,700–10,400 cal. BP. Core sampling in this and the 17490 core was every 10 cm, giving a resolution of ±360 years. With that level of fine-scale recovery, there has been suggestion that some of the cycles visible in both cores may equate to Heinrich events and D-O cycles in the North Atlantic (Sun and Li, 1999). Correlation between the timing of the palynological cycles and those revealed in the Greenland ice cores has also been proposed for earlier periods of the mid-Upper Pleistocene record in ISEA (Hunt *et al.*, 2007, 2012).

Data from ocean core 17490, also in the northeast part of the Sunda Shelf, show that a discrete pollen zone (Zone P2-b) commenced *c.* 13,280 cal. BP and concluded *c.* 11,600 cal. BP. This zone contained a rise in montane rainforest and conifer taxa, whilst broad-leaf and herbaceous taxa declined markedly, with the latter almost disappearing. This shift corresponds to the timing of GS-1 and suggests that although temperature depression was not nearly as marked as it was at more northerly latitudes, part of this region probably did experience changes in vegetation structure over this interval (Sun and Li, 1999; Sun *et al.*, 2000).

In Borneo, recent publication of a stalagmite sequence from Gunung Buda National Park provides an absolutely dated (by 75 U-series dates) oxygen isotope record for the last 27,000 years (Partin *et al.*, 2007). This record does show the development of drier conditions in the north of the island culminating at 16,300 ± 300 BP, coincidental with Heinrich event 1 (Rosseau *et al.*, 2006); however, there is no evidence of GS-1. Its absence led Partin *et al.* (2007) to propose that different mechanisms were operating between these two deglacial cold events, possibly with the latter being caused by a meltwater discharge coming largely from the Keewatin ice dome (central Canada), with subsequent pulses helping to maintain suppression of the thermohaline circulation through the stadial (*see* Tarasov and Peltier, 2006). Evidence from the high-resolution 18287-3 core from the South China Sea (5.65°N, 110.65°E) similarly echoes a muted stadial signal. Data from here suggested that while there is evidence of a cooling consistent with the timing of GS-1, the drop in temperature was only 0.2°–0.6°C (Steinke *et al.*, 2001).

Two comparatively long sequences are reported from the Sumatran Pea Bullock volcanic crater (*c.* 1400 m asl). Pollen evidence from Pea Bullock Core A indicates that a temperature minimum was reached at 25,651 ± 364 cal. BP (UB-3702) and that there was a possible decrease in temperature of *c.* 5.2°C. Rapid drop-off in the frequency of the cold indicator taxon *Vaccinium* after 21,498 ± 192 cal. BP (UB-3701) points to an early recovery here in the post-LGM period. In the Pea Bullock Core B, there is a final major peak in conifers of the genus *Dacrydium* at a point in the sequence dated to 12,537 ± 63 cal. BP (UB-3734). This falls within the GS-1 window and has been tentatively linked to a return to cooler conditions (Maloney and McCormac, 1995: 186). As noted, the Javanaese Situ Bayongbong core gives little indication that a GS-1 event had any major impact in the west Javan highlands.

Abrupt warming followed the end of GS-1 and marks the Pleistocene–Holocene transition. Measurements taken on the Greenland ice cores suggest that a significant part of the transition process may have taken place over a very short period of time, possibly decades only. Although this was in keeping with the established pattern of D-O cycles, unlike previously, the interaction of other factors, such as the elevated atmospheric CO_2 and decreasing glacial cover, meant that the comparatively warm conditions that had been initiated persisted. As elsewhere, in Southeast Asia, the transition to the Holocene was associated with further rising sea levels. From as early as *c.* 7.5 cal. BP and peaking *c.* 4.5 cal. KBP, these had reached and briefly exceeded modern values by 3–5 m (Geyh *et al.*, 1979; Hanebuth *et al.*, 2000; Tanabe *et al.*, 2006; Tjia, 1996), before stabilising at more or less their current position. There are still comparatively few clear indications about the type or scale of impact rising seas had on regional communities. Research in Oceania ties it closely to population migration (Dickinson, 2003). Within Southeast Asia, one of the few locations with high-resolution records comes from the very end of this period, at the rich open-air site of Khok Phanom Di (13.584725°N, 101.140644°E) in Chonburi Province, Thailand, a site excavated in 1984–5 by Charles Higham and colleagues (Higham and Thosarat, 2004a).

The thick cultural deposits of this site accumulated on top of Early Holocene marine clays from *c.* 3852 ± 80 cal. BP (ANU-5493) until *c.* 3542 ± 149 cal. BP (ANU-5482). Over the course of five or so centuries, the site's environs changed significantly. During its initial occupation, it lay near to the edge of a broad river estuary and mangrove forest, but as the sea level receded, these habitats were replaced by dry evergreen forest and swampy open woodland (Higham and Thosarat, 1998, 2004a). People did not abandon the site, though; rather, they adapted their activities to suit the changing conditions. A reliance on aquatic foods existed throughout the occupation, though mammalian game featured more prominently in the diet of these hunter-gatherers as immediate access to marine and riverine resources was progressively lost (Kijngam, 1991). The character of the strategies employed at Khok Phanom Di suggests that its inhabitants had inherited a mature maritime subsistence tradition and were the descendants of long-term coastal foragers rather than of recent arrivals from the interior (Higham and Thosarat, 2004a). While for a long time largely conjectural, genetic and archaeological evidence from the wider region is now beginning to provide additional if scattered glimpses of these early Southeast Asian coastal communities.

6.3 Genetic Evidence of Human Dispersal during the Last Termination

The great diversity of languages and cultures that exist in Southeast Asia has been long known. In recent years, it has become a spur to investigations seeking to understand the region's genetic history and its position in the global spread of humanity. While localized patterns of linguistic and cultural affiliation cannot always be taken as indicators of underlying genetic variability, that which is emerging in Southeast Asia well reflects this outward complexity; indeed, the demographic history of the region is turning out to be more intricate than anyone had thought. As a result, long-term *in situ* evolution of indigenous haplogroups and multifarious migratory pathways (e.g. Endicott *et al.*, 2006; Hill *et al.*, 2006, 2007; Macaulay *et al.*, 2005; Mona *et al.*, 2009; Soares *et al.*, 2008; Stoneking and Delfin, 2010; Tabbada *et al.*, 2010) are rapidly coming to characterize our understanding of the prehistoric populations here.

A significant component of this work has focused on dispersal within and outside of Southeast Asia. Based largely on linguistic evidence, it had been previously thought that the settlement of Southeast Asia (and Australasia) had been predominantly a two-phase process: an initial colonization by modern humans, representing the wider region's Australo-Melanesian heritage, followed by the arrival of a Mongoloid Austronesian population from Taiwan and China during the last 4000 or so years. When populations are considered at a genetic level, evidence for Holocene dispersal also duly emerges, but it seems to have been far from unidirectional (Tabbada *et al.*, 2010). By some margin the strongest signals of movement turn out to be internal rather than intrusive, and appear to have an ancestry extending into the late glacial (Hill *et al.*, 2007). Evidence for a significant intra-regional pulse has been linked to the effect that the flooding of the Sunda Shelf had on communities during the Last Termination.

The study by Soares *et al.* (2008) considered the most common and diverse indigenous ISEA haplogroup (haplogroup E) and plotted sequence variation in complete mtDNA genomes from 1704 samples from across the insular part of the region, 184 from West Papua and 962 from Taiwan. Haplogroup E is nested within M9 (itself thought to have a Southeast Asian origin *c.* 50 kbp) and is almost entirely restricted to Taiwan and Southeast Asia (Hill *et al.*, 2007), making it a particularly sensitive marker with which to track early population dispersal. It has an estimated coalescence age depth of *c.* 30 kbp and contains two subclades (E1 and E2), which themselves have estimated ages of 14,950 ± 3900 bp for E1 (maximum likely age) and 8000 ± 2050 bp (maximum likely age) for E2. These, in turn, subdivide into very regionally specific subclades: E1a and E1b, E2a and E2b.

The weight of the genetic evidence suggests that the homeland of haplogroup E lay somewhere in the northeast provinces of the region. All subclades of E are present in Borneo (a fact that has potential significance for the archaeology, as we shall see). There is the additional possibility that subclade E1a sequences, excluding an E1a1a variant, and E2a sequences may have arisen in adjacent islands of Wallacea. Soares *et al.* (2008) suggest that given that haplogroup E contains only three branches extending back to or beyond *c.* 12 kbp – E1, E2 and E1a1a – it is quite possible that a bottleneck event preceded the proposed dispersal phase. They intimate that the most likely cause was the rapid sea level rise linked to the GIS-1 transition (and mwp-1A), though robust calibration of the genetic dates is still required. Protocols employed by the authors to verify age estimations, crucial for such linkage, did also produce quite divergent results for some subclades, however. While presenting a very plausible cause, the timing is not yet accurate enough for us to posit a certain relationship to mwp-1A.

6.4 Technological Change

The inundation and following insularity that the Last Termination brought to Sundaic Southeast Asia affected changes in regional climate, weather systems, ocean circulation and, we can assume, habitat productivity. Human populations would have been displaced; some would have died out. Survival would have depended upon the ability of communities to adapt to the consequences of these challenges. While their most flexible device, the honed artifice and skill of living, is mostly lost to us, where that interface required a physical restructuring of the material world, we can still find traces of it (*see* Riddington, 1983). In light of this, for the remainder of this chapter, we focus on some of the key technological changes that emerged during this period; in Chapter 7, we examine some of the subsistence practices that these technologies enabled.

6.4.1 Bone Technology

In 1926, L. J. C. van Es, an employee of the Geological Survey, Bandoeng (Bandung), Java, was studying the Lower Quaternary fossils in the Madiun area of East Java. Reconnaissance by a local member of the Geological Survey, following up a rumour that bones had been discovered in the nearby cave of Gua Lawa, resulted not in the extinct species that van Es was looking for but in examples of modern taxa and a single worked piece of bone. Believing these to be from a site of early human habitation, in August of the same year, he opened a trial trench in this rock-shelter that would become the type site for an enigmatic early Southeast Asian 'bone culture'.

Van Es's trial excavation reached bedrock at −13.75 m, digging in 0.5-m spits, though all of the archaeologically productive levels came from fine-grained deposits in the upper 4.5 m of the profile. Following a visit by the then government prehistorian van Stein Callenfels in 1927, a joint excavation effort was mounted in the central area of the cave between 1928 and 1931. Although there was found to be some mixing of deposits due to commercial guano digging, a sequence of cultural material was successfully excavated. Among the artefacts recovered, implements of bone featured prominently.

Van Stein Callenfels claimed the existence of three cultural horizons at Gua Lawa, though there was never unanimous agreement about this (van Heekeren, 1957: 76). According to van Stein Callenfels' sequence, the deepest of these contained bifacially worked lithic points (termed 'arrowheads'; *see* Section 6.4.2) as well as ceramics. The horizon above this is said to have contained tools made exclusively of osseous material, and a range of types were proffered, including awls, daggers, fish hooks (four were identified, all from the trial trench between −1 and −1.5 m), edge-ground and 'spatulate' forms. The upper horizon included a mixture of iron and bronze objects together with Neolithic quadrangular stone adzes and ceramics.

With the exception of a single, partially flaked, stone arrowhead of questionable association, no other worked lithics and no ceramics were recovered from the middle horizon containing the bone tools; grinding stones and pounders, however, were described. There are also indications of a possible shift in the type of bone tools emerging with increasing depth: the edge-ground tools were predominately recovered from higher in the sequence, while spatuate pieces were found at lower levels (van Es, 1930: 335). At least a proportion of these implements are also described as having high polish over 'nearly the whole surface'. In all likelihood, these pieces had either been exposed to water abrasion post-depositionally or may have been shaped exclusively by this non-anthropogenic process.

Further survey and excavation in East Java subsequently established that facies of a bone tool industry comparable to that at Gua Lawa existed at nineteen other cave and rock-shelter sites. Horizons containing this technology appeared abruptly in an archaeological sequence, generally without struck lithics and generally underlying ceramic horizons. Van Heekeren considered the 'Sampung Bone Industry', as it was now termed, to represent the arrival of new technological innovations introduced to the region during the Early Holocene from Mainland Southeast Asia (van Heekeren, 1957: 85); van Stein Callenfels thought the industry an island derivation of the Hoabinhian (see later); Heine-Geldern (1936) saw it as something independent and perhaps of North Asiatic origin.

Although the work of early researchers drew attention to the fact that bone technology (a term that is used inter-changeably with, and that incorporates all forms of, osseous technology) was a not insignificant component of Southeast Asia's prehistory, when compared to the interest accorded the region's lithic record, detailed consideration was minimal. For example, Tweedie, in his *The Stone Age in Malaya* (1953), mentions their existence only briefly under the category of 'Anomalous Industries'. It is perhaps primarily because researchers had tended to view the recovery of bone tools, from the outset, as a discrete and exclusive entity, but one only sparingly represented in the region (as alluded by van Es, 1930: 337, and explicitly by Heine-Geldern, 1936: 32), that osseous implements elicited so little attention. Until recently, the only systematic classification was that of Harrisson and Medway (1962) made on the large assemblage of osseous artefacts (published, $n = 206$) from the Niah Caves, Sarawak. Their approach laid down strict guidelines on what should and should not qualify as 'worked', based on preciously early experimental analysis into bone fracturing (e.g. cf. Myers *et al.*, 1980; Johnson, 1985). They were also the first to question the bone culture hypothesis. The Niah evidence strongly suggested that this material was an integral part of tropical hunter-gatherer technology. Harrisson and Medway considered their schema to be applicable to that from another site, Gua Bintong (then on display at the Raffles Museum), and it was applied directly in the study of bone artefacts ($n = 50$) from Lobang Hangus, another of the entrances at Niah (Medway, 1966; *see* Chapter 7). The system, though, was never adopted any more widely than this, and while site-level reports continued quite frequently to include remarks about the presence of osseous artefacts, the position of this technology in Southeast Asian prehistory remained obscure until the last decade, when there was renewed interest on the part of several researchers, including the current author.

The regional development and deglacial distribution of bone technologies across the Sunda Shelf has been the subject of detailed study by the author (Rabett, 2002, 2005; Rabett and Piper, 2012). While such technologies have been a component of archaeological assemblages in Southeast Asia from *c.* 45 KBP, it was not until after the LGM that they appear to have ascended to become a significant component of forager toolkits. Across the area encompassed by the Sunda Shelf, many assemblages containing bone technology were also now found on or comparatively near to the coast. As a result, a degree of relationship to the inundation of the continental shelf was surmised (Rabett, 2005: 171), though it likely represents only one of several factors that influenced the pattern of distribution and use of such tools across the wider region.

This has always been a discontinuous technological record: many sites contain no evidence of bone tool use. Partly this will be linked to changes in local and regional environments and changes in the availability of prey – some connection to the abundance of arboreal game is apparent, though not universal (*see* Chapter 7). Site preservation and taphonomic factors, too, must be taken into account as tropical soils are mostly hostile to the long-term survival of bone. That said, it is equally true that there are sites with good bone preservation and suitable use contexts (e.g. the presence of arboreal fauna) but no bone tools.

The patchy nature of tool occurrence will also have been affected by the availability and choice of using other less durable organic materials – wood, rattan and especially bamboo – which very likely made up a large proportion of implements made by early foragers in this region (e.g. Mijares, 2008; Gorman, 1970; Pope, 1988; Raddatz, 2006; *though see* Bar-Yosef *et al.*, forthcoming). In terms of range weapons alone, while some authorities consider the blowpipe to be a purely Late Holocene innovation (e.g. Bellwood, 2007; King, 1995; Sloan, 1975), others, such as Oppenheimer (1998), have argued for a much deeper antiquity and with it potential impact on Pleistocene hunting practice. The two main types of blowpipe recorded ethnographically in Malaysia are a combined bamboo–wood variety and a purely hardwood type (Dunn, 1975; Evans, 1937; Williams-Hunt, 1952); none of the materials used to make them is very likely to survive archaeologically, though bamboo fragments have been recovered infrequently from Pleistocene contexts (e.g. Gorman, 1970, 1971). It has been suggested that bone points could have been designed to be used with a blowpipe (Olsen and Glover, 2004). While not impossible, this seems unlikely for a couple of reasons. Firstly, the redundancy of form that typifies blowpipe darts (*see* e.g. Yost and Kelley, 1983) – use of clay pellets has also been reported (Annandale, 1906) – contrasts with the variability of form in most bone point assemblages. Also, there is good evidence for the hafting of bone points (see later). Such a device is not required of blowpipe darts, which are often made from the leaf stalks of the sago palm (Skeat, 1902) – the species *Eugeissona tristis* is mentioned by Evans (1937). Soft pith, such as that found in the leaf bases of palms (Skeat, 1902: 129), is fitted behind the butt of the dart and suited to the diameter of the blowpipe bore to create an effective seal.

The bow and arrow may be of greater regional antiquity here (Barton *et al.*, 2009; Bellwood, 2007). The type that predominates in the Southeast Asian historical record – the 'simple' or 'self-bow' (Blackwood, 1950) – is fabricated on a range of wood species (*see* e.g. Blackwood, 1950; Pope, 1962; Radcliffe-Brown, 1948; Williams-Hunt, 1952), while varieties of bamboo feature prominently in arrow shaft manufacture (e.g. Griffin, 1997; Sillitoe, 1988; Vanoverbergh, 1925). Historically, arrowheads appear to have been made predominantly from wood, bamboo or bone – use of stone is not well documented – and individual types of arrows were designed for specific hunting conditions: more variation in form, less redundancy. Bow strings have been reported as made from materials such as twisted strips of the inner bark from *Ficus* sp. (Griffin, 1997), *F. laccifera* or *Anadendron* are mentioned by Radcliffe-Brown (1948). Williams-Hunt (1952) records the use of hornbill feathers for fletching. Cordage and bindings are made from fibres obtained from various kinds of plants. Again, most of this would not survive archaeologically except under exceptional circumstances, potentially leaving just occasional instances of bone armatures.

In recent years, a number of interesting discoveries have emerged from work carried out on an extensive suite of osseous tools from the Niah Caves (Barton *et al.*, 2009; Piper *et al.*, 2008; Piper and Rabett, 2009; Rabett and Barker, 2007; Rabett and Piper, 2012; Rabett and Piper, forthcoming) as well as assemblages from other sites elsewhere in the region (e.g. Olsen and Glover, 2004; Pasveer, 2004, 2006; Pasveer and Bellwood, 2004; Rabett, 2002; Velat, 2007; Figure 6.3). Collectively, these studies are bringing a new perspective to our understanding of this technology. The Niah Caves and Lang Rongrien currently contain the oldest examples of osseous technology in Southeast Asia. In both cases, the specimens from these sites date from the millennia leading up to the Late Pleistocene (*see* Chapter 5). Although it is not impossible that the two locales, 1700 km apart, are peripheral vestiges of a wide sphere of technological tradition now inundated on the Sunda Shelf, their broadly contemporaneous date does provide something of a starting point for what can be viewed as two separate developmental trajectories: one in ISEA, the other on Mainland Southeast Asia.

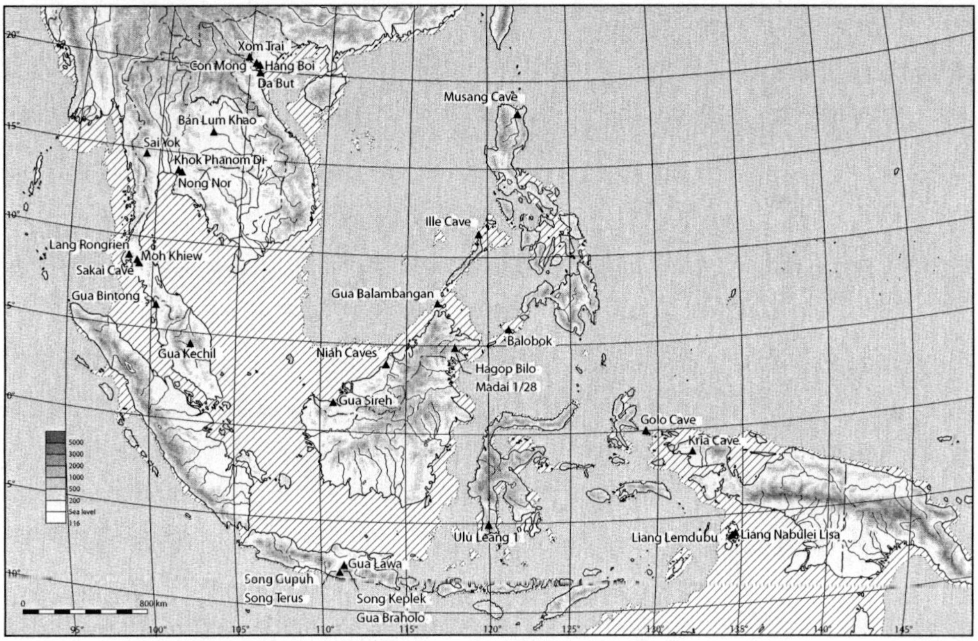

Figure 6.3. Map of Pleistocene Southeast Asia, showing the distribution of bone tool assemblages across the Sunda Shelf, Wallacea and northwest Sahul that are mentioned in the text. (Map drawn and prepared by D. Kemp.)

6.4.1.1 Island Southeast Asia

As noted, bone tools appear in ISEA from *c.* 45 KBP onwards, but it is a discontinuous record with no substantive evidence following the early finds at Niah until after the LGM. During the early deglacial, three artefacts appear at Liang Lemdubu, in the interior of Kobroor Island (6.152167°S, 134.371333°E), in the Aru Islands. Contexts immediately below the single and dual occurrences of here provide approximate dates for the implements of 19,720 ± 530 cal. BP (OZD-460; spit 17) and 15,487 ± 365 cal. BP (OZC-777; spit 26), respectively. There is also a single artefact from Hagop Bilo, Sabah, dated very broadly to between 17 and 12 kbp (Bellwood, 1984, 1988). Most evidence appears in the second half of the Last Termination and into the Holocene. This period sees the earliest evidence of bone implements in the western part of the Sunda Shelf: at Gua Braholo, in the Gunung Sewa area of East Java, from contexts dating to 13,873 ± 181 cal. BP (unknown lab code; Morwood *et al.*, 2008; Simanjuntak, 2000; Simanjuntak and Asikin, 2004), but it is in the eastern part of the shelf, again at the Niah Caves, where we see the earliest indication of an invigorated development in bone tool technology.

The excavations at Niah produced one of the region's largest assemblages of bone tools – now counted at 216 confirmed pieces (Barker *et al.*, forthcoming). The original assemblage (Harrisson and Medway, 1962; Medway, 1966) has been re-examined and supplemented by additional pieces during the course of analysis of the faunal remains from the cave, stored at the Sarawak Museum, or as a result of the NCP's (2000–4; Barker *et al.*, 2002, 2007) own excavations in the West Mouth (Barton *et al.*, 2009). The Phase II occupations here cover a period dating from 16,000 ± 125 cal. BP (OxA-15162) until 8400 ± 20 cal. BP (OxA-15161) and are subdivided into early and late episodes. On the basis of the original stratigraphic distribution of artefacts, most bone tool deposition beneath the rock overhang (Area A) occurred during Phase II (n = 29), with less (n = 13) appearing in later Holocene, Phase III contexts. Occasionally, artefacts show evidence of imposed tool form, though any effort

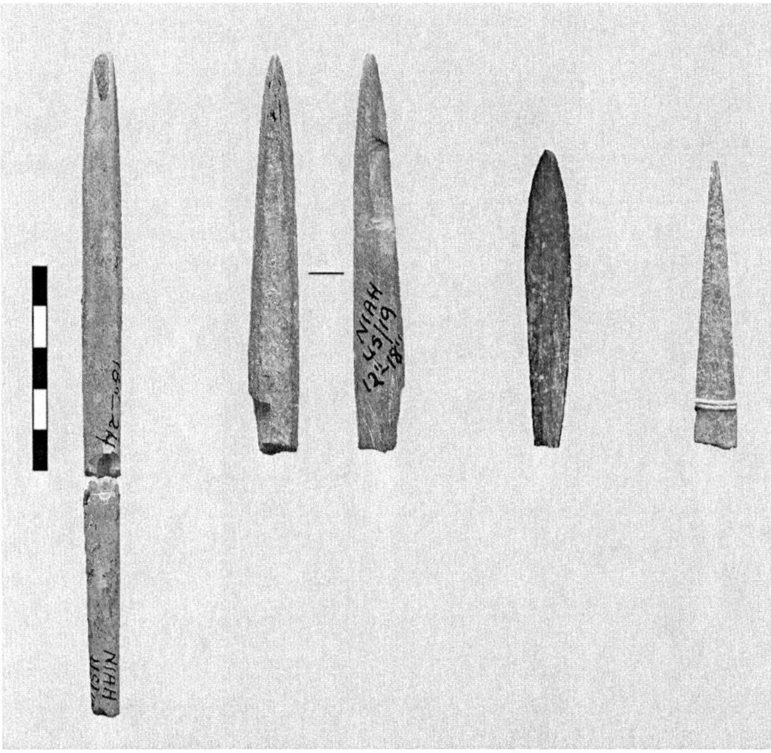

Figure 6.4. Examples of bone points from the Phase II occupation at Lobang Hangus, classified as likely to have been projectile armatures (scale in 5-mm increments). (Photographs by Ryan J. Rabett.)

towards standardization in shape is rare (Rabett and Barker, 2007). Projectiles and piercing tools both appear in the Phase II assemblage – indicating more than one use context for bone points – and the presence of worked fragments and off-cuts (bone tool manufacturing debris) lends support to the idea that on-site manufacture took place here during this and later times (Rabett and Piper, forthcoming).

Analysis and classification of the large ($n = 83$) bone tool assemblage – including worked bone, tooth and carapace fragments – on the east side of the Niah complex was undertaken by the author (*see* Piper and Rabett, 2009). Like the Area A archaeological sequence, once the Lobang Hangus stratigraphy was interpreted, three occupational phases were recognized: a Late Holocene phase (Phase III), commencing 4404 ± 50 cal. BP (OxA-13940); a Last Termination phase (Phase II), forming the larger part of total recovered material and dating from 14,484 ± 131 cal. BP (OxA-13936) to 12,373 ± 95 cal. BP (OxA-13939); and a very small pre-LGM (Phase I) assemblage dated to 38,058 ± 259 cal. BP (OxA-13938).

Eight bone tool fragments come from deposits above the transition between the ceramic and pre-ceramic phases; four are point-forms. Phase II (12″ to 48″ below ground level) contained the remaining 90.36 per cent ($n = 75/83$) of documented artefacts. Three of the 34 Phase II bone points that were available for direct study from Lobang Hangus carried modification that most closely paralleled that seen on ethnographic and experimental piercing tools (awls, needles etc). Seventeen fitted categories of modification most closely associated with projectile technologies (Figure 6.4), including self-barbed points (Figure 6.5), as determined through ethnographic comparison (Figure 6.4; Barker *et al.*, forthcoming). This finds additional support from two bone fragments found to exhibit surface modification

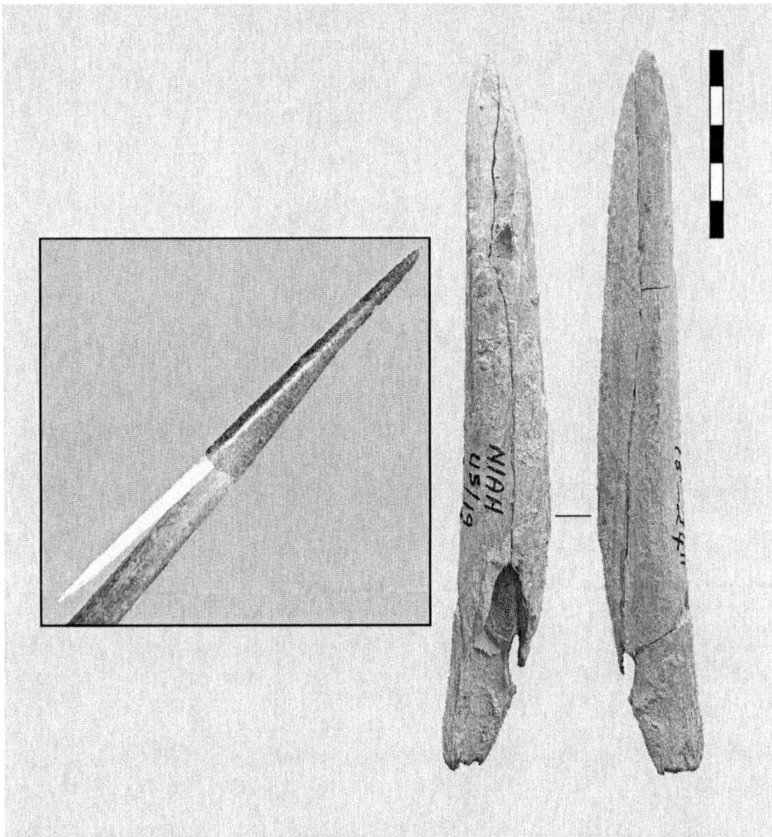

Figure 6.5. A 'self-barbed' bone point from the Lobang Hangus, Phase II assemblage (scale in 5 mm increments). The comparative piece is a bone-barbed spear from Australia, collected by A. C. Haddon, Cambridge University Museum of Archaeology and Anthropology, Accession No. Z 28493. (Comparative piece reproduced with kind permission of CUMAA; photographs by Ryan J. Rabett.)

that, on the basis of experimentation, is consistent with a hafting surface (Barton *et al.*, 2009). Neither line of evidence proves that the archaeological pieces were used as projectiles, but each does provide good grounds for supposing that this was so.

In terms of potential use context, ethnographic evidence from both Australia and the Andaman Islands describes how self-barbed bone points were often associated with spearing or the hooking of fish (Lampert, 1966; Pokines and Krupa, 1997; Thomson, 1936). A similar function may be extended to some of the Lobang Hangus pieces. The majority of points, though, comprises comparatively small, thin pieces as shown in Figure 6.4. Manufacturing traces on these are broken by fractures to tip or base, most likely the result of percussive impact damage of the kind inflicted on propelled projectiles (*see* Guthrie, 1983; Rabett, 2002). This is more indicative of the kind of damage one might find on the tip of either a light throwing spear or an arrow that would have been effective in fishing or taking arboreal game. The implication is that bone projectile technology may have become associated with more than one sphere of hunting and may represent a significant locus of innovation during the Last Termination and Holocene.

One of the Lobang Hangus bone tools studied by the author and an unknown number of the (n = 10) pieces identified by Harrisson and Medway (1962) as 'gouges' carry a level of surface modification that suggests a use other than as a projectile or piercing tool.

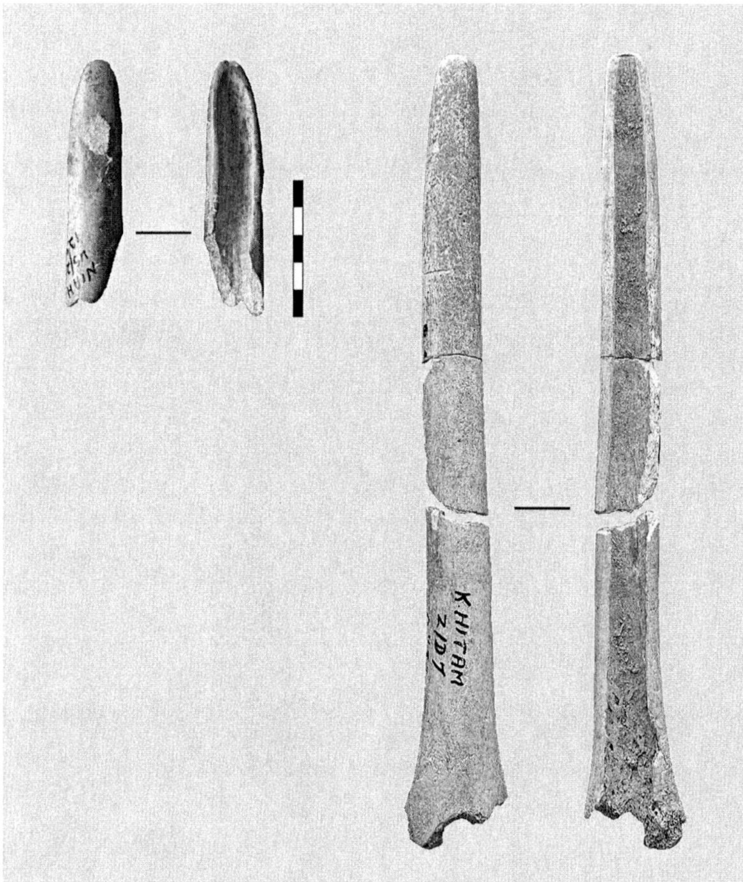

Figure 6.6. (left) An example of a longitudinally split short convex point, or 'gouge', with significant rounding and polishing to the utilised surface, Phase II, Lobang Hangus. The pattern and distribution of use-wear indicates that such a piece might have been used as a digging implement or coarse-weaving aid. (right) An example of a more complete specimen of this kind of implement from the Late Holocene levels at nearby Kain Hitam Cave, made on a Cercopithecidae femur and retaining the distal articulation as part of the handle (scale in 5-mm increments). (Right-hand image first published in *Asian Perspectives* 44(1): 154–79; photographs by Ryan J. Rabett.)

These pieces are usually longitudinally split diaphyseal fragments, with the curvature of the medullary canal retained. The 'tip' end of the fragment is very noticeably rounded in profile and carries a pronounced patina, extending down the dorsal surface and onto the flat fracture surfaces of the ventral, but not usually into the medullary canal. These fragmented specimens closely approximate in form and use-wear a single complete implement taken from the surface layers at the nearby cave of Kain Hitam (Figure 6.6) and ethnohistorical examples from Papua New Guinea, described as 'yam planters' (Blackwood Collection, 1937, Pitt-Rivers Museum, Oxford). This raises the possibility that some kind of digging activity may have been going on at Niah during this early period, with these tools being used to break up the earth. Experimentally (Rabett, 2002), digging rarely produced any fracture damage but exacted very visible changes in tool shape and reductions in length consistent with the wear seen on the archaeological pieces. It also resulted in the production of many longitudinal striations that were not seen, though one might presume that the development

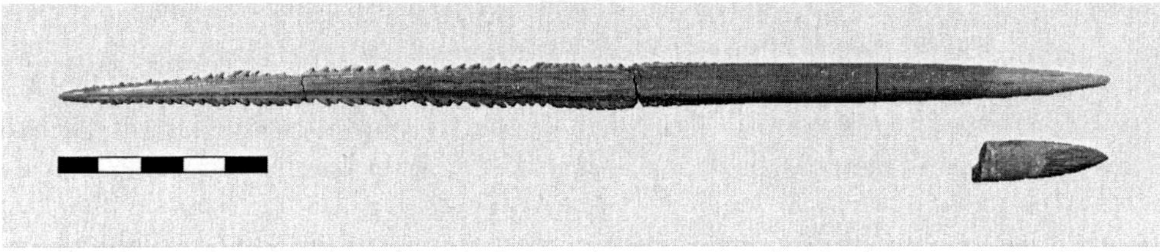

Figure 6.7. Stingray barb points (probably *Pastinachus sephen*) excavated by the Niah Caves Project in Area D of the West Mouth. The lower of the two fragments exhibits hafting fibres and resin adhering to the base (scale in 5-mm increments). (Photograph by Ryan J. Rabett.)

of these would depend on the nature of the substrate being dug. Hafted experimental edge-tools like these also proved to be very effective chisels when used with a hammer stone. Woodworking tended to depress the surface terrain of the exposed bone and produced a comparatively high polish. It also tended to result in pronounced fracture damage, and this appeared to correlate with the kind of work being carried out. Transverse breaks seemed to occur when the tool was being used to chip at the wood. Resistance at the leading edge, such as a knot in the wood, prevented the transfer of energy from the hammer blows delivered at the back of the tool. Diagonal fractures appeared when the tool was being used to lever splinters of wood away from the stake. They were formed when excessive rotational stress was caused by a combination of resistance in the wood and a twisting action of the haft. This kind of damage featured in the archaeological material from Niah, potentially implying that the effects of rotational stressing of these rounded-tip bone tools brought about by leverage. Although woodworking is another possible use context, the consistent tip form does not seem suited to this. A third alternative could be that such pieces were used as a weaving aid; comparable specimens are seen in other periods and contexts (A. Taylor, pers. comm.) – an interpretation that would account for the combination of rounded form, pronounced patina and damage induced by rotational strain, though further work will be needed before any one particular use scenario can be confidently assigned to these pieces.

Tool recycling and curation almost certainly featured in some parts of the Lobang Hangus assemblage, while some implements were probably also used in more than one activity. Finally, of the 83 modified pieces, 20 are likely to represent artefact débitage. This strongly suggests that manufacturing activities were taking place in the cave. The preservation of diagnostic articular ends provides another strand to this technological picture (Rabett and Piper, forthcoming). When such ends were preserved, they were exclusively from Cercopithecidae long bones – usually the femur. The straight and relatively thin walls of Cercopithecidae femora appear to have been ideally suited to bone tool manufacture. If this was the case, the limited presence of tools made on bone specifically from early periods of occupation may be related at least in part to the availability of these animals in the vicinity of the Niah Caves.

There is limited evidence of bone edge-form implements in the Lobang Hangus assemblage. Eight modified pig tusk fragments were, however, identified (*see* Rabett, 2004b). In each case, the original tusk was split longitudinally, the convex edge of the inside surface being subsequently worked to form a sharp edge. The outer enamel surface appears not to have been modified in any way. In their original analysis, Harrisson and Medway (1962) classed such pieces as 'knives'. Butchery is a potential use, but such wear and damage to the edge suggests resistance in the material being sliced. Either way, these implements were utilised on site in ways that the majority of bone tools was not.

One of the most significant new discoveries made during excavations at the West Mouth comprised 31 bone, tooth and cartilaginous (stingray and catfish barbs) artefacts from an area close to the limit of natural light penetration into the cave defined by the NCP as 'Area D' (Figure 6.7). This new collection of material (*see* Barton *et al.*, 2009) included two almost complete bi-points – one of them, made on bone, the other a worked stingray spine (Figure 6.7) – both displaying deliberate modification clearly for the purposes of hafting. That points in this assemblage functioned as composite armatures was further supported by the presence of a resinous material in crystalline form adhering to the back sectors of six pieces. On two occasions, this substance was found to contain bundles of transversely aligned fibrous material: the remains of binding strands. Spines from stingray made up almost half ($n = 15/31$) of the fragments within this assemblage. Comparison with specimens held at the British Natural History Museum indicates that the cowtail stingray *Pastinachus sephen* is a good candidate for the source of these spines, though other taxa that have brackish water tolerances and inhabit rivers and estuaries, for example, *Himantura* sp., cannot be ruled out. Whether the pieces were being employed in fishing activities is hard to judge; however, procurement of them, if not by exchange, does imply access to estuarine environments. Three use contexts were considered for these composite points: projectile heads, pronged spear tips or, potentially, fish hooks. The most likely, based on the arrangement of the preserved bindings, was that these were armatures to narrow-gauge single or multi-barbed projectiles – with light throwing spears (perhaps leisters) or arrows as the most likely candidates.

A feature common to both the bone and the spine points was the presence of red particulate matter adhering to the surface. No clear pattern of occurrence could be discerned. The presence of pigment is often assumed to be associated with explicitly symbolic behaviour. While this remains a possibility here, ethnographic evidence indicates how fishing nets may also be coated with ochre or bark-based solutions or be greased to remove their odour and make them less visible underwater (Satterthwait, 1986; Zaliznyak, 1998). A similar function may alternatively have applied in this instance.

AMS ^{14}C dates from directly associated contexts indicate that this technology appeared at Niah between 10,886 ± 148 cal. BP (OxA-12391) and 8793 ± 109 cal. BP (OxA-18358). These data currently comprise the earliest confirmed appearance of this kind of technology in the region and the first instance in which stingray spines have been used. Conditions did not permit the survival of resin and binding materials earlier than this at Niah, and armature technology almost certainly predates this; however, stingray barbs do not appear prior to this here, suggesting that it may well have been an Early Holocene innovation. Such pieces next appear at the then sub-coastal Thai site of Khok Phanom Di (Higham and Thosarat, 2004a). The existence and distribution of this technology during the intervening millennia are currently unknown.

A small number of other sites within Borneo have yielded a bone tool component. These include Gua Sireh, a cave site near Kuching, which has produced a couple of heavily worked pieces (a tanged arrowhead and eyed needle) dating to around 3441 ± 225 cal. BP (ANU-7047; Datan, 1993); two of the Madai caves (see later), MAD 1/28 and MAD 2, dating to 10.8 and 10.4 bp, respectively (Bellwood, 1984, 1988); and perhaps most notably Gua Balambangan (Rabett, 2005; Zuraina *et al.*, 1998b).

Gua Balambangan contained no evidence of bone technology before the Early Holocene, when a comparatively large assemblage of such implements ($n = 33$) appears between 11,453 ± 319 cal. BP (Beta-109141) and 10,007 ± 227 cal. BP (Beta-109140). A link to the exploitation of mangrove forest resources was postulated as being at least partly behind the rise in coastal associations (Rabett, 2005), a hypothesis based on parallel evidence from the Early Holocene in the North Arnhem Land area of Australia (Allen, 1986; White and Peterson, 1969; Woodroffe *et al.*, 1988). At Gua Balambangan, the possibility exists that the site was used as a kind of production camp: the item(s) being collected requiring the use of robust edged tools made on dense bone. Although further analysis of the faunal remains is needed, a cursory examination by the author did not reveal any immediate evidence of manufacturing debris, suggesting that these tool units were being brought to the site, where they were steadily used up with little visible effort expended on repair. The traditional argument that people turned to bone when quality stone was not available does not hold here, as mentioned (*see* Chapter 5), chert was both readily available (sourced to <1 km) and used by the site's occupants. A conscious choice was being made to use bone for certain implements. A comparison of the Balambangan and Lobang Hangus datasets (i.e. many edge-forms but no tusk pieces vs. few edge-forms and many tusk pieces) demonstrates how the rise in importance of bone technology in this part of the region was not necessarily tied to a single new strategy in resource extraction or manipulation but was being used in a number of different and in some cases pre-existing activities.

Northeast of Borneo, in the Philippines, the earliest record of bone implements comes from the site of Balobok rock-shelter, Sanga Sanga Island, in the Sulu Sea. Three cultural layers were identified here, all containing osseous artefacts (Bautista, 1999). The antiquity of the site has not been firmly established but points towards an occupation starting in the Early Holocene. More reliably dated are two heavily ground pieces and the distal end of a pig fibula ground into a point (use-wear traces on which suggest it was a perforating tool), from Ille Cave, in northern Palawan (11.196111°N, 119.505278°E). These pieces were recovered from shell midden layers dating to 6565 ± 57 cal. BP (OxA-16095; Lewis *et al.*, 2008; Ochoa, 2008).

In the municipality of Peñablanca, in northern Luzon, Musang Cave also dates to the Mid Holocene. Excavated 1976–7 by Barbara Thiel, this site lies *c.* 40 km inland, between the Cagayan Valley to the west and the Sierra Madra mountains to the east. Across eight (1 × 1 m or 2 × 2 m) trenches, culture-bearing deposits continued to a depth of *c.* 1.6 m. These deposits were split into main two phases of occupation. The uppermost (Cultural Level II) dates from 4110 ± 130 bp (GAK-7043) to 4980 ± 150 bp (GAK-7044), though these are on shell and of uncertain accuracy. Three bone artefacts are reported from these levels, as is a fragment of bone modified by the groove and snap technique (Thiel, 1990: 73), though the precise depths and dates of these pieces are not given. Bone technology does not appear in the lower horizons (Cultural Level I) at this site, which extends to the Pleistocene–Holocene transition.

To the southeast of Borneo, in southern Sulawesi, bone implements have been recovered from Ulu Leang 1, a rock-shelter *c.* 20 km from the modern coast. These are considered typical examples of pieces from sites across this island attributed to the 'Toalean' culture (see later; Olsen and Glover, 2004). These authors described an industry ($n = 134$ artefacts) that was composed predominately of bi-points ($n = 73$) and 'awls' ($n = 7$), in keeping with bone tool assemblages from other Toalean sites – though 'spatulae' are also reported in some instances (van Heekeren, 1972). The bi-point forms were produced by scraping a significant proportion of the surface on a fragment blank. Use damage was noted on several ($n = 13$) of these points, including tip crushing and the removal of longitudinal flakes from the tip, found to be consistent, experimentally, with impact damage. Hafting is inferred, and some evidence for the presence of a self-barb mounting system is noted on five points. Olsen and Glover (2004) considered it likely that these bi-points were used as projectiles. Pronounced surface incising is reported to occur on six pieces. The authors suggest that this may have been used for the application of poison (tests for poison residues proved negative, but preservation may be an issue in this respect). An alternative interpretation, which further supports their projectile hypothesis, is that these incisions, grouped around the base of a point, represent deliberate surface distressing to facilitate hafting (*see* Clark, 1954: 124), though markers of social identity would be a third possibility (e.g. Ford, 1959).

The 'awls' from Ulu Leang 1 appear in a variety of forms and were fabricated on a range of different bone elements. The only identifiable manufacturing modification is scraping to the laterals and in the vicinity of the tip. Use polish is not generally well developed on these pieces, suggesting a limited use-life. Dates obtained on freshwater shell and from charcoal at the site indicate that occupation can be bracketed to within the Early Holocene. The oldest date is 10,740 ± 50 bp (Grn-8648), but this was obtained on freshwater shell and deemed likely to be *c.* 1350 years too old. Bulbeck *et al.* (2000) have provided a corrected calibrated date at the 2-sigma range of 10,040–10,800 BP. Older levels do exist, but these lay just outside the drip line of the rock-shelter and lacked suitable dating material (Olsen and Glover, 2004: 282). Bone tools begin to appear in spit 11 – the underlying spit 12 is dated 7950–8360 cal. BP (2 sigma) – and continue to the top of the sequence; the uppermost dated spit is 3840 ± 172 cal. BP (PRL-230) on charcoal (Bulbeck *et al.*, 2000). Considered part of a separate but parallel development to the Toalean on the Minahasa Peninsula, the northern arm of Sulawesi, the Paso shell midden (Bellwood, 1976) has yielded radiocarbon dates of 8366 ± 478 cal. BP (ANU-1517) and 8178 ± 305 cal. BP (ANU-1518). A bone tool component is also in evidence from this site. Seventeen points were described as probably being awls or needles. The presence of a further five split and grooved fragments, also from the pre-ceramic midden deposits at Paso, suggests on-site manufacturing.

Moving from Sulawesi into the North Maluku province of Indonesia, Pasveer and Bellwood (2004) have reported on a total of 78 bone artefacts from the upper pre-ceramic phase at Golo Cave, Gebe Island. (A further 52 bone artefacts were recovered from the overlying final occupation.) As already noted (*see* Chapter 5), the pre-ceramic phase (20–80 cm below ground level) contained high frequencies of shell midden remains and lithic artefacts and has yielded an adjusted marine shell calibrated age of 7965–7743 BP (ANU-9449; CALIB 5.0.1) at 50–55 cm. A little higher in the sequence, at 45–50 cm, a charcoal sample produced a calibrated age of 3691–3244 BP (ANU-9448; CALIB 5.0.1; Szabó *et al.*, 2007). In their discussion of the bone technology, Pasveer and Bellwood (2004) consider that the particular abundance of bi-point forms at Golo Cave makes it likely that all stages of use-life occurred here (the recovered assemblage includes complete unused, broken unused and incompletely formed specimens). In terms of potential functionality, this category of artefacts, which are described as exhibiting considerable regularity in their proportions, is equated with small ethnographically documented fish gouges. The evident use-wear, however, was not found to be consistent with such utilization. Furthermore, fish remains are very scarce at Golo. These points were, nonetheless, probably part of a composite technology. Pasveer and Bellwood (2004) describe the bi-points as the most abundant bone tool form from North Maluku and consider that while shape, size and breakage pattern strongly support hafting, use as spears, arrow tips or barbs is unlikely. The high polish on many bi-points, combined with evidence of pressure damage to tips (most commonly involving small snap or step fractures), is considered more suggestive of a piercing technology; though the necessity of hafting a perforating technology of this kind is not clear.

A second set of implements, termed 'spatulae' and generally made on wallaby fibulae, formed a special class of uni-points that were comparatively large and carried a rounded rather than sharp tip. These are very similar to examples seen in the Niah material. They were found to be generally undamaged and modified by a patina that also extended down the shaft. The authors proposed that this pattern of wear would be consistent with that seen on perforating implements. An alternative function, as an extractive tool for removing meat from shellfish, was discounted as this has been shown experimentally (Webb and Allen, 1990) to leave deep striations and extensive chipping, neither of which was observed on the Golo spatulae. Use in weaving of cane or other stiff material was considered, though it was thought this too should have caused chipping along the tool edges. Compelling evidence for the repair of such pieces was recorded at Golo, and one might argue that this fact and the regularity of the proportions of these pieces are not functionally necessary features of a simple perforator.

The earliest excavated appearance of the bone technology in New Guinea comes from an assemblage ($n = 92$) of pieces at Kria Cave on the Bird's Head Peninsula, Indonesian Papua (1.251975°S, 132.333511°E), 325 m asl and is dated from 7680–7510 cal. BP (GrA-9103)

(170–175 cm) until 5210–4840 cal. BP (GrA-9100; Pasveer, 1998, 2004) using CALIB 4.3 (*after* Stuiver and Reimer, 1993; Stuiver *et al.*, 1998). Pasveer (2004) writes that despite the high occurrence of *Dorcopsis* sp. (wallaby) fauna at the site, and the oft-reported appearance of wallaby fibulae spatulate tools at other Australasian sites (see earlier), no similar tool was recovered from Kria Cave. Equally, there were no examples of pieces that could be classified reliably as repetitively or habitually used piercing implements. Bi-points and uni-points dominate the bone artefact assemblage in roughly equal numbers and are said to exhibit considerable uniformity in size and shape. The uni-points were almost invariably made on small to medium-sized mammal long bone fragments and retain the medullary canal on the completed piece; the bi-points were consistently made on splinters of cortical bone from larger animals. Use-wear analysis of both classes confirmed use as tools rather than as decorative pieces. Pasveer suggests that the bi-points were most likely hafted and, while not used exclusively as awls, were plausibly employed in some form of perforator role. Again, though, the redundancy of form and presence of hafting indicates a comparatively high level of manufacturing investment. Ethnographic evidence from Papua New Guinea and experimental work by the current author suggests that in most cases, perforators can be distinguished by low manufacturing input and over-exploitation (Rabett, n.d.-a; Sillatoe, 1988: 64). While use as perforators remains possible, other interpretations may be considered here.

In contrast to the findings at Kria Cave, bone artefacts are extremely rare at another cave excavated by Pasveer in the Bird's Head Peninsula. Toé Cave (1.284889°S, 132.218361°E) is situated in a low hill along the southern margin of the Ayamara Lakes. This site produced only three bone artefacts (all point-forms) and a single utilised *Dorcopsis* sp. incisor, dating broadly, but not with precision, to 3–9 cal. KBP. The stark contrast between these two otherwise technologically very similar caves (see later) highlights likely differences in the function of each during the Holocene (Pasveer, 2004: 352). It also rather draws our attention to the fact that while different technological media were used in subsistence, the two spheres of activity were not joined at the hip. As we saw in Chapter 5, at Golo Cave, the lithic assemblage gave every impression of being expediently produced, while the shell artefact assemblage was more standardized. At Kria Cave, a long-term emphasis on producing low-input, typologically ambiguous lithics existed again alongside a more standardized (both technically and in terms of use) and, if hafted, a more complex composite osseous technology.

In the Aru Islands to the south, Liang Lemdubu continues to feature during this period, while new evidence also appears from the nearby site of Liang Nebulei Lisa. Excavations at Liang Lemdubu yielded 37 bone artefacts. Pasveer (2006) describes it as comprising 21 short uni-points (or broken bi-points); two 'attenuated' (thin diameter) points; three pieces with rounded tips (spatulae); five mid-section (shaft) fragments with manufacturing traces; and six mid-section fragments with evidence of wear only (all from spit 1), which may constitute 'expedient' tools. Seventy per cent of artefacts were made on shaft fragments, with several recognisable to two species of wallaby. The majority of pieces ($n = 21$) came from spit 1 (0–5 cm) and the last 2000 years. Three pieces, as discussed already, came from deep in the deposits. The remaining 13 were recovered from spits 2–5. In terms of antiquity, spits 2–3 (10–15 cm) also have dates within the last 2000 years; spit 5 produced a much older date on a fragment of *Celtis* seed of 10,624 ± 65 cal. BP (OZF-356) at 20–25 cm and a calibrated date of 14,057–13,183 BP (ANU-10792) on shell (*Gelonia coaxans*) at 25–30 cm, indicating a hiatus in deposition during the Mid-Holocene. There are significant changes in the state and composition of recovered fauna between the earliest and latest use of the site, including, for example, increases in marine and estuarine molluscs in the Early Holocene deposits and especially so in the Late Holocene (O'Connor *et al.*, 2006). The increase in tool occurrence may be linked to these changes.

Work at Liang Nebulei Lisa on the same island (6.001867°S, 134.560450°E) produced a smaller sample of bone implements ($n = 15$) with 11/15 pieces coming from contexts dated to between 11,553–11,164 cal. BP (OZD-699) and 7754–7167 cal. BP (ANU-10905). The bone artefacts from this site reportedly exhibit considerable variability in size and shape. There is also a notable dichotomy between polished, intact uni-points and unpolished and damaged tips at Liang Nebulei Lisa. This is taken to suggest that either two modes of use are in evidence or that discard was taking place at different times in the tools' biography: sometimes soon after manufacture due to damage and sometimes later after a sustained period of use. No such disparity had been observed with the Liang Lemdubu material, which tended to be both polished and broken. Pasveer (2006) considers that the damaged, unpolished uni-points from Liang Nebulei Lisa may have been projectile points, while the remaining polished pieces were considered more likely involved in perforating or engraving activities. She concludes that the presence of predominantly broken pieces at both sites is consistent with them having probably been retooling locations.

With the exception of the late glacial date at Gua Braholo, East Java, bone tools in the western part of ISEA are confined to Holocene contexts and, therefore, are generally later occurring than in the east. Recent excavations at the cave of Song Gupuh (8.161111°S, 110.979939°E), also in East Java (Morwood *et al.*, 2008), have produced a large collection of $n = 80$ bone tools from the upper deposits of the site (in trenches N4 and P6), including 48 uni-points, 17 bi-points, 14 spatulate pieces and a further 11 artefacts from trench I2. No dates are available from the contexts where these pieces were recovered, though their first appearance coincides with a marked increase in the remains of *Macaca* sp. *c.* 11,410 ± 180-BP (Wk-14650; IntCal04), and they continue to occur through the Holocene. These implements were made on a variety of elements from several taxa, including cervid long bones and antler, bovid long bones, carnivora ulnae, suid lower canines and humeri, and cercopithecid canines, femora, fibulae and ulnae; the assemblage and range of raw materials was taken by the excavators as an indication of more intensive use of faunal resources. Although stratigraphic associations remain unclear at Gua Lawa, two radiocarbon dates have been recently obtained: 3242 ± 123 cal. BP (unknown lab code) and an earlier one of 9141 ± 232 cal. BP (unknown lab code; Simanjuntak and Asikin, 2004). While these are not reported with clear stratigraphic provenance, they do at least finally begin to put the sequence of this important site into some temporal context and bring it broadly in line with dates and developments seen elsewhere in ISEA and New Guinea.

6.4.1.2 Mainland Southeast Asia

On the mainland, several sites have been found to contain bone technology (Rabett, 2005). The frequency of reported pieces is mostly low (less than 25 and often less than ten per site) throughout the period, though accurate totals are not always available, and some sites do contain larger inventories.

At around the same time as the first discoveries of bone tools were made in Java, similar pieces were starting to appear in Vietnam from a number of rock-shelter sites, for example, Lang Bon rock-shelter in Thanh Hoá Province (Colani, 1930), and from open-air locales, most notably from Da But (*c.* 20.016667°N, 105.716667°E), a shell midden (*c.* 50 × 32 m) in Thanh Hoá Province. Etienne Patte, a French geologist, originally excavated Da But in 1926–7 (Bui Vinh, 1991; Khol and Quitta, 1978: 392). The site is now known to date from at least 6540 ± 60 bp (Bln-3509 II), a date taken on a *Corbicula* sp. shell 1 m into the 6–8 m that Patte excavated. Occupation may extend back to the Early Holocene, given that more recent work in the 1970s and 1980s failed to reach the base of the midden due to a high water table (Nguyen Viet, 2005). At Con Moong Cave (*c.* 20.3°N, 105.9°E) in northern Vietnam (Pham Huy Thong, 1980; Pham Huy Thong *et al.*, 1990), an undisclosed number of bone implements were uncovered from certain contexts within a 3.5 m excavated profile. Fifteen dates are available for this site (Nguyen Viet, 2000), ranging from 14,184 ± 133 cal. BP (Bln-3490/II) to 9500 ± 34 cal. BP (Bln-3482). Also in this part of Vietnam, the cave site of Xom Trai (*c.* 20.533333°N, 105.466667°E) yielded an undisclosed number of bone artefacts during excavations in 1986. Twenty ^{14}C dates from here all fall between 22,012 ± 211 cal. BP (Bln-3472) and 19,259 ± 114 cal. BP (Bln-3042), and bone technology appears to have featured from early on (Nguyen Viet, 2000). Of potential importance are the reported *c.* 105 bone tools (Ha Van Tan, 1976) from the rock-shelter site of Da Phuc, in Hoa Binh Province, though there is little information readily available about them. Ha Van Tan (1999) suggested that *c.* 250 bone artefacts had been recovered across 150 different prehistoric sites in northern Vietnam, but with 105 of that total coming from Da Phuc, the proportions per site for the remainder are likely to have been comparatively small (Matthews, 1966; Reynolds, 1990; van Heekeren, 1972). Such data clearly have the potential to bear on the mainland patterning of this technology, though, as with the case at Da Phuc, radiometric dates are currently unavailable from most sites. Excavations led by the author at two caves in Ninh Bình Province (Hang Boi and Hang Trống; *see* Chapter 7) do, though, affirm a generally low tool frequency. Single instances of bone points have been recovered from both sites. The example from Hang Boi comes from a context (5219) slightly younger than 13,601 ± 43 cal. BP (UBA-14887); that from Hang Trống was recovered from concreted shell midden deposits adhering to the wall above the main cave floor (Rabett *et al.*, ms. in prep.). These deposits are undated but, based on amino acid racemisation work on equivalent deposits at Hang Boi (*see* Rabett *et al.*, 2011), are of likely Early Holocene age.

In western Thailand, 15 bone points have been reported from the rock-shelter portion of the Sai Yok 1 site (see later). This site is undated, but the levels containing points are considered pre-ceramic and thought to be of Early Holocene age (Sørensen, 1988; van Heekeren and Knuth, 1967). In the northern half of the Thai–Malay Peninsula, at Moh Khiew Cave, the earliest date from a level bearing bone technology is 12,893 ± 128 cal. BP (OAEP-1284). The frequency of bone tools from this site also remains very low through the Last Termination and into the Holocene. The exact published number of tools from here and the Holocene site of Sakai (*c.* 100 km to the southeast, in the Palian District of Trang Province) is approximately a dozen (Pookajorn *et al.*, 1994), but of a total of 11 examined by the author, only 6 met formal tool criteria as defined in Rabett (2002, 2008). There are indications of a shift in subsistence focus at both sites, linked to their changing proximity to the coast, but there is no corresponding rise in the bone tool component at either. Isolated finds still occur as late as 4795 ± 203 cal. BP (OAEP-1290) at Moh Khiew, whilst at Sakai Cave, infrequent bone tools are found from Level 2 – 8421 ± 151 cal. BP (OAEP-1364) – to ethnohistorical evidence subsurface (Pookajorn, 1996).

On the Malaysian side of this peninsula, Gua Bintong is probably the most oft quoted of the sites with bone technology, though small numbers of tools have been found in other locales (Rabett, 2005). In 1936, H. D. Collings (assistant curator, Raffles Museum, Singapore) excavated residual culture-bearing deposits from the mouth of this cave in the northern Unfederated Malay state of Perlis. Located in Bukit Chuping (6.496456°N, 100.260931°E), one of the isolated karstic hills in this otherwise low-lying state, Gua Bintong had been largely stripped of its principal deposits by guano diggers. Collings' excavation uncovered part of a midden containing several species of coastal and freshwater molluscs, as well as abundant if fragmentary faunal remains, ceramics, lithics and 42 bone implements. Upon their original publication (Collings, 1937), these latter were compared to the Sampung material in Java and considered to be of 'Mesolithic' age (Erdbrink, 1954). No date is available for this important site – much of Bukit Chuping is now a quarry – but its cultural remains are conventionally included within the Malaysian Hoabinhian (e.g. Bellwood, 2007) of the Early to Mid-Holocene. A marine terrace 3 m above the current sea level in a neighbouring cave in the same massif has been dated to 5960 ± 224 cal. BP (unknown lab code; Haile, 1971), lending some support to an occupation of Gua Bintong towards the latter part of this period.

Of the few other larger assemblages of bone tools from the mainland that are dated, they too appear to be of Mid to Late Holocene age. The pattern of use at mainland sites seems to represent a continuation of that seen in the Late Pleistocene. From the Mid-Holocene onwards, as well as these occasional larger assemblages (if not always in conjunction with them), there is also an influx of much more invasively worked pieces alongside the minimally worked ones. These include fish hooks, tanged and barbed bone points (interpreted as arrow or harpoon heads) from a number of sites, for example, Gua Kechil in Malaysia *c.* 5457 ± 967 cal. BP (GX-0418; Bulbeck, 2003; Dunn 1964, 1966), Khok Phanom Di from 3852 ± 80 cal. BP (ANU-5493) to 3542 ± 149 cal. BP (ANU-5482; Higham and Thosarat, 2004a) and Nong Nor from 4440 ± 125 cal. BP (Higham *et al.*, 1998); and 'waisted' flat-bone implements at Ban Lum Khao in Thailand (Figure 6.8) from 3346 ± 49 cal. BP (Wk-4511) to 3188 ± 120 cal. BP (Wk-4509; Higham and Thosarat, 2004b). These may represent local innovations or introductions from elsewhere but perhaps not from ISEA.

In sum, there is considerable evidence from Island and Mainland Southeast Asia to support an early use of osseous procurement and processing technologies. In both areas, these almost certainly started out as a minor component of the toolkit, then proceeded along somewhat different developmental pathways. Between the Last Termination and the Late Holocene, tools made on these media appear to have become a prominent feature of assemblages in ISEA, and this may be linked to far-reaching changes in the nature of the landscape and intergroup communication. On the mainland, the frequency of occurrence in most cases remained low, with some indications during the Mid and Late Holocene of increases in inventory size and/or supplement by an infusion of pieces with much more evidently imposed and standardized forms. With the exception of equally late examples from Gua Sireh, this latter development does not seem to appear in ISEA. Otherwise, in both areas, fractured blanks tended to be augmented simply (Rabett and Barker, 2007), though tools were still tailored to a range of different use contexts. There is every reason to suppose that the many point-forms studied were projectile armatures for fishing or hunting (*see* Chapter 7). Other pieces suggest continued use in production activities, for example, perforating and perhaps weaving, while use as digging tools also remains a possibility.

Late glacial developments in bone technology in ISEA seem to have started in the north-east, so far most notably in Borneo *c.* 14.5 cal. KBP, and in close correspondence with the transition to GIS-1. Further occurrences come to light soon thereafter from other sites in the eastern portion of the region, with one early assemblage in the west. The timing seems to correspond with the intra-regional population dispersal Soares *et al.* (2008) postulate, though a direct link between the two is not demonstrated from this. Technological similarities have also been drawn for osseous (and lithic) implements between sites in Sulawesi and Australia, and farther west in ISEA itself during the Holocene (e.g. Olsen and Glover, 2004); the possibility that long-distance links were either already in existence or were being forged at this time has also been postulated (Bulbeck, 2008). The truth of any relationship between sites within ISEA and occurrences in Sahul during this period, however, remains to be proven.

By the beginning of the Holocene, evidence emerges again at Niah, and again currently earlier than elsewhere, for a new range-weapon innovation: the use of stingray-barbed pronged spears or arrows. An assemblage of edge-formed bone tools also appears at Gua Balambangan (*c.* 11.4–10 cal. KBP), where most came from a single hearth – a possible sign that retooling took place here, perhaps communally. Bone tool recovery in association with production and retooling sites appears to be something of a recurring theme and one that will need to be taken into consideration as we continue to explore matters of site function in the tropics.

The late glacial and opening millennia of the Holocene not only saw the emergence and growth of osseous technologies, it also saw changes in inorganic media. While some of

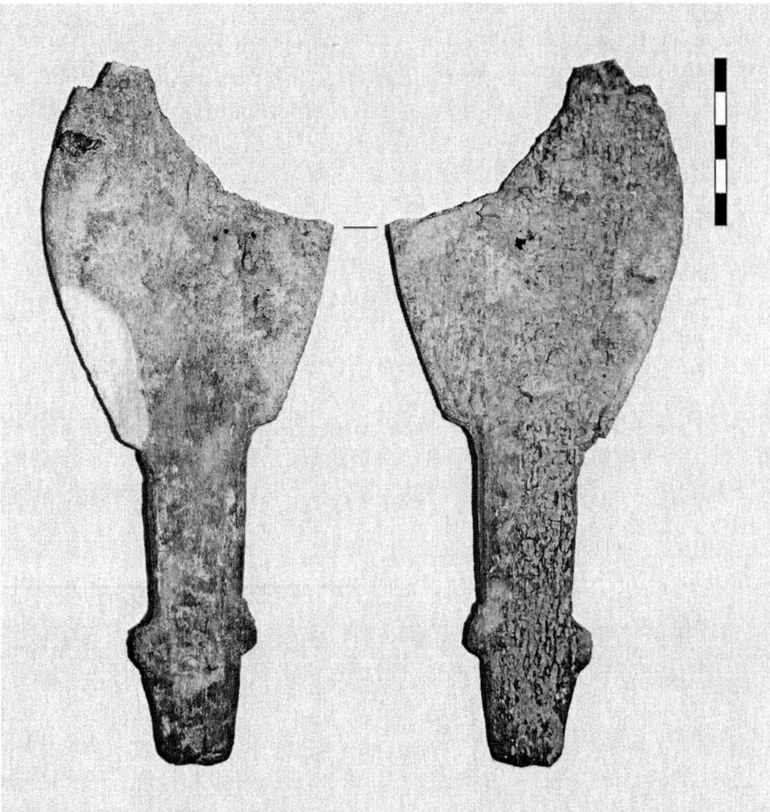

Figure 6.8. Evidence of extensively imposed form to create a hafting surface. Ban Lum Khao Catalogue No. 1380 (scale in 5-mm increments). (Photographs by Ryan J. Rabett.)

these developments occurred outside our period of interest (particularly the introduction of ceramics), many are tied in closely with it.

6.4.2 Inorganic (Lithic and Shell) Technologies

6.4.2.1 Island Southeast Asia

Late Pleistocene stone tool components persisted across the Pleistocene–Holocene transition in ISEA, but alongside them began appear often typologically distinct forms, including types of scraper, small flake-blades and bifacially retouched points. Current evidence has these appearing along an geographic arc stretching from central Sumatra, through the Sundas and eastern Indonesia, out to the islands around the margins of the Sulu Sea. Bulbeck *et al.* (2000: 96) noted that this distribution corresponds closely to that of obsidian and fine-grained siliceous stone resources, something that may have helped underwrite this development. It is worth noting, however, that the traditional un-retouched assemblage components were often also made from the same siliceous stone, making it unlikely that this technological theme was solely driven by raw materials. As Sémah *et al.* (2004: 47) point out, 'essentially the same raw materials feature commonly in the Pacitanian [pebble and flake] industry, among the artefacts found in the caves and rock-shelters, and on open-air Neolithic workshops'. Equally, in some instances, the new components were no less typologically ambiguous than before.

Sumatra. Much of Sumatra outside the northeast coastal margin (e.g. *see* Edwards McKinnon, 1991; van Heekeren, 1957) continues to have, for the most part, a largely unknown prehistory. In 1973, in an effort to begin rectifying this situation, a joint survey by the Indonesian Archaeological Institute and the University of Pennsylvania Museum was conducted to cover six of the seven provinces on the island. The principal discovery from this campaign would be the excavation of Tianko Panjang cave. Following test excavations at this site, Bronson and Asmar (1975) reported that the constituents of its lithic assemblage did not fall into readily definable types. Metrical analysis suggested against the flakes being referred to as blades or necessarily even 'microliths'. Although of the same general size and form as pieces recovered elsewhere in Indonesia (see later), the Tianko Panjang material is not considered morphologically comparable, suggesting that this was a distinct, local entity. The three available dates from the site indicate that the lower, pre-ceramic part of the sequence dates to between 11,988 ± 271 cal. BP (P-2250) and 10,381 ± 162 cal. BP (P-2248; 161–94 cm below ground surface). Given that only *c.* 5 per cent of flakes were found to have cortex and that only a single probable core was recovered, the authors concluded that preliminary flaking was carried out elsewhere in the landscape, with the closest, though not necessarily only, raw material source matching the physical character of the utilised obsidian on the Mesumai River, some 4 km from the cave.

Although prehistoric evidence from Sumatra remains distinctly under-represented 35 years on from the Tianko Panjang excavations, new expeditions are returning here. In 2001 an Indonesian–French project began pioneering work in southern Sumatra to attempt to establish a chronology for this part of the island. Initial reports look promising, with Palaeolithic artefacts being recovered from small river systems and caves in the Baturaja area of the South Sumatra Province. With respect to the Early Holocene, excavation work undertaken at Gua Pandan has also revealed the first examples of unifacially flaked pebble tools and retouched flake scrapers made, again, on a range of raw materials sourced to local rivers (Forestier *et al.*, 2006; Guillaud, 2006). Further work here can be anticipated in the future.

East Java. Javanese assemblages from the pre-ceramic period of the Holocene are often characterized by the presence of bone artefacts and flake tools, though in the context of otherwise considerable variability (Bulbeck *et al.*, 2000) and localisation. As pivotal to any discussion of lithic developments as it is to bone, is the rock-shelter site of Gua Lawa. The most-reported part of the lithic assemblage from the original excavations here was its typologically distinct bifacially flaked 'arrowheads', made of silicified limestone (Erdbrink, 1954). With the exception of a partially finished piece, these points were apparently recovered from layers underlying those associated with the type assemblage of bone tools. These distinct lithic points are said to have been recovered from greater stratigraphic depth than polished axes, which were also found here (van Es, 1930: 333), suggesting greater antiquity. Two types were reported. Stratigraphically uppermost were basally hollowed or 'winged' points; slightly deeper in the sequence were points with characteristically rounded bases.

Similar finds have come from surface discoveries elsewhere in east Java; thus far, however, only excavations at Song Perahu in the Tuban district (Forestier, 2007) have produced comparable pieces in context. At Song Perahu, their association with incised ceramics suggests them to be of Neolithic age. Overall, current evidence places the distribution of the round-base arrowheads and associated industries as coming from (and largely restricted to) the Tuban-Bojonegoro-Ponorogo-Pacitan districts, effectively forming a corridor of land connecting the north and south coast of Java between approximately 111° and 112°E (Forestier, 2007; Simanjuntak, 2000: figure 4). The potentially earlier, round-based arrowheads are also reported to be widespread among, but are also apparently confined to, sites in this part of eastern Java (Bellwood, 2007; Simanjuntak and Asikin, 2004). Surface and subsurface discoveries from mixed deposits in the Bandung Highlands region of West Java during the 1930s, though, did include a small proportion of obsidian-backed pieces and other retouched forms, considered by van Heekeren (1957: 104) to be of pre-Neolithic age, leaving the possibility of further finds and geographic revisions to come.

The existence of such distinct forms has encouraged typological analysis of some assemblages in this group of putatively related sites. While most of the components of these industries remain more difficult to classify typologically, detailed analysis is paying dividends, as subtle patterns of tool variability seem to be emerging. Forestier's (2007) study of the pre-ceramic lithics from Song Keplek (8.132608°S, 111.002214°E), a cave in the limestone karst of the Gunung Sewa range of mountains west of the city of Pacitan, illustrates this well. Dating to between 9.9 and 3.4 cal. KBP, this assemblage was made predominantly on local chert, with some additional limited use of fossilised wood and metamorphic stone. Forestier systematically defined a range of scraper forms (including side, serrated, concave and edge (or end) scrapers), an emphasis that has led him to suggest the assemblage to have an almost 'mousteroid' character (Forestier, 2007: 254). Other defined stone tools from this site include 'naturally backed knives', perforators, pyramidal pieces with convergent retouch, utilised flakes, cores and core tools, mortars and pestles, though these make up only a small percentage of the total lithic assemblage recovered (n = 2244/14,463, 15.5%), with the majority being unutilised pieces and presumably débitage.

As this kind of detailed analysis is extended to other sites, it will be interesting to see how well they continue to compare – all are currently grouped as 'Keplek' industries and are considered to follow representations at Song Keplek (Sémah *et al.*, 2003; Simanjuntak and Asikin, 2004). Equally interesting will be how these efforts help us begin to piece together differences in site function, which to date have generally not been sought (*though see* Morwood *et al.*, 2008). For example, the presence of grinding stones in some instances (Simanjuntak, 2000) may be pointing towards developments in plant processing, while the closely studied sequence at Song Terus, also in the Gunung Sewa district, shows a marked increase in the intensity of site use across the Pleistocene–Holocene transition (*c.* 12–10 cal. KBP; Morwood *et al.*, 2008). There is a noticeable emphasis on hunting cercopithecids at Gua Braholo, Song Terus and Song Keplek (Sémah and Sémah, 2012; Simanjuntak, 2000; Simanjuntak and Asikin, 2004), which also raises the possibility that there was growing emphasis on arboreal fauna (Rabett and Piper, 2012), as well as links to the coast (Simanjuntak, 2000), for example, in the collections of marine shells and sharks' teeth at Gua Lawa (van Es, 1930).

Flores (Lesser Sunda Islands). Moving eastwards to the island of Flores, the activities of *H. sapiens* at Liang Bua appear to be confined to no deeper in prehistory than the Early Holocene phase of this site's long cultural sequence (*see* Chapter 5). Moore *et al.* (2009) indicate that the arrival of modern humans *c.* 11 cal. KBP did herald certain changes in the island's lithic industry. Knapping practices came to include a greater reliance on the 'freehand' hard hammer reduction technique and unifacial flaking. There was an increased emphasis on chert raw materials, rising from *c.* 17 per cent in the Pleistocene to *c.* 60 per cent in the upper deposits. The first appearance of edge gloss to utilised (and mostly unmodified) flakes occurs during the Holocene. There is also an increase in the frequency of burnt lithic fragments. It is not thought that this formed part of any deliberate kind of heat treatment; rather that fires were being more frequently lit in the vicinity of where knapping was taking place. The lithic industry does not appear to be marked by the kinds of novel forms, such as points, that appear in other parts of the region from around this time.

Despite these differences, the basic lithic reduction sequence – involving a series of strikes, followed by a shift in the orientation of the piece or in the technique being applied to it and further strikes – shows no major change from when archaics were the sole fabricators to when *H. sapiens* were. In a regional context where stone tools have been found to be poor proxies for hominin species (Moore and Brumm, 2007; Reynolds, 2007), it is hard presently to judge the significance of this. The arrival of modern humans on Flores does, however, see the development of a range of behaviours not previously seen in the archaeological record from here, such as new ways of plant processing; formal treatment of the dead; the importation of coastal resources, notably marine shell from *c.* 40 km away; and the use of pigment (van den Bergh *et al.*, 2009).

East Timor. Glover (1977, 1986) reports on excavations carried out at four caves in East Timor from 1966 to 1967. The sites in question, Lie Siri and Bui Ceri Uato, are on the margin of the northern coastal plateau near Baucau and Uai Bobo 1 and 2, located in the central mountains.

At Lie Siri (*c.* 8.435489°S, 126.361992°E), we are chiefly concerned with evidence from horizons dated from between 7516 ± 110 cal. BP (ANU-171) and 8088 ± 160 cal. BP (ANU-236) and perhaps a little older (Glover, 1986: 55). The lithic assemblage from the site was fabricated predominately on fine noncrystalline flint from beach pebbles and nodules in local clay deposits. A small (<1%) component of the assemblage was also made on low-quality local obsidian. The industry from the levels in question consisted of 6103 fragments (including 5774 waste flakes and 329 worked pieces). Among the modified pieces, cores (n = 70) showed a strong emphasis for multi-platform reduction (n = 64/70). Utilised flakes, with either edge damage or silica (edge) gloss, were also present, as were various retouched pieces (n = 120) – an unquantified number of these had elongated, blade-like proportions. Predominantly, these latter were interpreted as various forms of scraper, notably side-scrapers (Figure 6.9), but a single burin was also identified from the same levels. Over and above the number of flaked stone fragments, three preparation tools were identified: a pounder, an anvil and a hammer stone. Most of the worked pieces appear to have been broadly 30–50 mm and occasionally larger. Glover (1986: 207) mentions that a tendency towards a reduction in size was noticed here and at Bui Ceri Uato as one comes forward through time – something that was not seen at the inland sites he studied.

Radiocarbon samples taken from the nearby site of Bui Ceri Uato (*c.* 8.446997°S, 126.359044°E) produced anomalous modern dates, leaving this site without an accurate radiometric sequence. On the basis of comparison to the other sites studied, the two basal horizons (I and II) were extrapolated by Glover to be in the region of 7000–9000 years old. Naturally, the site produced the richest lithic industry of the four, with a total of 39,585 pieces, with Horizons I and II yielding 8007 pieces. Most of these were, again, waste flakes (n = 7361, 92%); cores (n = 185, 2.3%) with multi-platform cores once more predominant; utilised flakes (including those with silica gloss) totalled 265 pieces (3.3%), while retouched tools numbered 147 (1.8%), with the majority (n = 79/147, 54%) being side-scrapers, with lower frequencies of 'thumb' and 'nosed' scrapers and burins. As at Lie Siri, some had blade-like proportions (Figure 6.10). Two anvils and 18 pounders were also recovered from these horizons.

On the eastern scarp of Hatu Ariana mountain, at *c.* 600 m asl, the two caves Uai Bobo 1 and neighbouring Uai Bobo 2 (*c.* 8.636519°S, 126.433997°E) are a little over twice the altitude of Lie Siri (240 m asl) and three times the elevation of Bui Ceri Uato (175 m asl), yet this does not seem to have had any appreciable impact on the character of the technology. The oldest radiocarbon date at Uai Bobo 1 is 3734 ± 118 cal. BP (ANU-414; Horizon IIIa). This is not the base of excavations, however. Age-depth relationships observed by Glover with Uai Bobo 2 provided an extrapolated age for the basal two horizons (Ia and Ib) at Uai Bobo 1 of *c.* 7000 years, though based on the character of the lithics, that age, he speculated, may be closer to between 7500 and 9000 years (Glover, 1986: 132). Thirteen archaeologically

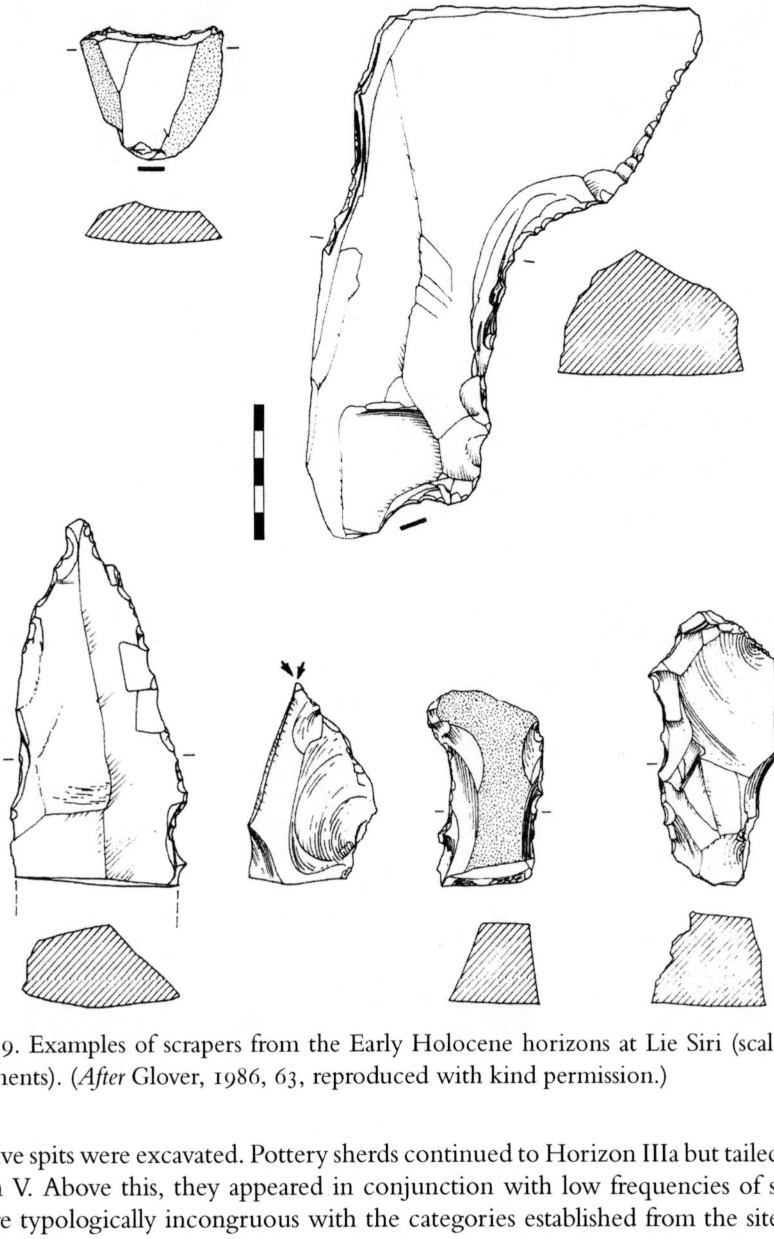

Figure 6.9. Examples of scrapers from the Early Holocene horizons at Lie Siri (scale in 5-mm increments). (*After* Glover, 1986, 63, reproduced with kind permission.)

productive spits were excavated. Pottery sherds continued to Horizon IIIa but tailed off below Horizon V. Above this, they appeared in conjunction with low frequencies of stone tools that were typologically incongruous with the categories established from the sites near the coast. The middle part of the sequence (from Horizon V down to IIa) contained the bulk of the total lithic assemblage (n = 7586/7856, 96.5%). The most represented were, again, waste flakes (n = 6721/7586), followed by utilised flakes (n = 536/7586, including those with silica gloss), side-scrapers (n = 156/7586) and cores (n = 73/7586, including five of the site's six blade cores). There were also isolated examples of nosed scrapers, adzes, burins and backed blades, with broken trimmed and miscellaneous pieces. Retouched pieces from these horizons accounted overall for n = 246/7856 (3.1%). Distinctive tanged points (n = 9/7586; *see* Figure 6.11) appeared in Horizon IIIc – where they were dated on associated charcoal to 2194 ± 110 cal. BP (ANU-237) – and above. The pieces from Horizons I and II were much more limited: only 43 waste flakes, 6 utilised pieces and 3 retouched pieces (all of them side-scrapers), together with a single multi-platform core.

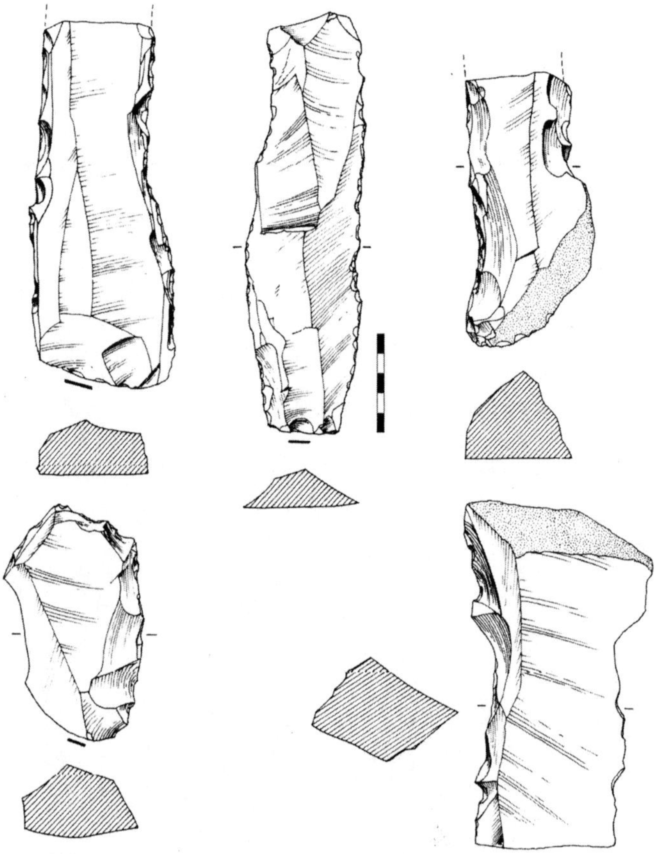

Figure 6.10. Elongated, blade-like scrapers from horizons I and II at Bui Ceri Uato, East Timor (scale in 5-mm increments). (*After* Glover, 1986, 103, reproduced with kind permission.)

An initial test pit (1 × 1 m) dug into a small cliff fissure 100 m to the south of Uai Bobo 1 produced the most finely and clearly stratified sequence of Glover's Timor excavations, giving him reason to expand this (now designated Uai Bobo 2) to a 3 × 2 m trench. Digging halted 4.9 m below the original fissure floor and had followed the receding walls of the fissure into a small subsurface chamber. In total, 13 horizons were identified (counting backwards from XIII at the top of the sequence) and anchored by four radiocarbon dates: 4096 ± 132 cal. BP (ANU-239; Horizon IX), 6307 ± 48 cal. BP (ANU-187; Horizon VII), 7838 ± 123 cal. BP (ANU-328; Horizon IV) and the much earlier 15,615 ± 666 cal. BP (ANU-238; Horizon I).

As at Uai Bobo 1, the flaked stone assemblage was recovered primarily from the middle part of the excavated sequence with very little in the way of lithics emerging from either the upper or lower part of the excavation. The earliest date was accompanied by a single utilised flake (without silica gloss). The material recovered from Horizons X to IV inclusive represented the bulk of classifiable lithic finds and to a time range from the Early Holocene to roughly the opening millennium of the Late Holocene (*see* Appendix), $n = 5905/5969$ (99% of the total lithic assemblage from the site). Waste flakes dominated ($n = 5289/5905$); utilised flakes (including those with silica gloss) accounted for $n = 332/5905$ (5.6%); and cores ($n = 46$), including two large specimens >10 cm in diameter; retouched pieces ($n = 231/5905$), made up 4 per cent of the assemblage – with the largest single category being side-scrapers ($n = 133$). As at Uai Bobo 1, one or two specimens were accorded to Glover's

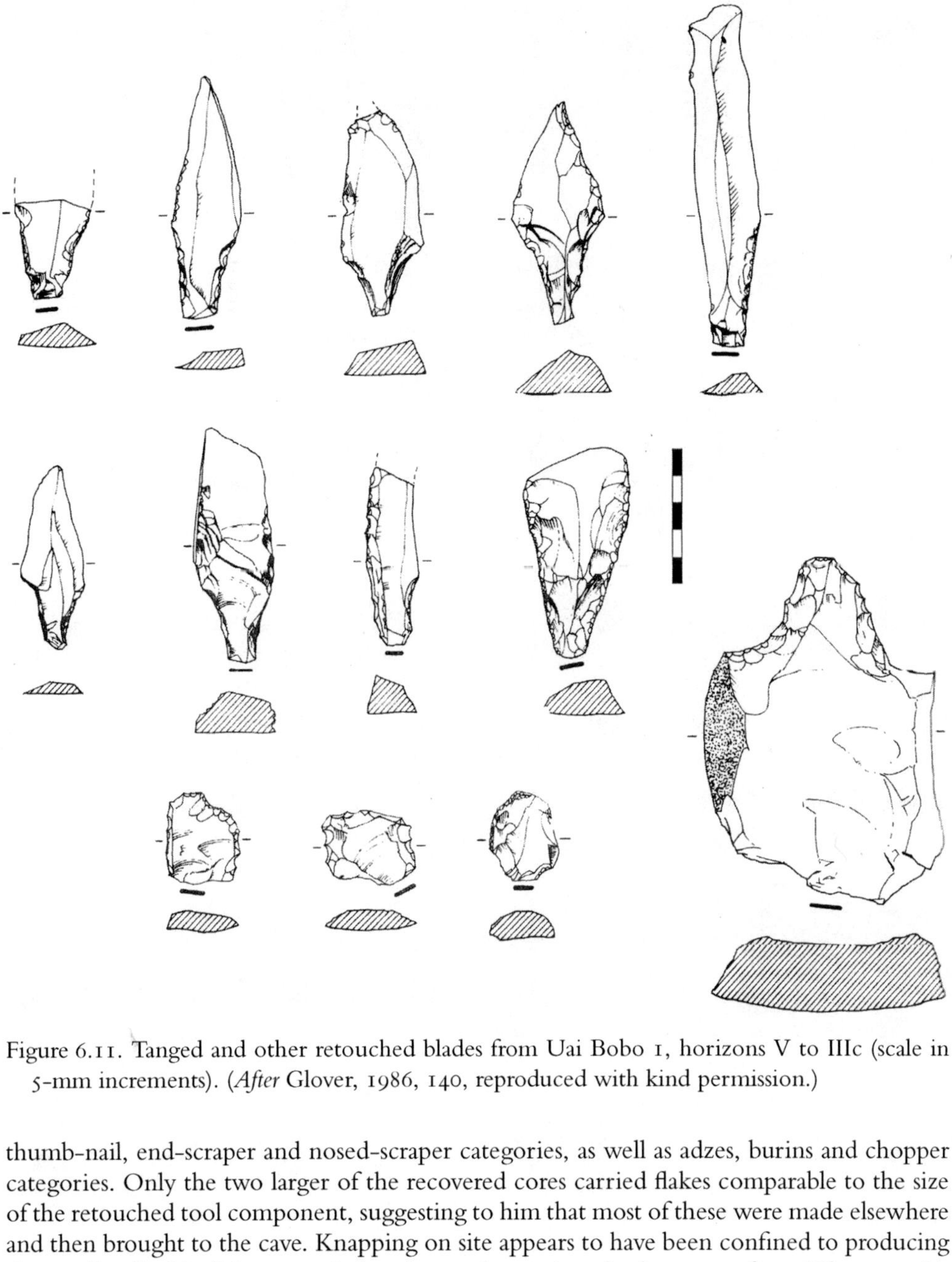

Figure 6.11. Tanged and other retouched blades from Uai Bobo 1, horizons V to IIIc (scale in 5-mm increments). (*After* Glover, 1986, 140, reproduced with kind permission.)

thumb-nail, end-scraper and nosed-scraper categories, as well as adzes, burins and chopper categories. Only the two larger of the recovered cores carried flakes comparable to the size of the retouched tool component, suggesting to him that most of these were made elsewhere and then brought to the cave. Knapping on site appears to have been confined to producing the small utilisable flakes recorded and to resharpening the larger artefacts (Glover, 1986: 171). Further to this interpretation, only three hammer and anvil stones were identified at Uai Bobo 2 (there were none found at Uai Bobo 1).

In sum, Glover's excavations established that occupation of East Timor extended back at least to *c.* 15 cal. KBP. His observations indicated little change in the character and composition of the assemblages (or recovered fauna) over the next 9000 years or so, other than trends towards a reduction in tool size at the coastal sites, something not apparent at the inland sites, and differences in the size of retouched pieces between the coastal and inlands sites (these artefacts were consistently larger inland). Utilised flakes, including some bearing silica gloss, suggestive of plant processing, featured prominently. Some of these pieces

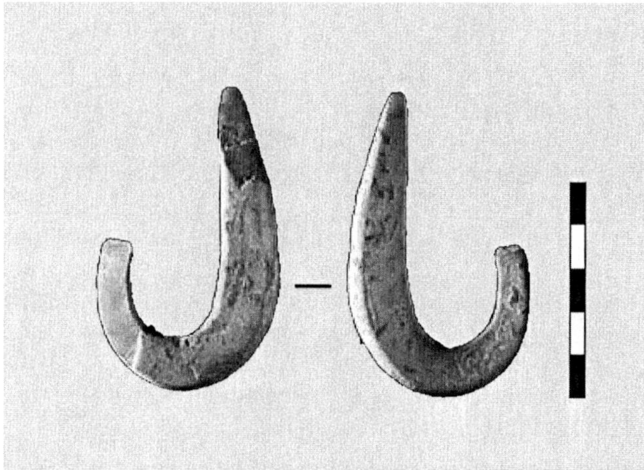

Figure 6.12. Fish hook made from *Trochus niloticus* (recovered from square F, spit 42), Lene Hara. (First published in O'Connor and Veth, 2005, figure 5, *Antiquity* 79: 249–56, reproduced with kind permission.)

were classified as blades, and some of the multi-platform cores were found to carry clear longitudinal ridges consistent with blade production (e.g. Glover, 1986: 135). The main typological class identified was various forms of scraper, particularly side and end-scrapers. At Uai Bobo 1, distinctive tanged points made on blades were added to the inventory late in the sequence *c.* 2.1 cal. KBP. As they appeared broadly during the same period as a marked widening of the faunal spectrum at these sites, the innovation may have necessitated, or at least been tied to, this shift in procurement strategies (Glover, 1977, 1986). Overall, when considering the place of these lithic assemblages in relation to others in ISEA, Glover saw the Timor material as an 'almost completely local development' emerging over a long period within a single cultural tradition (Glover, 1986: 210) – one with a predilection for long, blade-like elements.

Work undertaken much more recently in East Timor at the site of Jerimalai (O'Connor, 2007) produced a chert-based core and flake industry containing pieces with simple retouch but no discernible type forms. This industry was also marked by its continuity without significant change between the site's Pleistocene and Holocene levels (*see* Chapter 5). By comparison, two partial but typologically comparable fish hooks (*see* Figure 6.12), made on the operculums of large marine *Turbo* sp., were recovered from this site at levels dated to the Early Holocene. Another shell fish hook, made this time from the shell of a marine gastropod (*Trochus niloticus*), recovered from the nearby site of Lene Hara (*c.* 8.403472°S, 127.294394°E) has been dated to 9741 ± 60 bp (NZA-17000) from a secure stratigraphic location (O'Connor and Veth, 2005), confirming an Early Holocene age for this technology. Fish, including pelagic (open-sea) species, also feature strongly in the faunal assemblage from Jerimalai. These lines of evidence were taken to indicate that visitors to the cave had access to water-craft and the ability to fish in the deep waters off the coast (O'Connor, 2007). They also show that the imposition of form in technological systems *was* present here but, as at Golo Cave, was confined to shell working. Although shell fish hooks are a rarity in Southeast Asian archaeology (O'Connor and Veth, 2005), and even if manufacture was influenced by natural form, extensive modification of shell has a deep ancestry in this region. Just as lithics are poor proxies for hominins in Southeast Asia, lithic standardization (or the absence of it) is seemingly a poor indicator of overall technological sophistication.

Borneo. The locale of principal concern in Borneo is the Niah Caves. Full publication of the Late Pleistocene to Early Holocene cultural material excavated and analysed during the NCP is in progress (Barker *et al.*, forthcoming); only a broad outline of the character of that evidence is provided here.

Certain changes in the lithic assemblage of the West Mouth at Niah have been observed to take place between the first and second half of the Late Pleistocene. Reynolds (2007) relates how there is a paucity of pieces from the deep *c.* 40 cal. KBP contexts ('Hell'), including an absence of cores, little evidence of microdébitage and only a few confirmed flake tools made of predominantly, though not exclusively (e.g. Zuraina, 1982), local materials. The data suggest that limited lithic reduction was taking place somewhere nearby, though not necessarily in the cave. It also suggests that most of any incoming toolkit was also being taken away again when foragers moved on from the site (Barker *et al.*, 2007). No formal types were evident.

Although the deposits stratigraphically above these are known to have been rich in occupation evidence (Harrisson, 1957), they are currently most well understood from the immediately adjoining part of the West Mouth in Area A, from further into the West Mouth, in Area D, and from Lobang Hangus. In addition to the increased presence of an osseous tool component from these contexts, some changes also become apparent in the lithic industry. A few formal types begin to appear in the dominant flake assemblage, including burins and scrapers, while the curation of flakes is still a feature (Reynolds, 2007). New forms, such as pounding and ground stone components, seem to first appear at Niah during the first part of the Last Termination, and by the Early Holocene, stone pounders, various types of grinding stones (including for bone tool preparation), hammer stones, pebble tools, modified shell, and flaked and polished stone are also present, many of them likely used for processing plant matter.

In Sabah, two limestone massifs have received detailed archaeological work: Baturong and Madai (Bellwood, 1984, 1988). The location of particular interest to us in Bukit Baturong is Hagop Bilo. This locale comprises two sites, BAT 1 and, 35 m east of it, BAT2 (4.714442°N, 117.970014°E), just two of several sites that appear along the base of the precipitous south-facing cliff of this massif. The archaeological deposits at BAT 1 extended to *c.* 0.6 m and were divided into three stratigraphic units. Layer 1 comprised modern materials; Layer 2 contained Early Metal Age ceramic sherds, particularly in its upper 10 cm, below which the incidence of cultural material tapered off. Layer 3 contrasted noticeably to those above it, in that it included a recognisable shell midden accumulation together with animal bones and lithics. This layer also sat on archaeologically sterile sediments of presumed lacustrine or alluvial origin (Magee, 1988). Charcoal was not recovered from Layer 3, leaving radiocarbon dates to be obtained on freshwater shell – tests on a modern shell from the nearby Binuang River indicated that ages obtained would be in the order of 500 years too old and need to be adjusted accordingly (Bellwood, 1988). Four (uncorrected) radiocarbon dates spanned a period from 12,050 ± 120 bp (ANU-2558) back to 17,900 ± 200 bp (ANU-2560), giving a corrected and calibrated age range of broadly 20.7–13.3 cal. KBP and human occupation starting in the late LGM, and continuing through much of the Last Termination. The BAT 2 deposits appear to have originally followed essentially the same stratigraphic sequence, though they were disturbed by secondary burials.

Excavations were also undertaken by Bellwood *c.* 20 km east of Bukit Baturong, in the Bukit Madai (*c.* 4.719803°N, 118.164086°E). Between late 1979 and early 1980, digging took place in three separate cave mouths here: Agop Atas (MAD 1), Agop Sarapad (MAD 2) and Agop Alag (MAD 3), all within an inter-connected system known locally as the Madai Caves. The last of these produced a cultural sequence that confined to the modern era; the other two sites produced materials and dates that are much more relevant to the present discussion.

A total of 12 layers were identified at MAD 1. All of those down to Layer 11 are of Late Holocene age; Layer 12 is dated on the basis of two charcoal samples to between 8792 ± 370 cal. BP (ANU-2397) and 8205 ± 264 cal. BP (ANU-2398) and contained lithics but no organic material. Although left unexplored because of logistical constraints, Bellwood could not rule out the possibility that cultural deposits of greater antiquity may still be *in situ* (Bellwood, 1988: 113).

A secondary excavation to MAD 1, MAD1/28, undertaken *c.* 20 m to the west, comprised three additional 1 × 1 m squares and exposed a maximum depth of *c.* 2 m of deposit. In the absence of a clear stratigraphy, this was divided into four arbitrary layers. Although the uppermost of these returned a radiocarbon date of 9150 ± 100 bp (ANU-3086) on shell, the presence of ceramics indicates that considerable mixing had taken place. The layers below this seemed to produce material more consistent with the freshwater shell dates that were obtained from them, 7690 ± 100 bp (ANU-3087) to 10,800 ± 120 bp (ANU-3088), though the bottom two dates are inverted due to localised disturbance, and again all required correction for the effect of old carbon uptake into the shells. Lithics were recovered throughout the sequence and were particularly concentrated in the cemented deposits that marked the interface between the upper mixed layer (Layer 2) and Layer 2a (*see* Bellwood, 1988) – probably as a result of post-depositional processes rather than changes in the intensity of site use. Little detailed information is available about this specific assemblage, but it included utilised flakes with occasional examples of possible secondary retouch and edge burnishing, as well as flaked pebble tools (Bellwood, 1988).

The neighbouring excavation, MAD 2, produced a partially concreted freshwater shell midden (Layer 2) *c.* 20 cm thick beneath a thin skein of old guano deposits (Layer 1). Underlying deposits (Layers 3 and 4) were increasingly concreted, though apparently discrete, if undated, accumulations of lithics in the upper portion of Layer 3 signal human presence at the site prior to the formation of the midden, which is dated on shell to between 10,450 ± 110 bp (ANU-2554) and 9520 ± 100 bp (ANU-2553). Taking the recommended subtraction of *c.* 500 years off these dates gives a calibrated range of *c.* 11.8–10.3 KBP. The only question mark here is that the comparatively small size of the midden seems out of keeping with the millennium or so occupation period indicated by the dates, suggesting that either visits were only occasional or that there may be taphonomic processes as yet unaccounted for at this site.

At both Baturong and Madai, the motivation in lithic tool production appears to have been to create sharp-edged flakes. There is little indication of typologically distinct pieces and virtually no unequivocal evidence of secondary retouch. Almost all of the flaked pieces from both areas were struck from water-worn pebbles of chert or chert nodules available in local stream beds. Two principal sources were identified: the Kuamut Formation (Baturong) and the Chert-Spilite Formation (Madai). Interestingly, there is no evidence from either locale that the distinctive variety of the Kuamut chert used for making the Tingkayu bifaces was ever utilised here (Bellwood, 1988). Some tendency towards the creation of sub-prismatic cores was observed on about one third of the cores from BAT 1, but most did not carry dominant platforms and, with the exception of a single group of blade-like flakes also from BAT 1, there is very little indication of formal standardization in flake morphology or thickness. Utilised pebble tools are not frequent but have been recovered from these assemblages, some exhibiting clear battering damage, notably within the MAD 2 midden. An absence of similar pieces from BAT 1 is considered by Bellwood to be a recovery bias. Use-wear of various kinds was evident on sharp flakes, including micro-edge chipping and low frequencies of silica gloss. Most are from either poorly dated (BAT 1) or superficial layers (MAD 2), though a single specimen from MAD 1 is securely dated within Layer 12.

Briefly, and still in Sabah, it is worth making passing reference again to Gua Balambangan. No specific details are yet available for the lithic assemblage recovered from the Late

Pleistocene and Early Holocene archaeological deposits at this site. It reportedly comprises mostly flake tools, made of locally available chert, in association with cores, débitage and hammer stones. On this limited evidence the implication is that at least some of the lithic industry was being fabricated on site (Zuraina *et al.*, 1998b).

The Philippines. Archaeological investigations on the island of Palawan between Borneo and the rest of the Philippine archipelago have produced important sequences of early human occupation, both from sites on the east coast at Lipuun Point (Détroit *et al.*, 2004; Dizon *et al.*, 2002; Fox, 1970) and inland of Sibaltan Bay on the north coast at Ille Cave (Lewis *et al.*, 2008). The most well-known site from Lipuun Point is Tabon Cave. The evidence of early hominin occupation at Tabon was covered in Chapter 5; here we shall focus on the later pre-ceramic evidence during the Last Termination and Early Holocene.

Fox (1970) referred to the lithic facies from the earlier part of this period as Flake Assemblage II, though no specific details are available about its composition. The associated cave deposits rest on a thick gypsum speleothem *c.* 0.5 m below the surface. This layer has now been dated by U-Th series to 20.2 ± 0.8 KBP and 19.4 ± 0.8 KBP and, given this age, is thought to signal a significant local hydrological shift, possibly linked to conditions during the LGM (Lewis *et al.*, 2007–8).

Lying above the Flake Assemblage II levels, in the upper part of the sequence, is evidence from the final phase of site occupation at Tabon. This period is anchored by a single radiocarbon date from 20 cm below the cave floor to *c.* 10,451 ± 338 cal. BP (UCLA-284; Fox, 1970). The cultural remains recovered from these upper levels included what Fox referred to as Flake Assemblage I-B. Unfortunately, no detailed description of this material is available either, though combined analysis with another contemporary Early Holocene assemblage from the site, Flake Assemblage I-A, by Patole-Edoumba (2009) gives a broad sense of its character. Utilised flakes and flake-blades are reported, as are retouched pieces (of which burins appear to figure prominently), cores and, rarely, core tools. The cave appears to have been abandoned at some point during the Early Holocene as there then follows a significant hiatus in remains until the Late Holocene, when the site was used as a burial site during the region's Early Metal Age. Fox (1970) proposes that this abandonment was tied to an encroachment of the sea.

Also at Lipuun Point is the multi-chambered Guri Cave. Fox saw the Tabon flake tradition persisting at this site well into the Holocene – a radiocarbon date on shell from 60 cm below the surface, towards the base of the sequence, provided a date of 4070 ± 80 bp (UCLA-698), together with three further assays: two obtained from different parts of the same shell – 8130 ± 180 bp (AERIK-20) and 7890 ± 90 bp (AERIK-21) – and one on charcoal of 4220 ± 140 bp (AERIK-19) or 4769 ± 189 cal. BP (Fox, 1978). While the Guri Cave industry was perceived as a development of that at Tabon, there was a tendency towards smaller flakes and an increased incidence of secondary retouch. Large blades are also reported from Guri, whereas they were not present at Tabon. The production of such pieces would have required core preparation of a kind that was also not seen at the older site. Fox proposed (with reference to Australian comparatives) that the Tabon lithics represented a non-hafted technology, whereas the small flake and blade tradition from Guri may well have been composite pieces. Association with a larger quantity and diversity of vertebrate fauna, compared to that seen at Tabon, and with marine molluscs, was taken by Fox (1970: 50) as indication that the Early Holocene in this area saw a shift towards more effective hunting and trapping technologies. Although an interesting and pertinent interpretation in terms of the current discussion, it is one that took no account of possible differences in inter-site function and, therefore, needs to be treated carefully, particularly if one considers Duryong Cave, 11 km north of Lipuuan Point, also in Fox's (1970) report. At Duryong, Neolithic and Metal Age levels are underlain by a small flake-blade assemblage dated to 7834 ± 231 cal. BP

(UCLA-288). These lithics were excavated from within a midden containing densely packed marine and brackish water molluscs but only a *small* component of vertebrate fauna. The lithics show no evidence of secondary retouch, though they do appear to exhibit use-wear. Along with the flake component of the assemblage were small prepared cores, from which blades were apparently struck, and chopping core tools made on chert and basalt (Fox, 1970: 59).

At the north end of the island, excavations commencing in 1998 at Ille Cave (11.196111°N, 119.505278°E) produced one of the first well-stratified and well-dated excavations in the Philippines, covering the Pleistocene–Holocene transition. The deepest deposits were found to be in the East Mouth part of the cave, where a suite of 20 radiocarbon dates has been obtained from a 3.6 m depositional sequence, giving a range from *c.* 6.4 cal. KBP back to at least 11.2 cal. KBP and perhaps earlier. Deposits underlying a 7–5 cal. KBP shell midden layer have been found to contain hearth features, a flaked lithic industry, faunal remains and a single cremated burial (Lewis *et al.*, 2008: 320). The small lithic assemblage ($n = 52$ pieces) recovered from these lower levels has been found to contain a comparatively high proportion of cores (12.5%), several of which have longitudinal flake scars indicative of elongated blade-like removals. Artefacts were made of either chert or obsidian, neither of which is yet firmly sourced, but the former is thought to be locally occurring; the latter exotic, though, is of unknown origin. The absence of primary core reduction flakes, of hammer stones and of anvils implies that initial knapping activities probably took place elsewhere, before partially prepared pieces, including flakes, were brought to the site. Use-wear analysis indicates that a high proportion of exhausted cores were recycled as tools themselves. Collectively, the evidence describes a toolkit that was used primarily for maintaining or manufacturing organic tools (Lewis *et al.*, 2008: 324).

At Musang Cave, in Peñablanca, Luzon, 303 of the 1172 lithics excavated from Cultural Level I were considered to have been utilised to some degree. This level was dated on shell (and potentially problematic) from 11,450 ± 170 bp (ISGS-496) to 9390 ± 280 bp (GAK-7046; Thiel, 1990). All utilised flakes were small (10–50 mm, maximum linear dimension), with less than 3 per cent carrying secondary retouch and none that could be allocated to any formal type. The majority of pieces (*c.* 70%) was made of chert, the remainder on andesite, basalt, agate, jasper and jade (Thiel, 1990: 71). Almost all of the lithics from Cultural Level I were recovered from a midden otherwise composed of riverine gastropods (*Thiara scabra*) and animal bone near the cave mouth. The frequency of material increased through this level, suggesting greater intensity of site use during the Early Holocene, but without significant change in the activities being undertaken. Use-wear analysis of the utilised flakes suggests that this included the working of bamboo and bone and plant processing (on the basis of silica gloss and edge rounding). Although there is no clear transition to the upper cultural deposits (Cultural Level II), these are marked by the appearance of ceramics, a further steep rise in the frequency of vertebrate faunal remains, and a corresponding decline in molluscs (and mollusc size), suggesting a pronounced shift in site use and possibly subsistence pressure in the Mid- to Late Holocene.

Sulawesi. Turning southeast from Borneo, the island of Sulawesi has produced what has been termed a 'tradition' of stone tools from the Early to Mid-Holocene that is probably the most distinctive in the region (Pasqua and Bulbeck, 1998). Between 1902 and 1903, an expedition by the Swiss naturalists Fritz and Paul Sarasin into the interior of South Sulawesi, near Lamontjong, yielded Indonesia's first examples of a prehistoric flake-blade industry that would come to epitomize this part of the country: the Toalean. In view of its importance, it is worth pausing briefly to summarise the history of investigation.

The term was coined by the Sarasins from the local word *Toala* ('forest people') in reference to the then (incorrectly) presumed descendants, who still frequented the caves

(Bartstra, 1998). Subsequent expeditions were carried out in the Maros district by van Stein Callenfels and van Heekeren in 1933 and 1936, respectively; and by the Australian archaeologist, McCarthy, accompanying van Stein Callenfels in 1937, to further elucidate the character and distribution of the Toalean (van Heekeren, 1957). Initial impressions about the cultural sequence were framed within the then prevailing unilinear evolutionary scheme. The supposed youngest expression, the Late Toalean, was attributed to the Neolithic and contained distinctive hollow-based 'arrowheads'. Later redefined as *Maros* points (Mulvaney and Soejono, 1970: 171), these artefacts were mostly made on silicified limestone and were considered distinct from related forms with serrated edges. Accompanying the points were many shell artefacts and the so-called *Muduk* bone bipoints. Beneath this (and attributed to the Mesolithic) was the Middle Toalean, whose most defining artefacts were considered to be blades and geometric microlithics. Finally, there was the Proto-Toalean stratum: this was considered truly Palaeolithic and contained a larger, flake-based industry, featuring pieces that were considered to be incipient Maros points, called *Pirri* points. Although this interpretation has been questioned owing to the typologically selective way the deposits were being codified and to its reliance on European and Australian cultural sequences and material (Bartstra, 1998), broadly, the succession of forms has survived more detailed scrutiny.

After the Second World War, Dutch involvement in the archaeology of South Sulawesi continued under van Heekeren until 1950. During this time, work by Hooijer determined that the fauna associated with these sites was exclusively extant. This supposed that the origins of the Toalean were unlikely to be older than the beginning of the Holocene. Following an extended period of local political unrest, a joint Australian–Indonesian expedition, led by Mulvaney and Soejono, resumed work in South Sulawesi in 1969. This had the intention of clarifying the state and chronology of the Toalean sequence, and to assess lingering ideas, originally raised by McCarthy (1940), about the possibility of cultural links between Sulawesi and Australia in prehistoric (and historic) times (Chapman, 1986; Mulvaney and Soejono, 1970, 1971). A key finding of this expedition and one that is a salient reminder for the region as whole was that Toalean assemblages showed marked inter-site variation and, occasionally, as at the expedition's largest excavation, Leang Burung, even at an intra-site scale (Mulvaney and Soejono, 1971: 31). The classic Toalean artefact forms were present at Leang Burung, but the site was heavily disturbed; dating evidence was deemed to be inconsistent (Glover and Presland, 1985), though broadly of Late Holocene antiquity.

A couple of kilometres to the north of Leang Burung, project excavations under Ian Glover were working on the much more promising Ulu Leang 1 (*c.* 4.989058°S, 119.668239°E; Figure 6.13). Preliminary results, particularly with reference to the lithics from this field season and follow-up excavations in 1973 and 1975, were published in Glover (1976). While significant post-depositional processes, such as slumping and guano digging, did make this another stratigraphically complicated site, a reliable suite of radiocarbon dates was successfully obtained from minimally disturbed deposits. These indicated that cave use spanned the Holocene, from *c.* 10 cal. KBP until within the last thousand years (Bulbeck *et al.*, 2000), but that classically Toalean artefacts were not present at the beginning of the sequence. Undisturbed deposits excavated towards the cave entrance and from the basal layers of trenches further inside the cave produced an unnamed assemblage containing utilised and waste flakes, cores, and retouched elements, together with spatulate bone tools (Glover, 1976). All of the lithics in the small sample studied (n = 30) were made of a cream-white chert that was distinctly different from that used to fabricate tools higher up. The retouched pieces included various forms of scraper. While some changes in the frequency of occurrence were noted for these components, all persisted through to the top of the sequence; the Toalean elements appear to have been grafted into a pre-existing industry. Shell scrapers made of the bivalve *Cyraena* sp. (Glover, 1976: 138) were also added to the

Figure 6.13. Ulu Leang 1 at the end of excavations, July 1969. (Photograph by I. Glover, reproduced with kind permission.)

repertoire of implements, along with bone points, sometime around 7–8 cal. KBP (Bulbeck *et al.*, 2000). This was followed by the appearance of several types of abruptly retouched ('backed') flakes, together with flakes bearing silica gloss. Geometric and other forms of microlith began to appear *c.* 6 cal. KBP. The latest expressions of this are considered to be the Maros points (now incorporating at least four types, including the edge-serrated forms), which did not become common until *c.* 4.6 cal. KBP (Glover and Presland, 1985: 192–3; Figure 6.14). No evidence has emerged to challenge the possibility that the Pirri points were anything more than opportunistically shaped flakes (Mulvaney and Soejono, 1971: 33), while the credibility of blade production within the Toalean is also a matter of opinion. Chapman (1986) felt that a small proportion of the pieces from Leang Burung qualified as blades, though no prismatic cores were found; Glover and Presland (1985) do not see the existence of blades within the Ulu Leang assemblage.

Today the Toalean is known from about 25 sites across South Sulawesi. Despite sustained surveying up to the northern half of the peninsula, no evidence for its existence there has so far been discovered, suggesting it to have been a geographically restricted tradition, but it was one that appears also to have had localised variability. Assemblages excavated from the eastern and central part of the peninsula lacked Maros points and had few backed elements; where Maros points did appear, along the south coast, the basal hollowing was limited or even vestigial; some sites did not contain bone or shell implements. Maros points have even been recovered from Selayar Island, 30 km offshore at Batang Mata Sapo on the southern coast of the peninsula (Bulbeck *et al.*, 2000; *see also* Chapman, 1986). The composition of the Toalean carries some intriguing similarities to Javanese evidence, and the credibility of cultural links to here and Australia continues to divide opinion when it comes to the inference of possible maritime links between these areas (e.g. Bellwood, 1987, 2007; Bulbeck, 2008; cf. Forestier, 2007).

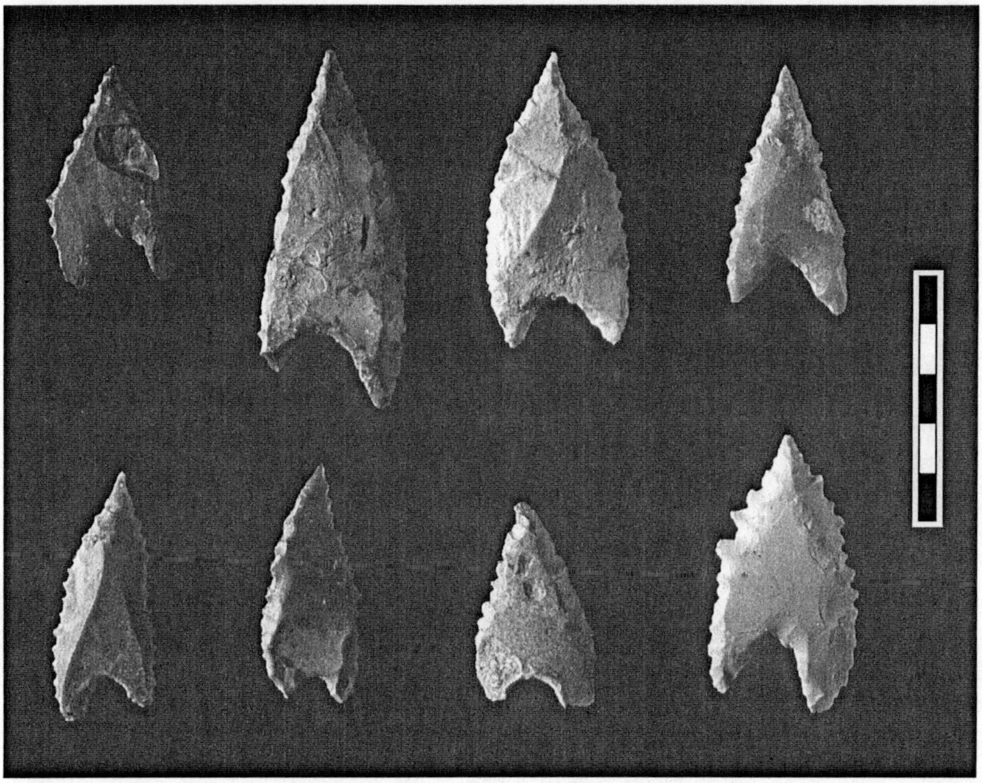

Figure 6.14. Examples of Maros points (scale in 5-mm increments). (Photograph by I. Glover, reproduced with kind permission.)

While the Toalean appears to be a geographically and chronologically confined entity, or group of related techno-facies, it does not represent the only evidence of Early Holocene occupation that has been discovered in Sulawesi. The pre-ceramic Paso shell midden (*c.* 1.206269°N, 124.854392°E) on the Minahasa Peninsula, in North Sulawesi (Bellwood, 1976), yielded a flake-based industry, made of locally sourced obsidian that is said to fit quite well into the general pattern of Early Holocene pre–flake-blade technology from ISEA (Bellwood, 1976: 248). Two charcoal dates are available for the site, 8366 ± 478 cal. BP (ANU-1517) and 8178 ± 305 cal. BP (ANU-1518), and these are taken to indicate a fairly short period of occupation of no more than around 500 years. The selection of flat-based, high-backed and steep-sided flakes for making tools at Paso is suggested to parallel similar tendencies in the pre-Toalean at Ulu Leang 1 (and is also considered not dissimilar to evidence from New Guinea and Australia), though not all of the tool types of South Sulawesi are present here (Whatrup, 1982).

Farther to the east of Minahasa and lying on the northern-most border of Indonesia, between the Celebes Sea and the Philippine Sea, are the 77 islands that make up the Sangihe and Talaud island groups, a regency of North Sulawesi. Investigations by Bellwood (1976) were centred on Leang Tuwo Mane'e (*c.* 4.505278°N, 126.815408°E) Karakellang Island in the Talaud group. This wave-cut rock-shelter formed in coral limestone and is situated 20 m inland of the coast. Two areas of the *c.* 150 m^2 shelter (designated H/K and P/Q) were excavated in spits, owing to an absence of visible stratigraphy, down to bedrock 1–1.5 m below ground. Neolithic, Early Metal Age and recent deposits were uncovered in both areas, but late pre-ceramic archaeology was only found in one area (H/K), below 0.8 m, and bore a single radiocarbon date on shell (*Turbo* sp.) of 4860 ± 130 bp (ANU-1717). All worked stone from the site was found to have been made almost exclusively of two kinds of locally sourced chert; the exception being a single obsidian nodule from Neolithic levels that was probably imported. The defining feature of the pre-ceramic industry is its high (*c.* 50%) proportion of formally defined blades, preferentially struck from one of the varieties of chert (grey-coloured). Flakes tended to be struck from the other, a brown/red variant, suggesting deliberate raw material selection. By Neolithic levels, the emphasis had shifted to only producing small, brown chert flakes. Although blades continued to be found in the site's Metal Age deposits, Bellwood (1976) surmised, on the basis of use-wear, these that later examples were not being freshly struck; instead, they were probably being recycled. Overall, elements of the older technological suite persist alongside the introduction of ceramic and other classes of artefact. Both the flake and blade components of the pre-ceramic industry at Leang Tuwo Mane'e are metrically quite small: 15–50 mm at their longest dimension. Prismatic-like cores were recorded here, though infrequently; similarly so, pieces with clear secondary retouch ($n = 6/655$ utilised pieces, 0.9%). Bellwood (1976, 2007) argued that the predominantly un-retouched pre-ceramic industry here, together with others from the Philippines, eastern Indonesia and Australia, represents one of two variants of a regional flake-blade technocomplex, the other containing the more heavily retouched components of the Toalean.

The oldest evidence of human occupation in this group of islands comes from Salibabu Island just to the southwest of Karakellang. The small limestone rock-shelter of Leang Sarru here lies *c.* 400 m inland of the eastern coast of the island (*c.* 3.932464°N, 126.701619°E). The original excavations carried out at Leang Sarru in the mid-1990s have been recently revisited (Ono *et al.*, 2010). Three cultural layers have been identified during the new excavations. These contained a large number ($n = 9465$) of lithic artefacts, comprising flakes and flake tools (retouched pieces), cores, hammer stones and débitage. All of the material appears to have been made of chert pebbles obtained from the island's watercourses and beaches. The small size of the flakes (mean, 25 × 10 mm) was recognised as a common feature of ISEA assemblages and probably representative of a restriction imposed by the choice of raw material. Although occupational evidence is most evident in Layers 1 and 2 (the latter dated to the Early Holocene, *c.* 8 cal. KBP), material also occurred below this in concentrations that would be consistent with intermittent site visits. Two conventional radiocarbon dates, obtained on the marine shell *Turbo marmoratus* from Layer 3, returned ages of 32–33 cal. KBP, in agreement with dates from the earlier excavation. Although some disturbance of deposits is indicated, the finds do further allude to the fact that people were already capable of travelling considerable distances by sea well before the LGM.

North Malaku Islands. In the North Malaku Islands, a small number of archaeological sites have yielded culture-bearing deposits from the Last Termination and Early Holocene (Bellwood *et al.*, 1998). On the island of Morotai, two are of note, both adjacent to the southern coastline. At Tanjung Pinang, a rock-shelter *c.* 140 m from the shore (*c.* 2.065208°N, 128.514578°E), 2.4 m of deposits span the period of human presence on the island. The top 30 cm of the excavation contained six Late Holocene secondary burials and associated pottery, which had been dug into pre-ceramic layers. These latter returned a sequence of five radiocarbon dates (on marine shell) from 3209–2959 cal. BP (ANU-7778) back to 9485–9260 cal. BP (ANU-7782), 70–80 cm below the surface. The lithic industry comprises pitted pebble anvils, which local ethnographic reference suggests may have been used for crushing *Canarium* nuts, large numbers of unworked pebbles collected from the nearby beach and un-retouched flakes made of the same local raw material. The *c.* 9.4 cal. KBP date came from the base of the uppermost of the two visually distinct layers at the site. Although two

Late Pleistocene dates were obtained from the underlying layer, no verifiable evidence of human activity could be tied to it.

Approximately 4 km to the west of Tanjung Pinang, one of three caves near to Daeo Village was investigated by Bellwood and colleagues in 1994. Daeo 2 (*c.*2.058892°N, 128.478525°E produced a very similar archaeological sequence to Tanjung Pinang. Late Holocene human remains and ceramics penetrated 30 cm into earlier cultural deposits, which extended to a depth of 70–80 cm before becoming archaeologically sterile. Although anchored by few dates, the period of site use here appears to exceed that seen at Tanjung Pinang by some margin. A radiocarbon date of 6405–6283 cal. BP (ANU-9452) was obtained on charcoal at a depth of 20–25 cm. This is out of sequence with a lower date at 55–60 cm but is considered by the excavators to be the more reliable of the two (Bellwood *et al.*, 1998: 246). This is immediately underlain (60–65 cm) by the much older date of 15,798–15,305 cal. BP (ANU-9450) taken from marine shell. Archaeological remains, including manuports and un-retouched flakes very reminiscent of Tanjung Pinang, and faunal remains, extend to at least 80 cm, suggesting that human occupation (albeit ephemeral) occurred even before this time. Two bone points (one of which is questionable) are reported from undated contexts between 30 and 50 cm. If the sequence dates are reliable, these are almost certainly of Early Holocene age. Besides evidence from Gebe Island (*see* Szabó *et al.*, 2007), this record is currently the oldest evidence of human occupation in North Malaku.

The only other location in North Malaku to have produced occupational evidence from the Last Termination and Early Holocene is Gebe Island itself. Three sites excavated on this island have produced pre-ceramic deposits: Golo Cave, Um Kapat Papo and Buwawansi. The earliest levels at Golo Cave were discussed in Chapter 5 and the bone technology has been discussed earlier, along with comparative comments made on the character of the lithics. Occupation continued through the Pleistocene–Holocene transition here, as attested by two dates on marine shell: 13,028–12,862 cal. BP (ANU-9512) and 11,925–11,121 cal. BP (ANU-9769). The pre-ceramic flaked stone industry was made on a wide range of metamorphic and siliceous rock and was accompanied by a series of shell adzes. Those dating to this middle period of occupation were made on giant *Tridacna* or *Hippopus* shells, while those from the Holocene were made from the lip of the large marine gastropod *Cassis* sp. (Szabó *et al.*, 2007).

At Um Kapat Papo (0.166667°S, 129.566667°E), the comparatively shallow cultural deposits (*c.* 70 cm) rest directly on bedrock with no underlying sterile layer. The upper *c.* 30 cm of ceramic horizons are underlain by contexts bearing an industry comparable to that seen at Tanjung Pinang, with flaked stone tools, putative plant processing stones and manuports, though at lower frequencies. The oldest date comes from spits 55–60 and is calibrated to 7198–7017 BP (ANU-9318).

The Buwawansi locale (0.016667°S, 129.333333°E), on the northwest of the island, consists of a sandy coastal terrace with rock-shelters cut back into ancient uplifted coral formations. Limited-scale excavations here in 1994 and 1996 by Bellwood and colleagues examined six locations (B1 and B3 were rock-shelters; B2 and B4–B6 were open-air sites). Of these, B1 is the only one of direct relevance here. Although cultural material included pottery through most of the sequence at B1, and no coherent pre-ceramic levels could be defined, basal layers dated on marine shell to 9182–8955 cal. BP (Wk-4628) did produce a single *Tridacna* shell adze similar to those seen at Golo Cave.

New Guinea. For the final examples in this section, we return to evidence from the western margin of the Sahul Shelf, firstly from the Bird's Head Peninsula and then from farther south in the Aru Islands. Juliette Pasveer's (2004) excavation of Toé Cave, in the Bird's Head (1.284889°S, 132.218361°E) produced a 1.1 m sequence that she divided into a Late

Pleistocene phase and a Holocene phase. Both span broadly 7000–10,000 years, with the former covering the period from 24,730–23,210 cal. BP (OZF-516) to 15,990–14,410 cal. BP (AA-33405). Initial occupation of the cave, however, dates to 25,940 ± 180 bp (OZG-063) on *Casuarius* sp. eggshell from 105–110 cm below the surface – a date that Pasveer notes should be taken as a minimum age owing to inconsistent age-depth determination in the site's lower deposits. A calibration of this would place initial human presence at the site *c.* 31 KBP. The Holocene phase spans the period from 10,210–9630 cal. BP (AA-33404) to 3320–2870 cal. BP (OxA-6201).

The Late Pleistocene levels at Toé Cave did not produce any *bona fide* cores, though several retouched flakes may have served this purpose and produced some of the 189 flakes recovered; of these, 15 had secondary retouch. Sixty per cent ($n = 9/15$) exhibited use-wear, while only 2 per cent ($n = 2/102$) of un-retouched flakes did. While utilization as cutting implements was noted, there was no evidence that any of these pieces were used for paring or shaving actions, suggesting that differences in tool production during this phase may have extended to differences in tool (and site) use.

The total excavated lithic assemblage from Toé Cave comprised 1775 fragments. Ninety-five per cent of these were made during the Holocene of poorly silicified chert, just 14.8 per cent came from Late Pleistocene contexts. Exotics were identified, including rounded pebbles of quartzite and metamorphic rock (but both at <1% occurrence). Some of the quartzite pebbles from the Holocene levels showed clear use as hammer stones and provided evidence that a degree of on-site manufacturing was taking place. This conclusion finds further support in the recovery of nine (mostly multi-platform and all but one chert) cores and an additional nine flake cores from the Holocene phase; and in the fact that the size of complete flakes was comparable to core flake scar dimensions: all of the ($n = 44$) retouched flakes from these layers could have been struck from cores such as the ones found.

The 1167 chert flakes recovered from Holocene contexts show a slight reduction in the dimensions of complete flakes compared to those from the Late Pleistocene; they also get thicker and heavier. These trends may reflect, at least in part, changes in the incidence of larger cortex flakes over time, something itself tied to the fact that the Holocene layers contained a larger proportion of a complete lithic reduction sequence. Otherwise no significant metrical difference in flake shape or retouched flake dimensions between the two phases was observed (Pasveer, 2004: 114). Among the 172 Holocene flakes that showed clear secondary retouch, the majority were also found to carry use-wear ($n = 157/172$, 91.3%), together with a much smaller proportion of the un-retouched flakes (*c.* 7%; *see* Pasveer, 2004: 116, for calculation criteria). A wide range of cutting activities best explained these traces, though shaving or planning activities were also indicated.

At Kria Cave (1.251975S, 132.333511E), on the north side of the Ayamara Lakes (*c.* 15 km away, in a straight line), Pasveer (1998, 2004) divided the 1.95 m excavated profile into five stratigraphic units. From these came a total of 1528 lithics. The industry recovered from levels below Unit I (dated to within the last 1800 years) comprised 88–96 per cent chert and seems to have similar if not identical properties to that used at Toé Cave. Other stones, including silicified tuff, limestone, quartzite and metamorphic rock, made up the remainder, all at low frequencies. The most immediately notable feature of the Kria Cave assemblage was that with the exception of a single and highly reduced specimen (made on silicified tuff), it contained no true cores and only three flake cores (and a further six potential ones) from Unit V: 7930–7590 cal. BP (OxA-6043) to 6890–6660 cal. BP (GrA-9101). Unit V is the oldest component of the site and the one on which we shall focus. Explanation for the near absence of cores is not immediately apparent, and Pasveer offers a range of possibilities, including the removal of cores off site due to scarcity of local flakable stone, to on-site destruction of cores (e.g. through disposal in a hearth), or that it represents a recovery bias and that knapping areas were spatially separate from the area excavated. The presence of a hammer stone from this same unit, and of small quartzite flakes in the assemblage (that possibly spalled off the hammer stone during use), is taken as indication that some degree of tool manufacture took place in the cave during this time.

Unit V was also found to contain 435 complete or fragmented flakes, including 42 retouched pieces. Using a ratio of width to length as a simple measure of shape, Pasveer determined there was little change in flake morphology throughout the sequence (a subtle narrowing and thinning of flakes and retouched flakes was noted later on *c.* 5.5 KBP, which may imply minor changes in technique). She drew from the overall similarity in the assemblage, however, that the basic stone-working techniques probably remained the same over the thousands of years of the occupation and compared closely to those seen at Toé Cave. The key difference between the two sites lay in the intensity of the reduction sequence, which was greater at Kria – as indicated by the predominantly smaller size of the struck flakes and lack of large retouched flakes. She suggested that this may be a reflection of less proximate raw materials (Pasveer, 2004: 127). Both kinds of use-wear identified during the Holocene phase at Toé Cave were also seen here. In Unit V (and right through the rest of the sequence), 100 per cent (n = 20/20) of the retouched pieces were shown to carry use-wear; n = 33/266 (12.4%) of un-retouched flakes exhibited use-wear or potential use-wear. This is higher than the value Pasveer quotes for the site as a whole (n = 56/659, 8.5%) and comfortably higher than that within any other of the site's units, implying perhaps further differences in utilization through time.

Pasveer concluded that both of these lithic assemblages are fairly typical of those recovered from most New Guinean sites: high levels of flaked débitage with little indication of formal types, even when pieces display secondary retouch. The basic reduction sequence itself remained the same and could perhaps be described as standardized, though with the same caveat noted with respect to Flores (see earlier). Some subtle differences and shifts over time were retrievable; the significance of these is less clear, though, other than the likelihood that distance and availability probably weighed on decisions about transporting lithic raw materials. There is nothing to indicate that ground stone pieces, such a prominent feature of Late Pleistocene and Holocene New Guinea (e.g. Bulmer, 2005), were employed by the people using either of these two caves – a fact that may have wider connotations about the spread (or not) of early landscape modification and cultivation practices on this island.

Aru Islands. Lastly, we turn once again to evidence from the Aru Islands along what would have been the coastal margin of the Sahul Shelf. The flake industry from Liang Lemdubu shows marked continuity in its make-up right through the site's sequence. In his analysis, Hiscock (2006) determined that although the overall size of recovered flakes is small (ranging from 2.7 mm to 45.4 mm), only *c.* 2 per cent of the assemblage consists of pieces <5 mm. Having discounted the possibility that this could be a recovery issue (i.e. bone fragments of this size were recovered during sieving), he concluded that primary knapping was probably being carried out elsewhere in the landscape – a position also supported by an absence of cores. This evidence, in conjunction with the low overall frequency of lithics from Lemdubu (total n = 527), suggests that the site was probably only ever visited for short periods of time throughout the Last Termination. Only four retouched tools were identified from the site, found across three spits (spits 7, 10 and 15).

Three changes were noted. Firstly, there was a shift in raw material preference that may mark changes in occupation at the end of the LGM. While chert dominates in the lower spits, from spit 11 upwards, it is gradually replaced by partially silicified limestone, though there is no significant evidence that this was accompanied by any change in manufacturing techniques or products (Hiscock, 2006: 223). Secondly, comparison between the occurrences of lithics to bone indicates that during the earliest occupation, a lot of bone was being deposited for comparatively few lithics. Thereafter, the ratios became more balanced, possibly marking a change in site function or patterns of discard around the end of the LGM. Thirdly, there is an observable increase in flake dimensions (particularly width) through time. This means that the degree of flake elongation declines through the sequence, reaching its weakest expression in the upper levels. In other words, there is no indication that the industry either contained or was developing towards a flake-blade component.

The stone tool industry from Liang Nabulei Lisa (*c.* 15–7.8 cal. KBP) is comparable to that from Liang Lemdubu. The assemblage of 42 artefacts is flake dominated, and once again, there is a marked shift in lithic raw material use. In spit 30 (11–10 cal. KBP) and below, chert was the preferred choice; spit 29 and above it was limestone, possibly reflecting the same shift in procurement strategies across the Pleistocene (Hiscock, 2006).

Summary. This survey of Island Southeast Asian inorganic technologies has revealed several recurrent themes from the Last Termination and Early Holocene. It is apparent that there is a marked tendency towards localised technological traditions and trajectories (e.g. as seen for the Javan, New Guinean, Timorese and Sulawesian data); sometimes, as in the case of the latter, these also have their own internal, possibly geographic, facies. Nonetheless, apparent similarities between some of the industries (especially between Java and South Sulawesi) continue to spark interest about possible cultural links within the wider archipelago. The presence of Maros points on Selayar Island, off the South Sulawesi coast, and the comparable industries from sites in the North Malaku Islands is certain indication that although they were localised, at least some of the societies of this period had maritime capabilities – a fact further supported by the discoveries at Jerimalai and Lene Hara.

Another feature is that lithic industries were predominantly flake based, with varying but generally low levels of retouched formal tool types. Of these, scraper forms are often identified, implying that many of these pieces were probably used in fabricating organic implements and artefacts. Blade-like flakes and formal blades do appear to have existed as a part of several industries, sometimes (as at Leang Tuwo Mane'e and possibly Hagop Bilo) with prismatic cores in evidence. Fine-grained siliceous stone is a dominant utilised raw material. The distribution and nature of cherts and obsidian through ISEA may well, as Bulbeck *et al.* (2000) proposed, partly underwrite the character of these industries and their tendency towards being microlithic. It is the case that formal types do emerge at this time (and subsequently, as we see with the points from Timor and Java, and the Maros points from South Sulawesi), but simply having access to quality flakable materials does not seem to have dictated this; amorphous components persist alongside more formally standardized ones. As Bellwood (2007: 198) observed, there is no wholesale replacement of flake industries by blade-based ones despite the availability and use of fine-grained materials. Also, if lithic reduction sequences are a poor proxy for the species that produced them in this part of the world, the absence of morphological lithic types cannot be taken as any representation of a community's level of technological sophistication (so often the assumption in traditional Palaeolithic archaeology). As the evidence from this region shows, groups may standardize production in one medium, such as shell (Timor and Gebe Island), yet maintain minimally standardized pieces in other media.

Amongst the developments that do occur during the period from the Last Termination to Early Holocene, some appear to have been linked to plant processing: silica gloss starts to be recovered from utilised flakes in many assemblages, grinding and pounding stones also. While plant exploitation will in and of itself not be anything new, such finds do allude to the founding of new techniques of harvesting and processing (be it for consumptive and/or fabrication purposes).

Other developments in the stone tool technology have been ascribed to shifting patterns of hunting (e.g. Fox, 1970: 50). Following an exhaustive study of the Toalean, Bulbeck *et al.*

(2000: 100) concluded that the differences they observed within this industry can be viewed as a microcosm of wider regional variability in flake-blade industries, involving divergent, localised solutions to improving projectile technology. Whether we can be as specific as this is a matter of opinion, though it would certainly seem that a significant technological realignment did indeed take place during this time, and it was one that was not only confined to lithics either. The overriding sense that one gets is that while these new developments were contingent on changing circumstances, they did not replace pre-existing technological elements; they developed from them or were added to complement them. People tended to revisit and reinvigorate parts of their existing technology to meet new local challenges, and although wholly new techniques (e.g. fabricating ceramics) were probably spreading by contact and trade as much as local innovation, these too were grafted into that existing ethic.

There are also numerous examples from ISEA of long-term continuity in basic lithic reduction techniques, often with different parts of the landscape used to conduct different portions of that sequence. Fox (1978) concluded that there were no fundamental changes in the production or identified tool types from the earliest levels to the final period of occupation at Tabon. A similar position has been adopted more recently by Patole-Edoumba (2009) following technological analysis of six flaked lithic assemblages from the islands of Palawan and Luzon. Glover (1986) saw this in the evidence from Timor, Pasveer from New Guinea, and Chapman (1986) for the Toalean. The basic forms of these industries appear to have remained locally consistent for thousands of years and, from the Flores evidence, seemingly crossed hominin species boundaries. Whilst on the face of it, the evidence from Mainland Southeast Asia will seem to tell a different story, the underlying themes will turn out to be surprisingly similar.

6.4.2.2 Mainland Southeast Asia

Across the Pleistocene–Holocene transition, human occupation of the peninsulas and interior of the Southeast Asian mainland appears to have undergone a shift from more flake-based industries (as seen, for example, at Lang Rongrien and Nguồm; *see* Chapter 5) to ones in which pebble tools featured prominently. What is not obvious at the moment is to what degree this shift has been affected by inconsistent recovery strategies and the hegemony of more typologically definable components over more amorphous ones. Taking the available evidence at face value, it appears that right across the mainland and possibly even beyond (e.g. Bowdler, 2006; Zöller, 2000) there emerge numerous lithic industries whose typological comparisons, range of tool components, raw material use and associated subsistence practices have resulted in them being portrayed as techno-facies of the Hoabinhian of northern Vietnam (Bellwood, 2007). This section sets out the basic criteria for defining the Hoabinhian and considers the significance of the variability that seems to exist across its wide distribution, with detailed reference to Vietnam, Malaysia and Thailand, where the mainland archaeological record from this period is best understood.

Vietnam. In 1923, whilst working for the Geological Service in French Indo-China, M. Etienne Patte discovered a flaked stone implement in a ford near the limestone massif of Bac Son in Lạng Són Province, along Vietnam's northeast border (van Stein Callenfels and Evans, 1928). Subsequent survey (1926–31) by Mansuy and Colani of the caves of this massif and in the neighbouring provinces of Hoà Bình, Ninh Bình, Thanh Hoá and Quãng Bình produced large numbers of stone artefacts in association with what they described as 'kitchen-midden' deposits. With their study arose the beginnings of a relative chronology of prehistoric occupation in northern Vietnam, the emergence of the Bac Son culture and its predecessor the Hoà Bình culture. Our chief concern here is with the latter because it is only this entity that has been described as extending far beyond the borders of Vietnam.

The Hoabinhian has been variously defined as a 'Mesolithic culture' (Matthews, 1966), a 'tradition' (Dunn, 1975), a 'techno-complex' reflecting a common ecological adaptation and basic 'stage' from which agriculture developed (Gorman, 1971, 1977) or broadly as a 'tradition' of upland foraging (Higham and Maloney, 1989). Most recently (1993–4), at a conference to commemorate Colani's work, the term 'industry' was adopted (Shoocongdej, 1996), or rather re-adopted, as this had also been the conclusion of the 15 participants at the First Congress of Prehistorians of the Far East, held in Hanoi in 1932, where it had been defined as follows:

> A culture composed of implements that are in general flaked with somewhat varied types of primitive workmanship. It is characterized by tools often worked only on one face, by hammerstones, by implements of sub-triangular section, by discs, short axes and almond shaped artefacts, with an appreciable number of bone tools. (cited in Matthews, 1966: 86)

Its technological component continues to be a defining characteristic of the Hoabinhian (Ha Van Tan, 1997; Marwick, 2008). Chief among its tool types is a unifacially flaked pebble core-tool class (Figure 6.15). It is with a slight amount of irony and testament to the perceived wider regional nature of this industry that this type tool is not named after a Vietnamese site or setting at all but after comparable finds made in the East Coast Province of Dutch Sumatra by van Stein Callenfels in 1925–6: the 'Sumatra type' (van Stein Callenfels and Evans, 1928: 149; or now 'Sumatralith'). It is defined as 'a flat, longitudinally-split pebble which has been centripetally flaked to produce a disc-like form' (Reynolds, 1990: 8), though this does not appear to have incorporated a deliberate intention to impose a prescribed shape. Ha Van Tan (1997: 37) has gone so far as to consider unifacial flaking of these implements as no less crucial to confirming affiliation. This criterion becomes an important issue, especially when one contemplates the Hoabinhian beyond the borders of Vietnam, as we shall see.

Another key tool type is the 'short axe'. These are described as unifacially or bifacially flaked pebbles, with the particular characteristic of a flat facet created through splitting the pebble along the transverse plane. What is less certain is whether these represent deliberate forms in their own right or whether they are broken fragments of larger implements – if the latter, though, it would seem to imply considerable redundancy in the circumstances of breakage to repeat the same form of break time and again. Additional core-tool classes include choppers and scrapers. These are variously labelled but usually distinguished by differences in edge angle. Flakes also make up a greater or lesser proportion of most assemblages (e.g. *see* Ha Van Tan, 1999). For a long time, the flake component was almost certainly left under-represented, creating a source of bias when it comes to comparing older with more recent excavation reports. Many flakes appear to have been utilised, but there is little or no evidence of morphological standardization, and secondary retouch is rare. Flakes with blade-like proportions do periodically occur, for example, as seen in the industry from Xóm Trại (see later). The final category of worked stone tool (as opposed to *a posteriori* implements such as hammer stones and pestles) are the so-called 'proto-neoliths' – a term that presupposes these to be forerunners to Neolithic fully ground stone pieces, which is by no means certain. These are partially or specifically edge-ground implements that, while common in the later Bac Son assemblages, do also occur in small numbers in some Hoabinhian ones and may have considerable antiquity. Trình Năng Chung (2008) asserts that edge-ground pieces recovered from Mai Da Điều and Xóm Trại caves push this preparation technique back to *c.* 16 kbp. Indeed such implements are listed as one of the defining criteria used within the tripartite system of subdividing the Hoabinhian that was also proposed at the 1932 congress. The Hoabinhian I was composed of large, crude flaked tools; Hoabinhian II was described as

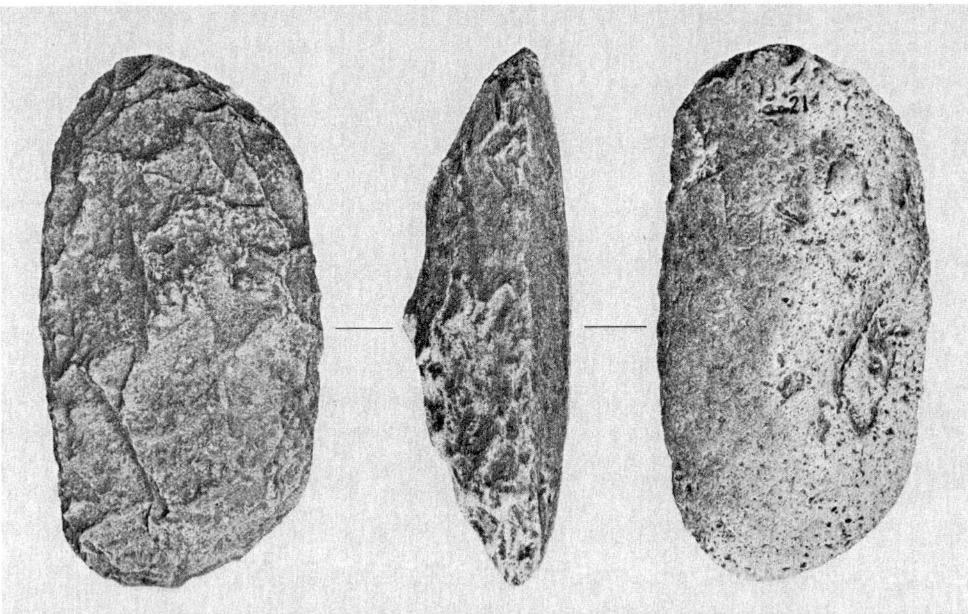

Figure 6.15. An example of a unifacial, centripally flaked pebble tool or 'Sumatra type', 16 × 84-mm, from Gunong Pondok, layer G, 3.56 m below datum. (Reprinted from van Stein Callenfels and Evans, 1928, plate LX, *Journal of the Federated Malay States Museums* 12.)

containing smaller, better-made pieces and proto-neoliths; Hoabinhian III carried tools that were smaller again and contained retouched pieces but no proto-neoliths. Chronological support for these substages, though, has generally not been forthcoming (Reynolds, 1990). It is complicated further by the fact that flaked implements have been found to persist alongside metal implements into the historic period (Davidson, 1979; Shoocongdej, 1996). Re-excavation and reassessment of many of Colani's sites by the burgeoning Vietnamese archaeological research community in the 1960s and the application of radiocarbon dating brought a degree of order to the country's prehistoric sequence, though the position of the Hoabinhian within that sequence remains fluid to this day. It is generally agreed to be dating broadly from the 'late to terminal Pleistocene to the early to mid-Holocene', but without any clear geographic boundaries apparently defined (Solheim, 1994: 10).

Technological trends in the late Sonvian (named after initial finds made near the Son Vi hamlet in Phú Thọ Province in 1968 and dated to between 23 and 13 kbp) seem to indicate that one or more facies of this industry gave rise to the Vietnamese Hoabinhian during the first half of the Last Termination (Ha Van Tan, 1976, 1997, 1999). The archaeological sequence at Xóm Trại Cave, located in the Muong Vang Valley of Hoà Bình, is considered to have produced one of the most classic examples of the Hoabinhian (Ha Van Tan, 1999). As noted earlier, twenty radiocarbon dates span a tight late LGM occupational sequence, including thick hearth deposits, from 22,012 ± 211 cal. BP (Bln-3472) to 19,259 ± 114 cal. BP (Bln-3042). A large proportion of the lithic industry from here is made up of partially flaked pebble tools, with a further proportion having all cortex removed or bearing ground edges. Large flake tools are also represented, as are appreciable numbers of small flakes (30–50 mm) and blade-flakes (Ha Van Tan, 1999). In the same valley, two other sites, Lang Vanh and Xom Tre, have comparable tool sets to Xóm Trại and have returned similar dates. Nguyen Viet (2000) has interpreted this as exemplifying the cool-adapted 'Typical Hoabinhian' phase. Also now seemingly belonging to this period is Tham Khuong Cave,

Lai Chau Province (*c.* 21.583333°N, 103.316667°E). Hoabinhian evidence from this site had been putatively and controversially dated to as early as 33,150 ± 2500 bp (Bln-1412) on land snail shell (*Cyclophorus siamensis*), which was questioned at the time (Kohl and Quitta, 1978). Re-examination of the site's record has produced a new, more consistent date of 15,800 ± 150 bp (HCMV-03/93; Ha Van Tan, 1997).

Following his climatic classification schema, Nguyen Viet (2000) proposed that the well-dated sequence at Con Moong Cave (*c.* 20.3°N, 105.9°E), in Cuc Phuong National Park – on the border between Ninh Bình, Hoà Bình and Thanh Hoá provinces – contains a late variant of the industry (one that was warm adapted) from 12,136 ± 132 cal. BP (Bln-3485) to 9500 ± 34 cal. BP (Bln-3482). Nine layers were differentiated within a 3.5 m series of deposits during excavations in 1975–6, purportedly containing cultural material of Sonvian, Hoabinhian and Bacsonian affiliation at different levels, with the Sonvian (Layer 9) dated to 11,755 ± 75 bp (Bln-1713) on land and freshwater shells (*C. siamensis*, *Antimelania siamensis* and *Lanceolaria* sp.). The accuracy of the age was confirmed by a second lab dating it to 11,090 ± 185 bp (ZK-379; Kohl and Quitta, 1978), though it is unclear and perhaps unlikely that either was corrected for carbon uptake bias, and also, not all authorities agree that the site contains three separate industries; Nguyen Viet (2000) suggested the entire sequence to be Hoabinhian. In addition to the bone points recovered from Con Moong, its Hoabinhian industry included freshwater shell scrapers, large numbers of flakes (some retouched), unifacially flaked pebble tools and a small number of edge-ground pieces (Ha Van Tan, 1999; Pham Huy Thong, 1980; Pham Huy Thong *et al.*, 1990). If these dates and interpretations are correct, the Hoabinhian existed alongside the Sonvian for some considerable time – something that must raise questions about the likelihood of the one giving rise to the other in any simple unilinear sense. It is interesting to note that renewed systematic excavations carried out at Con Moong in 1986 produced only two flaked pebble tools. The apparent emphasis on flakes at this site is mirrored by recent excavations in the near vicinity.

Since 2007, the current author has been excavating in Tràng An, a tourist park development centred around an isolated limestone massif, in Ninh Bình Province on the southwestern edge of the Song Hong delta. This park covers an area of 2400 hectares and is managed, in conjunction with the provincial Peoples' Committee, by the Xuan Truong Construction Company, both of whom have been sponsoring investigation into the early human settlement of this landscape. At its outset, archaeological work in Tràng An focused on the excavation of loose shell midden deposits at Hang Boi, a cave in the northwest corner of the park (20.258889°N, 105.888056°E) 78 m asl. Excavation was concluded here in 2010, when the base of the *c.* 4 m thick midden was found to overlie ancient roof fall. Laboratory analysis is ongoing, though radiocarbon dates obtained on charcoal through sequence indicate human occupation spanning the period from *c.* 13,601 ± 43 cal. BP (UBA-14887) to 10,620 ± 64 cal. BP (UBA-8371; Rabett *et al.*, 2009a, 2011; *see* Chapter 7).

Although the depositional context differs slightly, excavations (commencing in 2009) at the nearby cave of Hang Trống (20.250444°N, 105.890111°E) are providing an excellent parallel sequence that not only appears to overlap chronologically with that at Hang Boi but also now extends human activity in this landscape back to at least 18,549 ± 62 cal. BP (UBA-14886) and probably to *c.* 24,438 ± 93 cal. BP (UBA-17272; Rabett *et al.*, ms. in prep.). The latter is the deepest charcoal date yet obtained, taken from yellow colluvial deposits beneath the site's shell midden and still containing hints of occupation.

The archaeological evidence from both caves has been tentatively attributed to the Hoabinhian, though while small numbers of burnished pebble tools have been found (Figure 6.16), no classic lithic forms (such as Sumatraliths or short axes) have come to light at either site. Forty-nine stone pieces were found to exhibit evidence of deliberate modification in the 2007 study sample from Hang Boi (Rabett *et al.*, 2009a: table 5); 31 fragments came from

Figure 6.16. A pebble tool still *in situ* in the Deep Sounding of the Hang Boi excavation. (Photograph by L. Farr, reproduced with kind permission.)

the 2008 sample (Rabett *et al.*, 2011: table 5). Flakes predominate in each case ($n = 34/49$ and $n = 19/31$, respectively). Reduction appears to have been exclusively done by hard hammer technique; there is no evidence of secondary retouch and few signs of use-wear, though high magnification work has not yet been carried out. Edge damage indicative of trampling is present in some cases. Four artefacts, including two probable burin spalls, have blade-like proportions. The presence of burin spawls in the assemblage would suggest that some manufacturing activities took place at Hang Boi; a possibility made more likely by recovery of a worked bone fragment in 2008 and in 2009 by the site's only bone point (*see* Rabett and Piper, 2012). Among the few cores that have been recovered, almost all exhibit single striking platforms. The absence of micro-débitage (despite fine screening) does leave as unresolved the matter of on-site stone tool manufacture, though, as most of the lithic assemblage is made on limestone, distinguishing débitage from naturally occurring rock shatter does complicate this issue. The recovery of a single cortically polished flake near the top of the sequence, probably struck from a larger burnished stone (a recurrent feature at both sites), hints at further dimensions of tool curation, use and discard than are currently appreciated. On available evidence, there was no major retooling or manufacturing of stone tools at Hang Boi and probably an emphasis on expedient tool use (*see* Rabett *et al.*, 2011).

Preliminary analysis of the Hang Trống lithic assemblage appears to be telling a slightly different story. Artefacts recovered from the context dated to 18.5 cal. KBP include a core rejuvenation flake that carries at least two or three earlier hard hammer flake removals; a blade-flake, possibly struck using a soft-hammer technique; a large plain platformed thick secondary flake with slight rounding along the distal edge and evidence on its dorsal surface that many small flakes had been removed prior to this one (Figure 6.17); and a 'steep-edged' core (*after* Reynolds, 1989) that may also have been used as a pounder. Although only at a early stage of analysis, already these pieces are more securely suggestive of some on-site production, including basic core preparation and possible use of soft-hammer reduction.

This may indicate inter-site technological variability, though the behavioural mechanisms behind this difference remain to be considered in detail and could reflect spatial and/or temporal differences as well. The absence of classic Hoabinhian types and the limited emphasis on pebble tools in general does also lead one to ask whether either assemblage should be considered as a Hoabinhian affiliate at all. Both sites seem to accord well with other Hoabinhian cave midden faunal records, though the apparent homogeneity of this subsistence evidence may itself be something of a smokescreen (*see* Chapter 7; Rabett *et al.*, 2011). Comparison to the Con Moong assemblage may prove profitable and link the Tràng An industry to one of the facies there, although given the greater antiquity of the Tràng An sites, linkage to any particular climatic phase (*after* Nguyen Viet, 2000) would need to be considered carefully. Alternatively, the Tràng An industry may even represent a late-surviving example of a more flake-based industry.

This serves to illustrate that there is considerable variability between the assemblages from this period which might otherwise be subsumed under the heading of the Hoabinhian; something that archaeologists have yet to account for in a completely meaningful way. Are these facies of a single industry, or similar but not identical ones, and how much of the variability we see is linked to site function? Over the last decade, ten or so new open-air Hoabinhian sites have been discovered (Trình Năng Chung, 2008); further study of this new settlement component may well begin to elucidate these questions.

The possibility that there were several facies of a broadly similar stone tool industry existing concurrently in northern Vietnam between the Late Pleistocene and Early Holocene is one that Nguyen Gia Doi (2005) considers explicitly following three years' excavation at the Mai Da Điều rock-shelter in the uplands of Thanh Hoá Province. The 5.3 m sequence of deposits at this site produced one of the largest lithic assemblages in the country with *c.* 2000 pebble tools, including choppers, high-domed forms, pitted mortars, scrapers and Sumatraliths, and *c.* 10,000 flakes. While many flakes were apparently struck from large pebble cores, which were also present in the assemblage, a high proportion (84.4%) of the tool types usually assigned to the Hoabinhian are here made on flakes rather than pebbles, a characteristic reminiscent of what was seen at Nguòm (Ha Van Tan, 1999). There is some indication of changes in tool occurrence through time, including an early appearance of edge-ground pieces (Trình Năng Chung, 2008), though there are few specific details available in English-language papers. The sequence at Mai Da Điều is dated (on charcoal) from 23,534 ± 212 cal. BP (Bln 3542–86) to 8838 ± 142 cal. BP (Bln 3541–86), while the latest overall date (on land snail shell) has occupation concluding *c.* 4940 ± 90 bp (NU-1119-95). Another northern example is the Soi Nhu pebble and flake industry. This is recognised as contemporary with, yet distinct from, the Hoabinhian. Evidence of this industry has been recovered from 40 or so sites in the Ha Long and Tu Long Bay areas and adjoining provinces, where it is broadly dated to between *c.* 18 and 14 cal. KBP (Ha Huu Nga, 1998) though possibly as late as the Mid-Holocene (Ha Van Tan, 1999). There are also grounds for considering some kind of socioeconomic relationship and/or historical links between the Hoabinhian and the series of Mid-Holocene coastal occupations termed collectively the 'Dabutian', after the Da But type site, Thanh Hoá Province (Bui Vinh, 1991; Nguyen Viet, 2005; Rabett *et al.*, 2009a, 2011).

One of the most complete and probably most securely dated of the Vietnamese sites attributed to the Hoabinhian thus far is the recently (2004 and 2006) excavated cave of Hang Chổ, in Hoà Bình Province (20.835278°N, 105.652222°E; Yi *et al.*, 2008). Originally investigated by Colani, this site is situated *c.* 10 m above wide river floodplains at the base

Figure 6.17. (left, dorsal and ventral) A limestone blade-flake struck from the edge of the worked face of a core, possibly using soft-hammer flaking, small find: SF17, context (8016). (right, dorsal and ventral) A core face rejuvenation flake (hard-hammer strike) that would have removed a large chunk of previously damaged platform, SF15, context (8016). Both flakes from Hang Trống, Tràng An, Vietnam (scale in 5-mm increments). (Photographs by Ryan J. Rabett.)

of a limestone karstic tower and has produced an archaeological sequence containing an extensive freshwater shell midden, the deposits of which are now anchored against 44 AMS radiocarbon dates. These reveal site use from before 21,634 ± 268 cal. BP (SNU 04-233) until after 12,289 ± 378 cal. BP (SNU 03-156; charcoal dates) – radiocarbon dates on shell extend it to 8430 ± 180 kbp (SNU 04-A011). A test sondage into the base of one of the excavation trenches provided even earlier evidence of human presence at the site. A charcoal sample produced a date of 34,556 ± 247 cal. BP (SNU07-223; Yi *et al.*, 2008: 77).

In contrast to the Tràng An sites, the recovered lithic assemblage is dominated by flaked core tools. The authors do note a minor component of small flake implements, though they observe that these are mostly coming from the upper layers. Artefacts include an end-scraper and a couple of burins, none of which show much emphasis on standardization of form and are described as probably 'opportunistic' (Yi *et al.*, 2008: 78). There are few indications of major technological change through the duration of the sequence, but flake morphology does appear to become more elongated and slightly smaller towards the top of the sequence. Débitage flakes are present but in reportedly low numbers, something that suggests that tool manufacture may, for the most part, have been carried out somewhere else.

The overriding impression that one gets from the Vietnamese data is that while it is possible, maybe even justified, to call many of the Late to early post-Pleistocene sites 'Hoabinhian', there are definitely indications of technological variability at an inter-site level as well as changes through time. This is not a new observation (e.g. *see* Ha Van Tan, 1976: 153–8), though it still awaits a completely convincing explanation, as the true significance of these characteristics has yet to be fully appreciated. It seems possible, for example, that foragers in northern Vietnam developed independent, local facies out of an earlier, more predominantly flake-based industry, and that within these there existed varying degrees of attention to the importance of flaked core tools. Shifting proportions of core tool and flake elements is a recurrent theme, if one that is also affected by different approaches to archaeological recovery. It is certainly the case that tools were not consistently being fabricated on site, and from this it is also clear that any single site assemblage is only going to ever provide part of a wider set of activities being undertaken in the landscape. To achieve a more complete understanding of the Hoabinhian, it will be crucial for sites to be investigated as functionally integrated landscape loci rather than as self-contained archaeological islands, as has perhaps tended to be assumed previously.

Burma, Cambodia and Laos. In Burma (or the Union of Myanmar), although discoveries of prehistoric implements began in the 1870s, the first systematic exploration did not take place until the American South-East Asian Expedition for Early Man in 1937–8 (de Terra, 1938; de Terra and Movius, 1943). Following this, and with most archaeological commitments directed towards the exploration and conservation of ancient city sites, the next dedicated prehistoric expedition did not take place until 1969. Excavation through 1.22 m of deposits in cave A1 of the Padah-Lin Caves (*c.* 21.134169°N, 96.357811°E) provided archaeological evidence from the Late Holocene in Layer 1 (*c.* 0–0.14 m), back to 13,400 ± 200 bp (R-2547/5[B]) in Layer 4 (*c.* 0.77–1.03 m). Most dates, though, cluster between 6 and 7 kbp. The archaeological remains from this site (including 422 of 1600 lithic artefacts) were not apparently differentiated, suggesting that considerably greater clarity could be achieved with further excavation, but on typological grounds, they were ascribed to a Hoabinhian or Bacsonian affiliation (U Aung Thaw, 1971).

Despite considerable effort exerted in landscape survey, to date there is very little evidence for Pleistocene occupation in Cambodia, leaving its earliest certain evidence of settlement coming from the Holocene site of Laang Spean (*c.* 12.882878°N, 102.933203°E), Battambang Province (Stark, 2004). Three areas of this cave were excavated by Cécile and Roland Mourer between 1966 and 1968 (Mourer and Mourer, 1970, 1971; Mourer, 1977). Deposits in the central area of the cave proved the deepest, providing a dated sequence from 1120 ± 90 cal. BP (MC-270; on charcoal) back to 9860 ± 1154 cal. BP (LY-439) and spanning five cultural levels (and six stratigraphic ones; Mourer, 1977). The second of these dates, however, was taken on bone collagen and should be treated circumspectly, as Mourer notes (*see also* Walker, 2005: 31). The earliest charcoal date comes from the overlying layer (Layer 4): 7158 ± 91 cal. BP (MC-273). Interestingly, the lithic industry recovered from the base of the cultural sequence (Cultural Level I) comprised almost exclusively of unclassifiable chert flakes (Mourer and Mourer, 1970), possible evidence that flake-based industries persisted into the Holocene in some parts of the mainland, and further indication that raw materials did not slavishly dictate technological form in this region. The occurrence of chert declines as one ascends through the cultural sequence in favour of metamorphic rocks (chiefly hornfels), sandstone and quartzite, inferring that a shift in raw material access or procurement choice occurred during the Early Holocene, though there is no clear indication of what might have prompted it.

Cultural Level II (Layer 4) sees the appearance of a range of flaked core tools, including pieces classified as side-scrapers, Sumatralith-like pieces, short axes and end-scrapers, cores, discoid pieces and choppers. This level also sees the first instances of ceramic sherds appearing alongside this industry. Overall, the site's lithic assemblage is dominated by débitage ($n = 7725/8502$, 91%); flakes make up the next largest technological category ($n = 743/8502$, 8.7%), but among these, only one was found to carry pronounced (bifacial) secondary retouch, with an undisclosed proportion exhibiting traces of use-wear. Although typologically more identifiable, the pebble-derived core tools accounted for only $n = 34/8502$ (0.4%) of the total assemblage. The relative proportions of the different components, though, were taken to indicate that tool manufacture was taking place on site (Mourer, 1977).

After early exploration of the Annamite Mountains in northern Laos in the 1930s by Jacques Fromaget of the Service Géologique de l'Indochine, Hanoi, very little archaeological research was carried out in this country until as recently as the mid-1970s (Bacus, 2001). Fromaget's reports of finds in Hua Pan Province, in the country's northeast, certainly gave reason to believe that systematic cave exploration in the area would prove productive. One of the most promising sites here is Tam Hang, actually comprising a group of three rock-shelters beneath a limestone cliff – Tam Hang North, Central and South (Bellwood, 1979). Part of the sequence at Tam Hang South is reported to have contained up to 3 m of deposits with a lithic industry of Hoabinhian character, including unifacially flaked core tools. The site recently became the focus of renewed excavation by a joint Laotian–French team following its rediscovery in 2003. Already producing significant results in terms of human remains dating to the Last Termination and two distinct Pleistocene faunal sequences (Bacon *et al.*, 2008; Demeter *et al.*, 2004); future publications from this project are awaited with interest.

Malaysia. At the same time as discoveries were being made in Indo-China in the early decades of the twentieth century, British colonial officers were also beginning the first systematic excavations of the coastal middens, caves and rock-shelters of the Federated Malay States. Prehistoric investigations in Malaya date back to the early 1860s, with the mostly anecdotal observations of Earl (1863) into shell middens along the northwest coast in the then Province Wellesley. Van Stein Callenfels' (1936) analysis of three shell mounds at Guak Kepah, in the same vicinity, were considerably more scientific in approach: his geological assessment situated the sites on the shore of an ancient estuary, and some attention was given to assessing how the middens may have been formed. His chief interest, though, was with the cultural remains. These included grinding and pounding stones, 'Sumatra-types' and a bifacially flaked axe form, which may have been a short axe, and edge-ground axes. He concluded that the Guak Kepah industry was a variant of the Vietnamese Hoabınhıan – an affiliation, in part, influenced by the sites' proximity to Sumatra, where he had previously worked. This was 'as could be expected from the finds in the middens of East Sumatra, the Province Wellesley shell-heaps yielded the Malayan type of Hoa-Binhian culture' (van Stein Callenfels, 1936: 31). One of the shell middens on the Sumatran side of the Malacca Strait (Sukajadi Pasar III) has since been accorded a radiocarbon date partway through its sequence of 8162 ± 356 cal. BP (SUA-1107; Edwards McKinnon, 1991). Van Stein Callenfels' opinion, though, was also influenced by discoveries that he and I. H. N. Evans had made during their recent excavation of caves in the Gunung Pondok massif, Perak (van Stein Callenfels and Evans, 1928; Evans, 1928).

Interest in the archaeological record of the interior of the Malay Peninsula had also begun early. Between 1886 and 1891, the first curator of the Perak Museum, Taiping, L. Wray, had conducted excavations at Gunung Cheroh, outside the town of Ipoh, Perak, enthused by the surveys of Everett in Sarawak in the late 1870s and by his own initial discovery in 1880 of prehistoric habitation in Gunung Pondok (Wray, 1897, 1905). Organised investigations resumed under the guidance of Evans (Curator at Taiping from 1917 to 1932, e.g. *see* Evans, 1918, 1920, 1929, 1931). His collaborative work with van Stein Callenfels (1926–7), particularly at the rock-shelter of Gua Kerbau, in Gunung Pondok (*c.* 4.785861°N, 100.841083°E; Figure 6.18), marked the first systematic excavation carried out in the peninsula, reaching an eventual and surprising depth of 7.3 m (Evans, 1928). Their conclusions were tentative but included positive identification of Sumatra-type implements, proto-neoliths and other 'palaeoliths' (unpolished flaked pieces, including a unifacial point-form from 1.86 m below the surface as well as grinding and pounding stones, which occurred throughout the deposit). Notable also was an apparent distinction between the choice of raw material used in making the flaked implements and that used to make the Sumatra-types. The former was made on a 'dark grey rock' (the aforementioned point was made on biotite-hornblende schist), while the greater proportion of the latter were made of quartz. Ceramics were concentrated in the upper layers, with some fragments also appearing at depth, though animal disturbance was pronounced at the site and translocation of material would have been a distinct possibility.

The incidence of marine shells in the upper four of the seven layers excavated and their reduced presence and ultimate disappearance in the layers beneath suggests that the occupation may have spanned the Mid-Holocene high-stand (the Gunung Pondok is currently *c.* 25

km from the coast in a straight line). Van Stein Callenfels and Evans (1928: 158) concluded, based on this evidence and the discoveries made in Dutch Sumatra and Indo-China that 'a palaeolithic civilisation, making use of Sumatra-type implements, probably spread at a certain period through the south-eastern parts of the continent of Asia and even to Sumatra'. It is a curious historical fact that while this idea has been perpetuated in the interpretations of the wider continental distribution of the Hoabinhian, a similar contention, shared by themselves and others (e.g. Heine-Geldern, 1936: 29), that the Bacsonian also (as a 'palaeo-proto-neolithic stage') had spread overland, without reaching Sumatra, was never formally adopted outside Vietnam (*though see* Reynolds, 1992; U Aung Thaw, 1971).

The fledgling prehistoric research agenda of 1930s Malaya was intent on revealing reveal further traces of 'prehistoric migrations through the Malay Peninsula' (van Stein Callenfels and Noone, 1940: 119) by expanding on the small existing number of systematic excavations aside from Guak Kepah and Gua Kerbau (*see* e.g. Collings, 1936, 1937; Tweedie, 1936). It was felt that to achieve this necessitated excavation on both sides of the Titiwangsa Mountains (that extend 480 km south from the Thai border down the spine of the peninsula), and work undertaken at the rock-shelter of Gol Ba'it, Perak (*c.* 4.861111°N, 101.125°E) in 1936 opened this initiative (van Stein Callenfels and Noone, 1940). Two results of this excavation are particularly pertinent. Firstly, the incidence of burials in the upper, middle and lower parts of the sequence established for the first time in the Malay Peninsula and across the Indonesian archipelago affirmative evidence of changes in burial practice through time; with modern burials in the upper layers, secondary burials in the middle deposits and flexed burials attributed to the Hoabinhian *c.* 2.1 m below the surface (archaeologically sterile deposits were reached at a depth of 3.5 m, preceded by low frequencies of flakes).

Secondly, the excavation revealed indications of change in lithic raw material preference over time. Van Stein Callenfels and Noone (1940: 121) relate that '[those] who practised the custom of "secondary burial" preferred, as a material for their implements, fissile rocks such as slate and schist, whilst although these schist implements occurred lower down, the majority of the artifacts from the lower layers . . . were made of quartzite, chert and other non-fissile rocks'. A tanged schist 'chopper' is reported from the middle layers, for which the authors drew parallels to similar pieces they had excavated at another site on the peninsula and from Galoempang (= Kalumpang) in west central Sulawesi. The Kalumpang sites date to *c.* 3 cal. KBP and have themselves been indirectly linked on the basis of ceramic motifs to the Vietnamese Sa Huynh culture from approximately the same period (Bulbeck, 2004), raising an interesting, if currently tenuous, piece of evidence for intra-regional communication during the initial Late Holocene. The presence of worked cowrie shells (*Cypraea lamarckii* and *Cypraea caput-serpentis*) from Gol Ba'it in what look to have been the middle layers (0.76 and 1.31 m) was a feature also recorded at a number of other mainland sites (*see* e.g. van Stein Callenfels and Evans, 1928; Nguyen Lan Cuong, 2007; Rabett *et al.*, 2011), and certainly an indication of coastal links from at least the Early Holocene (Gol Ba'it is currently *c.* 60 km from the coast, though it probably would have been somewhat closer during this phase of its occupation). Shell modification for tool production, as evidenced, for example, at Gua Bintong (Cyprenidae; Collings, 1937), and a single instance at Gol Ba'it also seems to have been a Mid to Late Holocene phenomenon on the peninsula. Although published details about the full nature of the lithic industry from Gol Ba'it are scant, Sumatraliths were found in the lower layers of the excavation and were deemed associated material culture to the flexed burials, making them Hoabinhian, but with only a couple of exceptions, all of the other recovered stone artefacts (including choppers, flakes, grinding stones and pounders) were also linked to Hoabinhian types. This was despite the presence of ceramics to varying degrees of concentration throughout the deposits and the apparent changes in burial tradition.

Figure 6.18. An early stage in excavations at Gua Kerbau in 1926. (Reprinted from van Stein Callenfels and Evans, 1928, plate LVIII, *Journal of the Federated Malay States Museums* 12.)

Unfortunately, published details of the stone tool components from two other key sites excavated on the peninsula in following decades – Gua Cha, Kelantan (Sieveking, 1954), and Gua Kechil, Pahang (Dunn, 1964) – are also limited. In the case of Gua Cha, this was because the second half of a detailed site report (that containing data on stone tools and ceramics) unfortunately never materialised. Limited re-excavation of the site in 1979 (Adi, 1991) confirmed earlier preliminary remarks about the presence of morphologically varied bifacial pebble tools (i.e. Sumatraliths) in association with Hoabinhian deposits made by Sieveking, as well as the presence of débitage, but also noted that retouched flakes were numerous throughout the deposits. Adi (1991) did successfully allocate radiocarbon dates to the Gua Cha record. In so doing, he was able to show that the transition from Hoabinhian to Neolithic probably occurred here *c.* 3197 ± 284 cal. BP (ANU-2217). Dunn (1966) established the end the Hoabinhian somewhat earlier at Gua Kechil at 5457 ± 967 cal. BP (GX-0418) – at that time, only the second radiocarbon date to be published for Malaysia. The initial Hoabinhian occupation of Gua Cha predated 7207 ± 190 cal. BP (ANU-2218), an age that was obtained on charcoal from mid-way through the Hoabinhian deposits. Unlike Sieveking, Adi (1991) saw no significant break in deposition between the Hoabinhian and Neolithic occupations at Gua Cha, implying that observed changes marked indigenous cultural developments rather than being the result of an incoming population; probably taking place in the context of wider exchange networks, as evidenced by the presence of worked marine shells (*Mytilus viridis*) in Neolithic burials at the site (Sieveking, 1954: 88). With reference to more recent advances in Malaysian Palaeolithic archaeology and the place of technological change within that, it is useful here to briefly pause and consider the wider context in which this work was and continues to be conducted.

Archaeology in Malaysia has been said to have passed through four distinct research periods (Adi, 1987), and it is fair to say that perceptions have changed through those periods. This is not something by any means unique to Malaysia, but it so happens to be well exemplified and documented for here. The first three of these periods we have already briefly touched upon. Initial investigations were cursory and very much in the manner of nineteenth-century antiquarianism. The appointment of a succession of curatorial posts heralded a much more rigorous phase of inquiry and the first systematic excavations on the peninsula. The intellectual context during this period switched from curiosity to grand narrative and a perception that Malaysia was a thoroughfare of migrating, diffusing populations – aided by skeletal evidence that compared well with modern Melanesian groups (e.g. Tweedie, 1957) – on a unilinear evolutionary path towards neolithisation (Heine-Geldern, 1936). Incorporation of the later prehistoric Malaysian record, based strongly on its lithic component, into the Vietnamese schema followed from this as the main point of regional comparison; local innovation was rarely considered. After Malaysia's independence in 1957, this position began to change; focus shifted to be more closely concerned with internal trajectories of development as well as a much stronger scientific approach to the subject matter (*see* Dunn, 1964). At least in part, these changes were not unsurprisingly allied to a new sense of national identity and concern over cultural heritage. Over subsequent decades, interest in the cultural and ethnic origins of settlement of Peninsular Malaysia would become strongly instituted at governmental level (e.g. with the Antiquity Act of 1976 and the National Heritage Bill of 2005), while Malaysian archaeologists and the discipline as a whole also became established at a research and academic one (Adi, 1991). It has been in the context of this 'fourth phase' (Adi, 1987; *see also* Leong, 1999) that Malaysian archaeologists have applied themselves to a re-examination of Malaysia's Palaeolithic record. By the late 1980s, prehistoric archaeology on the peninsula was conducted primarily by research groups from the Archaeology Division of the Museums Department, Kuala Lumpur, and particularly from the then newly created Centre for Archaeological Research at Universiti Sains Malaysia (USM) in Penang (Leong, 1999).

Under the direction of Zuraina Majid, USM focused especially on tracing the early prehistory of the country. Working in Perak state, in the northwest of the peninsula, they have been conducting extensive survey and excavation work in the Lenggong Valley for the last 20 years (*see* Zuraina, 2003, 2005) – possibly the only such study of its kind in the region. Of particular interest here is the reassessment of Late to post-Pleistocene technological trends that has emerged out of this work. The revised model that Zuraina and her colleagues have proposed draws especially upon the technological character of the lithic assemblages from Kota Tampan (*see* Chapter 5) and from those recovered in major excavations at another nearby but much later site in Lenggong, Gua Gunung Runtuh (5.1175°N, 100.9675°E).

The Gua Gunung Runtuh (GGR) cave excavations (1990–1) are most widely known for the near-complete flexed male skeleton that was recovered from deposits here, bracketed to an age of between 9460 ± 90 bp (Beta-37818) and 10,120 ± 110 bp (Beta-38394; Matsumura and Zuraina, 1999; Zuraina, 1994, 2005), 0.6–1.0 m into the cave's deposits. Layers above the skeleton are undated until 0.2 m below ground level, where a Late Holocene age of 2620 ± 80 bp (Beta-37817) has been obtained. The base of excavations was reached 2.1 m down and at an age of 13,600 ± 120 bp (Beta-38338). All of the dates are listed as taken on freshwater shell (probably *Brotia* sp.) and do not appear to have been corrected for carbon uptake; thus a predicted calibrated range can only be given in the broadest sense, from about 15 to 2 KBP, with the burial taking place sometime around the Pleistocene–Holocene transition.

Analysis of the artefactual evidence from GGR was undertaken as two separate studies. The first was with reference to material excavated from within the same depth range and in the immediate spatial context as the skeleton. These were considered to be directly associated with the burial itself and labelled as 'mortuary objects', of which ten were lithics (three hammer stones, two 'slabs', one 'oval unifacial' pebble tool and four miscellaneous pieces). While the inhumation's right hand does appear to be grasping a small collection of animal bones, surety that the full collection of mortuary objects are indeed that is somewhat weakened by the absence of a recorded grave cut, the similarity of this material to lithics and faunal remains found elsewhere in the cave, and the comparable patterns in the vertical distribution of those remains. Such a burial is not, though, without precedent in the region, as a similar situation is reported with respect to lithic associations for two Last Termination inhumations at Tham Lod rock-shelter, northwestern Thailand (Shoocongdej, 2006; see later).

Owing to surface disturbance across the upper strata of much of the cave area at GGR, the bulk of technological (and faunal) analysis was carried out on finds made 6–14 m away from the inhumation but commencing at a similar depth of 0.5 m and extending to a maximum of 1.8 m. Excavation of the undisturbed deposits revealed discrete clusters of functionally related lithics (hammer stone, anvil, cores, débitage, flake and pebble tools) and, therefore, good reason to believe that tool production was taking place in this cave (Zuraina *et al.*, 1994: 154). The comparatively low frequency of material – $n = 622$ (including artefacts and débitage from disturbed and undisturbed areas of the cave) – suggests, however, that this was never an intensive on-site activity. Quartz and quartzite were the principal raw materials, accounting for *c.* 40 and *c.* 29 per cent, respectively, of material from undisturbed contexts away from the burial trenches. Slate, sandstone, limestone, granite, iron oxide and chert each composed less than 10 per cent of the total, and in the case of the final three, this dropped to less than 5 per cent. In most cases, the wider variety of raw materials was reserved for hammer stones and what were termed 'oval unifacial' pieces.

As no differences were discerned either spatially or stratigraphically, the GGR material categories were assessed as a whole. Approximately 27 per cent ($n = 18/66$) of the flaked pebble tools (total from undisturbed layers) were classed as 'oval unifacials'. Aside from variability appearing in raw material use, it was also observed in the invasiveness of flaking, which in some cases removed all trace of cortex from the one surface. These were described as being similar to a Sumatralith. 'Oval bifacial' pieces comprised a smaller proportion of the industry ($n = 6/66$, 9%). These were made of quartz, quartzite and slate only, though this exclusivity may well be a product of the small sample size. Two cores were identified, both made on quartzite and carrying a minimum of three flake removals. 'Palaeo-adzes', 'choppers' and 'perimeter-flaked' pieces and 'flake tools' were also noted but not described in detail. Hammerstones made up the greater part of the identified pieces ($n = 29/66$, 44%); most were made of quartz ($n = 19/29$), but other stone types were also noted. No flake tools were recovered from undisturbed contexts. Quartzite flakes dominated the débitage component of the overall assemblage ($n = 225/399$, 56% of débitage flakes) but were not quantified solely from undisturbed contexts.

The most distinctive trait of this assemblage and other contemporary assemblages excavated by USM in Lenggong was how closely they resembled the industry from Kota Tampan and how readily its components fitted into the morphologically and functionally based classification system devised at that site (*see* Zuraina, 1990: 80). This similarity led the excavators to conclude that there was no major change in lithic technology from *c.* 74,000 years ago until the Mid-Holocene. With the potential antiquity of similarly appointed industries at sites like Bukit Jawa and Lawin, in the Lenggong Valley, exceeding even this time depth, Zuraina viewed the term 'Hoabinhian' as of limited utility in describing the Malaysian record (Zuraina, 2003: 167). Such an emphasis on technological continuity will require further critical assessment, not least of which because it is extremely unlikely that all of the assemblages in question were being made by the same species of hominin – though, as mentioned earlier, this situation is not a unique to Lenggong. Given that the Lenggong evidence is based on such a discrete set of assemblages within a single valley, a more readily accessible question that also immediately arises is to what extent this record is representative of the wider Malay Peninsula. Work conducted on the eastern side of the Titiwangsa Mountains goes some way to providing an answer.

In 1990, survey work by the Malaysian National Petroleum Corporation (PETRONAS) in rural Pahang identified a series of limestone massifs along the Sungai Batu. Although the archaeological potential of this area had been noted as early as 1935 (Tweedie, 1937: 154), no systematic excavation had ever been conducted. The first of two caves, Gua Sagu, is situated several metres above ground level in Bukit Sagu massif and lies within sight of the western bank of the Sungai Batu (*c.* 3.980197°N, 103.146114°E) 20–30 m asl. The second site, Gua Tenggek, lies 4 km to the north-northeast of Gua Sagu (*c.* 4.014956°N, 103.160394°E) *c.* 40 m asl and was only discovered during the 1990–1 joint archaeological investigation by USM and the Department of Museums and Antiquities.

The depositional sequence at Gua Sagu bottomed out at *c.* 1.3 m (spit 13). Faunal remains ceased to appear in spit 11 (*see* Chapter 7), diagnostic lithics in spit 10. The earliest radiocarbon date comes from charcoal at 0.9 m and places initial occupation of the site at *c.* 16,991 ± 327 cal. BP (Beta-33181). The age of occupation around the middle of the sequence is measured by a date from spit 5 (at 0.43 m) of 2941 ± 95 cal. BP (Beta-36341), indicating a Late Holocene antiquity for the upper half of the sequence (verified by another date from higher in spit 2). Approximately 68 per cent (n = 134/196) of the recovered lithics come from this part of the sequence (spit 5 and above); however, it is unclear how the age-depth relationship plays out beneath this. Ceramic sherds do penetrate into the deeper part of the sequence, but it is considered likely that these are intrusive from above spit 6 (Zuraina *et al.*, 1998: 87). Most of the defined lithic artefact classes identified in spit 5 are still present in spit 7, after which both range and incidence drop markedly to leave only isolated examples of hammer stones, grinders and one case of an anvil. No cores are reported despite the presence of moderate frequencies of débitage.

The strongest case for technological continuity comes from the vertical distribution oval bifaces. These are present from spit 3 down to spit 9, and if one excludes flake débitage, they are the site's single highest recurring form (*c.* 24% of flaked implements are oval bifaces). Only two oval unifacial tools were identified in the entire sequence and only three flake tools: a situation that does give us cause to step back from considering it as arbitrarily 'Hoabinhian'. It is, however, slightly complicated by the presence of nine 'truncated' pieces in the upper half of the sequence. These are flaked at one end to form a use edge and have a straight opposing end (Zuraina *et al.*, 1998a: 72); in other words, short axes. There are no immediately apparent temporal patterns in raw material use: essentially the same range of materials was used throughout the site record. Pounders (elongated pebbles) were made exclusively of a form of slate; the oval bifacial pieces were made of a variety of different stone types, including basalt, slate and quartzite and limestone; the first three of these were also the most oft-selected raw materials at the site overall.

At Gua Tenggek, the upper 0.2 metres are undated, but the presence of pottery implies a later Holocene age. spits 3 and 4 are already of likely Pleistocene age, however, based on two dates: 12,494 ± 103 cal. BP (Beta-40968) on charcoal from spit 3 and 10,660 ± 110 bp (Beta-40969) on shell (uncorrected), indicating a significant depositional break of unknown cause. Occupation evidence from this site is also covered in detail in Chapter 7. Most of the site's small (n = 47) stone tool assemblage also occurs in spits 3 and 4 (n = 23/47, 49%). As at Gua Sagu, oval bifacial implements are strongly represented (n = 11/47), all but one coming from beneath spit 2, as are hammer stones (n = 14/47), with ten of the 14 appearing in pre-ceramic levels. Truncated tools are again present and span the chronological jump in the sequence, appearing in ceramic and pre-ceramic levels. No cores were recovered, though again, a small amount of débitage was (n = 41 fragments), probably from maintenance rather than production. Other notable absentees are oval unifacials (i.e.

Sumatraliths, *sensu stricto*) and flake tools; there is also only a single pounder in spit 1. The Gua Tenggek industry suggests only very ephemeral use of the site, but the character of the industry closely matches that of its near neighbour (Zuraina *et al.*, 1998a: 98).

The current state of the lithic evidence from the Last Termination and Early Holocene in Peninsular Malaysia suggests the existence of industries that were similar to the Vietnamese Hoabinhian, and as such, one can see why they were so readily attributed to it, but they are not carbon copies. Broadly, the Malaysian industries appear to show considerable technological continuity. At a tighter level, though, they also appear to have their own internal idiosyncrasies. The USM classificatory system has the laudable advantage of disassociating itself from the baggage of traditional type tools, though it cannot be said to have yet achieved a quantum leap analytically. While the Kota Tampan and GGR lithic industries can be usefully classified under the same system, the proportions of recovered artefacts in each case are very different, even though both appear to have been tool-producing stations. USM's Kota Tampan industry has an overriding predominance of débitage, with flake tools as the most common tool class, yet only ten oval unifacials are reported from the whole excavation (Zuraina, 1990: 88). At GGR, the picture is almost exactly the opposite: small quantities of débitage, no flake tools (from undisturbed contexts) and reasonably frequent oval unifacials. Compare this to the Late Pleistocene Bukit Bunuh assemblage and the proportions shift again: flake tools are the dominant group of implements, but 'choppers' and 'hand axes' occur at much higher frequencies than oval unifacials (25.3%, 22.9% and 8.4%, respectively; Mokhtar, 2006a: 63), and this composition is different again from that seen at Gua Sagu and Gua Tenggek. While formal differentiation between unifacial and bifacial pieces does permit analysts to begin to address the morphological variability that has long afflicted the Sumatralith (Reynolds, 1990), the full implications of inter-assemblage technological variability need to be explored in much greater detail before we can come to confident conclusions about the Malaysian record and how it relates to others in Mainland Southeast Asia.

Thailand. As elsewhere in Southeast Asia, the prehistoric settlement of Thailand was 'practically unknown' (Sarasin, 1933). Sarasin's initial excavations in the north (around Chiang Mai) and south (near Ratburi) yielded flaked stone implements from at least three sites. He compared this 'Siamian' industry favourably to the Hoabinhian discoveries that Colani and contemporaries were making in Vietnam, proposing that it appeared to 'belong to the same group of palaeolithic cultures' (Sarasin, 1933: 195–6). More than 70 years later, and despite containing the type site for the Hoabinhian outside of Vietnam (Spirit Cave) and one of the oldest, well-dated sequences in Mainland Southeast Asia (Lang Rongrien), there is still surprisingly little known about this period compared to those following in Thailand (e.g. *see* Higham and Thosarat, 1998).

After the chance discovery of six unifacially worked pieces by van Heekeren, whilst he was a Japanese prisoner of war near the Kwai Noi River in 1943–4, and a brief subsequent survey and collection of pieces by K. Heider in 1956, including Sumatraliths, a joint Thai–Dutch Prehistoric Expedition was formally sent to the area from 1960 to 1962. Although this expedition resulted in highly significant excavations of the Neolithic settlement at Ban Kao (Sørensen, 1999; Sørensen and Hatting, 1966), of particular interest here are its efforts at Sai Yok Cave, made as part of a general field reconnaissance of the Mae Kwae Noi and Kwae Yai river valleys, in Kanchanaburi Province (Kamminga, 2007; van Heekeren and Knuth, 1967). This location, 100 m back from the eastern bank of the Kwai Noi (*c.* 14.414625°N, 98.846803°E), consists of a rock-shelter and two caves within a small limestone massif. While the caves contained limited evidence of prehistoric habitation, the rock-shelter produced a deep cultural sequence, as did subsidiary excavations, especially the 'Sector X' trench, cut near the base of a talus slope that extended from the rock-shelter down towards the upper river terrace. The principal tool classifications were carried out by van Heekeren, who divided the recovered pebble-tool industry into two facies: a smaller, more 'finely flaked' one, which he considered reminiscent enough of the Vietnamese evidence to be called 'Hoabinhian proper', and an underlying generic pebble tool facies that was to his mind less sophisticated in its appointment and less typologically diverse. Though dominated by pebble tool forms, the 'Hoabinhian proper' also contained considerable numbers of flakes – Kamminga (2007) reports a figure of 34 per cent of the Sai Yok lithic assemblage, currently housed in the Department of Fine Arts, Bangkok – and a small subset of bladelets struck from pebble cores from the upper horizons of the sequence mostly, though not exclusively, of Neolithic affiliation. Neither the flakes nor the blades carried any evident secondary retouch. Bone points (mentioned earlier) and shell artefacts were also identified.

Kamminga's recent reassessment of the assemblage concluded that from its constituent parts – in which 'pebble adzes' (unifacially flaked pieces, aka 'Sumatralith' variants) were also prominent (40%) and appearing in various states of completion – the location was a manufacturing station. As concentrations of lithics increased away from the rock-shelter and caves, he proposed that the main activity area was on the open river terrace, close to the main source of utilised raw material. This conclusion finds credence through the fact that the terrace revealed very little in the way of habitation refuse (van Heekeren and Knuth, 1967: 100). Like Reynolds, Kamminga also favours interpreting transversely broken pebble adzes (short axes) as *a posteriori* artefacts – broken pieces rather than intentionally formed ones. He notes, 'it was evident that most of the transverse breaks were bending fractures . . . a common cause of production failure in stone tool making,' (Kamminga, 2007: 47). Assessing the true effect of tool biography on assemblage character remains an endeavour in its infancy with respect to the Hoabinhian but one that is starting to provide promising results, as in the case of Kamminga's work (*see also* Marwick, 2008; Reynolds, 1989, 1992; White and Gorman, 2004). Despite its importance, not only to our understanding of Hoabinhian techno-facies outside of Vietnam but also for the fact that it was the first systematically excavated prehistoric site in Thailand, Sai Yok remains undated but, as noted earlier, is accorded an Early Holocene age.

The first radiocarbon-dated prehistoric excavation in Thailand has probably cast the longest shadow of perhaps any from Mainland Southeast Asia: Spirit Cave (*c.* 19.567383°N, 98.111597°E) in Mae Hong Song Province, in the country's northwest. This excavation, carried out in 1966, was also conducted expressly as an attempt to create a local sequence for the Thai Hoabinhian, which could then be situated in relation to the emergence of food production (Gorman, 1969). The comparatively thin (*c.* 1 m) deposits contained discernible stratigraphy, which Gorman duly followed. All excavated sediment was passed through 1-mm sieves on site and all residues were retained, leaving there to be little doubt about the rigorous nature of recovery. Two cultural phases were discerned. Cultural Level I was characterized as Hoabinhian on the basis of albeit low frequencies of representative type tools, primarily unifacially flaked pebble tools (Sumatraliths) and grinding stones. The largest group of artefacts, though, comprised utilised and in some cases retouched flakes (occasionally this extended all along one edge), and it included a number of small pressure-flaked blades struck from large calcite flakes, as well as flakes struck on basalt, chert and coarse-grained quartzite. The latter of these was overall the most common lithic raw material used in tool manufacture; a material choice that persisted throughout the occupation sequence and which would have been readily available from the stream in the valley below the cave (Gorman, 1970). Edge damage was recorded on *c.* 30 per cent of flakes. This phase of cave use was dated to between *c.* 13,208 ± 573 cal. BP (FSU-315) and as late as *c.* 9520 ± 351 cal. BP (FSU-318; using the half-life convention of 5570 ± 30 years, rather than Gorman's corrected 5730 ± 40 years, which is no longer employed; Walker, 2005: 54).

Cultural Level II saw a continuation of the same technology but with the appearance of additional elements, including flaked and polished quadrangular adzes, small ground and four polished slate knives, together with cord-marked and burnished ceramics. Large quartzite cores are also mentioned, but it is unclear if these had also been present in Level I. Three examples of the new form of adze (all made of limonite, the only artefacts to be so in the excavation) were found in association with a datable hearth feature in the surface of Layer 2. Charcoal from here gave an age of 9546 ± 226 cal. BP (GaK-1846). This would be an early date for the appearance of ceramics in this part of the world and has raised some concern about the uniform antiquity of material from this level (Lampert *et al.*, 2003, 2004; White, 2004). At this juncture, it is worth noting that a single similar edge-ground slate knife was identified from deposits containing classic Hoahinhian tool types (including Sumatraliths, short axes and large numbers of flakes) and ceramics at nearby Banyan Valley Cave (also excavated by Gorman in 1972), dated to 6145 ± 141 cal. BP (GaK-4341); two other such knives were identified from Late Holocene deposits at that site, where pebble tools are absent (Reynolds, 1992). Details of a subsequent excavation season at Spirit Cave in 1971 are unfortunately unpublished; subsistence evidence from the 1966 work at the site is discussed in Chapter 7.

Since Gorman's excavations at Spirit Cave, there has continued to be only limited research into the Late Pleistocene to Early Holocene settlement of Thailand, producing less than a dozen well-controlled and well-dated sites (Shoocongdej, 2006). The earlier occupations at two of them, Lang Rongrien and Moh Khiew, have already been covered. Here we touch briefly again on these to consider their later deposits, before turning our attention to data from sites farther to the north, at Khao Talu, Lang Kamnan and, lastly, Tham Lod.

Excavations at Lang Rongrien emerged out of the first intensive archaeological surveys of southwestern Thailand, which took place between 1979 and 1983 (following a preliminary survey in 1974; Anderson, 1988). Below the disturbed Units 1–4 at Lang Rongrien (the 'Upper Levels', including Neolithic burials), Anderson identified a large number of thin horizontal layers, descending 1.5 m, which he grouped as Units 5 and 6 (and termed the 'Middle Layers'; Anderson, 1997). Dates obtained on charcoal from these units ranged between 11,055 ± 158 cal. BP (SI-6817) and 8541 ± 66 cal. BP (SI-6213), indicating them to be of Early Holocene age. A total of 206 artefacts were recovered from Units 5 and 6 (205 of which were lithics), and almost exclusively, they were made on locally available carbonaceous shale. Sixteen tool types were identified: non-utilised flakes (*n* = 123, 60%); bifacially worked pieces (*n* = 31, 15.1% subdivided into core, discoidal, elongate, irregular and biface fragments); unifacially worked pieces (*n* = 15, 7.3%); hammer stones, grinding stones and whetstones (*n* = 9, 4.4%); utilised flakes (*n* = 9, 4.4%); miscellaneous objects (*n* = 5, 2.4%); cores (*n* = 4, 2%); cobble spall scrapers (*n* = 3, 1.4%); choppers (*n* = 3, 1.4%); polished stone fragments (*n* = 2, 1%); short axes (*n* = 1, 0.4%); and pottery (*n* = 1, found in the uppermost part of Unit 5 and excluded from percentage counts here). Anderson described the assemblage as ostensibly Hoabinhian but in a vein that was characteristically more akin to the Malaysian examples of this industry than the Vietnamese. He attributed this primarily to raw material factors.

Early Holocene occupation at Lang Rongrien appears to have been comparatively intensive and probably included tool manufacturing activities. Technological changes from the Pleistocene occupation are apparent both in the make-up of the lithic inventory and in what it was made of. The Pleistocene industry contained a significant flake component (*n* = 28/45 artefacts, 62%, including evidence of retouch) – further demonstration that Mainland Southeast Asian industries have not always been pebble tool–dominated as once thought. The low incidence of lithic débitage and the much lower frequency of bifacial pieces (4.4%) may both signal a change in occupation habit between the Pleistocene and Holocene units at Lang Rongrien – a contention possibly supported by an observed shift in stone procurement (or preference). Although some of the objects from the Pleistocene units were made of locally occurring quartz, the greater part of the industry was made of materials like siliceous limestone, chalcedony and chert (Anderson, 1988, 1997).

The Late Pleistocene to Early Holocene levels at nearby Moh Khiew exhibit a similar shift from a flake-dominated industry to one with an increased pebble tool component. Pookajorn *et al.* (1994) relate that this too contained Hoabinhian types made of quartzite and chert and included unifacial and bifacial elements (though proportions are not published), while most of the struck-flake components were made of finer-grained stone such as chert and jasper. Débitage is reported to have been present in considerable quantities and was taken as indication of on-site tool production. As noted, bone tool use was also indicated – something not seen in the Early Holocene levels at Lang Rongrien. The dates for this period of site use at Moh Khiew range from the final millennia of the Last Termination, 12,893 ± 128 cal. BP (OAEP-1284), to 9439 ± 84 cal. BP (OAEP-1292).

The next two sites take us back to the vicinity of the Khai Noi River. The first of these, Khao Talu (*c.* 14.004167°N, 99.316667°E), was one of four caves excavated by Pookajorn between 1977 and 1979 (Pookajorn, 1984). A total of 3.2 m of deposits were excavated stratigraphically from the second chamber at Khao Talu, subsequently consolidated into six archaeological layers. Layer 1 contained modern, mixed deposits. Layer 2 (0.45–1.2 m) yielded a large number of pebble tools, some comparable to Sumatraliths, and débitage flakes but also ceramic sherds. This layer has a Late Holocene date (and, as with the other assays listed here, was derived from bone) of 3696 ± 485 cal. BP (OAEP-053). Layer 3 (1.2–1.7 m) once again produced pebble tools and débitage, though these were considered to have been manufactured using a less sophisticated technique than that seen in Layer 2. Layer 3 contained no ceramics and was dated from 5152 ± 550 cal. BP (OAEP-107) to 4771 ± 126 cal. BP (OAEP-054). In Layer 4 (1.7–1.9 m), 8377 ± 565 cal. BP (OAEP-056) to 5839 ± 573 cal. BP (OAEP-106), apparent short axes predominated the pebble tool component, together with hammerstones (Pookajorn, 1984: 34). Layer 5 (1.9–2.3 m) dated from 11,525 ± 790 cal. BP (OAEP-055) to 9390 ± 788 cal. BP (OAEP-111) and possessed a range of pebble tools according to six major types: massive high-domed unifacially worked scrapers (possibly cores, *after* Kamminga, 2007), large unifacial choppers, unifacial points, utilised flat scrapers, hammer stones and bifacially flaked points. Pookajorn (1984) compared this lower assemblage favourably to the Vietnamese Hoabinhian as well as the local Sai Yok and Spirit Cave industries – though it does not contain either Sumatraliths *sensu stricto* or short axes. The fact that there is some ambiguity as to its affiliation is also apparent when he intimates that some elements more closely fit those from sites in Peninsular Malaysia and Indonesia. The basal deposits (>2.3 m, Layer 6) were archaeologically sterile.

The second site is Lang Kamnan (13.981111°N, 99.42°E), a cave situated 110 m asl in limestone uplands *c.* 4 km to the north of the Kwai Noi River. The site was excavated by Rasmi Shoocongdej in 1991 following survey work in the Kanchanaburi province the previous year (Shoocongdej, 1996, 2000). Although the deposits were quite heavily disturbed in some areas of the cave (notably the cave mouth) to a depth of about 1 m, judicious placement of trenches ultimately produced a rich and well-dated (from 22 radiocarbon dates) cultural sequence of considerable antiquity from three of the five trenches excavated. Initial occupation at Lang Kamnan occurred prior to the LGM, *c.* 36,259 ± 766 cal. BP (GX-20068). As very little charcoal was recovered, dating was mostly carried out on land snail shell; correction for ancient carbon was not attempted as it was felt unlikely to have had a significant impact in this instance (Shoocongdej, 1996: 208). The generally low frequency of remains suggests that hunter-gatherers were visiting the cave only sporadically from this time until just prior to the Pleistocene–Holocene transition (Cultural Layers V–III). Shoocongdej (1996) labelled this span 'Period I' on the basis of the dates and internal homogeneity in the faunal and artefactual assemblages. Period II (Cultural Layer II) continued many of the characteristics of Period I and produced dates of 9308 ± 129 cal. BP (OAEP-1178) to 8347 ± 170 cal.

BP (OAEP-1178). Ceramics appeared at the top of the Period II sequence, though these were considered to have translocated from overlying Mid to Late Holocene deposits.

Technologically, the entire lithic assemblage (n = 874, tools and débitage) is described as not having employed flaking techniques to control the form of resulting flakes and, therefore, is considered to have been expediently produced. Possible applications are interpreted as including carcass processing or use in other forms of tool production, for example, on bamboo. The dominance of débitage flakes (n = 681/874, 78%) and the presence of hammerstones and waste cores suggested that most tool production took place on site, with some cores possibly being initially prepared here for transport elsewhere – based on a higher observed frequency of primary and secondary decortification flakes compared to tertiary flakes (Shoocongdej, 1996: 266). Evidence that both cores and flakes were utilised was obtained by microscopic examination of the assemblage. Few flakes showed any sign of retouch though. Locally (<5 km) obtained quartzite was consistently the dominant raw material used in tool manufacture (96.9%); the remainder was made up of small quantities of chert (1.93%), limestone (0.34%), chalcedony (0.23%), shale (0.11%) and quartz (0.11%). Shoocongdej (1996: 337) wrote, 'I observe no clear organizational changes in either tool technology [or] morphology between the Late and post-Pleistocene periods.' The evidence from Lang Kamnan suggests that at least a proportion of sites were occupied, and the technologies found there were created, on a very much *ad hoc* basis. This gave Shoocongdej reason to question the wider applicability of the 'Hoabinhian' when this may only apply to a subset of types and pays little attention to the evident variability seen in the region.

Between 2001 and 2006, the Highland Archaeology Project conducted detailed excavations at two rock-shelters in the Pang Mapha upland region, east of Spirit Cave: Ban Rai (Treerayapiwat, 2005) and Tham Lod. We will deal briefly with only the older of the two here: Tham Lod (*c.* 19.560519°N, 98.278997°E), which is located at the base of a limestone cliff just above the terrace of a small upland river (Shoocongdej, 2006), and 200 m from a much larger riverine cave of the same name. Sixteen radiometric dates (nine radiocarbon dates and seven thermoluminescence dates) have established that there was a human presence here from as early as 35,782 ± 226 BP (TL), making this comparable to Lang Kamnan and much farther inland (currently *c.* 290 km from the nearest coast) than either that site or Lang Rongrien, if one is pushing for the exclusivity of an early strand-lopping colonization of the region. Full details of this excavation have yet to be published; however, it includes four inhumanations with directly associated grave goods (lithic artefacts) recovered from Last Termination deposits within the 4.5-m sequence dug against the cliff wall (Area 1). These are dated between 15,877 ± 141 cal. BP (Beta-168224) and 13,891 ± 70 cal. BP (Beta-168223) on organic sediment (Pureepatpong, 2006; Shoocongdej, 2006). The Pleistocene–Holocene sequence does contain classic Hoabinhian types (notably pieces that can be classified as Sumatraliths and short axes) but also many less readily classifiable elements, all predominantly manufactured of local sandstone. A noteworthy preliminary finding is the existence of spatial variation in the composition of lithic assemblages between trenches (particularly between frequencies of flake and core representation). Had only one part of this site been excavated, the site as a whole might have been interpreted as characteristically flake dominated; if another area, as core dominated. In a region like Southeast Asia, where discerning technological patterning is already a challenge, accurate (and conservative) assemblage characterisation and horizontal as well as vertical trench exposure (where feasible) may be essential, as this example demonstrates.

Summary. Despite the influence that the Thai evidence has exerted at a regional level in the interpretation of early tropical hunter-gatherer activities, the allotment of sites dating to the Late Pleistocene and Early Holocene remains slim. Research here does have a renewed sense of purpose, though, particularly through the latest generation of archaeological students, and clearly has enormous potential. Three points stand out. Firstly, there is evidence that technological continuity was a feature of the Thai record during the Late to post-Pleistocene period. Older elements persisted while new ones were, introduced alongside them. While this certainly seems to hold true for the northern sites, the technological shift exemplified at Lang Rongrien is quite marked between Pleistocene and Early Holocene occupations. Potentially, though, even here the underlying process may be less so. The marginal role held by pebble tools in an otherwise flake-dominated industry could simply have become more central through time. Possibly these different local geographic trajectories were related to the differing impact of environmental changes in the two areas, even so, shifts in raw material usage and occupational activities at Lang Rongrien do indicate that a range of factors will likely have been involved.

Secondly, while the use of type specimens has the obvious advantage of making inter-site comparisons easier, the assemblage recovered from Lang Kamnan provides a salient reminder that there was more going on in these technological landscapes than simple adherence to transmitted knapping traditions. In Thailand, as elsewhere in the region, the record is complicated by our very sparing knowledge of early settlement systems – about where people were at different times of the year and the kinds of records they left at each locus in that network of movements. Clearly, lithics were being curated around the landscape. Initial knapping was sometimes occurring adjacent to or within habitation areas, but just as often, it was occurring somewhere possibly quite removed.

Thirdly, and following on from this, there is a genuine possibility that emphasis on type forms overdoes the importance of these at the expense of other variability within industries. Yet, paradoxically, some kind of standardization to our frame of reference *is* required to ensure effective and representative analysis. At least a portion of the mainland's inter-site variability is probably due to confustication between the numerous typological or technological stances that different researchers have adopted to describe it. The time is probably right to decide formally on a basic procedure for Mainland Southeast Asian lithic analysis, then each go away and assess local industries using it rather than continuing to fall back on antiquated typological approaches, which has never really been suited to the record of this part of the world anyway.

6.5 Conclusion

Climatic and environmental reconstruction for the Last Termination and Early Holocene in Southeast Asia is still in its nascence. The record remains patchy and discontinuous, but data are accumulating from a number of reliable proxies and sometimes to a sufficient resolution to begin revealing the temporality and impact that northern (and southern) hemispheric deglaciation had on this region. Throughout much of the period, temperatures were probably still depressed by 3°–5°C compared to modern values. In some parts of Sundaland, such as the Javanese highlands, conditions were possibly not that much drier than they are today, though the centre of the continent is believed to have been considerably more savannah-like. Similarly drier were at least some areas near to the eastern margin, as shown in Palawan, and conditions were more varied on the northeastern part of the shelf. By contrast, rainforest landscapes seem to have persisted in other parts of the east and west. We assume that rainforest contracted during glacial periods and expanded during interglacials; almost all of our models are built on this understanding. The recent work by Cannon *et al.* (2009) turns this on its head. Detailed computer modelling by these authors suggests that the dynamics of Sundaland forests – their contraction and recolonization phases – are divergent from those seen in other parts of the world, including equatorial Africa. Their model describes the current, interglacial and insular state of the rainforest in Southeast Asia as an expression of a refugial stage (when they are more vulnerable to destruction). Rainforest expansion in this part of the world, because of its conditions and geography, is more likely, they propose, to have occurred during glaciations, when there are much greater stretches of contiguous land. While this is diametrically opposed to the general and long-standing flow of interpretations, it offers a refreshingly original take on environmental changes here that should be readily testable.

The impact of the LGM may have been felt earlier in Southeast Asia than in Europe, based on data from the western side of the shelf, though further information is needed to confirm this. In Borneo, the Lake Pemerak results suggest more open vegetation towards the close of the LGM, followed by a mosaic character peat forest, incorporating montane taxa after 17.7 cal. KBP. These are still evident forest components *c.* 15.2 cal. KBP but have largely disappeared by 14.3 cal. KBP. The arrival of the GIS-1 warm substage appears to have occurred rapidly. In terms of temperature, habitat change and sea level rise – as indicated from the Song Hong delta – the turnover would have been well within the cross-generational memory of humans living in the Sundaic realm at the time.

Evidence for a return to cool conditions during GS-1 is much more equivocal. A cooling trend is noted in the northwestern and northeastern parts of the region with changing vegetation. It is much less evident farther to the south. No significant temperature change has been noted in Borneo, though the hiatuses in the Bornean sediment cores during this period may be an indication of associated depositional or erosional changes that we have yet to properly detail. Overall, while this was an event of considerable abruptness and significance at higher latitudes, GS-1 seems to have made a much more subtle impression on Southeast Asia. Shifts in forest composition and resource structure currently represent the most likely impact on human groups.

The effects of environmental convulsions during the Last Termination are now beginning to be tracked genetically in Southeast Asia. It is early days, but the mtDNA evidence does point to significant intra-regional population dispersal probably around the start of GIS-1. This movement may have been coming in the wake of a population bottle-neck caused by the rapid rise in sea levels at the interstadial outset. Whereas there is still some disjuncture between the developing genetic picture of *H. sapiens* dispersal out of Africa and the state of the archaeology, there are indications that both lines are in greater concordance during the deglacial period here, where supportive hints may be found in the bone technological evidence.

Bone tools have been a feature of early human occupation in Southeast Asia from before the base of the Late Pleistocene; the earliest appearance is currently from the 'Hell' deposits, Niah Caves, at *c.* 45 cal. KBP. Two early 'centres' of this technology can be conjectured regionally: one encompassing eastern Sundaland (chiefly Borneo) and Wallacea (northern Moluccas) to the northwestern edge of the Sahul Shelf (Aru Islands). The second, on Mainland Southeast Asia and currently covering an area from the northern half of the Thai–Malay Peninsula to northern Vietnam. In both areas, the frequency of this technology is initially low, but it appears on a variety of osseous media and likely was employed in a range of different ways. Although indeterminate, on the basis of the earliest Niah material, digging or production (weaving?) may have featured as one of the activities, butchery, potentially, as another.

On current evidence, the mainland centre appears to continue producing a low level of bone technology throughout the period of the Last Termination and on into the Holocene, with a late injection of more complex forms and larger assemblages. The situation seems to have been different in ISEA. Commencing with GIS-1, but continuing through (and seemingly not significantly affected by) GS-1, larger bone tool assemblages begin to appear, often containing a comparatively high frequency of point-forms. Juliette Pasveer and the current author have worked (separately) on a large proportion of the existing material. It is, therefore, with some degree of certainty that the conclusion we have both reached, independently, that many of these points were almost certainly hafted, has credibility. Our opinions diverge somewhat on the likely use contexts of this new emphasis on point-forms: Pasveer favours use as perforators, whereas I have been more inclined to think that many were projectile armatures, but with a smaller proportion probably used as piercing tools. There is room for both positions; for it is clearly a feature of the Last Termination too that bone technology was not confined to a single use context.

By the Early Holocene, this technology is linked to another regional innovation in the very strong evidence for the use of stingray barbs in composite range weapons from the Niah Caves (Barton *et al.*, 2009). From this time onwards, sites yielding comparatively large assemblages of bone tools appear from as far apart as New Guinea and now the southern half of the Thai–Malay Peninsula. The regional distribution cline of dates hints at a slight east-west trend, though further evidence will be needed to confirm this. The coincidence in the dates and what appears to be the main geographic centre of expansion is, nonetheless, intriguing. Despite its distribution, bone technology shows comparatively little evidence for transformation in form over time until the Mid to Late Holocene, and then, almost exclusively, evidence comes from the mainland – a change signalling the innovation or influence of other cultural trends.

Although the lithic technologies of the region are routinely divided into two quite separate technological provinces as well (a flake-blade tool tradition in ISEA and a pebble and flake tool tradition on the mainland), there are, in truth, some noticeable similarities between them. There is considerable evidence for continuity in the respective industries and the techniques used to produce them in both parts of the region. Also, Palaeolithic industries all across Southeast Asia seem to be intrinsically additive: they are prone to diversification, but newly developed, innovated or introduced components appear alongside persisting older elements. On the mainland, in particular, this fact is probably one of the reasons it has proved so difficult to categorically define the Hoabinhian.

Traditions of manufacture also seem to be predisposed to the development of geographically localised industrial facies, as communities addressed immediate needs, cultural preferences, and the conditions of subsistence. There may have been an underlying technological ethic that was, in many respects, very similar and deeply embedded in the way hominins adapted to the affordances and conditions of Southeast Asian landscapes.

Within this context, the differences that are apparent suddenly seem more contingent, more a reflection of those localised trajectories and adaptations than of the sweeping cultural norms that have tended (correctly or not) to epitomise European prehistory – and which made it the point of departure for the first researchers who started the business of tackling the Southeast Asian record. ISEA does show a greater propensity towards fine-grained lithic raw materials than is evident on the mainland. This is a fact of geology and geography, and it will inevitably have played a part in fashioning how those trajectories evolved. Neither area, though, appears to have been bound to raw material–defined controls. Flake-blades appear in small numbers in mainland industries made of coarse stone; ready access to siliceous materials did not result in a uniform shift in technologies towards Upper Palaeolithic blade-based industries in the Indo-Malay archipelago or Wallacea. It did give greater potential for the production of pieces like Maros points (or Tingkayu points) but shows that the same rules of technological standardization that mark other regions did not apply here. Indeed, it is very apparent that technological standardization in this region could exist in one medium and yet not in another, even at the same time and at the same site.

The overall impression one is left with is that the impact of deglaciation was felt, perhaps not unsurprisingly, most acutely in what were to become the insular provinces of Southeast Asia and along the coastal margins of the continent. It is here that we see most vividly how early Southeast Asian communities employed a particular set of behaviours in response to the environmental, demographic and probably social changes that were happening around them. In areas that were removed from the imperatives which this created, other similar but locally distinct adaptations continued under more enduring set of selection pressures across the Pleistocene–Holocene transition. At the same time as affecting its technologies, the inundation of the Sunda and Sahul shelves will have resulted in a major re-modelling of the environment for Southeast Asian hunter-gatherers. In the next chapter, we explore the effect this had on subsistence strategies.

Chapter 7

Tropical Subsistence Strategies at the End of the Last Glacial

7.1 Introduction

The greater part of global research into prehistoric hunting and gathering strategies has focused on human adaptation to the periglacial environments of northern Eurasia and America, or on the open savannah of southern Africa. Palaeolithic archaeology, in its analysis and interpretation, has drawn heavily on the detailed ethnographic studies of local modern forager populations currently or historically living in both of these environments. By comparison, and despite the existence of a diverse anthropological literature on modern tropical forager groups, detailed and specific treatment of prehistoric tropical hunting and gathering strategies has remained distinctly under-developed and under-applied. A small subset of tropical ethnographies has incorporated archaeological implications or had archaeological objectives. These have ranged from studies trying to assess the viability of early subsistence under rainforest conditions to those trying to reconstruct tropical site formation processes (e.g. Bird *et al.*, 2002; Engelhardt and Rogers, 1997; Politis, 1996, 1999; Politis and Saunders, 2002; Pookajorn, 1985) and, more rarely, global-scale syntheses (e.g. Mercader, 2003). At the present time, analysis of the Southeast Asian record, long one of the most neglected, is resurgent, both through innovative and systematic locally run projects and through collaborative international initiatives (e.g. Barker *et al.*, forthcoming; Barton, 2005; Barton *et al.*, 2009; Morwood *et al.*, 2008; Mudar and Anderson, 2007; Ochoa, 2008; Piper and Rabett, 2009, Piper *et al.*, 2008; Rabett *et al.*, 2011; Shoocongdej, 2000, 2006; van den Bergh *et al.*, 2009). Even so, many gaps remain in our knowledge, particularly in reference to our understanding of tropical settlement systems, site function and group mobility (Rabett and Barker, 2010; Shoocongdej, 1996).

Drawing on what are often criticised as 'laundry lists' of diverse faunal remains recovered from Southeast Asian caves (owing to a tendency to simply catalogue the many species identified), it had long been accepted that early tropical hunter-gatherers in this region economically targeted various habitats. A good deal of the logic for this model was derived from excavations made of the pre-ceramic levels (*c.*13.5–9.7 BP) at Spirit Cave (Gorman, 1969, 1970, 1971). Gorman's work not only opened a new dimension on Thai prehistory but also shaped the way researchers would perceive tropical hunter-gatherers for much of the last quarter century. Undisturbed cultural deposits that were well documented and convincingly interpreted made Spirit Cave a regional standard. At the time, this was the only site attributed to the Hoabinhian to yield botanical remains. While this situation has since changed (*see* e.g. Ceron, n.d.; Nguyen Viet, 2008), the potential for enhanced recovery of botanical remains using sediment flotation techniques is still to be fully realised. Part of the reason why Spirit Cave stood out was because sieving had been used only sparingly at many of the region's earlier archaeological excavations. Taking his new archaeological evidence in conjunction with local ethnographic material, Gorman suggested that five species of nuts, the bottle gourd, Chinese water chestnuts and leguminous beans (among others) were being collected or tended. Fragments of bamboo stem attested to the invaluable importance of this versatile species (*see* Raddatz, 2006).

Vertebrate remains from Spirit Cave, as well as from most other Hoabinhian sites, were (and continue to be) very fragmentary. Those that could be identified showed how despite the fact that it was a steep half-hour climb up to the cave from the valley below, people were making this climb and bringing back game from a range of different environments. Pig and deer came from terrestrial hunting; primates, civets, martens and squirrels came from hunting the forest canopy; fish, crab and shellfish remains from exploitation of local streams. The diverse collection of species suggested a subsistence economy in which multiple habitats were being targeted from one base camp with seasonal changes in occupation – based on the different species of molluscan evidence recovered. The evidence from Spirit Cave also gave a picture of continuity in ecological and cultural forms across the Pleistocene–Holocene boundary, a feature that did not match the Eurasian sequence. Although not the first to recognise the risk of superimposing the European divisions onto records from other parts of the world, the Spirit Cave sequence prompted Gorman to advocate the abandonment of the preferred European terminology for Southeast Asian prehistory (Gorman, 1970: 81).

In the event, the incipient agricultural model Gorman favoured for Spirit Cave has not been bolstered by any confirmatory evidence in the years since he proposed it; and some have argued that to consider these groups as somehow part of a wider Late to post-Pleistocene 'broad spectrum revolution' was unjustified. Subsistence in the tropics is likely to have always followed such a diverse pattern of exploitation as this is a direct reflection of the environment itself (Reynolds, 1990). The practicality of early human occupation in lowland rainforest in general was a hotly debated topic in anthropological circles during the late 1980s and early 1990s (e.g. Bailey *et al.*, 1989; Endicott and Bellwood, 1991; Headland, 1987; Rambo, 1985; Stiles, 1992). Some of the chief reasons for questioning the feasibility of early independent foraging included the fact that much of a rainforest's energy remains locked away in inedible tissue, either in wood cellulose or in foliage with high concentrations of toxins, thus making them too carbohydrate poor for human groups to exploit effectively. Yams – a contemporary staple – appear to have originally evolved under conditions of more marked seasonality, where a starch store is required. Wild varieties are smaller and more labour intensive in climax rainforest, making them (it was reasoned) unlikely props for independent foraging. Starch can also be obtained from plants like the sago palm, but there are differing opinions as to how laborious this palm is to harvest (Brosius, 1991; Headland, 1987; Robinson and Bennett, 2000; Townsend, 1974, 1990). Also, minimal amounts of energy go into producing flowers, fruit and seeds, and even these reproductive cycles can be easily knocked out of kilter by disruptions to annual receipt of rainfall (Bailey *et al.*, 1989). For example, in the lowland rainforests of Borneo, flowering, fruiting and leaf production can be irregular even between trees of the same species in different valleys. A four-year study of forest dynamics in lowland Peninsular Malaysia (MacKinnon *et al.*, 1997) revealed that 85 per cent did not flower or produce fruit at regular intervals, adding unpredictability to the issue of resource availability. Finally, there is often a wide distribution of individual members of the same species – plant and animal (the latter of which are also often arboreal and/or nocturnal) – meaning that acquiring food would involve potentially high travel costs for human foragers (Bailey *et al.*, 1989).

Overall, it was argued that many of the plant staples currently exploited by rainforest groups (and mostly these arguments have been about plants) may not have been present in their current densities before the development of swidden horticulture – given that plants such as yams and nuts are highly light dependent. These findings led to the revision of once prevalent views that modern-day rainforest foraging communities were static remnants of prehistoric hunting and gathering populations. Rambo (1985) asserted that most, if not all, modern hunter-gatherers in Southeast Asia are sufficiently involved in trade relations to the point where it is doubtful that they could survive if that trade were to be cut off. The logical conclusion to come out of this was that either prehistoric foragers did not live in Pleistocene rainforests or, if they did, that their livelihood had to be supplemented in some way beyond merely harvesting wild resources; otherwise, existence would have been marginal (Headland, 1987). Detailed archaeological research undertaken in Southeast Asia during the last ten years, however, has demonstrated that people were, in fact, successfully exploiting rainforest resources well into the Late Pleistocene, but with the important caveat that the structure of those forests did not necessarily match that seen in the Holocene (e.g. Hunt *et al.*, 2012); one might argue, scoring points for both sides of this particular argument.

In this chapter, we reconsider tropical subsistence, with particular emphasis on detailed examination of faunal assemblages from different parts of Southeast Asia, and use this as a means to understand early human responses to the profound environmental changes during the Last Termination and across the Pleistocene–Holocene transition.

The principal data for our discussion come from three areas of the Sunda continent: northern Borneo, the margin of the Song Hong river delta in northern Vietnam and eastern Peninsular Malaysia. The faunal assemblages discussed in this chapter are from a selection of sites and phases, which together cover each of the main climatic shifts of the Last Termination. Reanalysis of the Niah vertebrate fauna was carried out by the author and P. Piper; analysis of the faunal assemblages from Gua Sagu, Gua Tenggek and Hang Boi was carried out by the author.

7.2 Northern Borneo: Niah Caves

In 1958 the Sarawak Museum began a systematic excavation of deposits that lay beneath the small rock overhang on the northern side of the entrance to the West Mouth of the Niah Caves (Figure 7.1). Analysis of the archived vertebrate faunal remains from this part of the cave (Area A) was carried out as part of the Niah Caves Project (*see* Barker *et al.*, forthcoming; Cranbrook and Piper, 2007; Piper *et al.*, 2007a, 2007b, 2008; Piper and Cranbrook, 2007a; Piper and Rabett, 2009, forthcoming-a, -b; Pritchard *et al.*, 2009; Stimpson, 2009). The work involved examination of all bone fragments from 26 of the original 53 excavated trenches beneath the overhang (or 49% of the original excavated area). The trenches selected provided sufficient coverage of the whole area and range of depths to track major changes in spatial and temporal variation in represented taxa. Finer-scale variation was not possible due to the limitations imposed by the original excavation strategy to dig in coarse spits, the narrowest of which were 3″ (7.62 cm) thick and often 6″ (15.24 cm) or greater. The work was aided, though, by photographs and the copious site notes kept by members of the original excavation.

Of the 15,080 bone fragments recorded in Area A (Harrisson and NCP-recovered assemblages combined), 70 per cent of our sample was associated with Phase II (Last Termination and Early Holocene) activity, including 2071 identified specimens (*see* Table 7.2). Patterning within the total fragment counts, per Harrisson excavated spit, indicated that this bone assemblage probably accumulated in a shallow surface depression. The upper limit of

Figure 7.1. Photograph showing excavation work beneath the rock overhang (Area A, Niah Caves Project), probably 1965. (Photograph by Sarawak Museum Harrisson Archive, na868, reproduced with kind permission.)

the Phase II deposits lay at between 24″ and 30″ (i.e. material included for analysis comes from below 26″) below the 1958 ground surface; their maximum depth in the centre of the depression, where most material accumulated, was *c.* 60″. The visibility of this occupation unit is aided by the very low concentrations of animal bone in the 12–24″ levels above it. A similar hiatus also appears to have followed the preceding Phase I occupation. The available ^{14}C dates indicate that Phase II lasted for a period of approximately 7000 years: from *c.* 16 cal. BP until *c.* 8.4 cal. BP (*see* Appendix) – a spread of dates that suggests use of the cave spanned more than one climatic period. The early part of Phase II (60–48″) would have been deposited during the comparatively cool conditions in the wake of the LGM, anchored by one date 16,000 ± 125 cal. BP (OxA-15162) at 48–54″. The later part of the phase (48–24″) appears to have been deposited during the Early Holocene. Two dates anchor this part of the assemblage: 11,429 ± 109 cal. BP (OxA-15157) at 36–48″ and 8400 ± 20 cal. BP (OxA-15161) at 30–36″. This Late Phase II also corresponds with the first appearance of worked stingray barbs from another part of the West Mouth (Area D, see later). The absence of any dates within either GIS-1 or GS-1 sub-stages does raise some questions about on-site activity in the West Mouth during these periods. Although there is no visibly distinct evidence of hiatus in the sequence, a change in the rate of deposition may account for this gap.

7.2.1 Area A, Phase II

The overall taphonomic character of the Phase II bone is distinctive from both the underlying Phase I/Hell material and overlying Phase III/Neolithic assemblages and probably relates, in a broad sense, to localised environmental differences at the time of their respective depositions. Full details are presented in Barker *et al.* (forthcoming) and Piper and Rabett (forthcoming-a).

The greatest proportion of bone fracture types identified in the Phase II assemblage were transverse, longitudinal or columnar in form, often with slightly weathered and abraded fracture surfaces. These types of fracture are normally linked to dry, old bone rather than mechanically fresh bone (Johnson, 1985). In some cases it was possible to perform limited fragment refits between spits, implying that at least some of the assemblage had sustained reworking as a result of evident biostratonomic and post-depositional processes. Forty-sixty per cent of pieces, per examined spit, were found to exhibit slight rounding of fracture edges, combined with random surface scratching and pitting consistent with mechanical abrasion, possibly as a result of trampling (e.g. Lyman, 1994; Olsen and Shipman, 1988). The data reveal that bones accumulating closest to the cave lip sustained more extensive surface modification than those lying fully underneath the rock overhang. Another common surface feature was the appearance of deep longitudinal cracking of the cortical bone associated with bone surface flaking, erosion of fracture margins and exfoliation. These are all modifications that can be readily identified as coming from sub-aerial weathering in the tropics (Tappen, 1994). In this instance, the damage is consistent with the effect that would be expected from alternating periods of saturation and desiccation. In some cases, this damage was restricted to one or other surface – a pattern consistent with fragments having lain undisturbed on the ground surface. Manganese oxide speckling or more extensive staining was identified on a proportion of fragments. This brown-black chemical precipitates where conditions are moist or possibly as a result of sediment humification – an aftermath of human activity (López-González *et al.*, 2006; Marín Arroyo *et al.*, 2008).

The incidence of moderate and severe edge abrasion suggests that most human activity and foot traffic were probably concentrated at the northwest end of the overhang. In contrast to comparatively high percentages of charred and calcined bone that had emerged from the Phase I/Hell deposits in the West Mouth, burnt bone was almost entirely absent from the Phase II occupation (accounting for no more than 6% of fragments in any one trench and spit). The implication is that people were not living beneath the overhang nor, seemingly, were they dumping out hearth waste there. That said, the incidence of butchery traces to bone from Phase II clearly demonstrates that humans were the main agent of accumulation. Detailed analysis of carcass processing (element representation and butchery traces) was confined to three mammalian orders – ungulates, primates and carnivores – owing to the significant recovered numbers of each. Reference is made to the processing of other represented species from the site, but the low incidence of butchered remains belonging to, for example, sciurids and varanids precluded reliable quantification or interpretation in most cases.

7.2.1.1 Ungulates

From the trenches reanalysed during the Niah Caves Project, suids were found to be the most common species identified throughout Phase II (Table 7.1). These were almost certainly the bearded pig (*Sus barbatus*) on the basis of comparative geometric morphometric analysis

Table 7.1. Identified Taxa from the Early and Late Phase II Occupation of Area A, in the West Mouth of the Niah Caves (Excluding Human Remains)

Class	Order	Family	Taxon	Common Name	Area A (Early Phase II) 60–51″ NISP	Area A (Early Phase II) 60–51″ %	Area A (Late Phase II) 48–26″ NISP	Area A (Late Phase II) 48–26″ %
Reptilia	Testudines	Geoemydidae	Contains *Cyclemys/Heosemys*	Spiny hill/Asian leaf turtle	0	0	7	0.49
			Cyclemys dentata	Asian leaf turtle	3	0.47	10	0.70
			Heosemys spinosa	Spiny hill turtle	2	0.32	7	0.49
			Notochelys platynota	Malayan flat-shell turtle	5	0.79	14	0.97
			Geoemydidae spp.	Hard-shell turtle	101	15.96	297	20.65
		Trionychidae	*Amyda cartilaginea/Pelocheyls cantorii*	Asian soft-shell turtle/ Giant soft-shell turtle	0	0	2	0.14
			Amyda cartilaginea	Asian soft-shell turtle	3	0.47	22	1.53
			Dogania subplana	Malayan soft-shell turtle	2	0.32	2	0.14
			Trionychidae spp.	Soft-shell turtle	15	2.37	38	2.64
	Squamata	Pythonidae	*Python* spp.	Reticulated or Borneo blood python	7	1.11	0	0
		Varanidae	*Varanus* spp.	Monitor lizard	37	5.85	76	5.29
Aves	Galliformes	Phasianidae	*Arborophila* sp.	Partridge	0	0	1	0.07
			Phasianidae spp.	Wild fowl/pheasant?	3	0.47	2	0.14
	Falconiformes	Accipitridae	*Macheiramphus alcinus*	Bathawk	2	0.32	2	0.14
			Haliastar indus	Brahminy kite	0	0	5	0.35
			cf. *Accipiter trivirgatus*	Crested goshawk	0	0	1	0.07
			cf. *Ictinaetus malayensis*	Black eagle	0	0	1	0.07
			Spizeatus sp.	Changeable hawk-eagle	0	0	8	0.56
			Accipitridae sp.	Hawk	8	1.26	3	0.21

(continued)

Table 7.1 (*continued*)

Class	Order	Family	Taxon	Common Name	Area A (Early Phase II) 60–51″ NISP	%	Area A (Late Phase II) 48–26″ NISP	%
	Strigiformes	Strigidae	*Ketupa ketupa*	Buffy fish owl	0	0	1	0.07
	Coraciiformes	Bucerotidae	*Anthracoceros coronatus*	Pied hornbill	0	0	1	0.07
			cf. *Anthracoceros malayanus*	Black hornbill	0	0	2	0.14
			Buceros cf. *rhinoceros*	Rhinoceros hornbill	1	0.16	5	0.35
			Anorrhinus galeritus	Bushy-crested hornbill	0	0	1	0.07
			Bucerotidae spp.	Hornbill sp.	4	0.63	8	0.56
	Passeriformes	Corvidae	*Corvidae* sp.	Crow	0	0	1	0.07
Mammalia	Primates	Cercopithecidae	*Presbytis (presbytis)* sp.	Leaf monkey (not silvered langur)	4	0.63	14	0.97
			Presbytis sp.	Leaf monkey	3	0.47	14	0.97
			Presbytis cristata	Silvered langur	1	0.16	5	0.35
			Macaca fascicularis	Long-tailed macaque	6	0.95	10	0.70
			Macaca nemestrina	Pig-tailed macaque	1	0.16	2	0.14
			Macaca sp.	Macaque sp.	3	0.47	7	0.49
			Cercopithecidae sp.	Monkey	55	8.69	133	9.25
		Hylobatidae	*Hylobates* cf. *muelleri*	Gibbon	1	0.16	0	0
		Ponginae	*Pongo pygmaeus*	Orangutan	27	4.27	73	5.08
	Pholidota	Manidae	*Manis javanica*	Pangolin	3	0.47	0	0
	Rodentia	Sciuridae	*Petauristinae* sp.	Flying squirrel	0	0	1	0.07
			Ratufa affinis	Giant squirrel	1	0.16	3	0.21
			Sciuridae sp.	Squirrel	2	0.32	1	0.07
		Hystricidae	*Hystrix brachyura/Thecurus crassipinis*	Common porcupine	0	0	0	0
			Trichys fasciculata[a]	Long-tailed porcupine	0	0	0	0
			Hystricidae sp.	Porcupine	4	0.63	3	0.21

Carnivora	Ursidae	*Helarctos malayanus*	Sun bear	0	0	4	0.28
	Mustelidae	*Lutra sumatrana*	Hairy-nosed otter	0	0	1	0.07
		Martes flavigula	Yellow-throated martin	1	0.16	0	0
		Mustelidae sp.	Mustelid sp.	1	0.16	0	0
	Viverridae	*Viverra tangalunga*	Malay civet	1	0.16	1	0.07
		Paradoxurus hermaphroditus	Common palm civet	2	0.32	0	0
		Paguma larvata	Masked palm civet	1	0.16	0	0
		Arctictis binturong	Bear cat	4	0.63	10	0.70
		Herpestes sp.	Mongoose	0	0	1	0.07
		Viverridae sp.	Civet cat	7	1.11	7	0.49
	Felidae	*Felis badia*	Bay cat	1	0.16	0	0
		Prionailurus bengalensis	Leopard cat	0	0	3	0.21
		Felis/Prionailurus sp.	Cat	3	0.47	7	0.49
		Panthera tigris	Tiger	0	0	1	0.07
		Neofelis diardi[a]	Clouded leopard	0	0	2	0.14
Perissodactyla	Rhinocerotidae	*Dicerorhinus sumatrensis*[a]	Sumatran rhinoceros	1	0.16	0	0
Artiodactyla	Tapiridae	*Tapirus indicus*	Tapir	0	0	1	0.07
	Suidae	*Sus* cf. *barbatus*	Bearded pig	293	46.29	585	40.68
	Tragulidae	*Tragulus napu*	Greater mouse deer	2	0.32	5	0.35
		Tragulus sp.	Mouse deer	0	0	2	0.14
	Cervidae	*Muntiacus* sp.	Barking deer	1	0.16	4	0.28
		Cervus unicolor	Sambar	5	0.79	11	0.76
	Bovidae	*Bos* sp.	Cattle	6	0.95	10	0.70
			TOTAL (NISP and %)	**633**	**100**	**1438**	**100**

Note: Specific epithets are as follows: 'sp.' indicates confidence in attributing specimens to a particular genus but uncertainty with respect to which species within that genus they belong; 'spp.' indicates confidence that there is more than one species represented in the study sample but uncertainty as to which species those are; 'sp(p).' indicates that there is *probably* more than one species present in the sample.

[a] Renamed after Kitchener *et al.* (2006).

(Cucchi *et al.*, 2009). The presence of skeletal elements from all parts of the body suggests that complete carcasses were being transported back to the West Mouth to be processed. The same mix of juveniles, subadults and adults is found in Early Phase II assemblages, as was seen in the pre–Last Termination sample, suggesting a continuation of largely non-selective hunting techniques, with trapping a distinct possibility and, again, something that differs from modern hunting practice (Cranbrook and Labang, 2003). By Late Phase II, there is a slightly higher emphasis on taking young juveniles and subadults. This may reflect a shift to somewhat more targeted hunting or may reflect availability (Barker *et al.*, forthcoming). Either way, pigs continue to be a major draw for people coming to Niah and a food staple, quite likely related to persisting patterns in the migratory habits of this animal. Owing to this behavioural trait, *Sus barbatus* represents the only large ungulate likely to have been encountered in herds of significant numbers on a super-annual basis in the Bornean rainforests (Caldecott and Caldecott, 1985; Curran and Leighton, 2000, Dove, 1993; Pfeffer and Caldecott, 1986; Rabett and Barker, 2007) and a prime target for hunting. A similar interpretation, where there is also an exceptionally high dependence on pig, has been advanced for the Peninsular Malaysian site of Gua Cha (Sieveking, 1954). All other forest ungulates are more likely to have been encountered in small groups or as solitary individuals.

Despite its high incidence, there is only limited evidence of butchery to *Sus barbatus*. This fact might be in part due to the incidence of post-depositional bone surface modifications; however, while the relative frequency of surface-damaged bone is highest for pigs during the Phase II occupation, it still amounts to less than 1 per cent of the total recovered material (Piper and Rabett, forthcoming-a). The patterning looks genuine.

Evidence of butchery to the axial skeleton is scarce but does provide certain insights into processing habits towards these and perhaps other larger mammals. Despite the identification of 22 *Sus barbatus* mandible fragments in Area A, none was found to be cut-marked. As the skin is in close contact with bone on this part of the body, the lack of cut-marks suggests that neither skinning of the head nor tongue removal was a significant aspect of butchering this animal. There is, however, reason to believe that the brain was at least occasionally targeted as a resource during the Early Phase II occupation, as rejoining fractured pieces of the squamous portion of the temporal bone were identified in trench Y4, 54–60″, together with other *Sus barbatus* skull fragments including an occipital condyle; the implication being that the back of the skull was broken open to remove brain matter.

Turning to the appendicular skeleton, in his 1981 Nunamiut study (which still remains one of the most authoritative available), Binford observed that the proximal humerus is often flexible and can be broken open with sufficient applied leverage – an act that in large animals may take place before the scapula is itself removed from the carcass. Cuts are still employed in this act, though, as in that study (Binford, 1981: 122), none is apparent in the Phase II assemblage, leaving the actual process as yet to be determined; indeed, no cut-marks appear on positively identified upper thoracic elements at all. Butchery marks were also not found on femora, despite the confirmed presence of this element (i.e. nine proximal and 18 distal femoral fragments from Area A). A possible reason why there is such a dearth of traces here is that the pelvic limb of this and other large ungulates was one of the areas targeted for marrow extraction – as indicated by impact scars and rejoinable fractures. The lack of cuts may also, or alternatively, indicate that disarticulation was not attempted at the acetabulum; rather, the pelvis may have been split into two units, each with a limb still left attached on each side (*see* e.g. Bunn *et al.*, 1988; White, 1952). Disarticulation along the pelvic symphysis might have been easier than directly tackling the major femoral joint. Cut-marks to pig extremities consistently appeared on the calcaneus, phalanges and metapodia and were likely related to removal of the feet and skinning of these animals. All such evidence is restricted to Late Phase II, possibly signalling a change in processing routine (Piper and Rabett, forthcoming-a).

In terms of other ungulates, all three genera of deer that currently inhabit Borneo (*Cervus*, *Muntiacus* and *Tragulus*) are represented in this assemblage and occur at approximately the same frequency between the Early and Late components. Fragments of the Banteng (*Bos* cf. *javanicus*) were also occasionally recovered. Other large terrestrial mammals were restricted to a very low incidence during Phase II. The Malayan tapir (*Tapirus indicus*; Cranbrook and Piper, 2009) is represented by a single fragment, as is the Sumatran rhinoceros (*Dicerorhinus sumatrensis*; Cranbrook and Piper, 2007), the former only found in Late Phase II, the latter only in Early.

7.2.1.2 Primates

Primate taxa show an overall slight rise between Early and Late Phase II (*c.* 16% compared to *c.* 18%). Cercopithecidae (leaf monkeys and macaques) comprise the largest primate family in each part of the phase. This is because although some cercopithecid skeletal elements, such as the distal humerus, can be distinguished (Ki-Kydd and Piper, 2004), many cannot be identified higher than family. The remains of orangutan (*Pongo pygmaeus*) make up the largest proportion of fragments identified to genus or higher (*c.* 4% and 5%, Early and Late Phase II, respectively). Orangutans are predominantly arboreal and solitary primates and have a modern population density in lowland dipterocarp and mixed freshwater peat swamp of between two and four individuals per square kilometre (Morrogh-Bernard *et al.*, 2003). While these figures are likely to be lower presently than they were in prehistory, given the effects of historical hunting pressures (Corlett, 2007; Delgado and Van Schaik, 2000), it is still probable that this species was encountered far less frequently than troops of monkeys in the local environment around Niah.

In the Early Phase II assemblage, macaques are the next most frequent (1.58%); in Late Phase II, the leaf monkeys edge ahead (2.29%). As well as a general rise in primate bone between the periods, the increase in orangutan and leaf monkeys, which are also more strictly arboreal, could imply greater availability, capacity to access or shifting preferences towards them. Evidently, though, canopy rainforest existed near to the cave and was being exploited successfully throughout. The early presence of the Bornean gibbon (*Hylobates* cf. *muelleri*), another strictly canopy-dwelling species, lends support to this conclusion.

Body part representation does reveal cranial and appendicular parts during the Early Phase II, but the evidence for the transportation of whole carcasses from kill site to cave is much stronger during Late Phase II, especially for the cercopithecids. This is matched by an increase in observed cut-marks, rising from two to eight instances, including to both thoracic and pelvic limbs, extremities and mandible, with placement and type indicative of skinning and disarticulation, presumably in the cave. One fragment of *Presbytis* sp. distal humerus showed evidence of cuts to its posterior aspect, implying that disarticulation took place with the limb in the flexed position, possibly once *rigor mortis* had already set in (*after* Binford, 1981). Tooth eruption and epiphyseal fusion data suggest that the primate assemblage is mostly adults, with just one or two subadults and several juveniles recorded, a pattern consistent with a hunted (prime adult–dominated) population (*see* Piper and Rabett, forthcoming-a).

7.2.1.3 Carnivores

Carnivore remains accounted for NISP $n = 58$ of the total Phase II assemblage. Although there was a marked increase in absolute numbers noted between the Early and Late periods, the overall proportion of carnivore remains to that of other fauna actually declined somewhat, from 3.3 to 2.5 per cent. Certain species disappear from this record between the Early and Late Phase II: the Bay cat (*Felis badia*), yellow-throated martin (*Martes flavigula*), masked palm civet (*Paguma larvata*) and common palm civet (*Paradoxurus hermaphrodites*), while new ones appear, including the leopard cat (*Prionailurus bengalensis*), clouded leopard (*Neofelis diardi*), tiger (*Panthera tigris*; *see* Piper *et al.*, 2007b), hairy-nosed otter (*Lutra sumatrana*) and sun bear (*Helarctos malayanus*). Only the bear cat (*Arctictis binturong*) and Malay civet (*Viverra tangalunga*) persist throughout Phase II. Given that individual frequencies are low, however, it is impossible to say whether these changes reflect anything more than the vagaries of hunting success. Despite the fact that several cranial, mandibular and appendicular fragments from a range of carnivores were recovered from the Area A assemblage, none showed any signs of butchery treatment; though some evidence was found in probably comparably dated assemblages elsewhere in the West Mouth, making human predation the most likely cause of accumulation. These included a sun bear distal humerus, found to carry scrape-marks consistent with filleting of meat from the carcass (Piper and Rabett, forthcoming-a).

7.2.1.4 Reptiles, Birds and Rodents

A total of 650 reptile bones recorded in the Area A Phase II bone assemblage could be identified to family or higher taxonomic level. The majority ($n = 475$) comes from the Early Holocene part of the occupation, but their place in the proportion of the overall fauna only rises by a few per cent (27.64%–33.03%). The most frequently occurring single genus was *Varanus* (monitor lizards), but chelonid remains were the most numerous throughout, and the rise in their absolute and proportional values accounts for most of the observed increase in reptile numbers (*see* Pritchard *et al.*, 2009). In the final analysis, a total of 446 geoemydid (hard-shell turtles) fragments were identified in Phase II. These included three species of locally common turtles: *Cyclemys dentata* ($n = 13$), *Notochelys platynota* ($n = 19$) and *Heosemys spinosa* ($n = 9$). Each occupies a different freshwater and/or semi-terrestrial niche. The total trionychid (soft-shell turtles) is similarly distributed spatially beneath the overhang but fewer in number ($n = 84$). Two species have been identified on the basis of distinguishing carapace patterning (*see* Pritchard *et al.*, 2009, for criteria): the Asian soft-shell turtle (*Amyda cartilaginea*) inhabits slow-flowing, muddy freshwater rivers and peat swamps and spends most of its time buried in soft mud or sand; its principal periods of activity being restricted to dusk and dawn (Liat and Das, 1999). Hand collection of this species is reported ethnographically as achieved by wading through the water probing the riverbed with a barbed harpoon that will penetrate the soft shell (Piper *et al.*, 2008). The second species is the Malayan soft-shell turtle (*Dogania subplana*). By comparison, this tends to avoid the larger sediment-laden rivers; its preferred habitat is hill streams with stony beds (Pritchard, 1979). The collective range of habitats from which chelonids were being taken during this period is notably broad, including large and sluggish lowland rivers, ponds and river-edge terrestrial locations, as well as upland clear-running rocky streams; the NISP values suggest that hunting activities were particularly concentrated in lowland river and swamp contexts.

There is a caveat to this, though, involving the ease with which carapace and plastron fragments are identified. Unlike mammalian bone, not only are all shell fragments obviously chelonid (and usually readily distinguished between soft- and hard-shell variants), even the smallest pieces are identifiable to family – a degree of certainty that cannot be extended to other taxa under similar circumstances of fragmentation. This introduces a bias towards chelonids in any assemblage where they are present (e.g. *see* Barker *et al.*, 2010). While we can be certain that turtles featured prominently in subsistence activities at Niah, the true scale of that reliance remains to be clarified.

The only snake species that could be identified was the reticulated python (*Python reticulatus*). Its bones represent about 20 per cent of the total Squamata remains recovered from Phase II layers, but even so, snake bones are relatively uncommon in the assemblage and are one of the taxa (particularly for smaller species, such as cave racers) that form part of the cave's natural vertebrate community, so direct predation of them by humans must be viewed as less likely.

The remains of bird bones make a small but significant contribution to the Phase II assemblage. Ten species are represented from the 60 identified fragments recorded. These include four species of hornbill: the pied hornbill (*Anthracoceros coronatus*), black hornbill (*Anthracoceros malayanus*), bushy-crested hornbill (*Anorrhinus galeritus*) and rhinoceros hornbill (*Buceros* cf. *rhinoceros*). The Accipitridae are also represented by four probable species: the crested goshawk (cf. *Accipiter trivirgatus*), the black eagle (cf. *Ictinaetus malayensis*), the brahminy kite (*Haliaster indus*) and the bat hawk (*Macheiramphus alcinus*). A single species of owl – the buffy fish owl (*Ketupa* cf. *ketupa*) – was also identified. Of the seven bones of Phasianidae recorded, six were morphometrically consistent with jungle fowl (*Gallus* cf. *gallus*), while a complete right humerus was attributed to an indeterminate species of partridge (*Arborophilia* sp.). Some of the avian remains, such as those of the hornbills, could have been introduced by people. Others, in particular, those of the raptors and owls, were likely to have been naturally occurring, as the birds exploited the roosting and hunting opportunities provided within the cave (Stimpson, 2009).

Phase II also contained seven fragments of bone that were attributed to large squirrels, including four elements from the giant squirrel (*Ratufa affinis*) and seven fragments of porcupine (*Hystricidae* sp.). Rodents appear somewhat less frequently in the Late component than they do in the Early, but there is no clear explanation as to why this should be. The research of fish remains identified at Niah has not yet been undertaken, but preliminary observations show the presence of at least one genus and one family: *Arius* sp. and *Cyprinidae*.

7.2.1.5 Discussion

The taphonomic analysis of the Area A Phase II assemblage suggests that this deposit constituted part of a refuse midden, though, not one that seems to have included burnt hearth material. In one respect this is a curious fact, given the proximity of the cave lip and the ease with which materials could have been cast into the gorge as an alternative and easy discard strategy. Clearly dumping over the edge may still have occurred, but the use of caves as places to deposit food waste and not always as living areas may be related to the use of these places in the tropical landscape compared to cave occupations in colder climates.

The presence of water-eroded and weathered bone indicates that conditions around the overhang were periodically affected by poor drainage, and so made this probably not an ideal spot for camping in the first place, despite the cover afforded by the overhang itself from the likely steady rain of bat and swiftlet excrement from the cave roof. Nonetheless, the wide span of the West Mouth would have offered an open space for activities away from the tangle of the forest floor, protection from inclement weather and a viewing vantage point out over the gorge for watching and listening for game.

While a veritable spectrum of macrovertebrate fauna was deposited here, the primary food sources were much more restricted. The remains of suids, cercopithecids and (provisionally) chelonids account for the majority, and this shows no major change from Early to Late: 78.51 per cent compared to 81.13 per cent, respectively. Additional taxa, ungulates and small carnivores, notably, are present but in very low numbers. The fact that each of the three dominant taxa occupies a different ecological niche (terrestrial, fresh water and semi-arboreal) is, indeed, indicative of hunting strategies that were geared to the exploitation of multiple habitats, with each one placing specific requirements on requisite procurement techniques and technologies. On the basis of the age profile of the pig remains, terrestrial fauna were probably targeted through less selective methods, such as trapping, rather than through the more selective practices used in this area historically.

The presence of gibbon points toward the existence of tracts of closed rainforest canopy in the vicinity of the cave no more than three or so millennia after the close of the LGM. The Late Phase II occupation sees significant increases in all previously represented taxa and a slight widening of the overall diversity of recovered taxa (42 to 51), with some species dropping out of the assemblage and others added. The rise in the density of faunal remains during the Early Holocene seems to support a more intensive use of the cave by that time.

Element representation in the pigs and monkey data indicates consistent recovery of all major bone types thoughout phase II (Piper and Rabett, forthcoming-a). The Late Phase II occupation carries a much larger proportion of the butchered bone compared to the earlier part. Although this may be a statistical artefact (the earlier part of the assemblage is still numerically about half the size of the latter part), the difference in the butchery is much larger than might be affected by this discrepancy and, therefore, seems to mark a genuine shift in carcass processing techniques between these periods. What is also apparent is that the frequency of associated bone technology increases through time (*see* Chapter 6). The absolute frequency of bone points, specifically, also increases – something that may be linked to the rise in arboreal-adapted fauna. The range of bone implement types does not seem to alter greatly between the two subphases, suggesting that they continued to be used in established and defined tasks. A final observation about the Late Phase II fauna is that, perhaps as one would expect, it accords for the most part closely to the species represented in the other well-documented Early Holocene assemblage from the West Mouth, Area D. Positive support that these assemblages were probably contemporary comes from the closeness of the ^{14}C dates associated: Late Phase II (Area A), *c.* 11.4–8.4 cal. KBP, and Area D, 11.4–8.9 cal. KBP.

7.2.2 West Mouth, Area D

In 2001, renewed archaeological investigation in the West Mouth of Niah revealed an area of cultural deposits towards the back of the fenced-in archaeological compound (Figure 7.2). This location, Area D, became the focus of a small-scale excavation (2001–2) as part of the NCP (*see* Barker *et al.*, 2002b, forthcoming). The final trench measured 3 × 1 × 0.9 m, the northern half of which contained a complicated series of inter-cutting water run-off channels. These were orientated east-west and contained fine, loose silts and guano sediments. The upper channel fills were found to be much richer in fauna and cultural remains than the lower fills, though parenchymatous plant tissues were recovered throughout (Barker *et al.*, 2002b).

In the southern half of the excavated trench, calcareous silts dated from *c.* 8915 ± 103 cal. BP (OxA-11864) and older were found to contain considerable quantities of molluscs. These were dominated by gastropod species adapted to estuarine tidal environments *Neritodryas subsulcata, Neritina petitii* and *Neritodryas dubia* and the large bivalve *Polymesoda erosa*, suggesting that the Early Holocene waterways around Niah had become subject to salinization through rising sea levels. Underlying sediments comprised a series of sandy silts interspersed with thin, darker layers indicative of the development of more stable land surfaces. The earliest date from these lower deposits is 11,449 ± 137 cal. BP (OxA-11865; Barker *et al.*, forthcoming).

Analysis of vertebrate faunal remains showed no appreciable taphonomic difference between those recovered from the channel fills and those recovered from the layers: both were in predominantly very good condition, possibly indicating that sedimentation was occurring primarily under dry conditions and that the burial process was comparatively rapid. The survival of fragile elements and the lack of significant soil compaction suggest that human foot traffic was minimal this far back into the cave. Burnt bone was more frequently encountered than in Area A, but no *in situ* hearth features were found to suggest camping in the immediate area.

A similar range of species to those found elsewhere in the West Mouth was identified. These included *Cervus unicolor* and *Bos* cf. *javanicus* as well as tree-dwelling or arboreal-adapted

Figure 7.2. The archaeological compound of the West Mouth, Niah Caves. The rock overhang is left of frame; Area D is not visible in the picture but lies behind the excavated area in the centre of the frame at the edge of light penetration into the cave. (Photograph by Ryan J. Rabett.)

species such as *Ratufa affinis* and *Viverra tangalunga* – all at low frequencies of occurrence (mostly single, isolated fragments; Table 7.2). Reptiles, including varanids, geoemydids and trionychids, were also represented. As in the other archaeologically rich parts of the cave mouth, though, pig remains dominated and to a very similar degree (e.g. Area D = 40%; Area A, Late Phase II = 41%). The relative proportion of cercopithecids, though, is different to those seen in Area A. The remains of these animals form a notably higher proportion of the total assemblage than was found in Area A (Area D = 48.45%; Area A, Late Phase II = 12.93%). The contemporaneity of these two periods of site use suggests that this is possibly not due to environmental changes but rather to a short-term change in subsistence or discard strategies. The presence of species like the silvered langur (*Presbytis cristata*), pig-tailed macaque (*Macaca nemestrina*) and gibbon (*Hylobates* spp.) among the primate taxa confirms the continuing presence of mature, dense rainforest within the immediate vicinity of the cave during the Early Holocene.

While the overall spectrum of fauna represented in Area D appears to accord well with the Area A Late Phase II assemblage, certain other differences were also noted. Viverrid, mustelid and felid remains in Area A account for three times the proportion of the assemblage that they do in Area D (0.76% Area D, 2.29% Area A), whilst remains of *Pongo pygmaeus* are completely absent from the Area D assemblage and stingray barbs are absent from Area A. It is feasible that these differences may, again, have been linked to different activity (or refuse dumping) areas within the cave. It is impossible to say for certain if the same or different human groups were responsible for both of these middens. What can be said is that the evidence from this period in the West Mouth, collectively, exhibits notable if subtle differences to the slightly earlier occupation in Lobang Hangus.

7.2.3 Lobang Hangus, Phase II (GIS-1)

Details of the Lobang Hangus vertebrate fauna are presented in Barker *et al.* (forthcoming), Medway (1966), Piper *et al.* (2008), Piper and Rabett (2009), and Piper and Rabett

Table 7.2. Taxa Represented during the Early Holocene Occupation of the West Mouth, Area D

Class	Order	Family	Taxon	Common Name	Area D Late Phase II NISP	Area D Late Phase II %
Reptilia	Testudines	Geoemydidae	Geoemydidae spp.	Hard-shell turtle	2	0.76
		Trionychidae	Trionychidae spp.	Soft-shell turtle	2	0.76
	Squamata	Pythonidae	*Python* cf. *reticulartus*	Reticulated python	3	1.15
		Varanidae	*Varanus* spp.	Monitor lizard	12	4.58
Mammalia	Primates	Cercopithecidae	*Presbytis* (*presbytis*) spp.	Leaf monkey (not silvered langur)	7	2.67
			Presbytis spp.	Leaf monkey	6	2.29
			Presbytis (*Trachypithecus*) *cristata*	Silvered langur	2	0.76
			Macaca fascicularis	Long-tailed macaque	10	3.82
			Macaca nemestrina	Pig-tailed macaque	2	0.76
			Macaca spp.	Macaque	3	1.15
			Cercopithecidae spp.	Monkey	97	37.02
		Hylobatidae	*Hylobates* cf. *muelleri*	Gibbon	3	1.15
	Rodentia	Hystricidae	*Hystricidae* spp.	Porcupine	1	0.38
	Carnivora	Viverridae	Viverridae spp.	Viverrid species	1	0.38
		Felidae	*Prionailurus bengalensis*	Leopard cat	1	0.38
	Artiodactyla	Suidae	*Sus* cf. *barbatus*	Bearded pig	105	40.08
		Tragulidae	*Tragulus napu*	Greater mouse deer	1	0.38
		Cervidae	*Cervus unicolor*	Sambar	3	1.15
		Bovidae	*Bos* spp.	Cattle	1	0.38
				TOTAL (NISP and %)	**262**	**100**

Note: Specific epithets are as follows: 'sp.' indicates confidence in attributing specimens to a particular genus but uncertainty with respect to which species within that genus they belong; 'spp.' indicates confidence that there is more than one species represented in the study sample but uncertainty as to which species those are; 'sp(p).' indicates that there is *probably* more than one species present in the sample.

(forthcoming-a). This cave is the largest of the three main east-facing entrances to the Niah Caves system, though compared to the enormous West Mouth, it comprises a much smaller space – the cave floor area is *c.* 1500 m^2. Like the West Mouth, though, this cave also opens onto a steep gorge, with the valley floor lying *c.* 43 m below. The immediate environment outside the cave is currently seasonally flooded, mixed-swamp forest. There is no locally available palaeobotanical evidence to characterize the environment on this side of the Gunung Subis during the Last Termination. A Holocene core has been taken from the same gorge outside another of the east-facing entrances to Niah (Gan Kira; Hunt and Rushworth, 2005). This indicated that forest stands in the area were composed of relatively undisturbed back-mangrove swamp *c.* 6494 ± 95 cal. BP (Beta-193909), but whether the same can be said of earlier forest structure here is hard to ascertain at present.

Faunal remains from nine of the 31 10 × 10′ trenches excavated by the Sarawak Museum in 1959 and 1965 were included for reanalysis by the NCP. This amounted to a total NISP of $n = 9529$ macrovertebrate and bird remains; of these, $n = 3288$ (excluding human bone) could be identified to family or higher taxonomic level (*see* Barker *et al.*, forthcoming, revised from Piper *et al.*, 2008: table 2; Piper and Rabett, 2009; Table 7.3). This sample included all of the trenches excavated in 1959, when it was possible for all recovered materials to be transported to the museum for archiving. Less comprehensive procedures were employed during the 1965 dig, though material from a pit-like feature between trenches US/17a and US/21a was included when material from the site was restudied.

Taphonomic examination of the bone revealed abrasion to fracture edges on 10 to 30 per cent of fragments within any particular trench. For the most part, this was defined as 'slight' only, with small numbers of 'moderately' and 'severely' abraded bones (*see* Rabett *et al.*, 2006, for defining criteria). Some bone surfaces also showed evidence of superficial and chaotically orientated striations. This is interpreted as likely to be a result of foot traffic (*see* Olsen and Shipman, 1988). Such traces aside, bone condition at the site is largely very good, with limited weathering, minimal amounts of sediment or calcite encrusting and good preservation of cut-marks and bone working.

Patterning was observed in the fragmentation of long bones from intermediate-sized mammals (e.g. monkeys and civets) and larger ungulates. In the case of the former, this included the survival of articular ends and substantial lengths of the diaphysis, to the extent that it was possible to see that these had been repeatedly selected for bone tool manufacture (Rabett and Piper, forthcoming; Figure 7.3). By contrast, the large mammal long bones (including articular ends) were typically severely fragmented. Several instances of impact scars were recorded on long bone shaft fragments, and it seems very likely that these were deliberately broken to extract marrow (*after* Johnson, 1985). The vast majority of cut-marked bone identified in the Lobang Hangus material came from the Phase II period ($n = 74/77$ fragments, 96%).

7.2.3.1 Ungulates

As was seen in the West Mouth, pig remains form a large proportion of the Lobang Hangus Phase II fauna (NISP $n = 1095/3288$ or *c.* 33% of all identifiable fragments). All elements of the pig skeleton are represented, with the largest and best-preserved fragments being recovered from the pit-like feature in US/17a and US/21a. The implication, again, is that whole animals were being transported to the site for processing and, presumably, consumption. The fact that other identified large terrestrial mammal remains occur as isolated element fragments suggests that infrequent predation of these taxa may well have been linked to a different set of processing activities (perhaps dictated by body size), one that may have involved only portions of the carcass being brought back to the cave after a kill.

Table 7.3. Taxa Represented during the GIS-1 Occupation of Lobang Hangus

					Lobang Hangus 12–48″ (Phase II)	
Class	Order	Family	Taxon	Common Name	NISP	%
Reptilia	Testudines	Geoemydidae	Contains *Cyclemys/Heosemys*	Spiny hill/Asian leaf turtle	14	0.43
			Cyclemys dentata	Asian leaf turtle	18	0.55
			Heosemys spinosa	Spiny hill turtle	8	0.24
			Notochelys platynota	Malayan flat-shell turtle	28	0.85
			Geoemydidae spp.	Hard-shell turtle	297	9.03
		Testudinidae	*Manouria emys*	Asian brown tortoise	1	0.03
		Trionychidae	*Amyda cartilaginea/Pelocheyls cantorii*	Asian soft-shell turtle/giant soft-shell turtle	5	0.15
			Amyda cartilaginea	Asian soft-shell turtle	7	0.21
			Dogania subplana	Malayan soft-shell turtle	4	0.12
			Trionychidae spp.	Soft-shell turtle	29	0.88
	Squamata	Pythonidae	*Python* spp.	Reticulated or Borneo blood python	32	0.97
		Varanidae	*Varanus* spp.	Monitor lizard	404	12.29
	Crocodylia	Crocodylidae	*Crocodylus* cf. *porosus*	Estuarine crocodile	1	0.03
Aves	Anseriformes	Anatidae	Anatidae sp.	Duck	1	0.03
			Anser sp.	Goose	1	0.03
	Galliformes	Phasianidae	Phasianidae spp.	Wild fowl/pheasant?	4	0.12
	Falconiformes	Accipitridae	*Macheiramphus alcinus*	Bathawk	2	0.06
			Haliastar indus	Brahminy kite	2	0.06
			cf. *Accipiter trivirgatus*	Crested goshawk	1	0.03
			cf. *Ictinaetus malayensis*	Black eagle	2	0.06
			Spizeatus sp.	Changeable hawk-eagle	1	0.03
			Accipitridae sp.	Hawk	2	0.06

	Strigiformes	Strigidae	*Ketupa ketupa*	Buffy fish owl	3	0.09
			Strix leptogrammica	Brown wood owl	1	0.03
			Strigidae spp.	Owl	4	0.12
	Coraciiformes	Bucerotidae	*Anthracoceros coronatus*	Pied hornbill	2	0.06
			cf. *Anthracoceros malayanus*	Black hornbill	4	0.12
			Buceros cf. *rhinoceros*	Rhinoceros hornbill	5	0.15
			Anorrhinus galeritus	Bushy-crested hornbill	1	0.03
			Rhyticeros corrugatus	Wrinkled hornbill	2	0.06
			Bucerotidae spp.	Hornbill sp.	10	0.30
Mammalia	Scandentia	Tupaiidae	*Tupaia* sp.[a]	Tree shrew	1	0.03
	Primates	Cercopithecidae	*Presbytis* (*presbytis*) spp.	Leaf monkey (not silvered langur)	21	0.64
			Presbytis spp.	Leaf monkey	71	2.16
			Presbytis cristata	Silvered langur	7	0.21
			Macaca fascicularis	Long-tailed macaque	59	1.79
			Macaca nemestrina	Pig-tailed macaque	1	0.03
			Macaca spp.	Macaque sp.	46	1.40
			Cercopithecidae spp.	Monkey	532	16.18
		Hylobatidae	*Hylobates* cf. *muelleri*	Gibbon	4	0.12
		Ponginae	*Pongo pygmaeus*	Orangutan	128	3.89
	Pholidota	Manidae	*Manis javanica*	Pangolin	10	0.30
	Rodentia	Sciuridae	*Petauristinae* spp.	Flying squirrel	3	0.09
			Ratufa affinis	Giant squirrel	10	0.30
			Callosciurus spp.[a]	Tree squirrels	3	0.09
			Sciuridae spp.	Squirrel	49	1.49
		Hystricidae	*Hystrix brachyura/Thecurus crassipinis*	Common porcupine	2	0.06
			Trichys fasciculata[a]	Long-tailed porcupine	1	0.03
			Hystricidae spp.	Porcupine	10	0.30

(*continued*)

Table 7.3 (*continued*)

Class	Order	Family	Taxon	Common Name	Lobang Hangus 12–48″ (Phase II) NISP	%
	Carnivora	Mustelidae	*Aonyx cinerea*	Oriental small-clawed otter	3	0.09
			Martes flavigula	Yellow-throated martin	1	0.03
			Mustelidae spp.	Mustelid sp.	2	0.06
		Viverridae	*Viverra tangalunga*	Malay civet	2	0.06
			Paradoxurus hermaphroditus	Common palm civet	1	0.03
			Paguma larvata	Masked palm civet	7	0.21
			Arctictis binturong	Bear cat	133	4.05
			Arctogalidia trivirgata	Small-toothed palm civet	1	0.03
			Hemigalus sp.	Banded or Hose's civet	2	0.06
			Herpestes sp.	Mongoose	2	0.06
			Viverridae spp.	Civet cat	91	2.77
		Felidae	*Felis* cf. *planiceps*	Flat-headed cat	1	0.03
			Prionailurus bengalensis	Leopard cat	9	0.27
			Felis/Prionailurus spp.	Cat	34	1.03
			Neofelis diardi[a]	Clouded leopard	1	0.03
	Perissodactyla	Rhinocerotidae	*Dicerorhinus sumatrensis*[a]	Sumatran rhinoceros	2	0.06
	Artiodactyla	Suidae	*Sus* cf. *barbatus*	Bearded pig	1095	33.30
		Tragulidae	*Tragulus napu*	Greater mouse deer	14	0.43
			Tragulus spp.	Mouse deer	13	0.40
		Cervidae	*Muntiacus* cf. *muntjac*	Common muntjak	11	0.33
			Cervus unicolor	Sambar	5	0.15
		Bovidae	*Bos* cf. *javanicus*	Banteng	4	0.12
				TOTAL (NISP and %)	**3288**	**100**

Note: Specific epithets are as follows: 'sp.' indicates confidence in attributing specimens to a particular genus but uncertainty with respect to which species within that genus they belong; 'spp.' indicates confidence that there is more than one species represented in the study sample but uncertainty as to which species those are; 'sp(p).' indicates that there is *probably* more than one species present in the sample.

[a] Renamed after Kitchener *et al.* (2006).

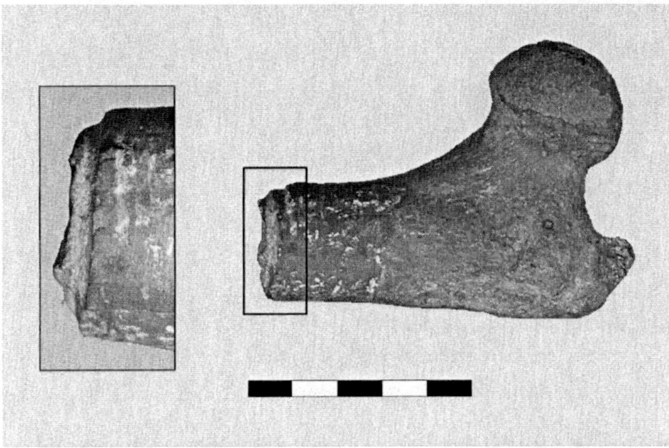

Figure 7.3. Fragments of Cercopithecidae long bone constituting manufacturing debris from bone tool production (scale in 5-mm increments). (Photographs by Ryan J. Rabett.)

Sixteen fragments of cranial and appendicular pig bone were found to be cut-marked (Piper and Rabett, forthcoming-a). Notably, there is evidence of butchery to the thoracic limb, including a butchered pig scapula, comprising two chop-marks close to the antero-medial edge behind the glenoid fossa – placement consistent with disarticulation of the joint from the front and separating the thoracic limb from the shoulder. There is also evidence for dismemberment around the elbow joint, with one example suggesting stiff-joint butchery. All are features not found in the West Mouth assemblages. By comparison, even though 17 distal femoral fragments were recovered from the Lobang Hangus Phase II assemblage, including one with a clear example of spiral fracturing, none was again found to be cut-marked. It appears here that the thoracic and pelvic limbs may have been treated in different ways, with the thoracic limbs being evidently sectioned and the pelvic limbs being removed as whole units or joints of meat, probably minus the foot.

7.2.3.2 Primates

Despite the numerical emphasis on pigs at Lobang Hangus, the assemblage actually contains a slightly higher proportion of arboreal-adapted game than it does terrestrial (*see* Piper and Rabett, 2009). While this was also noted in the Area D material, the relationship was reversed in the larger Area A Phase II assemblages, suggesting differences in prey availability and/or a differently weighted set of hunting strategies.

The remains of orangutan (*Pongo pygmaeus*) account for NISP $n = 128/3288$. This amounts to 3.9 per cent of the overall identified assemblage and a figure very comparable to that seen in Area A, potentially, therefore, more closely a reflection of their general availability in the local environment. The recovered orangutan material from Lobang Hangus consists primarily of extremities ($n = 89/128$), though most elements are present within the Phase II sequence.

Cercopithecidae are very well represented in the Lobang Hangus assemblage, comprising 22.4 per cent of the total and including, as in the West Mouth, the remains of *Macaca fascicularis*, *M. nemestrina* and *Presbytis* spp. Tooth eruption, wear and epiphyseal fusion data suggest that the cercopithecid assemblage here is dominated by young mature adults, with just one or two subadults and several juveniles recorded (Piper and Rabett, 2009). Such a pattern of occurrence suggests that foragers were using a selective hunting strategy and the means to effectively pursue it. Ethnographically, traps are employed to take some species of monkey, notably the typically inquisitive macaque (Piper *et al.*, 2008). It is not yet known what the impact of such methods may be on the age structure represented in this kind of prey; however, the very specific age structure at Lobang Hangus is inconsistent with non-selective trap mechanisms. The large number of bone tools and the likelihood that these included projectile armatures (*see* Chapter 6) presents an alternative demonstrable (as opposed to inferred, e.g. blowpipe) approach to hunting canopy game that permits this kind of deliberate prey selection. The four proximal ulnae of *Hylobates* cf. *muelleri* were identified during reanalysis of the assemblage, together with a further mandible and a cranial fragment recorded by Medway (1966: 209). All but the last of these were recovered from Phase II deposits. Given the canopy-dwelling habits of this agile ape, successfully hunting them by any means would have required considerable amount of skill and technological assistance. The use of range weapons seems the most probable option.

Butchery and disarticulation of the upper body of cercopithecids from Lobang Hangus were accompanied by the appearance of notable cut-marks to the anterior surface of the distal humerus, particularly a short distance above the medial epicondyle (defined as Hd-1; Figure 7.4). Cuts here dominate the small ($n = 5/6$) sample of primate distal humeri from this site and suggest a butchery *châine operatoire* that involved a preference for cutting the associated brachialis muscle during disarticulation. Significantly, cut placement of this kind does not appear in any of the Area A Phase II contexts, though was found in other West Mouth contexts dating roughly to the Middle to Late Holocene (Piper and Rabett, forthcoming-a).

Unlike the situation observed in the West Mouth, no cut-marks were found around the olecrannon fossa (posterior surface) of the distal humerus (Hd-4) of the Lobang Hangus assemblage. This distinction may be a reflection of differences in the timing between the acquiring and processing of primates as well as in the way the primate thoracic limb was severed in half at the two sites. In the West Mouth, there is only evidence of cut-marks to the distal humerus and no evidence of butchery to the radius; at Lobang Hangus, three examples of cut-marked proximal cercopithecid radii were identified. There are also hints that different species of Cercopithecidae were being butchered in slightly different ways, with a higher incidence of cut-marks appearing on macaque versus leaf monkey species (Piper and Rabett, forthcoming-a).

At a gross anatomical scale, all observed butchery traces on the proximal femur of the Cercopithecidae at Lobang Hangus, as in the West Mouth, represent primary processing: the removal of the pelvic limb. Again, although the numbers are never Figure, 7.5; large, and hence not statistically significant, it is notable that both the frequency and placement of cut-marks, and the muscles that were likely being cut, were found not to be the same. At Lobang Hangus, cuts to both the proximal and distal femur are much more evident than they are in the entire Area A Phase II assemblage. In addition to differences in frequency, butchery traces found on Lobang Hangus cercopithecid femora also showed a different placement to that seen in the West Mouth. While there are three instances of cuts to the area beneath the femoral neck (defined as Fp-1, one of only two places where cuts to proximal cercopithecid femora were found in the Area A sample), more variation in placement was noted in the Lobang Hangus sample. This sample is the larger of the two, and this will factor, but the suggestion is that although the upper and lower limbs were being removed during processing at both sites, the approaches to cercopithecid disarticulation may not have been the same: people were potentially using different operational sequences of butchery between two sites in the same cave complex. Whether this might reflect occupation by different forager groups will remain to be seen.

7.2.3.3 Carnivores

The Lobang Hangus assemblage includes a large proportion of semi-arboreal carnivorous species that have predominantly nocturnal activity cycles. These include the civets, *Arctictis*

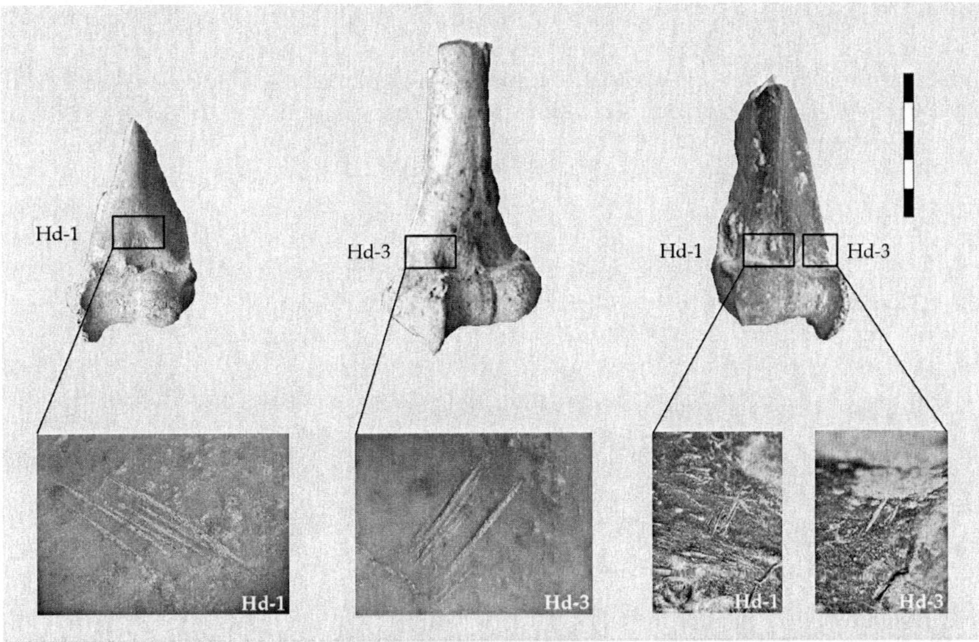

Figure 7.4. Examples of cut-marked distal humeri from Cercopithecidae from the Lobang Hangus, Phase II occupation (14.4–12.3 cal. KBP). (left to right) HLA-15322, multiple cuts to the anterior mid-surface, (L) *Macaca* cf. *fascicularis*; HLA-18496, multiple cuts distally of the medial supracondylar crest line, (L) *Macaca* sp.; HLA-17295, two sets of cuts: one to the anterior mid-surface, one in the vicinity of the medial supracondylar crest line, (R) Cercopithecidae (scale in 5-mm increments). (Photographs by Ryan J. Rabett.)

binturong; *Paguma larvata*, *Paradoxurus hermaphroditus* and *Arctogalidia trivirgata* and the felids, *Prionailurus bengalensis*, *Felis* cf. *planiceps* and *Neofelis diardi*. It is likely that traps rather than projectiles were used to catch such animals. Their collective high incidence here compared to what was seen in the Area A assemblages may well, in part, be down to differences in micro-habitat, but the extent of the difference makes it more likely that these animals were probably being more deliberately targeted at Lobang Hangus.

In the butchery evidence, the similarity in cut-mark placement on the calcaneus from both West Mouth and Lobang Hangus contexts suggests that the skinning of carnivores was commonly practiced, and seemingly this part of the reduction process exhibited low variability in the way it was executed from one species to another. In contrast, butchery of the thoracic limb showed a marked difference, especially with respect to the heavily muscled *Arctictis binturong*. This civet has a unique manner of locomotion: relying on its forelimbs to grasp, pull or brace itself, whilst its hind limbs work in coordination with its long prehensile tail as it climbs. Disarticulating the thoracic limbs at the shoulder is likely to be possible only with considerable difficulty; disarticulating at the elbow may well have been more readily achievable, though there would still be considerable muscle mass to deal with (Stinner, pers. comm.). This difference may go some way in explaining why butchery traces to the *A. binturong* thoracic limb occur almost exclusively to the distal humerus and at extraordinarily high frequencies. In all, 45 per cent ($n = 9/20$) of the *A. binturong* distal humeri in this assemblage were found to be cut-marked, a vastly higher incidence than seen on any of the principal taxa at this or any other site in the Niah complex. Significantly, this is also a frequency that is not even repeated for other viverrids (Figure 7.5). The cut-mark traces inflicted during the processing of this animal may be related to its comparatively large size (up to 20 kg) and unique adaptations; however, given such a high instance of butchery to one portion of one element, it is noteworthy that none of the ten examples of viverrid distal humeri identified from the Area A faunal assemblages (including two from *A. binturong*) was found to be butchered. Deliberate processing choices are in evidence here. These data suggest that the physiological differences between this animal and other viverrids were probably identified by foragers using Lobang Hangus, and they appear to have employed (as with the primates?) a distinctive butchery *châine operatoire* that was not used in the disarticulation of other, mostly smaller, viverrids. The evidence may even begin to reveal that this large civet was ethnotaxonomically perceived as somehow different to its Linnaean relatives.

Overall, while greater emphasis on nocturnal carnivores may relate in part to differences in micro-habitat, deliberate prey selection is a much more pronounced feature of hunting at this site compared to what we see in the West Mouth, making this perhaps the more likely explanation. Analysis of the carnivore (and especially the well-represented viverrid) data also revealed distinctive patterning in the way these animals were skeletally reduced during processing, patterning that was also not seen in the West Mouth.

7.2.3.4 Reptiles

The argument for selective hunting is further reinforced when one considers the herptofauna from Lobang Hangus. Reptiles comprise 25.8 per cent (NISP $n = 848/3288$) of the identified bone. Varanid remains dominate this total, while chelonid remains are also frequent (see earlier caveat), though it is interesting to note that the swampy habitat on this side of the massif and its propensity to flood (Harrisson, 1966) does not seem to have translated into appreciable differences in the acquisition of a species such as *Amyda cartilaginea*, which is adapted to such an environment.

One large fragment of a terminal phalanx has been tentatively identified as that of the estuarine crocodile (*Crocodylus* cf. *porosus*). Noted for its high tolerance of saline conditions, this crocodile can also be found in freshwater rivers and swamps (e.g. Stuebing *et al.*, 1994), further attesting to the likely existence of this kind of environment near the cave during the Last Termination.

7.2.3.5 Birds and Fish

The Lobang Hangus record (like that of the West Mouth) contains the remains of several families of large bird, including raptors (Accipitridae), owls (Strigidae), hornbills (Bucerotidae) and fowl (Phasianidae). While avian fauna may have been hunted for food or plumage, several of these species are known to frequent cave entrances, and the natural incorporation of their remains into the midden deposits is a further genuine possibility (Stimpson, 2009). Fish remains from this site have also not yet been studied in detail, though carp (Cyprinidae) and catfish (*Arius* spp.) were again identified during data entry (Piper and Rabett, 2009).

7.2.3.6 Discussion

To summarise the subsistence picture from the Last Termination and Early Holocene at Niah, the main terrestrial staple continues to be *Sus barbatus*, as it had been previously; possibly reflecting long-term proximity of the caves to the migration routes of this animal. Analysis of the faunal remains from Lobang Hangus indicates that the superficial similarity in species representation between this site and the Phase II occupation of the West Mouth belies considerable differences in prey population structure and game handling. Though pig is the single most dominant species at both sites, semi-arboreal taxa, such as cercopithecids and viverrids, make up a much larger proportion of the overall vertebrate assemblage at Lobang

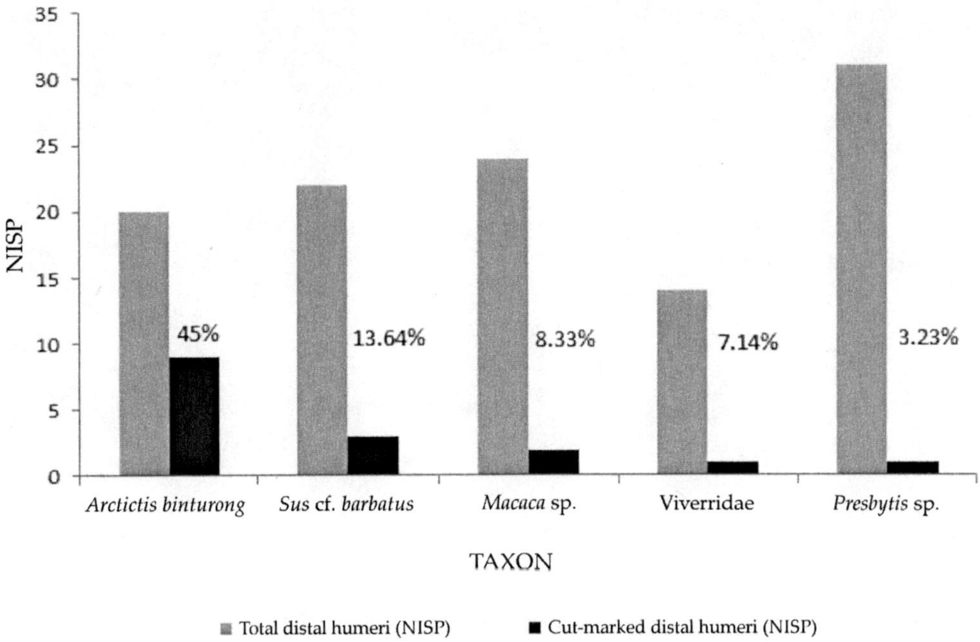

Figure 7.5. The NISP values and frequency of occurrence of butchery traces (%) to the distal humerus of *Arctictis binturong* compared to other key taxonomic groups from the Phase II cut-marked assemblage from Lobang Hangus.

Hangus, edging the terrestrial fauna into second place. This high incidence of semi-arboreal taxa coincides with the appearance of an extensive bone tool assemblage, including projectile armatures. The young adult-biased age structure of the cercopithecids from Lobang Hangus is indicative of deliberate prey selection, and projectile technology would be an effective way of achieving this patterning in the data. The presence of a comparatively high proportion of nocturnal species, suggests that trapping was also being used.

Local environmental differences almost certainly will have factored in the observed variability in prey occurrence between the Niah entrances. Temporal differences in occupation at the two entrances – early deglaciation (Area A, Early Phase II), GIS-1 (Lobang Hangus, Phase II) and Early Holocene (Area A, Late Phase II and Area D) – suggest that wider changes in climate may also be implicated. At the same time, though, there is good reason to believe that they were also shaped by the culturally informed choices in prey acquisition and processing of the people coming here and that some of those differences result from the cave complex being used by groups with subtly different habits. The occupations themselves also give us insight into the way caves may have been used and perceived by human groups in this part of the world.

Despite considerable evidence that humans were the primary accumulators of the Lobang Hangus midden, none of the Pleistocene layers there bear any evidence of hearths or ash dumps. As with Area A and Area D in the West Mouth, charcoal and burnt bone fragments are rare throughout the Last Termination (Piper and Rabett, forthcoming-a). The presence of trampling marks, butchered bone and bone tool manufacturing debris are good indication that people were active at both entrances, but precisely where they were camping is uncertain. If the records of Harrisson's work in the West Mouth and the NCP analysis of the preceding Late Pleistocene deposits are any indication, groups may have confined their site activities to the area along the cave lip only and did not venture regularly beyond the expanse of ground illuminated by daylight, but at least a portion of their camping refuse did find its way into the

Figure 7.6. (left) A view of the Upper Cave at Hang Boi looking east towards the archaeological compound; (right) access to the Lower Cave, with the opening to the cave mouth visible in the top right of the image. (Photographs by Ryan J. Rabett.)

cave mouth. The reasoning behind this choice, though, continues to elude and serves only to highlight how little we still know about cave occupation in the tropics despite more than a century of archaeological study focused upon it. Potentially, significance may have lain in the visibility of the steadily accumulating middens at these sites, less perhaps as territorial markers than as signifiers of social gathering points in the landscape and places where people went repeatedly to exploit locally abundant resources. Access to specific, perhaps more predictable food staples appears to have been an important factor in the repeated use of these sites and a characteristic that was not confined to Niah alone.

7.3 Northern Vietnam: Tràng An

As noted in Chapter 6, archaeological work in Tràng An began in mid-2007 under the author's direction and centred around the excavation of promising deposits at Hang Boi, a then recently rediscovered cave (Nguyễn Văn Lũ and Nguyến Cao Tấn, 2002; Rabett *et al.*, 2009) in the northwest of the park. This cave comprises an Upper Cave (the cave mouth) and a much larger Lower Cave, with several side chambers (Figure 7.6). The modern floor of the Upper Cave is slightly sloped from the back wall down towards the mouth and is made up of quite extensive midden deposits (greater than 200 m^2). This midden is dominated by land snail shells that have accumulated behind a rock-fall rampart at the front of the cave – almost certainly the result of a long-term process of roof collapse. Although the midden appears to extend across the entire floor area of the Upper Cave, as much as a third of it is obscured beneath a flowstone layer that also cements the surface to a depth of 10–20 cm. The excavated area began as a single L-shaped trench (Trench 1) covering nine square metres in the exposed northeast part of the cave floor. This was originally intended to extend from as near to the back wall as possible out to the drip line. Following the 2007 season, however, for the sake of practicality, the project concentrated on excavating a four-square-metre area of that in order to locate the base of the midden. The bottom of which was reached during the 2010 field season at *c.* 4 m below the surface; whereupon the loose shell and sediment was found to lay on top of further ancient roof-fall. Excavation beyond this was not attempted.

The Upper Cave also contains several additional crystalline midden and flowstone deposits. This appears to be a fairly common feature of Southeast Asian caves and one long reported

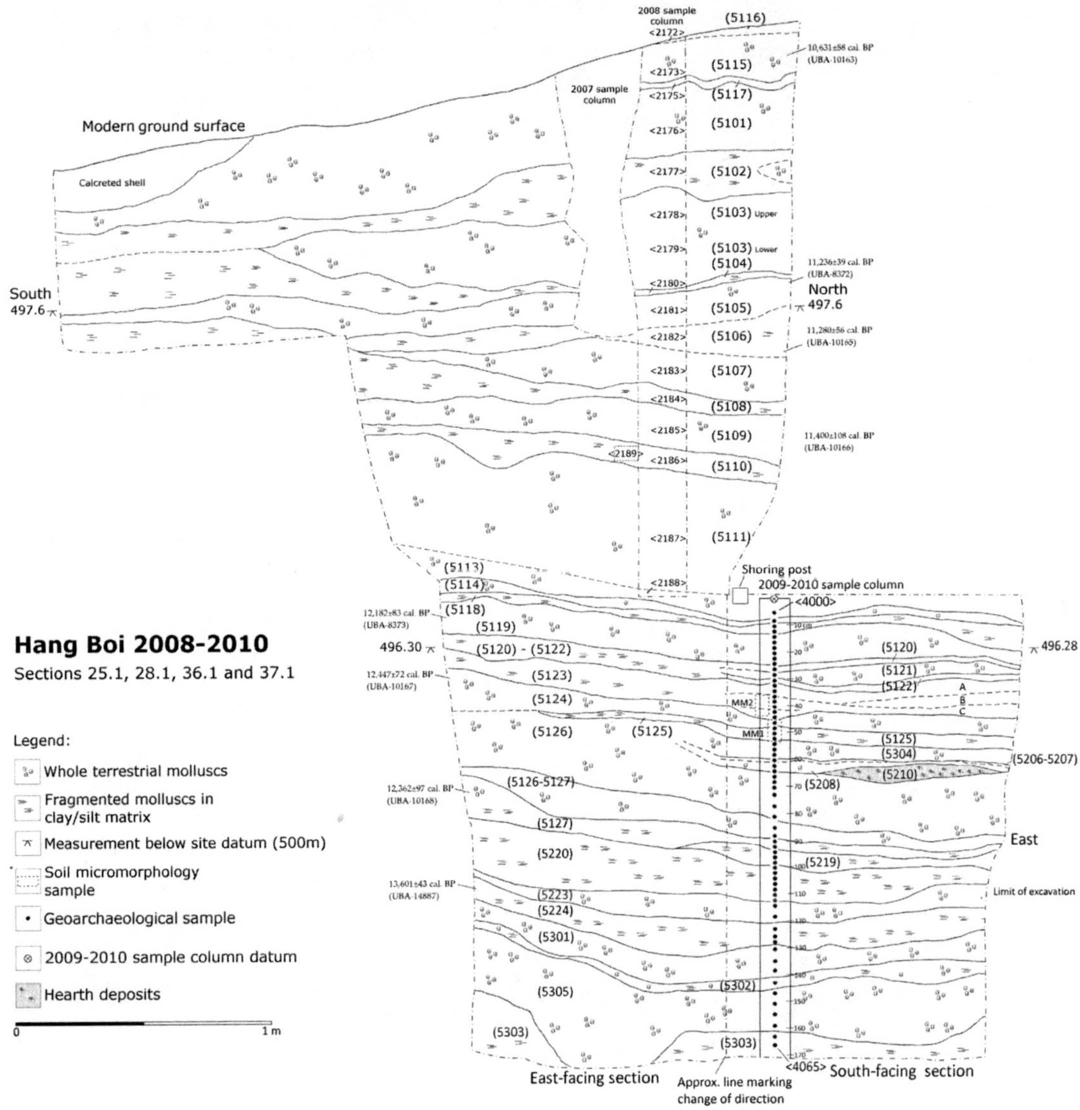

Figure 7.7. Contexts and sample columns in the east and south-facing sections of Trench 1 at Hang Boi. The figure is a composite compiled by the author of section drawings 25.1, 28.1, 36.1 and 37.1, covering excavations 2008–10, based on field drawings by L. Farr, J. Appleby, Nguyễn Cao Tấn and M. Morley.

(*see* e.g. van Stein Callenfels and Noone, 1940; Glover, 1976; Yi *et al.*, 2008). The assumption has often been that such deposits – because of their cemented state and that they are found adhering to cave walls – are of greater antiquity than the cave sediments. At Hang Boi, such deposits were found associated with cemented shell and bone remains, which we too had taken to be the remnants of older midden deposits that are no longer present in the cave (Rabett *et al.*, 2009a: 91–2); this turned out not to be the case.

Following preliminary work in 2007, a detailed stratigraphy of occupations was established and the age range of the sequence was clarified and extended by further radiocarbon dates. By the close of our investigations here, the site had nine AMS ^{14}C dates distributed throughout the excavated sequence (Figure 7.7; *see* Appendix). This established ages for the top (immediately beneath the flowstone cap), middle and basal deposits as well as specific dating of adjacent contexts to begin examination of depositional scheduling.

Amino acid racemization (AAR) dating on shell samples was undertaken by the NEaar laboratory, University of York (*see* Rabett *et al.*, 2011), with the aim of providing an independent measure from the radiocarbon chronology, to extend dating to the land snail shells that form the bulk of the midden deposits and, as part of this, to date the cemented shell and bone deposits on the cave walls. Although efforts to determine absolute ages ultimately proved unsuccessful because of high variability in shell decomposition, it was possible to establish relative age differences between the samples tested. This confirmed a trend towards increasing age with depth, without obvious reversal – lending indirect support to the ^{14}C dating programme. It also revealed that the cemented deposits were in fact most likely younger, by an estimated 3000 years or so, than the uppermost contexts of the midden (*c.* 10.6 cal. BP), rather than older. This had significant implications for our understanding of the history of occupation at the cave.

7.3.1 Midden Deposits from Hang Boi

The cave mouth midden stratigraphy at Hang Boi comprises alternating whole land snail-dominated layers containing very small proportions of fine silt matrix sediment with layers that contain much higher frequencies of crushed shell within a clay–silt matrix (*see* Rabett *et al.*, 2011). An example of this stratigraphy is shown in Figure 7.8.

The whole shell layers were found to exhibit little in the way of obvious internal depositional structure, though the upper *c.* 5 cm of individual whole shell units were frequently found to have proportionally more clay–silt matrix than the lower part of the same unit, indicating either some post-depositional translocation of fine sediment filtering downwards through the profile or spells of natural sedimentation between periods of site use. These units contained slightly more sediment closer to the front of the cave than they did farther back into the cave mouth, probably through a greater input of colluvial matter; they also contained comparatively higher quantities of vertebrate faunal remains, charcoal fragments and other cultural material. They appear to represent the discarded refuse of human occupation and are interpreted as probably accumulating comparatively quickly.

The clay–silt and crushed shell layers generally contain fewer cultural remains, and where faunal material was found to be present, it was often heavily encrusted with calcite, a consequence of diagenesis within these sediments. As was seen with the whole shell layers, there is an increase in the proportion of clay in these contexts towards the front of the cave. Internal bedding laminations were observed in some of the crushed shell layers, as was a gradual thinning of the layers towards the back of the cave mouth. These features suggest that the midden was periodically subject to low-energy water flow, almost certainly percolating through the cave roof. This process continues today after heavy rain, and it is evident in the midden as localised patches of dense calcite, particularly in the proximity of the modern drip line.

The clay–silt and crushed shell layers are thought to have formed through a mixture of anthropogenic and natural processes. The visual appearance of these contexts bears a close resemblance to the modern trampled ground surface of the cave (midden) floor, leading to the supposition that similar activity may have contributed to their more compacted appearance. The lower frequency of faunal and cultural material in these layers could be due to increased fragmentation and possible translocation caused by trampling, though substantial vertical movement of material does not seem to have occurred: there is no significant change in mean bone fragment length with depth, and also smaller elements, such as fish bones, are not confined to lower layers. In addition to human foot traffic, the higher proportion of sediment in these layers could mean that they formed over a longer period, an interpretation

Figure 7.8. The southeast-facing section of excavations at Hang Boi (2009 field season), illustrating the whole shell and sediment-rich crushed shell stratigraphy that has been revealed in the Upper Cave shell midden (scale in 0.5 m increments). (Photograph by L. Farr, reproduced with kind permission.)

that finds some support from fruit bat (*Rousettus* sp.) remains recovered in association. As fruit bats are sensitive to disturbance by human activity, their presence here may suggest that sufficient time had lapsed between forager occupations for roosting to begin in the cave. In essence, the suggestion is that the two principal structural components of the midden possibly accumulated at different rates.

Towards the base of the excavation, both whole and crushed shell contexts became considerably more clay rich than layers higher in the stratigraphy. This is a commonly reported feature of shell middens (Stein, 1992) and in many cases is thought to signal post-depositional basal saturation of the midden, whereby groundwater-hydrated clay and organic compounds within its lower strata and leached-out carbonates accumulate. Usually this feature is accompanied by a pronounced change in colour between upper (lighter) and basal (darker) midden layers. Many features of Hang Boi (rock-fall deposits, fine-grained mud and flowstone curtains and floors) are characteristic of a cave developing under vadose conditions – i.e. through water movement at or above the water table – which might support the interpretation of post-depositional saturation. There is, however, a large discrepancy between the young age of the midden deposits and the last time the cave was near to the local water table. There is also no clear colour distinction between the upper and lower portions of the stratigraphy (using the Munsell colour system). The change in sediment character is significant, nonetheless, and bone fragments from the lowest contexts exhibit demonstrable improvement in preservation, including a markedly reduced incidence of calcification. An alternative explanation is that the observed changes towards the base of the midden at Hang Boi could have formed because of increased erosion or degradation in the landscape outside the cave – a change that could have been climate-related. Detailed geoarchaeological analysis of these layers, currently under way at Oxford-Brookes and Cambridge universities, is aimed at resolving this issue.

In sum, the midden appears to comprise a series of gently sloping stable or semi-stable ground surfaces that were probably subject to periodic foot traffic, occasional low-energy water flow, incoming colluvial and other material (e.g. introduced by root systems) and phases of increased natural sediment build-up, occurring between periods of more rapid accumulations from human camp refuse (*see* Rabett *et al.*, 2011: figure 4). Typified by the accumulation of the predominantly whole shell units, these periods of activity also introduced vertebrate faunal remains and other archaeological material.

7.3.2 Faunal Remains

The invertebrate (crab) and vertebrate faunal analysis discussed here was carried out by the author on a one-square-metre column of contexts extending from the top to the base of excavations in 2008 (context 5127; *see* Figure 7.7). The results of that study are supplemented by preliminary identifications from the 2009 fauna.

7.3.2.1 Invertebrate Fauna (Molluscan and Decapod Remains)

Examination of a sample of the molluscan remains from Hang Boi that were collected and analysed by K. Szabó (Rabett *et al.*, 2009: table 3; Table 7.4) showed that *c.* 98 per cent of shells were land snails, with *c.* 91 per cent belonging to just two species: *Cyclophorus theodori* and *C. unicus* – a level of incidence that appears to persist through the excavated sequence. Although a well-reported feature of Hoabinhian deposits in this part of Vietnam (e.g. Hoáng Xuân Chinh, 1991), the status of land snails as a deliberately collected and consumed resource by humans has been rarely explored in detail. This is perhaps partly due to the fact that methods of distinguishing between collected and self-introduced assemblages are only now being developed (Lubell, 2004a, 2004b; Szabó *et al.*, 2003).

Multiple lines of evidence point towards the Hang Boi land snail component being predominately anthropogenic (*see* Rabett *et al.*, 2009a: 94–7). Firstly, the density of land snails in the midden (accumulating over comparatively short periods) and the geological and hydrological situation of the cave preclude the shells from being blown, washed or dropped into the cave. Secondly, the presence of such a large proportion of whole specimens is inconsistent with what would be expected from natural snail predators such as birds or rodents (though a small proportion have been found to have holes in them, and therefore some minor accumulation via this vector is possible). Thirdly, the biogeographical ranges of the snail taxa found in the environments beyond the cave indicate that local land snail diversity is rich (and seasonal). Although only a rudimentary survey has been undertaken as part of the work at the site, this demonstrated the presence of several different families. These included Cyclophoridae (including a range of genera such as *Leptopoma, Scabrina?, Japonia, Pterocyclos* and *Cyclophorus*), Camaenidae, Clausiliidae, Pupinidae, Zonitidae (Trochomorphinae), Subulinidae and Streptaxidae. The survey did not reveal any live populations of *Cyclophorus* spp. in the immediate environs of the cave, a fact that suggested against the existence of

Table 7.4. Identified Invertebrate Macrofaunal Remains from the Upper Cave Shell Midden (2007–8 Samples), Hang Boi

Order/Class	Family	Taxon	MNI	% MNI	NISP
Mollusca: Gastropoda	Neritidae	*Neritina* sp(p).	3	0.13	–
	Thiaridae	*Brotia* sp.	5	0.22	–
		Melanoides tuberculata	1	0.04	–
	Cyclophoridae	*Cyclophorus theodori*	1527	68.11	–
		Cyclophorus unicus	512	22.84	–
		Cyclophorus spp.	160	7.14	–
		Scabrina cf. *denudata*	1	0.04	–
	Vivaparidae	*Viviparus costatus*	5	0.22	–
	Pupinidae	*Pupina* sp.	1	0.04	–
	Cypraeidae	*Cypraea* sp.	1	0.04	–
	Streptaxidae	*Odontartemon costulatus*	1	0.04	–
	Camaenidae	*Camaena* sp.	16	0.71	–
Mollusca: Bivalvia	Corbiculidae	*Corbicula* sp.	4	0.18	–
	Unionidae	*Cristaria herculea*	1	0.04	
		Unionidae spp.	4	0.18	
Decapoda	Potamidae	*Villopotamon* sp.	–	–	401
		Potamidae sp(p).	–	–	696
		TOTAL (MNI/NISP)	**2242**	**100.00**	**1097**

Note: MNI = minimum number of individuals; NISP = number of identified specimens. Molluscan identifications are after Webb (1948), Abbot (1989) and Nguyen (2005). Values are updated from Rabett *et al.* (2011) to accord with the same sample criteria of the vertebrate fauna used herein. Molluscan analysis was carried out by K. Szabó; initial taxonomic identification of the decapod material was conducted by T. Naruse and D. Yeo. Specific epithets are as follows: 'sp.' indicates confidence in attributing specimens to a particular genus but uncertainty with respect to which species within that genus they belong; 'spp.' indicates confidence that there is more than one species represented in the study sample but uncertainty as to which species those are; 'sp(p).' indicates that there is *probably* more than one species present in the sample.

an immediately local population that would have been capable of self-introduction. The diversity, coupled with the relative natural abundances of individuals within these families, contrasts starkly with the more selective composition of the deposits at Hang Boi. Fourthly, the alternating stratigraphy of the midden between whole and crushed land snail layers and discrete patches of burnt shell are positive indicators of human presence as the chief accumulator. Finally, and in connection to this, there is also evidence from recent use of the cave by local hunters. This seems to follow a pattern very reminiscent of that seen in the archaeological deposits: small numbers of whole cyclophorids were brought to the site and eaten around a small fire (Figure 7.9).

Collectively, these features strongly point towards human collection and equally clearly towards a discerning selectivity on the part of the collectors for *Cyclophorus* land snails. Despite the site's proximity to freshwater, only very low frequencies of freshwater mollusc species were identified. These included the gastropods *Brotia* sp. and *Angulyagra* (= *Viviparus*) *polyzonata* and bivalves, including fragments of the large *Cristaria herculea*, the clam (*Corbicula* sp.) and freshwater mussel (Unionidae). In the 2007 studied sample, collectively, these genera accounted for only $n = 21/2239$ MNI, or 0.94 per cent. They come from a diverse range of freshwater niches. For example, *Brotia* sp. generally favours clean, running water; *Angulyagra polyzonata* is more silt tolerant and associated with calmer waters; while unionids are found

Figure 7.9. The remains of a hunter's small fire found on the open floor of the Upper Cave at Hang Boi in 2010. Newly introduced cyclophorid shells lie scattered around and within the charred wood (scale in 0.5-m increments). (Photograph by Ryan J. Rabett.)

buried in soft sediment of stream beds. If people were exploiting these niches for molluscs, they seem not to have been bringing them back regularly to Hang Boi in a quantity that would have made any significant calorific contribution (but see later).

Evidence that forgers using the cave were exploiting estuarine resources is currently almost non-existent despite the comparatively local proximity of this habitat during later occupation in particular (Rabett *et al.*, 2011). There is some evidence for contact with the coast during the last millennia of site use through the presence of a fragment of cowrie (*Cypraea* sp.) that was possibly worked and pigmented. The surviving morphological features of this shell preclude it from being *Cypraea moneta*, *Cypraea annulus* or *Cypraea caputserpentis*, though a positive identification could not be established from what remains (K. Szabó, pers. comm.). A small number of recovered shells do appear to have been used for ornamental purposes. This is shown by the recovery of three perforated Neritidae shells – probably either *Neritina (Dostia) cornucopia* or *Neritina (Neripteron) violacea* (*after* Qin, 1997; Siong and Clements, 2008; Nguyen Ngoc Thach, 2005; Figure 7.10). Both of the proposed species favour mangrove environments, and this could imply that these were carried back to the site from coastal estuary locations.

The high proportion of decapods (crabs) in the midden's fauna comes via large numbers of fixed and movable chelae (NISP). While one might expect to find lowland-dwelling parathelphusids (Ng and Kosuge, 1995) in such deposits, only potamids were found to be present, and only those belonging to a single, recently defined genus, *Villopotamon* sp. (Dang and Ho, 2003; Yeo and Ng, 2007; Figure 7.11). While it has not yet been possible to implement a comprehensive survey of the modern decapod fauna from Tràng An, on the basis of the limited work that has been undertaken (by Nguyêñ Van Son), modern specimens belong to a different family: Varunidae (D. Yeo, pers. comm.). This may indicate that there has been a turnover in decapod fauna since the site's occupation, or it may be further evidence of deliberate foraging selectivity. Either way, it is very apparent that people visiting

Figure 7.10. Three pierced Neritidae shells recovered from midden contexts at Hang Boi. The shell on the left is from context (5123) dated to *c.* 12 cal. KBP; the shell in the middle came from a test sondage in 2007 and is unsecurely dated, though based on its broad position in the stratigraphy, it is probably of Early Holocene age; the shell on the right came from context (5121), overlying (5123), and is thus of the same approximate age (scale in 1-mm increments). (Photographs by Ryan J. Rabett.)

Hang Boi were targeting local streams and rivers for crabs and bringing what they caught back to the cave. Either exploitation of freshwater molluscs did not figure in this activity or the greater part of those collected was generally consumed elsewhere.

7.3.2.2 Vertebrate Fauna

The presence of calcified sediment adhering to many of the excavated bone fragments, particularly from the upper part of the excavation, did effect bone identification and the study of surface modification. In spite of this, a small number of bones in the assemblage sample were found to exhibit evidence of chewing. The marks are not consistent with rodent gnawing and likely represent chewing by people or small carnivores. Evidence of butchery was noted, as was that of burning - represented by mostly charred bone and shell, as well as discrete lenses of charcoal and ash (*see* e.g. Rabett *et al.*, 2011: figure 6). Although overall levels of burnt bone were low, charcoal recovery was high for most contexts. Observed changes in species representation provide grounds for tentatively separating this assemblage into three different phases of activity between *c.* 12.4 and 10.5 cal. KBP.

7.3.3 Occupation Phases at Hang Boi

7.3.3.1 Late Glacial

Analysis of material from the earliest period of site use is still ongoing. From the 2008 field season, 285 fragments have been analysed from contexts (5120–5127) – 31 of these were identifiable to taxa – permitting only preliminary observations here. This is the only part of the sequence to contain a chronological reversal. Context (5123) is dated to 12,447 ± 72 cal. BP (UBA-10167), while context (5127), five layers stratigraphically below, returned a slightly younger age of 12,362 ± 97 cal. BP (UBA-10168; Figure 7.7). Although the standard errors do still overlap, the exact cause of this aberration is not known. There seems little to indicate significant reworking or post-depositional movement in any of the midden deposits excavated, but localised settling of materials cannot be ruled out and may have been the factor here.

Figure 7.11. (left) *Villopotamon* sp. Raffles Museum of Biodiversity Research at the University of Singapore. (right) Specimens of archaeological chelae from Hang Boi (scale in 5-mm increments). (Photographs by T. Naruse (left), reproduced with kind permission, and Ryan J. Rabett (right).)

Whole and crushed-shell layers are present right through the studied sequence (and indeed to the base of the midden), indicating that land snails were being collected regularly from the site's earliest occupation, as were freshwater crabs, confirming the long-term availability and reliability of these resources. As noted, this part of the midden is clay rich, and bone preservation is comparatively good, suggesting that the low frequency of bone recovery is a genuine feature of deposition rather than a product of post-depositional loss. The remains of fish and a single fragment of turtle further point to the exploitation of local waterways, while the presence of *Macaca* spp. remains can be taken as an indicator of local karstic forests. The proximal humerus of one cercopithecid specimen was clearly cut-marked (note, this is something not seen at all in the much larger Niah assemblage).

The only bone point to come from Hang Boi was recovered from these deposits (*see* Chapter 6; Rabett and Piper, 2012: figure 2). The piece in question was longitudinally scraped back to an old basal fracture. The extreme tip is missing, but the pronounced rounding and polish are more consistent with an implement used in a perforating function than as a projectile for taking arboreal game.

Carnivore remains are also in evidence from these late glacial layers. Most notably, this period sees the only occurrences of leopard (*Panthera* cf. *pardus*) recovered from the site. This animal is presently represented by two fragments: a fused distal fifth metacarpal that has been clearly burnt and a complete (L) second metacarpal (Figure 7.12). This species is renowned for its wide distribution globally, from sub-Saharan Africa to the Middle East and Asia (Meijaard, 2004). It has the broadest diet of large predators, and as a result body mass is highly variable between leopard populations living in different environments; both factors attest to its adaptive flexibility. Although its diet is catholic, when averaged across its whole range, the body masses of leopard populations have been found to be significantly correlated with those of prey species weighing 10–40 kg (Hayward *et al.*, 2006). Leopards tend to target prey that have a body weight less than their own, avoiding those which have the potential to inflict life-threatening injuries and favouring ones that can be stalked in dense, cluttered habitats. Evidence gathered in southern India (Ramakrishnan *et al.*, 1999) suggests that pigs may be too aggressive and dangerous for leopards to risk tackling; the same is thought to be true of sambar deer (*Cervus unicolor*). In one of the few studies of this animal to have been carried out in Southeast Asia, Grassman (1999) reported that leopards

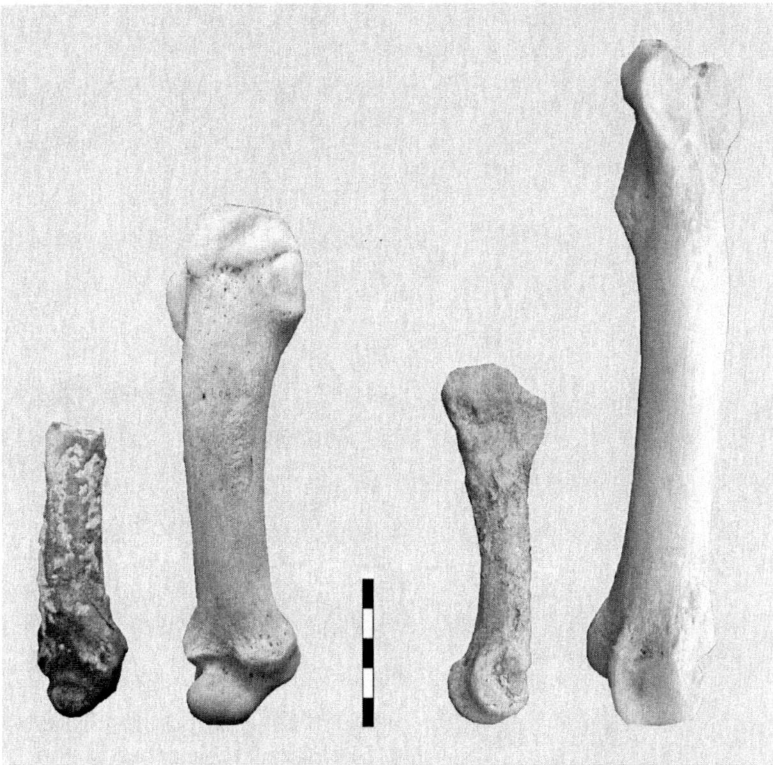

Figure 7.12. Leopard (*Panthera* cf. *pardus*): (far left) specimen (HBC-28505) distal (L) fifth metacarpal from context (5122), Phase I; (right of scale) specimen (HBC-28932) (L) second metacarpal, excavated in 2009 from context (5220), Phase I. Both pieces are compared against *Panthera tigris* (Grahame Clark Laboratory, University of Cambridge; scale in 5-mm increments). (Photographs by Ryan J. Rabett, comparative image reproduced with permission.)

from the Kaeng Krachan national park in northern Thailand were predominantly targeting hog badger (*Arctonyx collaris*) and muntjac (*Muntiacus muntjak*), though at least nine different taxa were identified from scats, including the remains of pig, cat, viverrid, porcupine and monkey (all at low frequencies).

The presence of *Panthera* cf. *pardus* at Hang Boi provides further suggestion that the upland landscape of this area was probably still forested. Given the leopard's wide dietary pallet, finding it does not demonstrate the local availability of larger forest game, but it does hint at their likely presence even if the people who were coming to this site were only infrequently hunting them. The 2009 find of a *Sus* sp. pelvic fragment from the midden's basal levels would seem to confirm this.

7.3.3.2 Pleistocene–Holocene Transition

An increase in both the incidence and range of identifiable fauna from context (5119) upwards to (5110) suggested a change in site occupation around the time of the Pleistocene–Holocene transition, based on two dates: 12,182 ± 83 cal. BP (UBA-8373) and 11,400 ± 108 cal. BP (UBA-10616). Alternating whole and crushed-shell components characterize this section of the stratigraphy. Freshwater crabs too feature strongly in the invertebrate assemblage. Freshwater shells were noted, but they were found only occasionally, and whether they appear during this phase for the first time will not be known until analysis is complete. A total of 410 fragments of vertebrate fauna were studied from this phase of site use in the 2008

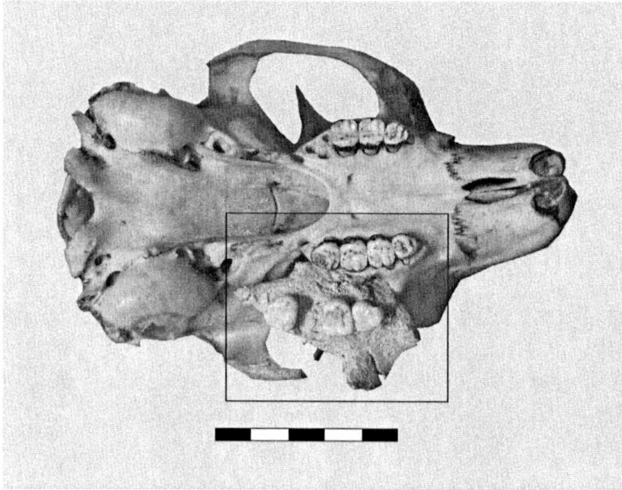

Figure 7.13. Archaeological specimen (HBC-28662) (L) maxilla fragment with PM^1, M^1 and M^3 *in situ*, context (5113), compared alongside *Ratufa* (= *Sciurus*) *bicolor* (University Museum of Zoology, Cambridge, Accession No. E.1572.A; scale in 5-mm increments). (Photograph by Ryan J. Rabett, comparative image reproduced with kind permission.)

sample; $n = 29$ permitted taxonomic identification. Within this, the proportion of primate remains drops compared to the late glacial, as does that of carnivores and fish. Deer, squirrel, snake and bird remains appear for the first time. The squirrel remains include a positively identified maxilla from a black giant squirrel (*Ratufa [=Sciurus] bicolor*; Figure 7.13). Whether a hunted food item or a natural death is unclear at present, though adult weight is *c.* 2 kg, and incidence of sciurid remains at the site may indicate predation of this shy species, which spends most of its time in the forest canopy (Lekagul and McNeely, 1988: 343).

Snake bone was largely composed of rib fragments and is, therefore, over-inflated relative to other taxa. Snakes may have formed part of the diet of foragers but, as with the squirrel, and bird remains, self-introduction is also a possibility. Fragments of pig continue to occur at low frequencies, as do chelonids – *Cyclemys oldhami* is positively identified for the first time. Comparatively intense carcass exploitation is hinted at during this phase by two instances where large mammal extremities had been split open to extract marrow.

Occupation during the late glacial and across the Pleistocene–Holocene transition took place at a time when the landscape beyond the massif was cut by a meandering river system and when the sea level was −60 to −50 m or lower compared to its modern value (Tanabe *et al.*, 2006: figure 4). The Gulf of Tonkin was open to the ocean, and at this point, Hang Boi lay *c.* 60 km from the nearest coast. The observed changes in vertebrate faunal composition do not exactly match known stadial and interstadial substages during the Last Termination but do chronicle human use of upland forest and freshwater habitats from GS-1, without apparent interruption but with subtle changes in site use. On-going post-excavation work from the Hang Boi excavation (including macrobotanical, lipid and pollen analysis, together with geoarchaeological work) will ideally bring greater resolution to these patterns of site use.

7.3.3.3 Early Holocene

The uppermost portion of the stratigraphy (5109–5100) dates to the Early Holocene, 11,400 ± 108 cal. BP (UBA-10166) to 10,631 ± 58 cal. BP (UBA-10163), and carries by far the bulk of the identified remains from the widest spectrum of taxa (*see* Table 7.5). Although the same caveat applies concerning chelonids, the fauna from this period contains a appreciable

Table 7.5. All Vertebrate Fauna Identified and Quantified from the Cave Mouth Midden at Hang Boi

					Hang Boi					
					Late Glacial		Pleistocene–Holocene		Early Holocene	
Class	Order	Family	Taxon	Common Name	NISP	%	NISP	%	NISP	%
Pisces	–	–	–	Fish	10	32.26	5	17.24	60	22.39
Chondrichthyes	Rajiformes	Dasyatidae	Dasyatidae sp.	Sting ray	0	0	0	0	[a]	0
Reptilia	Testudines	Geoemydidae	*Cuora (=Pyxidea) mouhotii*	Keeled box turtle	0	0	0	0	6	2.24
			Cuora cf. trifasciata	Chinese three-striped box turtle	0	0	0	0	[a]	0
			Cyclemys oldhami	Dark-throated leaf turtle	0	0	1	3.45	1	0.37
		–	Chelonia spp.	Turtle species	2	6.45	1	3.45	71	26.49
	Squamata	–	–	Snake	0	0	6	20.69	61	22.76
Aves	Galliformes	Phasianidae	Phasianidae sp.	Wild fowl/Pheasant?	0	0	0	0	1	0.37
	–	–	–	Bird	0	0	1	3.45	9	3.36
Mammalia	Scandentia	Tupaiidae	*Tupaia glis (=belangeri)*	Common tree shrew	0	0	0	0	1	0.37
	Soricomorpha	Soricidae	Soricidae sp.	Shrew	1	3.23	0	0	0	0
	Primates	Cercopithecidae	*Macaca arctoides*	Stump-tail macaque	0	0	0	0	[a]	0
			Macaca cf. *assamensis*	Assamese macaque	0	0	0	0	[a]	0
			Macaca nemestrina	Pig-tailed macaque	0	0	0	0	[a]	0
			Macaca cf. mulatta	Rhesus macaque	0	0	0	0	2	0.75

(continued)

Table 7.5 (*continued*)

Class	Order	Family	Taxon	Common Name	Hang Boi: Late Glacial NISP	Hang Boi: Late Glacial %	Hang Boi: Pleistocene–Holocene NISP	Hang Boi: Pleistocene–Holocene %	Hang Boi: Early Holocene NISP	Hang Boi: Early Holocene %
			Macaca sp.	Macaque sp.	1	3.23	0	0	1	0.37
			Cercopithecidae sp(p).	Monkey	11	35.48	5	17.24	18	6.72
	Rodentia	Sciuridae	*Ratufa (=Sciurus) bicolor*	Black giant squirrel	0	0	1	1.12	[a]	0
			Sciuridae sp(p).	Squirrel	0	0	1	3.45	6	2.24
		Muridae	*Rattus* sp.	Rat	0	0	0	0	2	0.75
		Hystricidae	Hystricidae sp.	Porcupine	0	0	0	0	1	0.37
	Carnivora	Mustelidae	*Arctonyx collaris*	Hog badger	0	0	0	0	[a]	0
			Mustelidae sp(p).	Mustelid sp.	1	3.23	0	0	1	0.37
		Viverridae	*Paradoxurus hermaphroditus*	Common palm civet	0	0	0	0	[a]	0
			Herpestes sp.	Mongoose	0	0	1	3.45	0	0
			Viverridae sp(p).	Civet	1	3.23	2	6.90	2	0.75
		Felidae	*Panthera* cf. *pardus*	Leopard	2	6.45	0	0	0	0
			Felis sp(p).	Felid sp.	1	3.23	0	0	1	0.37
		Suidae	*Sus* sp.	Pig	1	3.23	2	6.90	5	1.87
	Artiodactyla	Cervidae	Cervidae spp.	Deer species	0	0	3	10.34	19	7.09
		Bovidae	*Bos* sp.	Cattle	0	0	0	0	[a]	0
		TOTAL (NISP and %)	**31**	**100**	**29**	**98**	**268**	**100**		

Note: Chiroptera fauna has been identified but not yet quantified: genera included *Hipposideros diadema*, *Miniopterus* sp., *Rousettus* sp., *Rhinolophus* sp. and *Taphozous* sp. Quantities and relative proportions of other remains were determined from the complete 2008 1 m^2 study sample column; *Panthera*, *Felis* and *Sus* also include three individual finds from 2009. Specific epithets are as follows: 'sp.' indicates confidence in attributing specimens to a particular genus but uncertainty with respect to which species within that genus they belong; 'spp.' indicates confidence that there is more than one species represented in the study sample but uncertainty as to which species those are; 'sp(p).' indicates that there is *probably* more than one species present in the sample.

[a] Identification to genus or species made in the 2007 Early Holocene sample that did not appear in the 2008 sample from that period.

Figure 7.14. (left) Keeled box turtle (*Cuora mouhotii*). (Photograph by T. Blanck, reproduced with kind permission.) (right) Nuchal fragment (HBC-28579), as indicated on the comparative, identified to this species from context (5109), 11.4 cal. KBP (scale in 5-mm increments). (Photograph by Ryan J. Rabett.)

emphasis on Geoemydidae. At least two species have been positively identified from this phase: *Cuora mouhotii* (keeled box turtle; Figure 7.14) and *Cyclemys oldhamii* (dark-throated leaf turtle; Blanck, pers. comm.). The former is a predominantly terrestrial species that favours moderately moist habitats such as are found around the base of hills; the latter is commonly found in small water courses in upland environments (Bonin *et al.*, 2006). *Cyclemys oldhamii* replaces the previously reported identification (*Cyclemys dentata*; Rabett *et al.*, 2009a). *Cuora* cf. *trifasciata* was also tentatively identified in the 2007 sample, a species that inhabits clear or partly vegetated shallow upland streams and ponds, to an elevation in Vietnam and Laos of 200–800 m (Blanck *et al.*, 2006). In terms of element representation: limb elements, carapace and plastron were all identified, as were scapulae and pelvic fragments. Cut-marks to a femur and humerus from the Early Holocene portion of the assemblage confirms predation of these reptiles.

Terrestrial game included cervids and suids. Available comparatives precluded definitive higher taxonomic identification of either, though, based on size, *Cervus unicolor* are certainly and *Muntiacus* sp. (probably) represented. Both of these deer are forest browsers, with the former being crepuscular, nocturnally active and favouring wooded areas, while the latter is more active during the day and forages along the forest edge or within woodland clearings (Lekagul and McNeely, 1988; Payne and Francis, 1998).

Some differences in element representation and, therefore, carcass treatment were observed with regard to these taxa. Although the cervid remains predominately consist of extremities, some limb elements and cranial bones, including fragments of antler and teeth, were also recorded. One cut-marked fourth carpal was identified. The small number of *Sus* sp. remains comprised extremities, with the exception of two fragments of canine. Cut-marked pig bone from this period included a single fifth metatarsal and a cuboid. While at the very least, carcass portions of cervids were being brought to the site and processed, the picture for suids looks to have been different. The absence of any meat-bearing pig bones may indicate either that these were never brought to the cave or they were removed off site after disarticulation.

The presence of cercopithecids points to the continued existence of local forest stands during the Early Holocene. Dental remains (including material recovered as part of the 2007 faunal sample) have been ascribed to the rhesus macaque (*Macaca mulatta*) and Assamese macaque (*Macaca assamensis*) on the basis of morphological and biometric comparisons with Swindler (2002). The *M. mulatta* specimen included a partial piece of the left mandible that was cut-marked on the buccal surface – probably related to skinning or the removal of the mandible from the cranium. Cut-marks were also found on a fragment of cercopithecid distal humerus, close to the olecranon fossa. *M. mulatta* and *M. assamensis* are sympatric in modern northern Vietnam, where they are restricted to broadleaf evergreen forests (Sterling *et al.*, 2006).

In addition to these remains, three skeletal fragments from the pig-tailed macaque (*Macaca nemestrina*) and one from the stump-tailed macaque (*Macaca arctoides*) were identified in the 2007 sample from this phase (comparisons were made against specimens held at the Natural History Museum, London). *Macaca arctoides* currently inhabits Tràng An, deftly moving around the sheer faces of its karstic towers (Figure 7.15). This species is found throughout wooded upland areas of the country. While *Macaca nemestrina* tends not to inhabit coastal forests regionally, reports from Vietnam indicate that it may be found in mangrove and peat swamp forests and is less restricted in its habitat preferences than *M. mulatta* and *M. assamensis*. While co-occurrence of these different species of macaque is known, it is presently confined to an area between 14° and 17° latitude (Sterling *et al.*, 2006); Tràng An lies north of this, at *c.* 20° latitude. A modified Early Holocene distribution is feasible, but further analysis will also be required to completely clarify this. There is no evidence from Hang Boi to suggest that bone technology was being used in the acquisition of any of these primates or other arboreal taxa such as squirrels, which are today taken using traps hung in trees (Rabett, pers. observation).

Small carnivores are a further feature of the Early Holocene sample assemblage. Felids, mustelids and viverrids are all represented in low frequencies. The presence of cut-marks on two felid elements indicates that these animals were definitely being targeted as prey. Unlike the situation at Lobang Hangus, however, they do not appear to have been a major component of the diet. Also recorded from this phase are fragments identified as tree shrew *Tupaia glis* (= *belangeri*) and of murid. These are unlikely to have been introduced by people – though, potentially, the latter may have been attracted to human food waste.

This phase of occupation at Hang Boi was capped by a flowstone layer that was initially thought to represent the end of sustained site use. If the relic midden material adhering to the cave walls above this level does indeed date to the Mid-Holocene, as is now suspected, isolated fragments of Da But pottery found in the Lower Cave may have been deposited during that continued use rather than later, after a hiatus (Rabett *et al.*, 2011). The extensive nature of flowstone in the cave mouth suggests that there was either a significant alteration in the cave's hydrological system or an extended interval of increased rainfall and water percolation during the Early Holocene. The latter seems more likely.

In the landscape beyond the cave, this phase of site use was associated with a time – from *c.* 11.6 to 10.2 cal. KBP – when the area around Tràng An was dominated by salt marsh and lagoon habitats, as determined from analysis of the ND-1 sediment core (Tanabe *et al.*, 2003a, 2003b, 2006), which lies 29.5 km northeast of Hang Boi. Correlation between data in the ND-1 core (Unit 2.2) and regional sea level rise during this period (Tanabe *et al.*, 2003b) indicates a sea level rise from approximately −33 to −25 m.

Notwithstanding the general proximity of coastal habitats, only one questionable piece of evidence suggests that food resources acquired there were being brought back to the cave. This is a single, small fragment of stingray (Dasyatidae sp.) spine recovered in 2007 from a context dated to 11,236 ± 39 cal. BP (UBA-8372), the early part of this phase, but it is unknown if this is an estuarine or freshwater ray. Despite increasing proximity, groups camping at Hang Boi in the Early Holocene appear not to have been exploiting these habitats. At least four possibilities may be entertained to explain this situation.

First among these, the paucity of evidence may have been due to practical limitations of accessing the coast through the dissected terrain of the massif itself. Despite its elevated location, the surrounding lowlands almost certainly would not have been visible from the

Figure 7.15. One of a population of *Macaca arctoides* (stump-tailed macaque) that currently inhabits Tràng An Park. (Photograph by Ryan J. Rabett.)

cave or its immediate environs. Under current conditions, sub-coastal lowlands lie less than 1 km away to the northwest of Hang Boi and no more than 2.5 km away to the east. The landscape away from established paths and open dolines within the massif, however, comprises broken and jagged limestone outcrops covered with dense secondary forest. To give a sense of the current difficulty involved in traversing this kind of environment, taking the most navigable route to a second known upland cave, just over one kilometre from Hang Boi (with guides and a GPS), involved a *c.* 4 hour round trip just to reach the foot of the hill where this cave was situated (*see* Rabett *et al.*, 2009b). If the local terrain was comparable to the present day during the occupational periods at Hang Boi, this may have affected movements around the landscape and prompted meandering rather than direct routes of travel and may have limited daily movements beyond the internal network of dolines.

A second possibility relates to how far inland coastal resources are likely to be transported. Erlandson (2001) has suggested that a coastal foraging zone should be tied to the distance a group can travel in a day to collect food: 5–10 km, a figure in keeping with the standard foraging radius of ethnographic hunter-gatherers. Claassen (1998: 43) quoted a distance of *c.* 11 km based on a long-term ethnographic study along the coast of South Africa. Drawing from studies of 14 tropical forager groups, Shoocongdej (1996: table 2.7) also indicated that the range of daily forays from residential camps was in the range 2–10 km. If group movement were more 'logistical' (*after* Binford, 1980), involving same-day, overnight or multiple-day forays, these distances ranged from 3.2 km to as much as 30 km (Shoocongdej, 1996: table 2.8). Although perhaps not easily accessible, coastal habitats would probably have lain within a day's travel from Hang Boi by this time and would certainly have been reachable if that travel involved an overnight camp *en route*.

Increasing distance from the point of acquisition increases transportation costs, and this is likely to have a direct bearing on what is brought back to a residential camp on the inland edge of a foraging range. In terms of large mammals, this is often expressed through on-site biases in element representation that may reflect field butchery of a carcasses and portage of preferred portions: the 'schlepp effect' (*after* Daly, 1969). A similar logic has been proposed through ethnoarchaeological research in the Torres Strait for the transportation of collected marine molluscs (Bird *et al.*, 2002). These authors concluded that where locations of procurement and consumption are different, the faunal assemblage of residential sites may be heavily biased against the archaeological recovery of bulk (shell) remains from central prey species such as large clams. The flesh of these is liable to have been removed and the shells discarded in the field before returning to the residential camp as a way to maximise collecting returns. With reference to ethnographic evidence among native coastal communities of the Joaquin and Sacramento valleys of California, Cook (1946: 51) similarly noted that while some molluscs were shipped inland in the shell, it was far more economical to export the meat after it had been dried. What is clear for Hang Boi is that groups were fishing and collecting, and returning with, large numbers of freshwater crabs and other riverine resources during this period. The same schlepp effect model might have greater value in explaining the low incidence of freshwater molluscs: what appears of these taxa in the midden could be residual to what was collected and mostly consumed elsewhere.

The implication in the current context is that the distances involved between Hang Boi and any exploited coastal resources may well have left a bias against finding evidence of marine mollusc remains at the site. Taking the return rates of different species as a guide, Bird *et al.* (2002: table 1) predicted at what distance field processing becomes the more economic option. While the transportation of certain species could clearly be affected by this molluscan schlepp effect and become archaeologically invisible at Hang Boi, the shells of other collected species included in this study (such as neritids) were found to be carried much farther inland, *c.* 5 km, one way. The presence of the neritid and the cowrie shell fragment in the Early Holocene occupation at Hang Boi supports the idea that people coming to the site did have some degree of contact with the coast. Indeed, perforated neritids begin to appear on site *c.* 12.4 KBP, suggesting such links were already in place at a time when the coast was even farther away. This notion finds support in another recent and currently unique find in the region: a Hoabinhian female burial where cowrie shells (*Cyprea arabica*) were placed in the eye sockets (Cuong, 2007). No radiometric dates are available for this find; however, the current distance of the site (Phia Vai) to the modern coast is *c.* 500 km, indicating that whatever the actual age, these shells were being transported or traded a considerable distance inland. At Hang Boi, shells were acquired as part of group movements or through an exchange system, but still the evidence that this also involved procurement for subsistence is not strong.

A third scenario that might explain the situation at Hang Boi is that despite apparent coastal orientations elsewhere in the region, such as along the northwest coast of Peninsular Malaysia or the northeast coast of Sumatra (Endicott and Bellwood, 1991; Edwards McKinnon, 1991), in Vietnam, the Hoabinhian is a characteristically inland phenomenon. Ha Van Tan (1997) reported of the existence of a single open-air site (out of more than a hundred), dated to 5184 ± 110 cal. BP, which has a maritime focus: Bau Du, in Quang Nam Da Nang Province. If this characterisation is correct, then it is possible that Hoabinhian subsistence was broadly geared towards the exploitation of inland resources – an illustration of the fact that proximity to coastal resources does not automatically mean they are immediately adopted to become pre-eminent in forager economies (*see* e.g. Szabó *et al.*, 2003). While this may fit the evidence from Hang Boi, as noted, the cultural remains from the site can be considered as only provisionally Hoabinhian.

A fourth explanation for the absence of maritime resources at Hang Boi is that local shoreline exploitation was simply not economic enough to justify the entailed transport costs of acquiring them. The productivity of maritime habitats would have been strongly linked to the rate of coastal inundation. A detailed sediment study into the response of salt marsh habitat to climatic fluctuations (specifically, Heinrich events) has been carried out for the Late Pleistocene in Venezuela (Gonzalez and Dupont, 2009). The evidence presented indicates that only pioneer species are able to cope with the stressful high rates of change and erosion during the early stages of coastal succession. It is only once the rate of sea level rise has slowed that vegetation can become more firmly established, forming sediment traps and raising productivity. Woodroffe (1990) calculated that inundation rates exceeding *c.* 0.8–1 mm/yr (0.08–0.1 m/100 yr) tend to preclude mangrove from becoming established. Mangrove forests did not appear in the vicinity of Tráng An until *c.* 9 cal. KBP (Tanabe *et al.*, 2003a). Between 8 and 6 cal. KBP, the shoreline remained stable in the vicinity of nearby Ninh Bình town, creating marine notches in the surrounding karstic limestone uplands, including Tráng An. It is still to be determined but this may indicate that nearby coastal habitats were being subjected to high inundation rates during the Early Holocene occupation at Hang Boi, that this hampered the establishment of coastal habitat succession and with it, perhaps, the desirability of this habitat for foraging.

The material evidence of contact with the coast or with coastal communities does raise two further possibilities: that access to the coast may have been socially constrained or else formed part of seasonal movements. The former is hard to demonstrate but should not be overlooked. Potential support for the latter can be more readily found, as Nguyen Viet (2004) has indicated, in the fact that cyclophorids are still actively collected during the wet season (April–October). Observations by the author concur that these snails are much less noticeable in the landscape during the winter dry season. Under this scenario, marine (and other) resources may have been collected and consumed at a different time of the year and entirely separately from the occupation of Hang Boi. On present evidence, any one of these scenarios (or a combination of them) could account for the position at Hang Boi, and this serves to highlight that sites in this region need to be considered as a consequence of forager mobility rather than as self-contained settlement entities.

The Early Holocene occupation was the most intensive and diverse uncovered at the site. There are strong indications that people were using this location to exploit upland and local freshwater resources. Hints of contact with the coast are found, but there is little to suggest that visits to coastal habitats formed part of subsistence activities carried out here at this time. The limited recovery of freshwater molluscs may reflect transportation choices that resulted in most being consumed or processed away from the cave.

Evidence at other later sites in this area of northern Vietnam suggests that this economic picture would change as the Holocene progressed, and the emergence of the Da But culture took shape along the coastal lowlands of the Gulf of Tonkin. The Da But is regarded as providing clear evidence for human adaptation to the Mid-Holocene sea high-stand (Nguyen Viet, 2005). Evidence from the type site indicates that by *c.* 7445 ± 45 cal. BP (Bln-3509 II; but dated on *Corbicula* sp.), a complex economy existed at this open-air locale that was drawing on resources from adjacent uplands, freshwater swamp, lake and riverine habitats and extending out into the sea (Nguyen Viet, 2005). The history of the relationship between the coastal uplands and the coastline itself remains ambiguous. That said, given the Early to Mid-Holocene estimate for the cemented shell deposits at Hang Boi and the fragments of Da But pottery in the Lower Cave here, people were still coming into this upland landscape for some considerable time after *c.* 10.6 cal. KBP.

7.3.4 Discussion

Occupation during the late glacial appears to have focused on the lowest range of species. Although all species then present, except the *Panthera* cf. *pardus*, would persist through the entirety of later site use, the relative importance of some (e.g. primates and carnivores), in terms of overall frequency, shows signs of declining through time, whilst the importance of others (notably chelonids) rises significantly. Large terrestrial ungulates were probably to be found in the landscape though appear not to have been primary prey. The relative proportions of *Sus* sp. do not actually change significantly at any point, though the targeting (and perhaps the availability) of cervids does seem to be something that becomes more pronounced over time. Element representation gives the impression that pig carcasses were being butchered somewhat differently to deer. There are also hints that large game was being more intensively processed during the second phase of activity, around the Pleistocene–Holocene transition, when species numbers and diversity first increase, perhaps signifying periodic incidences of food stress. Despite the exploitation of arboreal fauna from the earliest use of the site, there is currently only limited evidence of bone technology and no direct indication that it was employed in a hunting context.

Freshwater molluscs are present throughout the sampled sequence but always at far lower quantities than the land snails, which are consistently the dominant component of the midden and probably the mainstay of consumption activities here – and something which appears to happen, at least periodically, in the present day. These differences in occurrence might signal that freshwater molluscs were being consumed elsewhere; though this does not sit comfortably with the high incidence of freshwater crab remains (including some burnt specimens), which were clearly and consistently being brought back to the cave from presumably similar habitats.

None of the faunal evidence gives any indication that local salt marsh and lagoon habitats were being exploited from this cave during the Early Holocene, when it was being most intensively occupied. Yet there is evidence of coastal contact from before the Pleistocene–Holocene transition in the form of pierced neritid shells. Either, groups coming to Hang Boi were also travelling to the coast periodically (perhaps seasonally) or they were engaged in some form of exchange system with communities living along it.

A few centuries before the sea began to lap around the edges of the massif itself (Tanabe *et al.*, 2006: figure 10a), the midden was capped by a layer of flowstone – seemingly, related to a time of regionally elevated rainfall. This, though, does not seem to have signalled the end of site use. Indications are that people continued to visit Hang Boi until the Mid-Holocene, but all of the midden deposits associated with this activity, save vestiges cemented to the walls, have since been scoured from the cave by unknown means, leaving only hints of this period of occupation, including Da But pottery fragments that found their way into the Lower Cave. Salt marshes were still a prominent feature of the landscape beyond the massif during this late period, before they were replaced with classically deltaic sediments and estuarine habitats between 9.8 and 9.2 cal. KBP (ND-1 Unit 2.3; Tanabe *et al.*, 2003b, 2006).

Macrobotanical remains, including nuts, seeds and parenchyma, have also been recovered from the midden deposits at Hang Boi (Ceron n.d.). Allied with this is a study of plant macromolecules (specifically, lipids) that have been extracted from the cave sediments. The on-going studies of these and pollen evidence will incorporate another essential stratum of subsistence data into this picture, as has been demonstrated by Nguyen Viet (2008) for selected other northern Vietnamese Hoabinhian sites.

In an attempt to move away from broad-brush site characterisations, Hang Boi has been described as a 'shell-bearing midden camp' (Rabett *et al.*, 2011: 167), adapting Widmer's (1989) nomenclature. While mostly made up of land snail shells, the midden also contains the remains of many other animals and plants as well as material cultural, hence the term *shell-bearing midden*. The 'camp' designator refers to the probable episodic nature of the occupation; people were the primary but not sole agents of accumulation here.

The impression from the fauna is that land snails provided a reliable probably seasonal staple that was supplemented by resources from a range of local freshwater, terrestrial and arboreal habitats. Some of these foraging forays involved trips of sufficient distance that they may have warranted off-site processing. If hunting and gathering routines involved trips to the coast, no evidence of such excursions survives at Hang Boi. The low incidence of freshwater molluscs might be a product of transportation decisions and underscores the likelihood that Hang Boi gives us only a partial picture of the economy of the early human groups living in this area. It represents just one component of a subsistence landscape. The evidence from this site also cautions against assuming that proximity to maritime resources will mean that they are immediately exploited (see earlier). It grants an insight into the existence of a possible time-lag between access to such resources and their full economic integration into site use. On current evidence, the upland karstic landscape of Tràng An continued to prove a strong attractor and reliable larder in the face of considerable environmental change well into the Holocene.

7.4 Eastern Peninsular Malaysia: Gua Sagu and Gua Tenggek

The 1990–1 excavation at Gua Sagu by Universiti Sains Malaysia (*see* Chapter 6) was undertaken in 10 cm spits across 14 trenches, most of which were 2 × 2 m. All but one of these trenches (C5) was contiguous, and deposits are considered to have been laid down in a near-horizontal stratigraphy (Zuraina *et al.*, 1998a). Six radiocarbon dates were obtained during the excavations at Gua Sagu and neighbouring Gua Tenggek. The uppermost dates at Gua Sagu came from spit 2 and spit 4 and indicate occupation during the period 1164 ± 111 cal. BP (Beta-32486) and 2941 ± 45 cal. BP (Beta-36341), respectively (Zuraina, 1998). From the evidence laid out in the original site report, and supported by findings in this study, spit 5 seems to correspond to a transitional period, equating broadly to the Mid-Holocene. Prior to this, there is a considerable gap in the dated deposits at Gua Sagu. On available data, it is difficult to say much about the temporal nature of the occupations here or whether there were any hiatuses in that occupation. Faunal and cultural remains were found in every excavated spit, but this does not necessarily equate to a sustained habitation. That a small number of human remains was found in the upper levels at Gua Sagu raises the possibility of disturbance through the digging of later (probably Neolithic) grave cuts. The small amount of evidence from the ceramics, and from the fauna slightly higher up (spits 4 and 5), also hints at some admixture of deposits. The two dates from the bottom of the excavation are out of sequence, raising some uncertainty about the integrity of these as well. The date for spit 9 is given as 16,991 ± 327 cal. BP (Beta-33181), whereas that for spit 10 is given as 14,843 ± 225 cal. BP (Beta-42121).

At Gua Tenggek, stratigraphic relationships are of greater potential concern, both because the 15 excavated 1 × 1 m trenches were dug in four isolated groups within the cave that could not be linked and because the excavators noted the cave floor to be ‘uneven and sloping’ (Zuraina *et al.*, 1998a: 95). Quantification of bone (and other) remains, and the interpretation of them are subject to the uncertainty that surrounds the original depositional character of this site. There are no date reversals, but only the two dates: 12,494 ± 103 cal. BP (Beta-40968) and 10, 660 ± 110 bp (Beta-40969; shell, uncorrected). Owing to the positions where these samples were taken and the known slope in the cave floor, these may actually be dating the same deposit (*see* Zuraina *et al.*, 1998a: 105). Unlike Gua Sagu, where on-site disturbance is enigmatic, here it was rather more evident. Unfortunately, there is no way of knowing how old the upper levels in the sequence are (spits 1 and 2). These hold half of the total recovered fauna ($n = 192/383$). However, the presence of marine shells and a concentration of pottery fragments in spit 1, and to a lesser degree extending into spits 2 and 3, would seem to place them to sometime within the last 3000 to 5000 years, as the excavators proposed. But probably spit 3 in some areas of the cave was excavated into older deposits than in others. Holding these issues to one side, analysis of the faunal remains from these two sites (weighted towards Gua Sagu) has nonetheless revealed new details about early human occupation in the Malay Peninsula and has reinforced others.

7.4.1 Bone Preservation

The existence of localised vertical movement of materials is evident in at least one probable incidence at Gua Sagu, where bone fragments likely belonging to the same element were identified across two spits. The fragments themselves do not rejoin; however, the bone morphology, condition and colour strongly suggest them to have been from the same element. Burrowing activity cannot be discounted. A low frequency of porcupine (Hystricidae) remains ($n = 4$) was identified in the assemblage, though there is clear evidence that sometimes porcupines were being brought to the site as a food item, based on cut-marks to a fragment of proximal femur. The monitor lizard (*Varanus* sp.), a noted scavenger, does occur at both sites, but numbers are again low: for Gua Sagu, $n = 9$, and for Gua Tenggek, $n = 1$. Furthermore, no examples of gnawed bone were identified in either assemblage, which might otherwise indicate the presence of secondary agents moving or consuming midden material.

Several fragments of bone from towards the entrance to Gua Sagu were found to be heavily weathered, exfoliated and insect bored; numbers of weathered bone dropped significantly for interior contexts. There were also examples, albeit infrequently, of differential weathering and mineral staining to the cortical surface of bones. This suggests that debris sometimes remained partly exposed to the elements and undisturbed for considerable lengths of time, hinting that human visits were sporadic.

At Gua Sagu, trenches adjacent to the cave mouth contained numerous bone fragments which exhibit considerable manganese oxide build-up. This is perhaps as one might expect: material deposited in this area would be subject to periodic water run-off and short-term accumulation of pools below the entrance drip line; it would probably have also been a focus of human activity. Manganese was further observed on bone fragments from trenches deeper inside the cave. In those instances, it is more likely that water was percolating through fissures in the limestone, probably in the days after heavy rainstorms. The presence of mineralised bone in that area also supports this interpretation.

Although the incidence of manganese-stained bone remains largely constant throughout the sequence of excavated spits at the cave mouth, in the interior, there is a hiatus in incidence (equating to spits 7 and 8), when this part of the cave may well have been much drier. There are no dates pertaining to these levels, but the change may be an indication of a temporary shift in local climate towards less intense precipitation. Water appears to have once again begun to percolate into the back of the cave during the period when the deposits in spit 5 were accumulating (also undated). A small number of bone fragments from just inside the cave mouth also show evidence of root etching across their surfaces, suggesting that plants also grew periodically here. Occasional instances inside suggest that some light also penetrated into the cave.

In the Gua Sagu bone sample, the degree of edge abrasion (none, slight, moderate or severe) was recorded for 834 fragments. Fragments that were wholly encrusted with sediment or those with recent damage were excluded. Across the site as a whole, bone with no edge abrasion was found most commonly ($n = 600/834$). The proportion of unabraded to abraded bone did, however, show some variation by trench and spit. Bone from those trenches closer to the cave mouth exhibited a higher incidence of abrasion than material found farther inside. For example, in spit 7 of Trench X3, 48 per cent showed some degree of edge abrasion, compared to spit 5 of Trench B4, towards the back, where only 11.4 per cent was abraded. This suggests that some areas of the cave were more prone to foot trampling than others.

Compared to Gua Sagu, and even allowing for differences in assemblage size, the overall occurrence of manganese staining to bone at Gua Tenggek is markedly less: 5.7 per cent ($n = 22/383$ studied fragments) compared to 17.7 per cent ($n = 202/1140$ studied fragments).

This suggests that the environment within this cave during its occupation was generally drier (though this need not reflect wider climatic trends). The spatial distribution of manganese suggests that both of the main areas of bone accumulation (near the entrance and towards the centre of the cave) were slightly damper than elsewhere, but the difference was not substantial. The presence of root castings in the sediment rind on several fragments from spits 1 and 7 in two trenches near the entrance testifies to localised plant growth just inside the cave; evidence on fragments from the central area shows that there was sufficient light for plants also to grow here.

A total of 210/383 fragments were recorded for abrasion at Gua Tenggek (again, the figure excludes recent breaks and encrusted fragments). The results suggested that bone was subject to minimal post-depositional mechanical damage, and this included bone in the area around the cave mouth, a finding that is consistent with transitory site use.

7.4.2 Evidence of Burning

The interpretation by Zuraina and her colleagues, that a significant part of these two bone assemblages was the product of human hunting and collecting was supported by the presence of burnt fragments. Separating burning from manganese staining required close examination, though distinguishing the crystalline pattern traced by manganese growth from the more destructive cracking and charring wrought by heating was readily achievable under a low-powered light microscope. Across the whole sample from Gua Sagu, reanalysis yielded an incidence of 12.5 per cent (n = 143/1140) charred and calcined fragments of bone. These results broadly concur with Zuraina *et al.* (1998a).

Of the three discrete concentrations of burnt bone at Gua Sagu, the most consistent through time was at the cave mouth; though between spits 2 and 4, the higher concentrations of burnt bone appeared at the very back of the cave. In spit 5, burnt bones were spread more evenly between the front and back, but from this level downwards, to spit 8, burnt fragments again had accumulated at the front of the cave almost to the exclusion of anywhere else. Overall, the evidence suggests that while the front of the cave was the main focus of activity, with either a hearth or hearth dumping area probably lying nearby (though none was formally identified), the back of the cave may also have been periodically used.

At Gua Tenggek, the distribution of burnt bone showed a marked concentration at the cave entrance in spit 1. Otherwise, there appears to have been a repeated, if ephemeral, deposition of burnt material in the central part of the cave through time (i.e. from spits 1–6), possibly indicating a shift in activity from the interior to the cave mouth late in the cave's use.

7.4.3 Invertebrate Fauna

At Gua Sagu, molluscan remains predominantly comprised of the riverine gastropod *Brotia spinoza* (85.7%), frequently displaying removal of the apex, a feature interpreted as evidence of human consumption (Zuraina *et al.*, 1998a). The remaining 14.3 per cent of recovered shell was described as (unidentified) marine shell, a fact which in and of itself provides an important aspect to the occupational story from this site. Given the distance of the cave to the modern coast (*c.* 28 km), this provides an interesting parallel yet contrasting picture to that seen at Hang Boi. The number of recovered marine shells (av. n = 20 NISP) remains noticeably consistent throughout the first five excavated spits; none occur below this level. As one comes up through the depositional sequence, in spit 5, there is a sharp increase in the numbers of *Brotia Spinoza*, and after spit 4, there is a correspondingly significant drop in the frequency of vertebrate faunal remains appearing at the site. This apparent subsistence shift also correlates with the arrival of ceramics in the upper spits. It seems likely that these changes related to a noticeable and sustained subsistence shift – spit 4 is dated to 2,941 ± 45 cal. BP (Beta-36341) – one that is mirrored in the results of the vertebrate faunal analysis here. At Gua Tenggek, *Brotia spinoza* make up more than 99 per cent of the recovered shell, a significant rise which appears in spit 4. Three unnamed marine shells were found at the site: all came from the Holocene-aged but undated spit 2.

7.4.4 Vertebrate Fauna

All of the material from the small Gua Tenggek bone assemblage ($n = 383$) was reanalysed by the author. Time constraints meant that only a *c.* 50 per cent sample studied from the Gua Sagu assemblage could be studied. This comprised 45 per cent randomly sampled and a further 5 per cent judgementally sampled, giving a total sample of $n = 1140$ fragments (Rabett, n.d. -b). The bones studied from Gua Sagu were drawn from 11 of the 14 excavated trenches. A cursory look at most of the remaining material did not suggest that the relative distribution of fragments would change significantly if the sample were expanded to include the entire assemblage; the overall impression of the Gua Sagu sample is that it is representative.

The reanalysis of the vertebrate faunal remains from Gua Sagu and Gua Tenggek has made it possible to expand the number of represented taxa at both sites from the original publication (Table 7.6). Some of the new species were confined to levels above spit 5 at Gua Sagu and above spit 4 at Gua Tenggek and were of later Holocene age. They are, therefore, not covered in detail here. Of principal concern are the lower spits dated, or thought to date, from the Last Termination or Early Holocene. This, unfortunately, imposed considerable interpretative limits owing to the small sample sizes. The ratio of identified fauna to unidentified fragments for this period at Gua Sagu is $n = 210/800$ (spits 5–11); at Gua Tenggek, it is a tiny $n = 13/134$ (spits 4–7).

During this time, geoemydids and trionychids were the most numerically and taxonomically dominant animals found in the fauna from either site (though bear in mind the earlier caveat to chelonid identification). It is reasonable to assume that these freshwater turtles were not self-introductions on account of butchery traces to several fragments of bone, carapace and plastron. They were probably caught in nearby rivers or streams. Adult *Cyclemys dentata* (Figure 7.16) are almost entirely terrestrial, whereas *Dogania subplana* tends to avoid the larger, sediment-laden rivers frequented by the larger trionychids (such as the specimen that appears in spit 1 at Gua Tenggek) in favour of hill streams (Pritchard, 1979). In this respect, it is perhaps noteworthy that trionychids do not appear in the deeper levels at either site. The earliest they occur is spit 6 at Gua Sagu, whereas *Cyclemys dentata* occurs throughout both sequences – a feature which presumably stems from a change in availability or hunting strategy. At Gua Sagu, a large concentration of chelonid remains found in Trench Y3 (spit 8; $n = 39$ NISP) may represent a discrete butchery event that took place close to the cave entrance. The NISP values of turtle remains in the levels below spit 5 are approximately double what they are in the levels above it, further strengthening the likelihood of a shift in procurement emphasis around this time – which is undated but assumed to be of Holocene age. Similarly, at Gua Tenggek, almost all of the identified fauna from the Pleistocene and Early Holocene layers was turtle ($n = 11/13$). Spatially, chelonid bone was found mostly in the central part of this cave, with four instances from trenches adjacent to the cave mouth.

The remains of cervids, including sambar deer (*Cervus unicolor*), and of pigs (*Sus* sp.), cattle (*Bos* spp.) and mouse deer (*Tragulus* sp.) make up the principal terrestrial mammals targeted as food sources, though none occurs in large numbers. The presence of a rhinocerotid in the assemblage is attested on the basis of a single (butchered) phalanx 1. At Gua Sagu, the majority (77%) of remains occur in the X and Y trenches near the cave mouth, the main focus of activity. Only a solitary fragment of *Bos* spp. was recovered from the early levels at Gua Tenggek, and all of the pig remains were confined to levels above spit 4.

Table 7.6. Identified Vertebrate Fauna from Pleistocene and Early Holocene Levels at Gua Sagu (spits 5–11) and Gua Tenggek (spits 4–7)

					Gua Sagu (Spits 5–11)		Gua Tenggek (Spits 4–7)	
Class	Order	Family	Taxon	Common Name	NISP	%	NISP	%
Pisces				Fish	8	3.81	[a]	–
Reptilia	Testudines	Geoemydidae	Contains *Cyclemys/Heosemys*	Spiny hill/Asian leaf turtle	0	0	1	–
			Cuora cf. *ambionensis*	Malayan (= East Indian) box turtle	[a]	–	0	–
			Cyclemys cf. *dentata*	Asian leaf turtle	7	3.33	[a]	–
			Siebenrockiella crassicollis	Black pond (= marsh) turtle	1	0.48	0	–
			Geoemydidae sp.	Hard-shell turtle	139	66.19	9	–
		Trionychidae	*Dogania subplana*	Malayan soft-shell turtle	5	2.38	1	–
			Trionychidae sp.	Soft-shell turtle	0	0	[a]	–
	Squamata	–	–	Snake	3	1.43	1	–
		Varanidae	*Varanus* sp.	Monitor lizard	9	4.29	[a]	–
Aves	–	–	–	Bird	[a]	–	0	–
Mammalia	Primates	Cercopithecidae	*Macaca nemestrina*	Pig-tailed macaque	[a]	–	0	–
			Presbytis sp.	Leaf monkey	2	0.95	0	–
			Cercopithecidae sp.	Monkey	0	0	[a]	–
	Rodentia	Muridae	*Rattus* sp.	Rat	0	0	[a]	–
		Sciuridae	Sciuridae sp.	Squirrel	[a]	–	[a]	–
		Hystricidae	Hystricidae sp.	Porcupine	3	1.43	0	–
	Chiroptera	Hipposideridae	*Hipposideros* sp.	Roundleaf bat	2	0.95	0	–
		Vespertilionidae	*Miniopterus* sp.	Bent-winged bat	[a]	–	0	–
		Pteropodidae	Pteropodidae sp.	Fruit bat	1	0.48	0	–

(*continued*)

Table 7.6 (*continued*)

					Gua Sagu (Spits 5–11)		Gua Tenggek (Spits 4–7)	
Class	Order	Family	Taxon	Common Name	NISP	%	NISP	%
	Carnivora	Viverridae	Viverridae sp.	Civet	1	0.48	0	–
		Felidae	*Panthera* sp.	Leopard (or tiger)	1	0.48	0	–
	Perissodactyla	Rhinocerotidae	*Dicerorhinus sumatrensis* (?)	Sumatran rhino	[a]	–	0	–
	Artiodactyla	Suidae	*Sus* sp.	Pig	6	2.86	[a]	–
		Cervidae	*Cervus unicolor*	Sambar	7	3.33	0	–
			Cervidae sp.	Deer	11	5.24	[a]	–
		Tragulidae	*Tragulus* sp.	Mouse deer	[a]	–	[a]	–
		Bovidae	*Bos* sp.	Cattle	4	1.90	1	–
		Capridae	Capridae sp.	Goat	[a]	–	0	–
				TOTAL (NISP and %)	**210**	**100**	**13**	**–**

Note: Specific epithets are as follows: 'sp.' indicates confidence in attributing specimens to a particular genus but uncertainty with respect to which species within that genus they belong; 'spp.' indicates confidence that there is more than one species represented in the study sample but uncertainty as to which species those are; 'sp(p).' indicates that there is *probably* more than one species present in the sample. The frequency of occurrence values was not calculated for Gua Tenggek owing to the very few taxonomically identifiable pieces.

[a] Present at the site but confined to the upper, ceramic-bearing levels.

Figure 7.16. (left) Asian leaf turtle (*Cyclemys dentata*; Accession No. 7490, Chelonian Institute, Florida); (right) specimens of third and fourth (L) peripheral dermal plates of this species from Gua Sagu Trenches X2 (30–40 cm, spit 4) and X3 (60–70 cm, spit 7), left–right (scales in 5-mm increments). (Photographs by Ryan J. Rabett.)

In the absence of any significant attention by scavengers at Gua Sagu, element representation contains notable absences for some parts of the skeleton. Most identified elements are either metapodials (mostly metatarsals) or hind limb long bone fragments; phalanges are very few ($n = 4$) given the number of these per whole skeleton. There is no evidence of humeri and only two fragments of scapula and one of ulna to attest that the fleshy parts of forelimbs were consumed on site. The lack of humeri is particularly significant as, in the absence of other agents, large mammal humeri frequently survive in archaeological assemblages from this period in Southeast Asia, for example, they are found at the Niah Caves and more locally at the site of Gua Cha, in the west Malaysian state of Kelantan (Rabett, 2007).

Equally surprising is that there are only three isolated teeth and no other mandibular or cranial bones, with the exception of some cervid antler fragments at Gua Sagu. Although these exhibit polishing and facet wear, under examination, it is with considerable certainty that these were a natural product of wear and damage during the lifetime of the animal (*see* Olsen, 1989). The suggestion is that throughout the occupation sequence at this cave, people were not bringing complete carcasses of large mammals back to the cave, or else they were selectively dumping remains elsewhere. The skeletal element picture for ungulates is identical at Gua Tenggek: isolated fragments appear at a range of depths.

Evidence for the exploitation of arboreal fauna at Gua Sagu and Gua Tenggek is limited. Only two fragments of leaf monkey and two other unidentified primate fragments (not listed) came to light from the early period of occupation. All other incidents of primate remains ($n = 11$) are confined to deposits above spit 5 at Gua Sagu. At Gua Tenggek, only two fragments of primate were identified: one from the early and one from the later periods of occupation. These findings contrast with evidence from late glacial and early post-glacial hunting practices in the lowland rainforests around the Niah Caves or in the karstic uplands around Hang Boi.

The tiny collection of primate bones makes element representation somewhat moot, but of those elements represented, examples include extremities, meat-bearing long bones (one femur and one distal humerus) and cranial and mandible elements – including half a set of maxillary teeth of a *Macaca nemistrina* (spit 4 at Gua Sagu). Although one can only be very tentative about this, the probability is that whole monkey carcasses were being brought back to the site and processed. Although primates (and other semi-arboreal taxa) seem to have played a comparatively small role in subsistence activities at Gua Sagu at any time, increased frequencies in the later period do signal a shift towards the exploitation of more arboreal-adapted game during the Mid-Holocene. This is in accordance with the shift in hunting strategies proposed by Bulbeck (2003), based on evidence from several other fauna-bearing sites in the Malay Peninsula and one reiterated herein (*see also* Piper and Rabett, forthcoming-b), but one that apparently came some time later than it did in lowland Borneo.

Carnivores also played a very minor role in subsistence at both these sites, with all examples confined to the early part of each sequence. At Gua Sagu, aside from one identified viverrid right proximal femur, the only other piece of note was the proximal end of a first phalanx (unknown digit) of a large carnivore in spit 10. This spit is dated to 14,843 ± 225 cal. BP (Beta-42121), placing it close to the start of the GIS-1 interstadial. It is likely to have been from a large felid, *Panthera* sp., though it seems too small and gracile to be *Panthera tigris* (*contra*. Zuraina *et al.*, 1998a). The proximal depth is 10.20 mm, and the proximal breadth is 12.53 mm. The second measurement matches it more closely to *Panthera pardus* (10.95 mm) than the *Panthera tigris* average of 19.5 mm (*see* Piper *et al.*, 2008).

Among the fish remains, which appear across a range of levels at both sites, all recovered elements were either dentary/vomer fragments or spines; strangely, there were no vertebrae. This is unusual when compared to fish remains recovered from similar assemblages in the region (e.g. Rabett *et al.*, 2009a). Dry sieving of sediments was definitely used during the original excavation, precluding loss because of recovery strategies. Although the grade of mesh is not recorded, the fact that microvertebrate fauna were recovered suggests that it should have been fine enough to have caught at least larger fish vertebrae. This absence cannot currently, therefore, be explained. Microverterbate fauna are, nonetheless, quite scarce, with one exception: Trench A5 (spit 9) from Gua Sagu, where the numbers of whole or partial elements are disproportionately high ($n = 41$ NISP). This spit is dated to 16,991 ± 327 cal. BP (Beta-33181). Most significant among these remains are the bats (Chiroptera).

Hipposideros sp. (the roundleaf bat) probably occurred at Gua Sagu throughout the depositional sequence, on the basis of evidence in spits 2, 4, 5 and 9. Spatially, occurrence of this genus was fairly confined, appearing only in the middle of the cave. This may well indicate the presence of a natural roost in the roof here, making this a natural death assemblage. The fact that it appears to have endured for as long as it did could also imply that human presence or disturbance was not excessively disruptive. During this early period, there were also other genera of bat in the cave, which have not been recovered at any point higher in the sequence. The first is another insectivorous bat, *Miniopterus* sp. (the bent-winged bat). These tend to fly above the tree canopy, taking insects on the wing, and roost in caves in near-total darkness. The second taxon is from the family Pteropodidae, which are frugivorous bats. Their sensitivity to disturbance may account for their later disappearance. Gua Tenggek does not seem to have served as a roosting site at any point throughout the excavated sequence.

7.4.4.1 Butchery

A total of 13 fragments of cut-marked bone were identified within the Gua Sagu sample; five further pieces were found in total in the Gua Tenggek assemblage. If one considers only incidence from the late glacial and Early Holocene, the frequencies are 4/770 = 0.52 per cent (Gua Sagu) and 2/134 = 1.49 per cent (Gua Tenggek). To put this into some context, evidence at the other sites examined herein was as follows: Hang Boi, 0.25 per cent ($n = 14/5556$), the Niah Caves, Sarawak (Area A, Phase II, West Mouth), 0.53 per cent (56/10,513 NISP) and Niah Cave (Lobang Hangus >12″) 0.94 per cent (83/8767 NISP). The values are broadly similar despite the differences in, sample size, geographic location, taphonomic factors affecting the bone assemblages or the kind of elements that

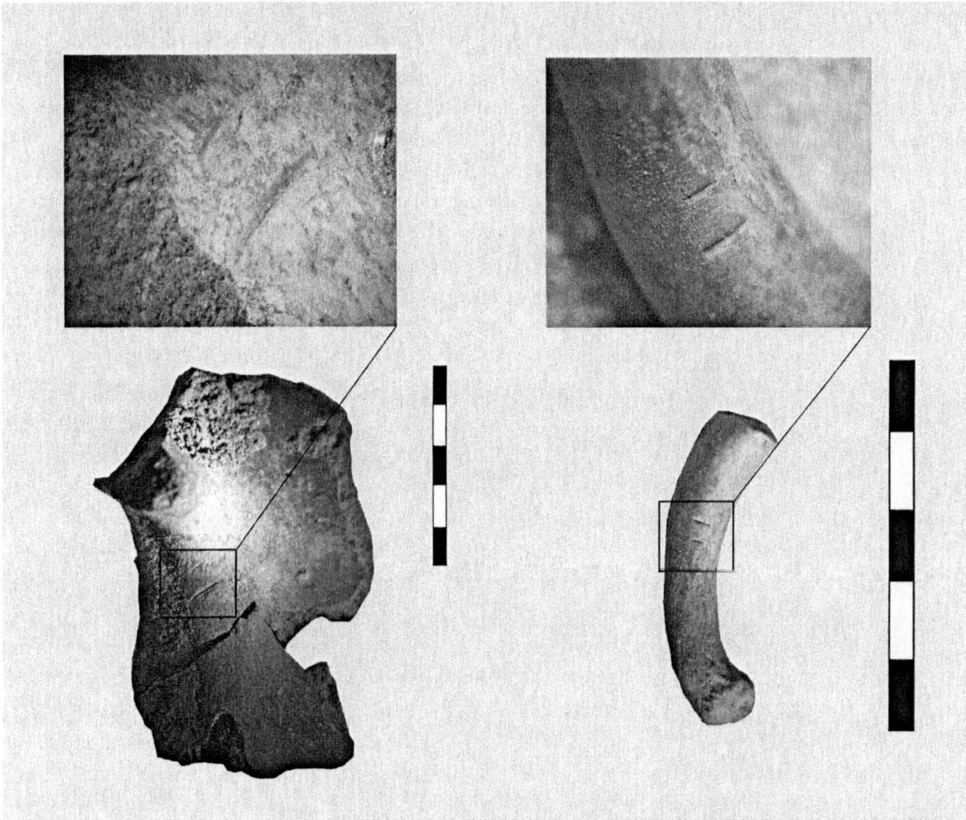

Figure 7.17. (left) Cut-marks to the inside of the hyo- or hypoplastron dermal plate of a chelonid, Gua Sagu Trench X3 (0–10 cm, spit 1); (right) cut-marks to a chelonid femur or humerus mid-shaft Gua Sagu A5 (80–90 cm, spit 9; independent scales, both in 5-mm increments). (Photographs by Ryan J. Rabett.)

occur within a given assemblage. Yet close comparison of where cut-marks were located reveals site-specific patterning.

Intermediate and large mammal butchery at these two Peninsula sites accords quite closely with that seen at Niah: cut-marks appear on metapodials and around the articular joints of long bones of *Sus* sp. and Cercopithecidae. Despite comparatively high levels of chelonid recovery at Niah Cave (Area A, Phase II; $n = 560$) across a similar range of species (Pritchard *et al.*, 2009), none was found to bear evidence of butchery. By contrast, at Gua Sagu, there are three instances, one occurring at the top and two from the early part of the sequence (Figure 7.17); at Gua Tenggek, there is one case of chelonid butchery, if from near the top of the sequence. In both cases, these instances are drawn from much smaller assemblages than Niah. This would seem to suggest that chelonid processing techniques were similar between Gua Sagu and neighbouring Gua Tenggek, that such techniques were enduring and that they appear to mark a different operational sequence compared to that employed habitually at Niah. There is also no evidence for porcupine butchery at Niah while, as already noted, one example occurs at Gua Sagu from the earlier deposits. Conversely, there is a distinct lack of small carnivores in either the Gua Sagu or Tenggek assemblages, while there are frequent occurrences at Lobang Hangus which, as we have already established, was probably down to more than simple availability.

7.4.5 Discussion

As the evidence from Gua Sagu is more numerically and stratigraphically robust, interpretation of the data is probably more secure and so forms the central part of this concluding discussion about the situation, history and use of these sites.

The earliest phases of human occupation of Gua Sagu began in the millennia immediately following the LGM and during a period of rapid regional sea level rise from *c.* 14.6 to 14.3 cal. BP (Hanebuth *et al.*, 2000). Around 15 cal. KBP, the coast would have been approximately −100 m below its modern value and more than 500 km away on the eastern edge of the still almost fully exposed Sunda Shelf (*see* Sathiamurthy and Voris, 2006: figure 5). By *c.* 12.6 cal. KBP, the sea level was −60 m, and the coast would have been *c.* 70–80 km away (*after* Voris, 2000).

At Gua Sagu, the existence of local forest stands is attested by the presence of *Presbytis* spp. remains in spits 10 and 7 and by the existence of a small colony of Pteropodidae, a specialist canopy-feeding species of bat in spit 9. Thus, on the eastern side of the Malay Peninsula, the vegetation looks to have been more closed than is presently envisaged for the western side of the Titiwangsa mountain chain (e.g. Bird *et al.*, 2005). It is possible that increased human activity at the cave after this time may have disrupted some of the roosting bat communities here, as remains from *Miniopterus* sp. and Pteropodidae spp. did not reappear higher in the sequence.

Although cercopithecids were hunted from the time of the site's initial occupation, the very low frequency of remains suggests they were not a significant dietary component. Freshwater courses probably lay nearby to Gua Sagu and immediately outside Gua Tenggek. It appears to have been from these upland rivers and streams that people were drawing most of their food in the form of freshwater molluscs, fish and turtles – as well as collecting the raw materials for stone tools (Zuraina *et al.*, 1998a: 106). Large land mammals, including *Cervus unicolor*, *Sus* sp. and *Bos* spp., also featured in people's diet from early on. Their significance is hard to gauge; element representation does not seem to indicate that whole carcasses were being brought into the caves. Unlike the primates, these animals may have been hunted and disarticulated elsewhere, with only certain sections of the body being introduced onto site.

The environment inside Gua Sagu seems to have been comparatively dry during this early period, as indicated by the presence of only limited manganese build-up on bone fragments. A comparatively high occurrence of edge-abraded bone was found at the cave entrance, the area probably experiencing the most foot traffic. This area also seems to have been the principal location for the accumulation of charred and calcined bone fragments – almost to the exclusion of anywhere else at times. The earliest evidence of burnt material and cut-marked bone, though, comes from farther inside. It seems likely that initial occupation was focused in the middle part of the cave and that activities were probably limited in nature owing to minimal levels of bone abrasion. This is echoed by the low incidence of lithic artefacts from these deepest levels (Zuraina *et al.*, 1998a: 83). Subsequently, the cave mouth appears to have been the focus of habitation.

From spit 7 upwards, conditions in the cave changed. The build-up of manganese staining on the bone assemblage is indicative of increasing levels of moisture or humification of the sediments. To begin with, this was mostly confined to the cave mouth and may represent the effects of run-off from rainfall, leaving the cave interior still reasonably dry. By spit 5, however, the bone evidence suggests that water was percolating more persistently into the back of the cave or that human refuse was accumulating there. Much of this period, from spits 7–8 to basal spit 5, broadly accords with the onset of the Holocene and possibly elevated levels of precipitation.

Although interior conditions may have been generally more humid by this stage, people continued to come to Gua Sagu. In fact, this time seems to have heralded the most intense usage of the cave (on the basis of the volume of recovered material), though again, the actual timing of occupation can only be guessed at. The front of the cave continued to be a focus for the deposition of burnt material, though the amount of edge abrasion drops noticeably and picks up instead at the very back of the cave. Either this indicates another occupational move to the back or, alternatively, this area became a refuse dump for activities at the cave mouth. Between spits 5 and 4, two trenches at the back of the cave also accumulated almost half of the total butchered remains from the site ($n = 6/13$). This could be used to support either proposed scenario, though the relocation of refuse is probably the more likely.

By spit 5, other changes were taking place. Pottery may have already begun to be brought to the site by this point. From spit 5 onwards, it begins to appear in quantity; there are also increases in the frequency and types of lithic artefacts and a shift in subsistence practice. The incidence of freshwater molluscs now begins to increase significantly, and for the first time, marine shells start to be brought into the site from a coast that is likely to have been *c.* 30 km or less away. Fishing continues, as does the capture of turtles, though the range of species begins to widen, with instances of *Cuora ambionensis* and *Siebenrockiella crassocollis* and the Trionychidae turtle *Dogania subplana*, which appear from spit 6 upwards. The butchery traces on different chelonid elements compare sharply with the general dearth of evidence on these at other contemporaneous sites and hint at a locally specific set of practices for processing these animals.

The first appearance of marine molluscs at Gua Tenggek occurs higher in its sequence (spit 2) and numbers are very low ($n = 3$). This might be an indication of a parallel Mid-Holocene phase here, though there would have had to have been a hiatus in occupation, given that spit 3 is dated to *c.* 12.4 cal. BP. In the final period of occupation (spit 1) at Gua Tenggek, people deposited parts of a large trionychid at the cave. As these turtles tend to favour the more sluggish waters of sediment-laden lowland rivers, its presence and the limited showing of marine shell suggests that whatever the date of this final period, people coming to the site were moving quite widely over the landscape.

Terrestrial mammals continued to feature in the diet of people at both caves; however, their frequency of occurrence broadly declines as exploitation of freshwater recourses increases. There is also some reason to believe, based on dental evidence, that hunting strategies may have changed to emphasise the predation of adult individuals, but the pattern of element representation does not seem to have changed. More conspicuously, at Gua Sagu, the frequency of arboreal and semi-arboreal mammalian fauna rises, with 11/15 fragments of Cercopithecidae and indeterminate primates identified coming from spit 5 and levels above. Given that the overall frequency of vertebrate fauna has dropped by this time, this does mark a significant change. It is possible that it signals an increase in wider forest cover (or forest productivity) in light of higher rainfall and reduced seasonality during the Holocene, but there is no evidence for the existence of bone technology at either site that might suggest changes in the means of procurement. Additionally, and unlike the situation at Niah, at Gua Sagu, there is no increase in the exploitation of other semi-arboreal taxa such as cats and civets. The presence of human bone extremities in the uppermost spit at the site hints that perhaps one of the final acts to take place at the cave was a burial site. Not long after this, the cave appears to have been abandoned.

7.5 Inter-Site Patterns of Subsistence through the Last Termination and Early Holocene

Subsistence evidence during the earlier part of the Last Termination is represented by the Early Phase II vertebrate faunal assemblage from the West Mouth, Niah, and from the basal deposits from Gua Sagu, both dating from 17 to 16 cal. KBP. The Niah assemblage

(n = 633/2071, 30.5% of total Phase II) exhibits a continued emphasis on non-selective hunting of *Sus barbatus* seen in the pre–Last Termination sample. This was undoubtedly the principal game animal here. As whole carcasses were being brought back to the site, kills were probably being made locally, perhaps during migratory movements of this animal. Evidence of how pigs were butchered is largely missing, though, despite reasonable bone surface preservation. Breaking open the back of the skull, probably to remove the brains, is indicated but appears as an isolated case. Absolute numbers of these and other large ungulates increase between the Early and Late phases, but their proportions remain very similar – possibly a reflection of fairly stable availability. Primates formed a secondary component, with both strict- and semi-arboreal species recorded as present, testifying both to the local existence of closed-canopy rainforest and also reliable, if unknown, means of procurement – bone points are present but do not appear to have served as projectiles. Available age-profile data on the cercopithecids show these to have been mostly adult; whatever method was employed, it was capable of taking game selectively. Both axial and appendicular element fragments were found, suggesting the transportation of carcasses back to the cave, though the evidence for this is stronger later. Carnivores feature in both components of Phase II, but there are changes in composition. Although the proportion of carnivores is slightly higher in Early Phase II, numbers are always few, and it is hard to draw any firm conclusions with respect to this change. Among the well-represented reptile fauna are members of both Geoemydidae and Trionychidae families, indicating that a range of freshwater habitats was also being targeted; how they were being processed is not known.

The Gua Sagu remains from this period appear to be few and of limited diversity (of the total assemblage, they comprise n = 170/1140 fragments, or 15%). Without dates on the layers between these lower spits (9–11) and those higher in the sequence it is impossible to be certain about the total depth of related deposits. The represented taxa include specimens from semi-arboreal, terrestrial and aquatic habitats – with chelonids well (and perhaps over-) represented – suggesting that mainland groups were also routinely taking game from each of these settings. The identified remains included cut-marked fragments of a *Presbytis* sp. mandible and chelonid long bone, testifying that human agency was responsible for the accumulation of much of this small assemblage and hinting at differences in butchery practices between here and Niah.

Pigs were clearly the staple source of protein and fat at Niah, whereas freshwater resources appear to have been more central at Gua Sagu. Arboreal game were taken at both sites but formed a larger component of the diet at Niah. Although no firm prediction can be offered, the comparatively lower occupation intensity at both sites may be an indication of lower population density or increased mobility compared to later periods. New data emerging from Hang Trống (Tràng An), which also has contexts dating to this part of the Last Termination and earlier, will provide a useful additional point of comparison on this matter in the near future.

Moving forward in time, subsistence patterns during the GIS-1 interstadial are exceptionally well represented by the large, Phase II assemblage at Lobang Hangus. The identified vertebrate fauna from here still show a very strong emphasis on the exploitation of *Sus barbatus* and a continuing important reason for coming to the Gunung Subis massif. Whole carcass transportation is again evident, but now cut-marks appear across a range of elements. On this basis, there are indications that thoracic and pelvic limbs may have been sectioned differently, and there is one example of stiff-joint disarticulation – a marked contrast to the Early Phase II material in the West Mouth. Why this should be is uncertain, though it is definitely not a taphonomic or element representation issue. It could be an indicator of changing approaches to butchery or that the West Mouth and Lobang Hangus entrances were visited by different forager groups.

While pigs were also a major focus of hunting during GIS-1 at Niah, arboreal and semi-arboreal game now surged. This probably, at least in part, marks an increase in forest development and the availability of these taxa within it. Invigorated interest in arboreal prey was accompanied by an overtly selective attitude towards hunting them. People using Lobang Hangus had a very clear sense of which arboreal and semi-arboreal fauna they wished to target. The age structure identified in the cercopithecid remains emphasised young mature adults (Piper and Rabett, 2009), and this was coupled with an apparently deliberate preference for small carnivores, particularly viverrids and especially the *Arctictis binturong*. Butchery techniques applied to both the monkeys and the bear cat show signs of being peculiar to this occupation. In the case of the latter, this even distinguishes it from the treatment accorded to other viverrids. On this evidence, the possibility that we are dealing with more than one community of people visiting Niah looks plausible. While those coming to Lobang Hangus had the same intention of hunting pigs, they appear to have had or arrived with a somewhat different knowledge base and set of subsistence preferences to those seen in the Early Phase II occupation of the West Mouth. This also included a resurgent emphasis on bone technology and possible reliance upon it in securing at least a portion of the much more prevalent arboreal prey.

Occupation during the thousand years of the last stadial was covered by data from Gua Tenggek and Hang Boi. As was the case for Gua Sagu, without a clear sequence of dates from Gua Tenggek, it is not possible to determine exactly the time span and hence range and occurrence of fauna dating GS-1. We can only be reasonably certain that people were here around this time. The fauna is very sparse but demonstrates a continued exploitation of freshwater resources. Fish and chelonids are represented; freshwater molluscs (*Brotia spinoza*) are also reported in appreciably larger quantities than in preceding levels. A single fragment of pig was identified, giving some indication that terrestrial game was also being taken if infrequently.

Greater clarity comes from the Hang Boi sequence, where, although not yet completely analysed, occupation commenced towards the end of GIS-1. Based on the sample studied, the range of taxa is considerably reduced compared to that seen here in the Early Holocene, but as at Gua Tenggek, we find good evidence that freshwater habitats were being routinely exploited. The low incidence of freshwater molluscs might be a consequence of transportation choices, but in light of the other riverine resources that were consistently finding their way to the site this is probably not the whole story. The highest frequencies of cercopithecid and carnivore remains (including the large felid, *Panthera* cf. *pardus*) also appear during this period and tail off in importance later on, in apparent contrast to the picture seen elsewhere. The single bone point from this depth does not look to have formed part of whatever technology was being deployed to hunt these animals; other methods must have been available. Although bone point technology and increases in arboreal game are concomitant in the region, this association seems to be most prominent in ISEA and possibly does not hold for the mainland (*see* Rabett and Piper, 2012).

Cave occupation during the Early Holocene was more intensive and involved a wider array of species, with some previously minor dietary components rising in importance and new ones being added. Evidence for this period came from the Late Phase II occupations in the West Mouth and again from Hang Boi.

At Niah, the hunting of pigs was still central to subsistence. Almost identical proportions of this animal were recorded in both the Area A and Area D assemblages (*c.* 40%). Evidence for its processing now appears in the form of cut-marks to extremities, an indication that primary butchery – skinning and probably removal of the feet – was definitely taking place on site, though other details continue to be scant. There are noticeable expansions in some taxa, such as chelonids, with several species of turtle now more strongly represented. Cercopithecidae frequency rises slightly in Area A but does so dramatically in Area D. This distinction may be a product of spatial patterning in midden formation, given the close contemporaneity between these parts of the site, but could mean they were quite discrete accumulations in their own right.

At Hang Boi, the Early Holocene sees a pronounced increase in both the quantity and diversity of vertebrate fauna represented, suggesting this to have been the most intensive period of activity at this site. Chelonids, in particular, show a comparative surge in number, as do fish remains. Cercopithecids form a less significant part of the diet than previously, but a greater diversity of species was being taken. Large terrestrial mammals, especially cervids, rise in relative significance but, despite their larger meat utility, seem to remain ancillary. Cyclophorids remain the primary midden component, as they had since the site was first occupied. Although the per-unit calorific value of these snails is low, their greater local availability looks to have been a key attractor that drew people to this location over the long term (Rabett *et al.*, 2011).

7.6 Conclusion

In this chapter, a selection of recently studied faunal assemblages from the Last Termination and Early Holocene has been discussed in some detail, with evidence drawn from Island and Mainland Southeast Asian settings. These studies have demonstrated that behind the traditional picture of broad-spectrum tropical forest subsistence, which was based on species identification alone, there lies an altogether more polychromatic image. Within it can be found a wealth of information on trends and variation in resource reliance, prey selection and processing habits, on forager mobility, and ultimately, on the processes behind human adaptation within this region.

The data show that certain components of these assemblages remained consistent throughout this period and beyond, and did so despite the effects of climatic and environmental change. These resources were notable for the reliability with which they could be accessed, and the sites that we have in the landscape appear to have been tied to them. These may be considered to have been food staples. The fact that they continue to account for the similar proportions of the prey assemblage across climatic boundaries could suggest that these changes in conditions were significant but not so profound here as they were elsewhere – echoing one of Gorman's early contentions (Gorman, 1970: 101). An alternative explanation could be that these were species best able to tolerate environmental change. It is not yet clear how many of these central resources a community would have relied upon, but given the nature of the environment, we can assume there were probably several, scattered at points across each community's foraging range. Movements were probably cyclical (including seasonal) and timed to take advantage of local peaks in the availability of a given staple.

Other faunal taxa appearing in both the Island and Mainland assemblages looks to have been more opportunistically taken as a supplement to these core foods. Changes in the representation of these suggests that cooler interludes during the Last Termination were marked by contraction of the faunal base and perhaps a shift to a higher level of forager mobility compared to that seen during the last interstadial and into the Early Holocene. Under these latter conditions, the assemblages came to contain greater quantities and a broader diversity of exploited fauna, signalling more intensive occupations, or perhaps a burgeoning population.

These periods of expansion in site assemblages did involve the introduction of new resources, but there is a strong sense that forager groups emphasised intensifying their existing subsistence base, raising previously minor components to greater prominence. This has particular reference and relevance when we consider the promoted availability of coastal resources to previously inland sites. The Hang Boi evidence suggests that increasing proximity to the coast need not always lead to an immediate preference towards foraging in coastal habitats from such sites.

One of the regional trends to emerge across the Pleistocene–Holocene boundary was a rise in the occurrence of arboreal and semi-arboreal fauna. In no small part, this probably reflects forest expansion and enhanced availability, but there also exists evidence for quite deliberate targeting of these taxa also. There are indications that this shift was shadowed by technological developments and innovation in the sphere of bone tools, though other non-preserving range weapons may have existed for some time, and these (or other systems such as traps) seem to have remained preferable, particularly in Mainland contexts.

Lastly, this chapter has also demonstrated that tropical subsistence strategies combined tangible threads of continuity with considerable inter-site variability. At times, this variation in practice is even apparent between chronologically associated assemblages at neighbouring sites and extends to contrasting patterns in carcass reduction, sometimes even in relation to the same taxon. Subsistence practices appear to have carried a distinctly local character in Southeast Asia – an observation that brings us to the final chapter of this volume, one that will consider specifically how early humans colonized this region's Pleistocene environments, how they adapted to the changes wrought within them and, perhaps most fundamentally, how regional trajectories here, and globally, may have helped shape the development of what has come to be current modern human behaviour.

Chapter 8

Ex Levis Terra

8.1 Introduction

The identity of European stone tool industries emerged out of a chronological system devised originally by Gabriel de Mortillet (1821–98), which had marked progressive epochs in human evolution along a unilinear trajectory. Each epoch was named after a type site – following geological convention – and comprised material culture thought to reflect the character of its developmental stage (Chazen, 1995). Such an overtly progress-driven perspective has long been officially consigned to history, with only the naming convention retained, and other conventions have been introduced (e.g. Clark, 1971). Arguably, though, the search for human behavioural origins, which featured so strongly in twentieth century Palaeolithic archaeology, represented – if implicitly – a continuing expression of that same principle: a search for the path of development leading to the present. The Palaeolithic sequence of western Europe formulated in this way and influenced by this perspective became the standard against which archaeological records in the rest of the world would be judged. This singular point of geographic and cultural reference has come to be actively challenged only since the turn of the millennium, starting with the seminal study by McBrearty and Brooks (2000) on developments in the African Middle and Late Stone Ages. Although it has been known for some time that early cultural records outside of Europe often differ from those seen within it, the notion that such regional developments should represent something more than mere shadows (or lesser versions) of the European sequence has amounted to a paradigm shift in interpretation affecting both sides. Finlayson and Carrión (2007) have recently proposed, for example, that the technological innovations of the European Upper Palaeolithic would be best understood as responses to living in vast open spaces rather than the consequence of a cognitive revolution. Rather than a marker of human modernity, in their view, the repeated turnover in stone tool industries in the European record may be consequential of the way hominins were responding to oscillating conditions in the geographic environments and shifting population densities that existed there. In Asia – and

Chapter title translation: 'Out of the unstable Earth'.

in terms of the current study particularly Southeast Asia – a different range of conditions was encountered and a different set of responses would have been required. As new sites and new data continue to emerge from regions beyond Europe, the traditional model of the emergence of human modernity, with its emphasis on a singular evolutionary trajectory to the present and behavioural unity cast in the likeness of the European, looks increasingly untenable.

On current evidence, *H. sapiens* were almost certainly present in Southeast Asia by 60–50 kbp, and perhaps slightly earlier, while our species had reached central and northern Asia by at least the beginning of the Late Pleistocene. No satisfactory explanation has yet been proposed, however, to account for why the adaptations that early modern human migrants to Asia were sufficiently parallel to those already existing among populations of archaic hominins as to make their initial arrival often so technologically ambiguous. Addressing the cause(s) of this enigma is also beyond the scope of this book. An attempt can be made, though, drawing on the various findings of the current study, to shed light on one of the key mechanisms behind the development of regional behavioural trajectories among early modern humans: the phases and processes involved in colonizing new and transformed environments – an investigation that may also begin, perhaps, to provide a revised perspective on what Upper Pleistocene 'cultures' actually were.

8.2 The Colonization of New Environments and the Recolonization of Changing Ones

Traditionally, developments in modern human behaviour have often been closely attributed to responses to climatic deterioration and the challenges that faced a tropical hominin entering colder mid and high-latitude environments. Niche competition between hominin populations in these new areas aside, the demise of anatomically archaic humans is often linked to their apparent inability to adapt to these worsening conditions, whilst hominin expansion, globally, has been linked to climatic amelioration. Many, though, find it hard to accept that a species like *H. neanderthalensis*, so well adapted to the rigours of existence in periglacial environments, could be rendered extinct by conditions at the end of the last glacial across their entire range, particularly given that the climatic deterioration during late MIS-3 and MIS-2 that culminated in the LGM did not mark the deepest glacial event across the entirety of that range (e.g. Chlachula, 2001: 136). Nor is it yet compelling to propose that an even earlier hominin species (*H. erectus*) that was capable of farther-reaching geographic dispersal, the ability to apparently rebound from the regional impact of major geological events (the Toba OTT eruption and the MN-AT impact), and potentially even island colonization, should have been stopped in its tracks by Pleistocene rainforests, particularly since these latter forests were not even structurally analogous to those we see today. Following the analysis laid out in this book, one can propose that the extinction of these archaic hominins and the regional developments in modern human behaviour were less tied to either deterioration or amelioration *per se* and more to the pace and extent of fluctuation between those states. The demands placed on hominins to adapt to increasingly rapid and extreme swings in climate, occurring down to a frequency of centuries or decades by the Late Pleistocene, would push our genus to the edge of extinction. Anatomically modern humans have gone on to populate, and in many respects dominate the planet, but the survival of a single species of *Homo* might hardly be considered an evolutionary success story in the traditional sense.

The adaptive trajectories taken by different populations of early humanity would have been dictated by a range of variables, including, but not confined to, regional differences in resource structure and availability, differences in population and social dynamics and the effects of sympatric occupation of the landscape by other hominin species. The coarseness of the Pleistocene archaeological, environmental and chronological records, however, means that if one is to attempt reconstruction of these trajectories and interactions, the most resolved period to study is currently the Last Termination (22–11.7 cal. KBP). Climatic and palaeoenvironmental shifts during this period are amongst the most pronounced of the Pleistocene; archaeological remains are widely available and preservation is generally good, whilst error margins on radiometric dating are consistently at the sub-centennial scale; precisely that necessary to examine correlations between human and environmental sequences of change with a reasonable degree of accuracy.

The data on technological change and subsistence variability during the Last Termination in Southeast Asia give only a sense of what was transpiring in that region and only the first hint of what was happening across the expanse of Asia as a whole. The conclusions reached here are presented in the knowledge that they are only one step down a long analytical road. Nevertheless, from these data it is possible to construct a first picture of the short-term process of human adaptation to palaeoclimatic shifts as part of the much larger story of varied adaptive trajectories around the Pleistocene world. One of the useful ways that these adaptive strategies can be assessed is to consider them as forms of colonization.

Over the past 40 years, many models have been devised to account for different episodes of modern human dispersal. These are particularly well developed for the early palaeo-colonization of the Americas (e.g. Erlandson and Moss, 1996; Fix, 2002; Martin, 1973) but have also appeared in recent years with respect to the recolonization of western Europe after the LGM (e.g. Housley *et al.*, 1997) and the original spread of modern humans out of Africa into South and Southeast Asia – the Southern Route hypothesis (e.g. Fix, 1999; Lahr and Foley, 1994; Oppenheimer, 2004, 2008). The simple Wave of Advance models that have featured in many interpretations of this process (driven, essentially, by increasing population density) have now been superseded with sophisticated computer modelling with multiple variables of geography and predictive maps of resource distribution (e.g. Alroy, 2001; Anderson and Gillam, 2000; Field *et al.*, 2007). Fine-grained archaeological analysis into the process of early human colonization on the ground, though, is still largely lacking due to the overall coarseness of the Pleistocene archaeological record. The knowledge that does exist strongly suggests that the colonization of new landscapes often did not happen in one push but rather as a process. This perspective has been explored in detail by Davies (2001) for the spread of the Aurignacian into western Europe. However, the patchiness of the Pleistocene settlement record and chronological resolution for that time frame makes it difficult to identify which part of the colonization process is being recorded at any one site or if evidence in a particular region marked sustained occupation or only temporary immigration (e.g. Davies, 2007; Housley *et al.*, 1997).

Periods of colonization and acclimatization to new environments involve significant behavioural adjustment (Gamble, 1993). Studies examining this span a wide range of fields and periods, for example, from the initial colonization of the New World (e.g. Kelly and Todd, 1988) to that of the earliest English colonists at Jamestown, Virginia (Blanton, 2003), to polar and space exploration (e.g. Launius and McCurdy, 2007; Mocellin and Suedfeld, 1991) and even modern recommendations about adapting to, rather than simply attempting to mitigate, current climate change (IMecE, 2009). Additionally, psycholobiological analysis of group interaction under conditions of prolonged remoteness, which are likely to have characterized early human colonization phases (Kelly, 2003), observes the presence of striking similarities between the psychological challenges facing groups in modern-day polar, underwater and space environments (e.g. Ursin *et al.*, 1991), providing further avenues that can be pursued in the reconstruction of early, historical and future human diaspora.

Blanton (2003) suggested that the Jamestown experience, while acknowledging it is not representative in its detail of all colonization circumstances, may nonetheless be used to help model this process at a broader scale. He defined three stages to the process of successful colonization based on the Jamestown evidence. Firstly, he proposed that there may be a phase of resistance and maladaptation. This represents the steepest part of the learning curve, as survival depends on the speed and success in gaining local knowledge while a fledgeling community is at its most vulnerable. As part of on-going research into reproductive isolation and speciation, evolutionary biologists have recently proposed a very similar principle, 'immigrant inviability' (Nosil *et al.*, 2005). This derives from the concept that the chance of survival among immigrant populations of organisms is compromised when they colonize a foreign habitat that is ecologically divergent from their native one. Low population density or a sudden reduction in effective population size can exacerbate this maladaptative phase to its logical conclusion: localised extinction.

Through the following couple of generations, there is consolidation though, and the pioneer population is augmented and may become more gender balanced (Blanton, 2003; Tolan-Smith, 2003; Ursin *et al.*, 1991). Initial successes are furthered by deliberate subsistence experimentation. Again, a similar pattern can be discerned over shorter periods of time with respect to modern expeditions to extreme environments (e.g. Mocellin and Suedfeld, 1991). Finally, the cumulative effect of a successful adaptation will be the development of a 'creolized' culture – a locally specific version of the parent culture with its own discernable identity.

Culturally and technologically specific conditions will obviously have factored in the experience of the Jamestown colonizers. Responses to a new environment will be affected by how suddenly (and permanently) a group is confronted by it. Blanton (2003: 193) noted that 'the age of sail thrust colonizing parties into strange environments very abruptly'. The extent and regularity with which contact can be (or is chosen to be) maintained with the parent society will be a further feature that will also vary between circumstances (Rockman, 2008). In addition, the extent of adaptive resistance within a colonizing community during the first phase will depend upon prevailing cultural mores and the degree of familiarity (by experience or tradition) with what is entailed in meeting such a challenge. Mobile hunter-gatherers may be expected to be more adept at such changes than groups from settled and more socially stratified societies. The model presented herein, which draws on Blanton's insightful work, is intended to create a basis from which we can begin accounting for the kinds of variables that might be expected to affect the process of generational-scale adaptations to environmental change in settled regions and new territories. This process to colonization is examined through four descriptive (and inter-related) phases, and with particular reference to the circumstances for early populations in Southeast Asia. These phases are 'Transference', 'Diversification', 'Innovation' and 'Divergence'.

8.2.1 Transference

The initial exploitation of a new environment will be subject to the prevailing traditions and approaches to life and exploitation of the home environment. In essence, transference represents the imposition of old habits onto new settings. Only once people are acting and living within the affordances of a new environment – what NASA calls '*in situ* resource utilisation' (e.g. Murphy *et al.*, 2009) – will that imposition become challenged or qualified. The period of transference takes little account of the actual appropriateness of pre-existing views to new situations; one starts by doing what one knows. As a thought process, this will also be, in part, driven by the notion of recognition, namely seeing in the new environment features or resource characteristics that are apparently consistent with memories or experience from the old one. Additionally, initial colonization may involve the repeated use of selected points on the landscape for resource extraction, not necessarily because these represent the best-positioned or the highest-quality resources of a particular type but because they represent 'known quantities' within an otherwise largely unfamiliar setting. For example,

Beaton (1991) has referred to the exploitation of 'mega-patches' (e.g. coastlines, rivers, plains or lake-shores) in relation to early Palaeo-Indian colonization: basically, coarsely defined but preferred habitats. Such locales will probably be few and form crucial nodes within a rudimentary economic system covering the bare essentials of survival. In essence, they form the first efforts at risk reduction within a new landscape (Kelly, 2003).

As 'landscape-learning' (Kelly, 2003; Rockman, 2003) develops over time and groups gain greater experience of a new setting, this rudimentary system will tend to be augmented and altered. One might characterize this as a shift from using predominantly 'transported knowledge' to relying increasingly on locally 'acquired knowledge'. This process of familiarization may have been more effective if the small, incoming groups remained in close contact with one another within a comparatively restricted geographical area rather than co-residing in fewer but larger groups. The former scenario would help facilitate the speed of landscape learning within the 'nursery area' without the resource depletion that larger groups would inevitably cause, particularly to a rudimentary economy. Communities emerging from this preliminary phase are more likely to leave signatures in the archaeological record, simply through their enduring presence and through innovative technological forms. Even so, this opening phase is not without retrieval potential archaeologically.

Take for example the distribution of Clovis technology across Late Pleistocene North America. States to the east and south of the Great Lakes, where Clovis and other related fluted point traditions predominately occur (Anderson and Faught, 2000) are believed to have experienced conditions favouring complex disharmonious mixes of modern and extinct fauna under a fairly extreme climatic regime during the Late Pleistocene (Dincauze, 1993). By the time of the Clovis cultural horizon, during the last couple of millennia of the Pleistocene – long-held as the most reliable evidence of human dispersal across the continent; though for changes to that perspective *see* e.g. Beck and Jones (2010) – the environment was experiencing dramatic and persistent changes in habitat distribution, seasonality and growing season. Biomass that had been previously quite high was dropping rapidly, and species that could not either adapt or move on were becoming extinct. If humans were entering or, as is perhaps considered more likely on current evidence, expanding within North America around this time, the environments they encountered would have been in a profound state of flux. Refinements to the ^{14}C age range for the spread of the Clovis tool industry (Waters and Stafford, 2007) indicates a remarkably tight window for its appearance and spread through North America. Using the INTCAL04 calibration curve, a maximum age range for Clovis, based on only the most well dated contexts, was calculated by these authors as 13,250–12,800 cal. BP. What is more, if this is correct, the Clovis expansion was occurring right on the boundary between the end of GIS-1 and the onset of GS-1. This brief colonizing pulse is not only one of the most archaeologically visible phases of a post-LGM expansion within the Americas (Fagundes *et al.*, 2008) – it was taking place during climatically deteriorating rather than ameliorating conditions. Kelly and Todd (1988) proposed that a high degree of environmental fluctuation would have placed considerable strain on these groups in terms of food acquisition and the predictability of resource timing and distribution. Faced with this scenario, they suggested that migrant hunter-gatherers would have needed to be highly mobile and probably would have been able to most successfully adapt by targeting game of comparable size to that which they will have hunted in the past. This would also have maximised the value of transferable skills in terms of animal procurement and carcass processing, increasing their adaptive potential. In other words, the emphasis might have rested less on knowledge about landscapes as a whole and more on knowledge (including knowledge transfer) about specific resources (Kelly, 2003). Transferable knowledge about target resources would be one of the ways through which key nodes within any preliminary knowledge system would be formulated. More intimate knowledge of a landscape, as well as its cultural significance, would come through refinement of that initial point-to-point understanding of the newly colonized landscape – an understanding that is itinerant rather than spatial (Gamble, 1998: 441). This familiarity and deepening knowledge are brought about through habitual activities and increasing experience of places across seasons and across years – what Ingold (1993) has referred to as the emerging 'taskscape'.

Returning to the Jamestown experience, Blanton writes that early subsistence evidence from that settlement includes a lot of small mammals and reptiles because these were easier to catch by the inexperienced English settlers than the preferred large game. In a similar way to the earlier example of dispersal in the Americas, this would also have been a period of elevated risk and dietary stress because of the lack of knowledge held by the settlers (Blanton, 2003: 194). As Winterhalder (1980: 148) has observed, 'the pathway to adaptive change need not bear any straightforward relationship to normative qualities of the relevant habitat'. One might argue that the distinction in this case is that the English settlement would have been made up of translocated people. These were individuals who had arrived by sea with no real possibility of gradual acclimatization *en route* and who were disposed towards approaching colonization from a particular set of social rules, including an inherent belief in cultural superiority over the locals they met. The Late Pleistocene arrival of humans in the Americas via the Bering land bridge may have eased the physical transition through gradual acclimatization. If the coastal colonization models currently in favour (e.g. Dixon, 2001; Erlandson and Braje, 2011; Mandryk *et al.*, 2001) continue to gain support in the field, this may have to be reconsidered, but under either scenario, the transference of attitudes towards colonization during that period is itself a largely unexplored but fascinating area of research.

If the transference of hunting and gathering strategies marks the initial colonization of a new environment, it may also feature in the adaptation to ecologically rapid shifts within an existing one. The duration of this transitional period, then, will be affected by a range of variables, including the compatibility of the resource character and structure between environments at the point of origin and those at the destination, the level of socioeconomic predilection towards dealing with such conditions and the size of the incoming population. However, the duration of this 'maladaptive' period may be extended where destination or occupied landscapes are themselves being re-modelled by selective pressures acting differentially on different members of the resident biotic community.

Although dependent on the resilience of resident animal species to abrupt external impacts (e.g. such as the sudden introduction of a new hunting technology), in many cases, biotic communities at the point of debarkation will have been living under different and potentially lower levels of community-wide selection pressures compared to those acting on the incoming colonists. The response of resident human groups to rapid environmental change introduces certain additional differences to standard models of colonization. These latter tend to be dominated by factors associated with the effects of translocation on human groups. Where the entire environment is in a state of flux – as it frequently was during the Late Pleistocene – strong selective processes will have been affecting resident, as much as incoming, human groups in a region. Such change would also have been acting concurrently and differentially on individual elements of the food web on which those human populations depended. Success at maintaining or reformulating a viable subsistence base is likely to have been hampered by this turbulent state. Depending on the flexibility of the human cultural-economic systems, this will affect the length of the maladaptive period.

Responses by resident early hunter-gatherer groups to climatic swings, such as the deterioration in conditions during the late glacial of northern Europe, have tended to focus on visible signs of population contraction into more southerly geographic refugia (e.g. Mellars, 2006b). This interpretation of the data does, though, tend to perpetuate the image of community-wide responses to change – a position that ecologists have now largely abandoned. As a shortcoming in archaeological theory, it is also now being highlighted by studies that have sought to compare *H. sapiens* to *H. neanderthalensis* adaptive strategies as a way towards explaining their differential survival (e.g. *see* Marean, 2007). This issue may prove equally instructive when we are considering localised variation in the adaptive responses of early *H. sapiens* groups to climatic change. Although the Late Pleistocene European steppic and tundra-like landscapes were more productive than similar environments are today, the subsistence economy of modern human groups may have differed less from that of Neanderthals than it was from that of other modern humans living in other regions. The resource structure to which modern humans were obliged to adapt in Europe selected for novel cultural and economic mechanisms that could ensure survival through monitoring access to and distribution of those resources. It is generally accepted that in the humid tropics, a comparatively large proportion of the material culture of early people living in this environment was probably perishable and, therefore, has been lost to us archaeologically. I think, we would be wrong, however, to assume that this means that no more can be said or to assume that the continuation of predominantly cobble- and flake-based technologies should be explained away by inadequate lithic resources, cultural retardation or loss of 'complexity' through cultural drift. The emerging Southeast Asian Palaeolithic record suggests that there is much more that can be retrieved about early human adaptation in these and other neglected environmental settings if we would but look.

8.2.2 Diversification

The increased level of risk associated with early colonization may result in population contraction, a population bottleneck or, in extreme cases, extirpation. If a colonizing population has successfully established a foothold in a new landscape, they will have begun to settle into the adaptive setting that it offers – rather than simply sustaining themselves with the knowledge and technologies they brought with them. We can illustrate this by considering how the structure of tropical and periglacial environments will have presented different sets of constraints and opportunities to successful settlement.

Palaeoenvironmental reconstruction indicates that during the second half of the Upper Pleistocene, during the period of its colonization by modern humans, Europe experienced of brief episodes of temperate forest development. Although persisting woodland stands in sheltered riperian and southern peninsular refugia provided regional sources for this expansion, the climatic instability of the period exerted a strong selective pressure against more sustained sylvan colonization of the wider landscape. The tempo of change and turnover of conditions left insufficient time for many species to adapt. The dominant biome was the vast, open, arid but productive steppe (Guthrie and van Kolfschoten, 2000) populated by a novel faunal community comprising large nomadic herds of herbivores, such as horse, reindeer, bison and mammoth, that ranged widely and of species, such as wild boar, that have very wide ecological tolerances.

Given that *H. sapiens* are also particularly capable of rapid adaptation to changing conditions through behavioural realignment and cultural augmentation, such ecological instability may actually have aided colonization of this landscape by mobile bands capable of exploiting wide-ranging fauna. Some scholars see this reflected in evidence for hunting specialization during the Upper Palaeolithic sequence. Mellars (2004b) reports that 47 per cent of 166 faunal assemblages from southwest France, spanning the Aurignacian to the late Magdalenian, show greater than 90 per cent of identified ungulate remains to be those of reindeer (*Rangifer tarandus*). The stark contrast he draws to evidence for Mousterian specialized hunting (one of 95 sites studied) gives good reason to suppose differences in procurement preferences and strategies. Other studies, though (e.g. Britton *et al.*, 2011; Grayson and Delpech, 2002), find evidence that the interception of migrating reindeer herds was a feature not only of modern human groups but also of Neanderthals, leaving this matter unresolved and potentially more a reflection of species abundance and availability, and about acquiring knowledge of their nature and movements.

Modern studies and ethnographies indicate that reindeer (and caribou) may be readily intercepted at certain points in the landscape, such as river crossings, or where ice lingers into the spring as part of insect avoidance (e.g. Hagemoen and Reimers, 2002); however, as a key resource, they are prone to instability. Herd numbers may crash suddenly or alternate migration routes might be adopted because of changing access to grazing and other environmental factors and predation (Gunn, 2003; Joly *et al.*, 2007; Urquhart, 1989). Heavy reliance on such a fluctuating resource will have had a significant effect on how risk avoidance strategies developed among European Palaeolithic hunter-gatherers. How these strategies were articulated among Neanderthal groups is still to be determined, but it should not be assumed that they were less attuned; among modern humans, such dependency may have fostered an enhanced reliance on long-distance communication and social networking to capitalise on periods of abundance and offset periods of scarcity.

The picture in the tropics of Southeast Asia is quite different. Modern zooarchaeological studies from a number of sites in the Southeast Asian record, including those herein, are finding that individual components of early economies seem to persist as site-specific (or location specific) staples – supplemented by more opportunistic hunting and foraging. For example, ungulates are comparatively rare in some instances, such as the Late Pleistocene–Early Holocene sites of Lang Rongrien (Thailand) and Hang Boi (Vietnam), where emphasis lay on procuring turtles and land molluscs, respectively (*see also* Nguyen Viet, 2004, for further Vietnamese examples). By comparison, ungulates dominate the terrestrial faunal component from Late Pleistocene deposits at the Niah Caves (Borneo). The bearded pig (*Sus barbatus*) formed a key resource here from the beginning of occupation *c.* 44 kbp (age beyond current calibration) right through until the Mid-Holocene. The strong migratory and herding behaviour of this animal places it in the unique position among tropical mammals of being both numerous and comparatively predictable during its mass movements from one valley to another as it takes advantage of mast-fruiting events (Curran and Leighton, 2000). As indicated earlier, the exploitation of pig herds may have been one of the particular attractions to foragers on the coastal lowlands of northern Borneo in visiting the Niah Caves. There is still a long way to go in reconstructing Pleistocene rainforest settlement systems, but the data that we do have are emerging as broadly consistent with the basic nature of adaptations reported in the local ethnographic literature.

The resource structure and scheduling within a modern rainforest environment are considerably different from that found at higher latitudes and, though similar to those experienced by early modern human arrivals during the Upper Pleistocene, are unlikely to match them exactly. As already stated, there is an emphasis on species diversity over species quantity, and while seasonality may be less pronounced, it is marked by rainfall maxima related to monsoon cycles and strengths. Residential and hunting strategies today are adapted to account for wet and dry season differences in terms of the different restrictions and opportunities they pose (e.g. Eder, 1988; Griffin, 1989; Griffin and Griffin, 2000). In Borneo, irregular super-annual cycles caused by the El Niño–Southern Oscillation (ENSO) are closely tied to peaks in forest resource availability that are separated by sometimes long intervening periods where forest food resources may become scarce (Chan, 2007: 201; Curran and Leighton, 2000; Dove, 1993; Leighton and Leighton, 1983). One of the principal arguments put forward to explain why forager groups came to practice mixed economies was the paradoxical dearth of edible and accessible products, in particular, carbohydrate staples that closed-canopy rainforest apparently offered, making husbandry a necessity. While this is no longer supported by the archaeological evidence, which strongly indicates long-term (pre-agricultural) occupation of the Bornean lowlands (e.g. Barker *et al.*, 2007; Barton, 2005; Rabett and Barker, 2007), probable differences in the structure of these forests between the Late Pleistocene and those that had developed by the Late Holocene (and Anthropocene) do lend some support to a scenario that sees resource scarcity as one of the factors that may have helped promote the widespread adoption of swidden-based agriculture during the last few millennia.

Modern small-scale indigenous communities in Southeast Asia tend to practice a strategy of switching from one component of a mixed economic system to another as needs arise and changes in circumstance demand (e.g. Eder, 1984; Endicott, 1984; Endicott and Bellwood, 1991; Griffin, 1985; Noone, 1936; Sellato, 2007: 82; Figure 8.1). Being able to constantly switch between parallel strategies ensures that return rates are maintained. In the past, as in the present, in basic terms, people appear to have visited sites to exploit spatially restricted resources out of a range of common staples, probably at particular times of the year. As needs demanded, previously minor components of the economy might also be raised to greater significance. In discussing why the Batek (a dialect group of the Semang in Peninsular Malaysia) favour a frequently shifting economy, Endicott (1984) proposed that it was more efficient than following any one practice. He noted that when collecting wild foods, the Batek do not behave as if those resources were scarce but supposed that they could subsist on them indefinitely. He explains how each of the components that make up the Batek economy (foraging, hunting, shifting cultivation and forest collection) have productive and lean periods. Being able to switch between them ensures that return rates are maintained. Individual components of an economy persist or are deliberately retained where they are compatible with or help support alternative and newly adopted components. The flexibility of this kind of 'additive' economy – that is, new components are added to, rather than replacing, existing ones – provides its own buffer against risk and promotes socioeconomic and cultural continuity (Meijaard *et al.*, 2005: 10). Furthermore, although tropical forager groups globally (and routinely) possess extensive knowledge of their surroundings, a detailed understanding tends to be accorded only to certain sections of it – those that have the greatest immediate socioeconomic importance (e.g. Dunn, 1975; Rival, 1996). Resource utilisation at any one time may be focusing on only a small fraction of what could be exploited (Sellato and Sercombe, 2007: 44). By these means, additional risk mitigation may be built into this kind of economic system by shifting attention to different resources, previously ancillary (but not necessarily secondary in calorific terms) and raising them in importance.

Details about how this system emerged as a general economic strategy among most rainforest communities in Southeast Asia are complex and themselves related to responses to colonization during the region's Neolithic and, historically, to state and economic forces. That said, the inherent characteristics of the Holocene rainforest itself, taken together with the many abrupt changes affecting communities within it – such as the rapid alteration caused by logging (Meijaard *et al.*, 2005), which can result in the complete physical obliteration of a 'learned landscape' (Brosius, 2007; Langub, 1988) – appear to have been similarly strong forces of selection in the development of extremely malleable and situational economies. As such, this characteristic may be adopted provisionally as part of what one might expect to see in the colonization process of highly diverse environments. How early human groups undertook different adaptations is reflected in the kinds of innovative behaviour each came to exhibit.

8.2.3 Innovation

Innovative behaviour appears to have strong links to demography. A higher population density is considered conducive to more rapid change as there is a larger pool of interacting learners to generate that change (Henrich, 2004). These situations provide crucibles of innovation. The European evidence seems to contain just such pulses of cultural invention during the Late Pleistocene from within areas and periods when hunter-gatherers appear to have congregated at generally higher population densities (e.g. Conard and Bolus, 2003;

Figure 8.1. View of wet rice fields just prior the transplantation of rice seedlings, near the village of Pa' Dalih in the Kelabit Highlands of Sarawak. Bordering the rice field in the left of the picture are fruit trees and a large hill sago palm (*Eugeissona utilis*). The pith of the sago trunk is the staple food of the nomadic Penan. Even though the Kelabit are intensive rice farmers, the shoots of the sago palm are eaten. The palm's leaves were also traditionally used for thatching. (Photograph by L. Lloyd-Smith, reproduced with kind permission.)

Gamble, 1991; Jochim, 1983; Mellars, 2006b). Demographic circumstances have also been cited in the innovative first appearance in Palaeolithic Europe of humans employing material culture as a vehicle for transmitting identity and alliance. Gamble (1982, 1986, 1991) has proposed that under circumstances of low population density and existence in elevated risk, such as ahead of the LGM, day-to-day living is likely to have involved extended periods of group isolation and infrequent small or large-scale meetings between people. Ethnographically, such environmental circumstances appear to be met with 'open systems' of group interaction; essentially, systems that emphasise mutual support and rapid information flow as means to offset times of stress – particularly resource stress. They contrast to 'closed systems', where trade and social interactions are more exclusive and greater emphasis may be placed on the identification and possession of territorial rights. Gamble has argued that open systems would have been of significant adaptive advantage to communities in the Late Pleistocene northern Europe as material culture can provide a medium through which a visual system of communication could be transmitted over and above spoken language or face-to-face communication, augmenting these forms with symbolic markers of identity and history into relationships even if personnel changed through time (Gamble, 1998, 1999; *see also* Finlayson, 2004). In essence, such a system would have worked by rapidly bridging social distance between people and groups (even when personnel in those groups changed) and would have helped administer access to resources in periods of abundance or scarcity.

Whereas elevated population densities appear to have been more the exception than the rule in periglacial environments, closer to the equator, the reverse may be true. Ethnographically, modern hunter-gatherer population density – measured as persons per 100 km^2 – in seasonal and wet tropical forests is considerably higher (an average is 32.8 from a sample of 18 groups from around the world, but excluding the Semang, which is listed as having variable densities from 5 to 19 persons per 100 km^2) than that reported for sub-Arctic and cold forest environments (an average of 4.14 from a sample of 32 groups). Both values are calculated from Kelly (1995: table 6-4). This difference is related to the carrying capacity of an environment and its productivity but is also moderated by other important variables such as competition for resources, foraging return rates and forager impact. With respect to the last of these, it has been a key finding of recent ethnographic (Puri, 2005a, 2005b; Posey, 1985; Rival, 1993) and ethnoarchaeological (Politis, 1996) studies that tropical hunter-gatherer mobility has a discernable effect on the distribution and productivity of rainforest resources. Previously abandoned camps provide concentrated patches of desirable and harvestable foods (which also attracts game) and can be exploited during timed return visits. While the relationship is not a simple one, tropical environments appear to be capable of supporting a comparatively high density of hunter-gatherer communities.

That this should have been the case in the remote past as well as historically is now gaining support. Although tropical palaeodemography is still a nascent science, work using mtDNA diversity as a proxy for population size has found that between approximately 45,000 and 20,000 years ago, over half the world's modern human population may have lived on the Indian subcontinent and in the vicinity of the Thai–Malay Peninsula (Atkinson *et al.*, 2008), and thus possibly Sundaland. Archaeologically, population density is hard to track in the tropics because of preservation and sampling issues, and the limited existing understanding of settlement systems (Rabett and Barker, 2010); however, several lines of evidence from Southeast Asia indicate that while exchange and trade networks certainly existed by the Mid-Holocene, and probably earlier (Bulbeck, 2008), an emphasis on parochial behaviour is a consistent feature. We have already seen in the preceding chapters that there are indications for localised lithic and bone technological trajectories, and methods of faunal processing in this region during the late glacial. This pattern also occurs in more recent millennia. For example, Chia's (1997) study of ceramic material sourcing in the Lenggong Valley, Peninsular Malaysia concluded that the production of vessels recovered across a sample of sites (n = 29) scattered between eight limestone massifs was focused on localised clay sources and with no indications of bartering and exchange of products among the sites studied, though inter-site similarities in firing methods, decoration, shape and tempers were taken to indicate that knowledge exchange between communities may have occurred (Chia, 1997: 126–7). Holocene burial practices at Niah parallel those seen elsewhere in the region, and there are hints that group members were exchanged beyond the immediate area. Even during what seems to have been this more expansive phase of regional interaction, though, locally distinct burial traditions are still apparent; in the case of the Neolithic, some of these distinctions occurred between communities using caves only a few hundred meters apart (Barker *et al.*, 2011). Collectively, the evidence does begin to align itself with what one might expect were there a prevalence of comparatively more closed social systems – themselves a feature of higher demographic density (*see* Wobst, 1976).

Varying degrees of social closure are found among the region's modern indigenous groups. For example, it is seen among the Penan, Semai and Melayu Asli groups in Malaysia (Brosius, 1993, 2007; Fix, 1995) and on the Andaman Islands, for example, between the Jarawa and Onge (Endicott *et al.*, 2006), though it is not always present. Although the Semai, Melayu Asli and a third group, the Semang, share a common ancestor, the last is overtly exogamous and maintains open social networks (Fix, 1999). It is unique mtDNA lineages found in the Semang (particularly in M21) that are believed indicative of initial colonization of the region prior to the Late Pleistocene (Macaulay *et al.*, 2005). Since patterns of residence and sociality have been found to impact on genetic variation, the modern Semang system could have very deep roots (Stoneking and Delfin, 2010). This presents a challenge to the proposed demographic model. While one might anticipate that initial colonization would be at low population densities and require open systems of communication, its persistence under tropical conditions seems out of step with the mass of other evidence. Although it represents only one possible explanation, it is worth noting that the social and economic character of modern indigenous Southeast Asian groups need not be representative of past formations (Oota *et al.*, 2005). This not only reminds us against assuming antiquity in the behavioural customs of these small-scale societies, it is in reality completely in keeping with a predilection towards switching strategies according to need – something the Semang practice as much as other indigenous groups here. Complexities in Semang history may go some way to explaining the contradictions, though a more comprehensive explanation does need to be formulated.

The challenges do not end there, though, for if population densities have been comparatively high in Southeast Asia since the Late Pleistocene, why is this not reflected in conspicuous levels of innovation? Part of the explanation for this may have been the specific cultural and economic perceptions about the affordances of raw materials. In Southeast Asia, allowing for the fact that the scale of reliance on non-preserving media (a common recourse when trying to explain the character of Southeast Asian technology) remains a largely unknown quantity, the archaeological evidence suggests a propensity towards taking existing natural forms and materials from the environment and applying only minimally invasive modification to them before use; basically, a predilection for expedient forms of technology. This generally more minimalist emphasis on manipulation or standardization contrasts with something like the European Upper Palaeolithic, which can be said to have embodied a tendency for imposing a much greater diversity of forms onto a more restricted range of raw materials (Rabett, 2011; Rabett and Barker, 2007).

A history of early innovative behaviours is known archaeologically from this region. For example, in the Niah Caves record, this includes an early use of pigment (*c.* 40–45 kbp) or use of pits, probably to detoxify forest products (34–29 kbp) and the development of composite harpoon or leister technology (Early Holocene). The Tingkayu lanceolate points and knives (Sabah) or pseudo-Levallois points of Leang Burung II (Sulawesi) would be other examples among numerous ones discussed herein. When these inventions occur, however, they appear to be used to complement existing technologies rather than to supplant them. With few exceptions – such as the possible spread of bone projectile technologies, itself possibly linked to a significant post-glacial intra-regional diaspora (Rabett and Piper, 2012) – such innovations also seem to have remained localised. In light of this, it is possible that social closure tended to restrict inter-group interaction under normal circumstances. This could have reduced the effective population pool for the transmission of newly generated knowledge. The social setting between these two regions may have been quite different, with a commensal impact on innovation: traditionally low population densities with open networks and pulses of population packing on the one hand, compared to traditionally higher population densities with greater long-term social closure on the other.

In essence, the environmental, demographic and social situation in Southeast Asia may have meant that innovation in material culture and how it mediated interaction developed in different ways to that seen outside the tropics. A statement by Gamble (1982: 101) comparing *H. neanderthalensis* and *H. sapiens* social networks in Europe has equal credence here with respect to inter-regional populations of modern humans: 'The differences between ... populations and their distinctive material culture was not derived from differences in biology and intelligence. Rather it stemmed from differences in their societies, the structure of their alliances, that resulted in the potential information of material culture being used in alternative ways.'

8.2.4 Divergence

One of the consequences of the colonization process is the evolution of existing knowledge domains into more locally pertinent ones. Knowledge systems become tailored to the particular behaviour and ecology of the new resource spectrum, and there is divergence from the old. There is the potential loss of information about some taxa and the creation of new information about novel forms; a divergence of cultural identity, and even language, and greater adaptive and social specialisation to the affordances of a new (or transformed) environment. With reference to later periods, Voeks (2007: 283) has observed that the ethnographic subsistence niches associated with foraging economies, when compared to farming, present different essential details about plant knowledge. These differences translate into diverging paths in ethnobotanical knowledge acquisition and its cultural retention (*see* Brown, 1985).

Although divergence need not apply equally to all spheres of knowledge or practice, broadly, one might expect to see behavioural differences emerging between early and more established settlement phases during colonization. As stated, the initial exploration of a new environment tends to be undertaken not by large residential groups, but by small, often logistical, ones. This appears not to be something confined to our prehistory; we see the same pattern time and again, as in the European colonization of North America and in the exploration of extreme environments such as the Poles and in space exploration (e.g. Arnould, 2001; Ursin *et al.*, 1991; cf. Hienz *et al.*, 2007). Where initial colonization translocates such groups either seasonally or for brief periods of time, or rapidly over a wide geographic area, there is little evidence of regional variation. MacDonald (1998) suggests that these kinds of widely distributed 'cultural phenotypes' may imply the maintenance of inter-regional contact between colonizing groups – a position he advances via the long distances (up to *c.* 500 km) that lithic raw materials used in the production of Palaeo-Indian Folsom projectile points often moved. Equally, if there is constant change confronting a pioneer population – relentless movement into uncharted territories rather than settlement consolidation, for example – there may also be little scope for behavioural or cultural divergence to occur and limited potential for consequential regionalisation paralleling Blanton's third phase of colonization: *creolization*.

Blanton (2003) characterizes creolization as the adoption of subsistence, technological and other elements from more than one contributing culture. The anthropological use of creolization presents complex issues about exactly what is being identified (*see* e.g. Jourdan, 2001). For example, is it the process or the result of the structural and cultural modification of a population? Given that all ethnographic societies are in a state of change (to a greater or lesser degree), what makes the changes associated with creolization specifically different from other forms? Does creolization represent a transitional or an achieved state? While an emergent cultural creole may embody a sense of stability in the way that Blanton (2003) suggests, this does not necessarily mean that a situation of long-term adaptive stability or viability has also been reached. Even if it is visibly the case that a new social formation has been created (signalling a departure from what went before or compared to a contemporary but geographically removed parent society), it represents the attainment of a state whose stability may itself be transient or unsustainable. For example, Riddington (1983) writes of the contact dynamic between the Athapaskan people of northern Canada and industrialised society. 'In order to obtain the exchange medium of furs and meat demanded by traders, the Dunne-za [people] were compelled to alter their own relationship to the animals and even to contribute to an overall decline in animal population' (Riddington, 1983: 62). Jourdan (2001) allows that despite its ambiguities, the term *creolization* has gained value in anthropological discourse by virtue of its positive attributes: as a way of examining rapid cultural change; by the fact that it foregrounds cultural heterogeneity over homogeneity; and as it focuses on the forces of fragmentation rather than integration, process rather than stasis, and the multiple and mutable rather than the singular and normative in terms of cultural identities.

In archaeology, the term is also gaining credence in the study of pre-ethnographic spheres of colonization – for example, Webster (2001) has proposed that the Roman provinces might be more usefully described as 'creolized' rather than 'Romanized'. With respect to Palaeolithic archaeology, it could be a valuable means of characterisation in the study of interaction between hominin species. The term is adopted tentatively in the current context. The *Oxford English Dictionary* (OED) indicates with historical reference that the term *creole* can be applied to a person (plant or animal) born and naturalized to one country but not of indigenous origin. A potential alternate term could be *hybrid*, which the OED defines specifically as 'anything derived from heterogeneous sources, or composed of different or incongruous elements'. While applicable, this latter word carries predominantly biological overtones; *creole* presents greater semantic space for biological, cultural and linguistic reference, making this the more preferable term here.

At present, the closest approximation to creolization entertained in Palaeolithic research is the often vague use of the term *acculturation* (*though see* Davies, 2007; Tostevin, 2007). Most commonly, this is seen as a product of discourse between modern humans and Neanderthals and is generally assumed (realistically or not) to have been unidirectional in terms of what was imparted culturally. There are a number of ways in which detailed consideration of emerging creoles might be productively applied to the European Upper Palaeolithic sequence in the form of testable hypotheses. For example, one might examine the material components held in common spatially between contemporaneous cultural complexes and their antecedents. Currently such components tend to be recognised simply as evidence of common adaptations or evolutionary carry-overs, when they could be tested more specifically as possible trails of technological transfer between parent and daughter populations.

Applying greater sophistication to existing models of European Upper Palaeolithic colonization would have further beneficial consequences. Significant departures observed in the character and composition of grouped site assemblages would no longer be simply typologically compartmentalised, leaving them characteristically isolated from one another and static; they would gain a viable mechanism of change in early modern human dispersal across this continent. To implement such a model, one would have to address a number of issues, paramount of these being temporal resolution (Davies, 2007; Douka, 2009). Other issues, though, would include distinguishing between changes in inter-site functionality from wider cultural change, and determining what should, in fact, constitute 'rapid cultural change' in the Pleistocene – does one compare backwards against the monolithic lithic industries of the Middle and Lower Pleistocene or forwards against the mercurial, short-lived cultures of the recent past? With such matters considered, however, the Upper Palaeolithic of Europe suddenly begins to grind into motion. Cultural turnover becomes linked to an interactive process between groups and the environment rather than the unilinear evolutionary development it has tended to be portrayed as. Each new cultural expression 'momentarily' emerges in the wake of a dispersive or innovation-driven pulse where adaptation to new sets of conditions or niches has been successful. In other words, these cultural formations may become visible in the archaeological record only when (and if) there has been sufficient time for divergence (creolization) from their parental forms and if they have not been found to be maladaptive.

Within the European sequence, one could also assess what level of archaeological evidence might be equated with maladaptive responses; how might the so-called transitional industries such as the Szeletian, Uluzzian or Châtelperronian fare when viewed from this perspective? Could they be interpreted as maladaptive forms whose makers died out or as emergent but short-lived creoles? If the 'Developed' Aurignacian, as defined by Davies (2001), is distinct enough to separate it from a 'Pioneer' phase, one might consider classifying this as a creole both in terms of the consequence of cultural dialogue (including inter-hominin) and the successful passage of an incoming population through adaptive misalignment towards local conditions and naturalization within a new territory. At the same time, though, if no evidence indicating a maladaptive phase can be identified within an industry like the Aurignacian, which is presumed to be a primary colonizing phenomenon, what does this say about the credibility of this enduring interpretation of it? Finally, one might also examine the pattern of correlation between Palaeolithic cultures in the European sequence against what is known about the patterns of climatic and ecological variability in the North Atlantic region. The Developed Aurignacian emerged under the increasingly rapid fluctuations of a climate regime steadily sliding towards the LGM (Davies, 2001). From then on, the region experienced a period of major cultural diversification, innovation and change (not to be confused with a notion of incrementally increasing cultural complexity), features in line with the expectations of the colonization thesis presented here. In no small part, however, was this trajectory affected by the particular combination of ecological conditions and kinds of social interaction peculiar to Europe – a combination that was not repeated verbatim in other parts of the world.

In light of the foregoing, we may begin to place some kind of formal definition of a hominin 'palaeo-culture' (aka the 'Palaeolithic cultures', 'industries' or 'techno-complexes' of Palaeolithic archaeology). These may have been phases of quasi-stable adaptation, expressed through a visible divergence in material remains and social and economic activity from geographically or chronologically removed group(s) – particularly where that separation can be identified as from a 'parental' community. These cultural states emerge following a period of adaptive stress and economic re-alignment and diversification (*see* Latinis, 2000, for a parallel argument). They are defined as 'quasi-stable' because their durability, however long it may be, is still much more transient than generally seen for most organisms. This is by virtue of the fact that hominin behavioural and cultural adaptation are by their very nature more plastic than biological adaptation, and in *H. sapiens*, have arguably attained their most malleable state yet. The emergence of a new human palaeo-culture meant that a certain degree of adaptive stability had been reached and, with it, a tendency towards specialization. This situation would persist until such a time as external (e.g. environmental) or internal (e.g. social) changes tested that palaeo-culture's resilience and a new cycle of transference, maladaptation and so on began. If it failed, population collapse and extirpation would ensue. This was possibly where the late archaic hominin groups faltered. The palaeo-culture system of *H. neanderthalensis*, for example, was specialized to a range of Eurasian settings and conditions, and while changes during the Late Pleistocene (including the arrival of *H. sapiens*) may have initiated adaptive responses within some enclaves (as seen e.g. in the Châtelperronian techno-facies), the species ultimately did not survive the high frequency of environmental resets during this period. Among the late-surviving hominins only *H. sapiens* evolved palaeo-cultures that became capable of adapting fast enough and successfully enough to keep pace with the conditions of the Upper Pleistocene.

Across much of Pleistocene Eurasia, successful human adaptations were geared towards coping with the forces of unpredictability associated with the abundant but comparatively narrow range of animal and plant taxa available to be exploited. Tropical subsistence, by comparison, may have long been adapted to dealing with resource diversity. The flexible subsistence practices of historically recorded hunter-gatherer societies living in such rich environments may be only the latest expression of a much older principle. The chief concern of Late Pleistocene and Early Holocene hunter-gatherers in the tropical lowland forest of Southeast Asia was probably not how to mitigate the risk of not gaining pivotal provisions (*see* e.g. Bird-David, 1990; Endicott, 1984; Endicott and Bellwood, 1991). Rather it lay in how to map their activities onto peaks of abundance in a variety of staples – peaks that followed a range of productive cycles, some of them seasonal, some not (such as ENSO) – and what was the most appropriate artifice and technology for securing provisions within such diversity (*see* e.g. Griffin and Solheim, 1990). The expeditious manner in which colonizers appear to have adapted to the wide range of habitats that would have confronted them across Sahul has been linked to the possibility that such groups possessed 'an economy structured into different strategies' (Bowdler, 1990: 340) – just the kind of flexibility the evidence suggests we might expect of colonizers from Southeast Asia.

8.3 The Last Termination in Southeast Asia

The broad plains and upland margins of Sundaland were spared the extreme temperature swings of the Last Termination experienced at higher latitudes but were nonetheless subject to fluctuations in the extent and composition of landscape cover. Forested areas in particular, then as today, comprised a wide diversity of flora and fauna. Economic resources will have been predominantly dispersed and varied, though a long-term hominin presence will have affected changes to the clustering of edible resources; something that will probably have imparted a guiding influence on mobility and been an aid to the successful settlement of this landscape.

For reasons that are still to be fully appreciated, all species of hominin occupying the region through the Pleistocene appear to have employed broadly similar facies of a low-input lithic technology that confounds traditional typological classification. This has sometimes created a considerable challenge when it comes to determining who was responsible for a particular industry. The enduring nature of this emphasis on amorphous technologies may have been influenced by the kinds of lithic raw material that were available – e.g. in ISEA – though the significance of this has been overstated in the wider literature. Even when siliceous materials were utilised, this did not result in a widespread shift towards the manufacture of blade-based toolkits. Whereas traditionally, this fact has been equated with cultural impoverishment or a loss of skill sets in Asia (and especially Southeast Asia), the current study has argued to the contrary. The skills to produce such forms *are* present, knowledge of the mechanical properties of a range of technological materials *is* evident; what is different is that demographic, cultural and economic circumstances seem neither to have encouraged nor required the production of lithic toolkits other than those that we see.

Another long-held perspective on the generally low input character of Southeast Asian technologies argues that people needed their tools to be flexible enough to be suitable across a range of subsistence and organic tool manufacturing contexts (e.g. Hutterer, 1976). One could also hypothesise that they may have been less central in creating and maintaining identities and social networks to buffer against resource scarcity. Their role in carrying messages of group affiliation or to help bridge social distance between groups may have been less pronounced than that seen elsewhere.

While something like tool standardization is to be found, it seems to have had only a limited impact on the manufacture of implements in stone or bone until late, possibly more so in shell, but it does give one cause to question whether standardization is a valid marker of technological sophistication as traditionally assumed. Cultural artefacts made on durable media appear to have been born out of a distinct approach to the affordances of raw materials and may have been used in different forms of social negotiation to those seen elsewhere under other demographic and environmental conditions. We can be reasonably certain that non-preserving media will have formed a substantial portion of foraging accoutrements and that taphonomic processes in general are probably exerting a greater influence over archaeological patterning than we have hitherto supposed (e.g. *see* Langley *et al.*, 2011), leaving persistent but not insurmountable gaps in our knowledge.

In Late Pleistocene Southeast Asia, unlike in Eurasia, Australia or North America, although climatic instability instigated novelty in vegetation structure – for example, episodically combining montane and lowland forest taxa – in most cases, this did not extend to the establishment of novel zoological communities and the presence of megafauna (*though see* Piper *et al.*, 2007a); fauna was largely extant. At a superficial level, this pattern of diversity has been a primary feature of site assemblages from the region. For a long time, and in the context of limited zooarchaeological research in this region, it had been assumed that this indicated an emergent 'broad spectrum' subsistence pattern (Gorman, 1970), though rather than setting the stage for the arrival of agriculture, as such practices appear to do in western Eurasia, in the tropics, it is deemed likely a feature dictated to a significant degree by the structure of this environment (Reynolds, 1990). When compared closely (in terms of species representation and abundance and, where possible, patterning in butchery strategies), it becomes clear that tropical subsistence activities, as with genetic, linguistic, lithic and ceramic traditions, were noticeably local in character; providing grounds for suggesting that early tropical foragers may have maintained relatively closed social networks and settled into this landscape at comparatively high population densities (*see* Table 8.1). Both are features that will have affected trajectories of development in material culture. This is not the whole story though, as there is considerable genetic and archaeological evidence of changes during late to post-glacial times, when contact between populations, the translocation of species and the existence of exchange systems become quite pronounced (*see* e.g. Bulbeck, 2008).

For example, the increased prominence of bone technologies, including projectiles, from GIS-1 onwards marks a stream of technological innovation continuing into the Early Holocene, with new composite projectiles using stingray barbs. Although a robust connection remains to be demonstrated, the pattern of bone tool occurrence across the region seems to coincide with genetic evidence for a phase of population constriction and then intra-regional dispersal out of northeast Sundaland and western Wallacea (Rabett and Piper, 2012). This rise also correlates with greater exploitation (and probable availability) of arboreal taxa in Holocene forests. Projectiles, though, are not the only kinds of osseous implement that begin to appear in quantity; piercing and possible digging implements are among others documented, as are fish hooks made of bone and marine shell. Developments in range weaponry were only one part of a selective technological re-organisation taking place at this time. When taken in concert with other observed developments in this and other spheres, the indication is that it formed part of a phase of wider socio-economic change.

Nonetheless, certain subsistence practices were transferred across periods of climatic and landscape change during the Last Termination. Particular components of faunal assemblages remained consistent throughout different phases of occupation. It has been proposed that these elements may be considered dietary staples and that sites were perhaps placed for the exploitation of these reliable resources. Changes in subsistence practice, when they did occur, seem often to have supplemented rather than replaced existing ones. At Niah, the shift towards arboreal prey did not signal the end of the long-standing exploitation of the bearded pig; rather, it may be taken as an act of maintaining (or transferring) existing habits whilst instigating or expanding others: diversifying as part of adapting to post-glacial conditions. Through the Vietnamese data, we also begin to get a sense of how sub-coastal groups were responding to sea level rise and increasing coastal proximity. Here, despite the encroachment of salt marsh, lagoon and mangrove habitats, resources obtained from them were not immediately incorporated into the exploitation strategies of groups camping in Tràng An. Existing subsistence patterns were first intensified. The process of innovation appears to follow a recurrent pattern: low-importance, long-present components of the technological and faunal package were taken and 'cultivated' to prominence. People were emphasising approaches that gave them new ways of unlocking the environment by expanding pre-existing, but little used, parts of their knowledge base.

Table 8.1. The Broad Adaptive Circumstances and Strategies of Tropical Hunter-Gatherer Communities during the Last Termination, Southeast Asia

Criterion	Adaptive Circumstances and Strategies
Environmental character	High diversity of species encountered in low quantities (with certain exceptions, most notably the bearded pig);
Vegetation character	Fluctuating (timing largely correlates with North Atlantic climate shifts); mosaic of montane and lowland species; enhanced seasonality; the existence of rainforest refugia was probable, though whether such refugia should be seen as glacial or interglacial phenomena is an emerging point of discussion;
Temperature variance	Muted (the highest variation was likely to be at an elevation of *c.* 10° at the beginning of GIS-1); very minor to no temperature change recorded for GS-1;
Landscape change	Mixture of steady and pulsed inundation of the Sunda Shelf; some changes in forest composition linked to GS-1; structure of modern lowland rainforest is likely to be a Holocene phenomenon;
Inter-hominin interaction	Low (*H. floresiensis* was apparently extinct by *c.* 17 KBP), though certainty on this matter does not yet exist;
Inter-group interaction	Links between parts of the region are indicated, especially in the Holocene, but social networks may have been relatively closed, particularly before the latter half of the Last Termination;
Population density	Potentially high compared to Upper Palaeolithic hunter-gatherer densities in Eurasia; linked to differences in environmental structure;
Mobility patterns	Still largely unknown; generally assumed to be high, though preference for local raw materials may indicate mobility was lower after than it had been prior to the LGM; movement was possibly prescribed to a degree by utilisation of old camps as concentrated resource patches; intra-regional maritime dispersal and routes of contact probably featured from quite early in the period of regional inundation and further developed thereafter;
Subsistence regime	Localised practices utilising reliable staples often supplemented opportunistically; 'strategy switching' in accordance with the diverse nature of the environment and according to circumstance (with low emphasis on buffering against single resource scarcity); a tendency to intensify and diversify an existing system in response to changing circumstances; a possible time lag to the exploitation of new resource habitats;
Technological regime	Generally low input in nature; mostly expedient but locally distinct technologies that were flexible across subsistence strategies; where previously innovations or traditions had tended to remain localised and brief, this period sees the beginning of a trend towards greater dissemination; projectile technology was apparent, with bone armatures evidently used, though osseous materials had a number of applications and various non-preserving procurement systems were also (probably extensively) utilised; there is limited emphasis on the use of durable technology as a medium for cultural or group signalling; technology was additive: new forms were added or developed, but old forms endured; standardization is found in certain media but not others until quite late.

Southeast Asia's late glacial prehistory has been shaped by a fundamental re-organisation of its landscape, by probable population crashes and by dispersal and innovation within its interior landscapes and along its coasts, but above all, it has been shaped through the successful economic and social attunement of incoming groups to this tropical realm, and it to them.

8.4 Discussion: Hominin Dispersal, Climate Change and Behavioural Evolution

Palaeo-climatic research has drawn particular attention to the importance of the MIS-11 (*c.* 428–352 kbp) interglacial. For some, this has been because of its potential as an accurate analogue to the Holocene (and to aid in predictions about future climatic conditions) (Loutre and Berger, 2003) for others, because it marks the boundary between two different global climate regimes. The preceding was characterized by comparatively subdued climatic oscillations, during which early species of hominin were evolving, dispersing and adapting. The post-MIS-11 climate is characterized by a series of glacial–interglacial cycles that would exhibit increasingly frequent and high-amplitude changes that would become most pronounced during the Upper Pleistocene (MIS-5e to MIS-2). MIS-5e would signal both an increasingly steep slide towards high-frequency, high-amplitude climate changes and also the first dispersal from Africa of the last Pleistocene species of *Homo* – *H. sapiens*. It is coincidental, but in light of the thesis advanced here, of potential importance, that the onset of this trend in growing climatic instability can be traced back to *c.* 230 kbp (Dansgaard *et al.*, 1993), a period closely tied to the first emergence of defining morphological traits for *H. sapiens* (*though see* Bräuer *et al.*, 1997).

There is currently disagreement about the extent of the role played by climatic conditions in the extinction of late-surviving archaic hominin groups in Eurasia. Two positions may be identified: one focusing on climate as a secondary cause, the other on it as a primary one. Work carried out by d'Errico and Sánchez-Goñi (2003) into the late surviving Neanderthals in the Iberian Peninsula concluded that while climate changes associated with Heinrich event 4 (*c.* 39.5–38 cal. BP) almost certainly did impact on already dwindling Neanderthal populations in what was a likely ecological refugium, the effects of this climatic downturn may also have impeded the southward spread of *H. sapiens* populations (*see* Sepulchre *et al.*, 2007). D'Errico and Sánchez-Goñi (2003) propose that the final extinction of the Neanderthals in southern Iberia came not from climatic causes but from competition for resources with immigrant modern humans. By comparison, Jiménez-Espejo *et al.* (2007) use proxy marine sediment core data to demonstrate that there were significant palaeoenvironmental changes taking place in the western Mediterranean area at the time of the Neanderthal extinction. They propose that climate variability during the cold phases of the MIS-3 could have had a detrimental effect on late Neanderthal populations in Iberia, with the definitive disappearance of the species from Europe occurring within the time frame of the MIS-3 to MIS- 2 transition and the extremely adverse conditions of Heinrich event 2 (*c.* 25.5–22.5 cal. BP). Comparison of cultural, genetic and palaeoenvironmental lines of evidence has led others (e.g. Finlayson and Carrión, 2007; Giaccio *et al.*, 2008; Hetherington and Reid, 2010) to propose that the effects of Late Pleistocene rapid climate change were of paramount importance to the fate of human groups.

Where both secondary and primary positions seem to agree is in the existence of a link between the decline of one population (Neanderthals) and the expansion of another (modern humans) in the context of the high-amplitude climate fluctuations during the late glacial. If climate did have a significant effect on the ultimate extinction (or extirpation?) of Neanderthals here, Jiménez-Espejo *et al.* (2007: 848) propose that variability during the cold episodes may have acted as a 'territory cleanser' that would have favoured subsequent colonization of southern Iberia by modern humans. The full dynamics of such a mechanism have yet to be explored in detail, but the link between climate change, population bottlenecks and diaspora in hominin evolution – and particularly the evolution and spread of *H. sapiens* – is a recurrent theme.

Recently published data from the Early Holocene are now beginning to describe links between abrupt climatic and environmental change, and population movement and cultural development. Turney and Brown (2007) suggest that a sea level rise of 1.4 m played a major role in the spread of migrating Neolithic peoples into Europe from the Levant, linking land loss with mass migration and regional scale cultural change. Berger and Guilaine (2009) argue that an abrupt climate shift caused by the sudden discharge of Lake Agassi, in North America, *c.* 8.2 cal. BP precipitated a pronounced cooling in the North Atlantic realm and also affected the development of the Neolithic in the Mediterranean region.

From their study of 86 mtDNA genomes belonging to all five of the Native American haplogroups (A, B2, C1, D1 and X2a), Fagundes *et al.* (2008) concluded that there was a moderate bottleneck event seemingly concurrent with the LGM and probably within Beringia. This was followed by rapid demographic growth in the millennia immediately following the LGM – a scenario that, they argue, is consistent with a major range expansion that probably extended initially down the Pacific coast before extensive penetration into the continental interior. Soares *et al.* (2008) saw population constriction preceding expansion in ISEA in the context of coastal inundation. The recent publication of new dating evidence corroborating the late survival of *H. erectus* in Java to between 60 and 40 kbp (Yokoyama *et al.*, 2008), linked with changes in palaeoenvironment (as marked by the Ngandong–Punung faunal turnover) and the arrival of modern humans in this region, will make a future interesting line of inquiry into this demographic mechanism. Whether the late survival and ultimate extinction of archaic forms and expansion of modern humans in Southeast Asia can be linked to the same underlying process remains to be seen, but one may present the hypothesis that the scale of climatic and environmental change during the late glacial was such that the ability of their respective bio-cultural repertoires to adapt quickly enough had been finally surpassed. Population collapse followed, reducing the coverage of communities, opening habitats to immigrant groups and exacerbating vulnerability to factors such as resource competition, disease or even inter-group hostility. Arguably, the Late Pleistocene population turnover between archaic and modern humans in western Eurasia may also be treated in this way (in form if not in specifics); the recolonization of Europe following the downturn of the LGM already has been (Blockley *et al.*, 2006). In Africa, late MSA technological and population developments may be linked to adaptive responses to situations imposed by a rapid succession of climatic and palaeoenvironmental changes at the transition between MIS-5 and MIS-4 (Mellars, 2006c). Some authors also now propose that the original dispersal of *H. sapiens* out of Africa (Carto *et al.*, 2009; Walter *et al.*, 2000), as well as earlier phases in hominin evolution within Africa (deMenocal, 2004), may have been affected by climate instability.

From the evidence presented in this volume, it appears that the greater part of Pleistocene hominin diaspora was not constrained by specific periods of climatic deterioration or even necessarily advanced by its amelioration. The dispersal of hominins was linked to adaptive capacities that emerged under a long-term regime of climatic and ecological instability – one that was of increasing magnitude after MIS-11 and markedly so from MIS-5e to the end of MIS-2. Recognition of the importance of this relationship is not new. Over the course of several publications, Richard Potts (1996a, 1996b, 1998a, 1998b, 2003) has laid out an hypothesis that increasing climatic and environmental variability over the last 3–5 million years has selected for the evolution of organisms, including hominin taxa, that are able to adapt to significant long-term ecological disruption. This has been defined as the Variability Selection hypothesis (Potts, 1996a: 231). Selection for adaptive flexibility in early hominins (beginning with bipedalism, encephalisation and tool use) permitted habitat expansion (Potts, 2003). Alongside these bio-behavioural traits, some archaic hominin species appear to have evolved adaptations tailored to specific environmental situations – that is, physiological specializations. Examples of this include the development of the massive craniodental architecture of robust Australopithecines in East Africa to help masticate tough fibrous savannah plants, the facial and postcranial adaptations of *H. neanderthalensis* to the glacial environment of the Eurasian steppe or the diminutive form of *H. floresiensis* through what seems to have been long-term island isolation. The frequency and amplitude of ecological changes were largely within the reach of physiological adaptations with varied levels of non-biological augmentation (i.e. early forms of material culture).

Arguably, however, a threshold was reached during the Middle (or perhaps early Upper) Pleistocene when the frequency of climatic and environmental shifts became such that, among hominins, physiologically dominated adaptation could not respond with sufficient speed to keep pace with the constant resetting of ecological conditions (*see* Winterhalder, 1980: 138). This gave impetus for an adaptive response that operated over much shorter timescales, probably not only in *H. sapiens* but also in other then-surviving hominins. Potts (1996a: 203–4) writes, '[c]ultural behaviors, in the modern human sense, evolved in the midst of nature's grandest alterations . . . Neither stability nor directional change can be credited for defining these elements of human behavior . . . honed by their role in buffering fluctuations in the conditions of survival.' Where the conclusions presented here differ slightly from the Variability Selection hypothesis is not in the long-term effect of inconsistent adaptive settings to the development of adaptive versatility over many generations (Potts, 2003) but rather with the nature and importance of short-term adaptations to that process.

Potts (1998a: 111) relates how hypotheses about the importance of short-term variability tend to emphasize 'habitat variation that an organism is likely to experience and adapt to in every generation' – what Dean (2000) refers to as 'high frequency processes'. Examples operating at this scale during the Holocene include seasonal or multi-annual cycles of variability such as herd migrations, the monsoon, annual changes to the position of the ITCZ and the periodicity of ENSO (*after* Dunbar, 2000). By virtue of the comparatively high frequency at which these phenomena occur, an individual will likely live through many such cycles in the course of his or her lifetime. Cumulative personal and community experience will amass with respect to most of the range of expressions that these kinds of mechanisms will assume in any given year – even if they are not strictly predictable year to year and embody multiple scales of cyclic variation, as is the case with ENSO. In relation to this latter, Anthropocene climatic anomalies linked to ENSO have been observed as having a noticeable effect on communities and ecosystems in just the last 30 years (Dunbar, 2000). This phenomenon may have an influence that extends beyond multi-annual change. A study by Anderson *et al.* (2005) has matched the episodic pattern of initial island colonization in the Pacific during the later Holocene with periods of reversed wind direction associated with ENSO. Confirmation of ENSO-like activity during colder phases of the Pleistocene between late MIS-6 and the Last Termination using annually banded corals from Papua New Guinea (Tudhope *et al.*, 2001) also raises the possibility that its wider effects were not confined to the recent past.

Taken together, experience of these annual and inter-annual events may provide a wellspring for adaptive resourcefulness to expected (and possibly unexpected) climatic fluctuations, but it is not this kind of short-term variability with which we are chiefly concerned here. The kinds of climatic shifts we have discussed for the period of the Upper Pleistocene occurred for the most part less often, at generational or decadal scales. As a result, they may have been less predictable when considered against an existing knowledge base and, therefore, more likely to cause adaptive stress. These shifts fall within, but further down, the continuum of what Dean (2000) has referred to as 'low frequency processes' – that is, lengths of time equal or greater than a single human generation of about 25 years. Even when these were cyclic phenomena (such as D-O events), complete cycles would probably have been beyond the experiential grasp of Pleistocene communities, taking as they did many generations to complete. As Dean (2000) goes on to point out, knowledge about the periodicity of environmental variation will be related to the level of information storage within a society; in essence, recurring experience brings cultural reinforcement. While the potential depth of oral traditions (particularly ritualised ones) should not be underestimated (e.g. Echo-Hawk, 2000; Minc, 1986), a conservative estimate for secular histories is put at *c.* 200 years. This is considered to be the kind of temporal depth to which particular observations retain 'verbal durability', for beyond a certain point, information about behavioural responses to environmental change will, as with other details of life, be distilled into abbreviated pneumonic forms (Westley and Dix, 2006). Culturally constructed and locally contingent, these schemata or condensed versions of social memory form valuable experience-based guides of deeds but no longer recall all the specifics of the situations from which they are derived (e.g. Dean, 2000: 107; *also see* Minc, 1986).

At the upper end of low frequency processes, some climatic modulation during the Upper Pleistocenemay has happened over the course of as little as a decade. Talking in broad terms, Burroughs (2005: 101) puts the scale of inter-decadal variance prior to *c.* 12 kbp at approximately ten times the level currently experienced. With such variability, one can see that there may have been little scope to predict the changing course of conditions with any certainty, potentially heightening stress. Though change itself can become routine: for example, the rate of coastal inundation in Southeast Asia included periods of steady rise punctuated by pulse events. During phases when gradual encroachment was the norm, coastal flooding may have become a familiar and expected part of a community's life, and lodged in its history even if, at a global scale, these interludes were part of a far less predictable mechanism.

Emergent versatility in behaviour is predicated on the repertoire and flexibility of adaptive mechanisms available to an organism for it to respond to changing circumstances. This will be at least partly dependent on the rate of generational turnover and the level of experiential transmission between generations. Responses to Upper Pleistocene climatic fluctuations at short and longer-term scales are likely to have had considerable significance during this phase of our species' history (Burroughs, 2005) – just as it appears to have done for other species over the same period (*see* e.g. Hewitt, 2003; Roy *et al.*, 1996). The difference in the case of *H. sapiens* is that this would take a particular, highly flexible and ultimately species-defining form.

Whereas possibly all zoological species are to a degree what could be called *trans-somatic* (i.e. perception and survival mechanisms go 'beyond the body', e.g. through territorial behaviour), humans (and probably other hominins) have a propensity not only to incorporate their surroundings into our existence (in the same spatial and physiological sense) (*see* e.g. Everden, 1993; Fox, 1990), but also, to varying degrees, to reshape the material world to conceptual specifications. We use it as well to define and physically extend or enhance the human body in its interactions with other people, things and the environment. To communicate about, perpetuate and manage our lives, humans depend upon this 'materialised culture' – what we could call *cyber-somatic* devices, that is, devices (either found in the world and used in their natural state or modified from that state) that enable us, that embody information both about our activities and sometimes about other spheres of meaning that are connected to them. Until now, this materialisation of culture has been almost exclusively physically external but socially integral to modern human activities and identity at group and individual levels; this may change in the future to even more fundamental integration (Launius and McCurdy, 2007). As such, though, archaeological artefacts are more than the sum of their physical attributes; they are the 'fossilised' remains of bolt-on parts that past humans used in the actuation of their humanity. One can liken materialised culture with the use of prosthetics. At first thought, the analogy may seem an imperfect one: prosthetics replace or augment disenabled body parts; however, in a sense, this is exactly what materialised culture does. It is not that we are disenabled; rather, in our translocation into contexts (physical or social) where the required speed of adaptation outpaces biology, materialised culture extends and augments those aspects of human physiology that are most relevant to realigning us to the contingent demands of the situation. An amputee uses an artificial hand to progress his or her interaction with the world, but at the same time, that artificial appendage becomes part of that individual's body, part of what it means to be that person. In the archaeological record, what survives when the people are long gone is, effectively, these prosthetics, the keystones for archaeological reconstruction and interpretation.

One may propose with some confidence that most, if not all, hominin species, and especially the late-surviving representatives, possessed this quality to a greater or lesser degree, probably irrespective of other evolutionary (physiological) trends in body or brain size, as the behavioural capacity of *H. neanderthalensis* and possibly even *H. floresiensis* suggests. In the end, however, and in the face of selective pressure to respond to increasingly variable conditions, it appears that only one species was able to change quickly enough. The later Pleistocene history of humanity, its movements, behaviours and regionalism, can be thought of in terms of evolutionary selection for increasingly short-term, localised and material culture–dominated adaptations in the face of increasing and compounded environmental disequilibrium.

8.5 Conclusion

This book has ultimately sought to draw together three inter-related strands of evidence – climatic instability, hominin regional dispersal and hominin adaptive responses to both new and transformed environments – as a way to explore the regional development of modern human behaviour. The character of Southeast Asian archaeology, the principal case study used herein, is particularly fitting as it has long inspired debate about this very topic. When considering the emergence of modern human behaviour, the assumption, and indeed the emphasis, has always been on establishing the earliest incidence of a particular cultural trait. One does not look to the Neolithic to learn about the emergence of human modernity. This is because it is also assumed that all those changes that make us what we are were long since established by the time ceramics and agriculture began to appear. Researchers have sought to locate the core area of the world where this modernity first took shape and model its ensuing diffusion or acculturation into other regions; firstly, this was taken to be Europe, currently, many argue that Africa should be our primary focus. The mental capacity for the kinds of sophisticated behaviours we attribute to ourselves is increasingly manifest from an early date in the MSA record of Africa (McBrearty, 2007). We assume, though, that it stops there: that everything that comes afterwards is simply an expression of that capacity; that behavioural modernity is a catalyst rather than a process that is itself open to change. While I am in complete agreement with McBrearty and others that a Eurocentric approach to Palaeolithic archaeology is no longer tenable, I am not convinced that change to an Afrocentric (or indeed an Asiocentric) stance is breaking open a new conceptual paradigm. I have argued here that we possibly should no longer be searching for this kind of emergence at all.

The adaptive diaspora of our species that started in the Middle Pleistocene has not stopped; the journey is not over – nor is that journey simply about our global spread. Who we are today is the sum of all of the most adaptive and, therefore, bio-culturally inherited processes that have occurred along the chain of our ancestry over the last 200,000 or more years, in Africa and beyond it. The long-term pulses in hominin dispersal into (and adaptation to) new environments, and survival through changing conditions in the world's different biographic regions, were and continue to be intimately tied to the process of establishing the emergent character of our species. By dint of the timescales involved, most of those characteristics emerged during the Pleistocene, though with the heightened pace of change experienced among human populations since then, there will come a time, perhaps sooner rather than later, when this is no longer the case.

The evidence presented here is more environmentally deterministic when it comes to human adaptation in the Pleistocene. It places climatic and ecological instability, the 'flickering switch' (Taylor *et al.*, 1993) of the Pleistocene, as a central though not sole driver behind the way the regional expression of that character developed. Like each step taken before, the situational behaviours that emerged through humanity's regional adaptations during our Pleistocene history would become the basis from which we would draw up to meet the next round of ecological and social challenges. Indeed, for a species whose behavioural characteristics had been honed in the context of such fluctuating conditions, we were next to be confronted by a period of the most stable climatic conditions for 420,000 years (Petit *et al.*, 1999). This could cast the development of human civilisation during the Holocene in rather a different light. While the external stimulation of rapid climate change was largely suspended during the Holocene, the only hominin adaptive mechanism that had proved adept enough to meet it has continued to evolve, now by predominantly internal rather than external stimulation (i.e. more socio-culturally deterministic). The result is that our adaptive interface has itself become a self-transforming and specialized system that creates its own rapid turnover of pseudo-environments in cultural space for us to adapt to (*see* e.g. Laland and Brown, 2006). Arguably, this may be one of the reasons for the brevity of post-Pleistocene cultures and the exponential rise in cultural diversity that we have seen globally during this period. If this scenario is even remotely correct, there is a real chance that palaeo-cultures were significantly different. We should not expect to find the same attenuated cultural entities in the Pleistocene, or even perhaps the same fragmented sense of individualism that so permeates particularly Anthropocene societies (and much of modern archaeological theory). The irony is that our actions during the Holocene have contributed unintentionally towards putting us back onto a path to globally capricious climate conditions that may not be wholly dissimilar to those under which our species evolved. This is not something that should be read as any kind of endorsement of our role in causing the present state of affairs but instead as a prompt that we should perhaps begin to consider the way we respond to these changes with a view to a much wider and deeper behavioural context.

The manner in which global changes at the end of the last glacial impacted Southeast Asian forager communities was in large part shaped by the structure and affordances of a tropical environment: its high species diversity, low species density and often pronounced vertical resource stratigraphy. These communities were characterized by localised subsistence and technological systems, perhaps indicative of higher population densities and closed social systems; an intimate knowledge of plant-processing techniques – from the removal of toxins to the creation of pigments; a tendency to situate sites near to reliable food staples; and a long-term use of predominantly low-input flexible stone, bone and shell technologies, where standardization was limited, where innovations were mostly additive and, until around the time of the Pleistocene-Holocene transition, often expressed only briefly and not widely disseminated.

In closing, this study has attempted to expand our understanding about what drove inter-regional differences in early modern human behaviour and in particular those specific adaptations pertaining to the Southeast Asian record. I would argue that the reason for inter-regional disjuncture between this and other parts of the world goes beyond and much deeper than previously proposed reasoning. Tropical populations were adapting to Southeast Asian landscapes in different ways to those living in and adapting to the Eurasian mammoth steppe, African savannah, montane or desert settings. How much and in what way each set of regional adaptations contributed to the pool of behavioural capacities we see in ourselves today is largely an untouched field of investigation. What can be said, though, is that the explosive cultural diversity of post-Pleistocene *H. sapiens* emerged out of our species' many varied, successful and geographically differentiated adaptive strategies over the preceding 200 or so millennia. Indeed, the deep propensity towards dispersal and behavioural heterogeneity may itself have been profoundly important for our species' survival, when other Late Pleistocene hominins – in the case of *H. neanderthalensis*, in particular, with only slightly more specialized adaptations (Marean, 2007) – ultimately met with extinction. If dispersal and diversification in our acculturated adaptive strategies can, at a broad level, be considered as similar to speciation events (cf. e.g. Moyle *et al.*, 2009; Schemske, 2010; Sobel *et al.*, 2010), even if not so mutually exclusive, regionalisation in Pleistocene human behaviour, emerging across its extraordinarily wide geographic domain, may itself have enhanced the evolutionary fitness of *H. sapiens* in the face of increasingly unpredictable conditions: a single species carrying myriad possibilities, a Pleistocene 'explosion' of new life ways.

Appendix

Radiocarbon and Calendar Dates for Last Termination Tool Assemblages from Southeast Asia

Pleistocene/ Holocene Epoch	Site (Island Southeast Asia)	Decimal Lat. and Long.	Location	Uncalibrate Date (years bp)	Reference(s)	Calibrated Date (years BP)	Lab Code
Pre–Last Glacial Maximum	Niah Caves (Hell)	3.813908N, 113.781414E	Borneo (Sarawak)	41,200 ± 400	Barker *et al.* (2007)	44,941 ± 329	OxA-15630
	Niah Caves (Area A)			36,470 ± 250	Barker *et al.* (2007)	41,675 ± 278	OxA-15164
				35,890 ± 250	Barker *et al.* (2007)	41,164 ± 281	OxA-15163
	Niah Caves (Hell)			35,690 ± 280	Barker *et al.* (2007)	40,984 ± 305	OxA-V-2076-16
				35,000 ± 400	Barker *et al.* (2007)	40,344 ± 415	OxA-15126
	Niah Caves (Lobang Hangus)	3.813228N, 113.785544E		32,660 ± 210	Piper and Rabett (2009)	38,058 ± 259	OxA-13938
	Leang Sarru	*c.* 3.932464N, 126.701619E	Sulawesi (Talaud Islands)	28,760 ± 150	Ono *et al.* (2010)	32,426–33,079[f]	TERRA-070407a03
				28,460 ± 150	Ono *et al.* (2010)	32,223–32,832[f]	TERRA-070407a04
	Callao Cave	*c.* 17.7N, 121.816667E	Philippines (Luzon)	25,968 ± 373	Mijares (2005)	31,195 ± 430	Wk-14881
	Toé Cave	1.284889S, 132.218361E	New Guinea (Bird's Head)	25,940 ± 180	Pasveer (2004)	Uncalibrated	OZG-063
	Song Gupuh	8.161111S, 110.979939E	East Java	23,750 ± 475	Morwood *et al.* (2008)	26,300 ± 950[d]	ANUA-32719
Last Glacial Maximum	Toé Cave	1.284889S, 132.218361E	New Guinea (Bird's Head)	20,310 ± 110	Pasveer (2004)	24,730–23,210[c]	OZF-516
	Hagop Bilo	4.714442N, 117.970014E	Borneo (Sabah)	17,900 ± 200	Bellwood (1988)	Uncalibrated	ANU-2560
	Liang Lemdubu	6.152167S, 134.371333E	Aru Islands (Kobroor)	16,570 ± 510	O'Connor *et al.* (2006)	19,720 ± 530	OZD-460
	Gua Balambangan	*c.* 7.200725N, 116.8531E	Borneo (Sabah)	16,530 ± 160	Jeffrey (2005), Zuraina *et al.* (1998b)	19,651 ± 180	Beta-109143

Last Termination				13,930 ± 50	Zuraina (1998)	16,235 ± 129	Beta-109144
	Niah Caves (Area A)	3.813908N, 113.781414E	Borneo (Sarawak)	13,745 ± 55	Pritchard *et al.* (2009)	16,000 ± 125	OxA-15162
	Daeo 2	*c.* 2.058892N, 128.478525E	North Malaku (Morotai Island)	13,930 ± 140	Bellwood *et al.* (1998)	15,798–15,305[e]	ANU-9450
	Uai Bobo 2	*c.* 8.636519S, 126.433997E	East Timor (Hatu Ariana)	13,400 ± 520	Glover (1986)	15,615 ± 666	ANU-238
	Liang Lemdubu	6.152167S, 134.371333E	Aru Islands (Kobroor)	13,300 ± 300	O'Connor *et al.* (2006)	15,487 ± 365	OZC-777
	Liang Nebulei Lisa	6.001867S, 134.56045E		13,130 ± 80	O'Connor *et al.* (2006)	16,292–14,817[a]	OZF-518
	Niah Caves (Lobang Hangus)	3.813228N, 113.785544E	Borneo (Sarawak)	12,500 ± 50	Piper and Rabett (2009)	14,484 ± 131	OxA-13936
	Gua Braholo	*c.* 8.081664N, 110.754894E	East Java	12,060 ± 180	Simanjuntak and Asikin (2004)	13,873 ± 181	None
	Hagop Bilo	4.714442N, 117.970014E	Sabah (Baturong)	12,050 ± 120	Bellwood (1988)	Uncalibrated	ANU-2558
	Liang Lemdubu	6.152167S, 134.371333E	Aru Islands (Kobroor)	11,700 ± 130	O'Connor *et al.* (2006)	14,057–13,183[a]	ANU-10792
	Padah-lin Cave (1A)	*c.* 21.134169N, 96.357811E	Burma (South Shan State)	13,400 ± 200	U Aung Thaw (1971)	Uncalibrated	R-2547/5[B]
	Golo Cave	*c.* 0.008975S, 129.391692E	North Malaku (Gebe Island)	11,480 ± 70	Bellwood *et al.* (1998)	13,028–12,862[e]	ANU-9512
	Agop Atas (MAD 1/28)	*c.* 4.719803N, 118.164086E	Sabah (Madai)	10,800 ± 120	Bellwood (1988)	Uncalibrated	ANU-3088
	Golo Cave	*c.* 0.008975S, 129.391692E	North Malaku (Gebe Island)	10,540 ± 70	Bellwood *et al.* (1998)	11,925–11,121[e]	ANU-9769
	Niah Caves (Lobang Hangus)	3.813228N, 113.785544E	Borneo (Sarawak)	10,450 ± 45	Piper and Rabett (2009)	12,373 ± 95	OxA-13939

(*continued*)

Pleistocene/ Holocene Epoch	Site (Island Southeast Asia)	Decimal Lat. and Long.	Location	Uncalibrate Date (years bp)	Reference(s)	Calibrated Date (years BP)	Lab Code
	Agop Sarapad (MAD 2)	*c.* 4.719803N, 118.164086E	Borneo (Sabah)	10,450 ± 110	Bellwood (1988)	Uncalibrated	ANU-2554
	Tianko Panjang Cave	*c.* 2.005956N, 101.936047E	Sumatra (Jambi)	10,250 ± 140	Bronson and Asmar (1975)	11,988 ± 271	P-2250
Early Holocene	Gua Balambangan	*c.* 7.200725N, 116.8531E	Borneo (Sabah)	9960 ± 190	Jeffrey (2005), Zuraina *et al.* (1998b)	11,453 ± 319	Beta-109141
	Niah Caves (Area D)	3.813908N, 113.781414E	Borneo (Sarawak)	10,000 ± 55	Barker *et al.* (forthcoming)	11,449 ± 137	OxA-11865
	Niah Caves (Area A)	3.813908N, 113.781414E	Borneo (Sarawak)	9995 ± 40	Pritchard *et al.* (2009)	11,429 ± 109	OxA-15157
	Song Gupuh	8.161111S, 110.979939E	East Java	9961 ± 60	Morwood *et al.* (2008)	11,410 ± 180[d]	Wk-14650
	Liang Nebulei Lisa	6.001867S, 134.56045E	Aru Islands (Kobroor)	9870 ± 70	O'Connor *et al.* (2006)	11,553–11,164[a]	OZD-699
	Lene Hara	*c.* 8.403472S, 127.294394E	East Timor (Tutuala)	9741 ± 60	O'Connor and Veth (2005)	Uncalibrated	NZA-17000
	Ille Cave	11.196111N, 119.505278E	Palawan (Sibaltan Bay)	9740 ± 75	Lewis *et al.* (2008)	11,170 ± 82	OxA-14163
	Tianko Panjang	*c.* 2.005956N, 101.936047E	Sumatra (Jambi)	9580 ± 130	Bronson and Asmar (1975)	10,921 ± 222	P-2249
	Niah Caves (Area D)	3.813908N, 113.781414E	Borneo (Sarawak)	9560 ± 60	Barton *et al.* (2009)	10,886 ± 148	OxA-12391
	Agop Sarapad (MAD 2)	*c.* 4.719803N, 118.164086E	Borneo (Sabah)	9520 ± 100	Bellwood (1988)	Uncalibrated	ANU-2553
	Liang Lemdubu	6.152167S, 134.371333E	Aru Islands (Kobroor)	9400 ± 50	O'Connor *et al.* (2006)	10,624 ± 65	OZF-356
	Tabon Cave	9.275497N, 117.979975E	Palawan (Lipuut Point)	9250 ± 250	Fox (1970)	10,451 ± 338	UCLA-284

Tianko Panjang Cave	*c.* 2.005956N, 101.936047E	Sumatra (Jambi)	9210 ± 130	Bronson and Asmar (1975)	10,381 ± 162	P-2248
Agop Atas (MAD 1/28)	*c.* 4.719803N, 118.164086E	Borneo (Sabah)	9150 ± 100	Bellwood (1988)	Uncalibrated	ANU-3086
Toé Cave	1.284889S, 132.218361E	New Guinea (Bird's Head)	8860 ± 70	Pasveer (2004)	10,210–9,630[c]	AA-33404
Gua Balambangan	*c.* 7.200725N, 116.8531E	Borneo (Sabah)	8930 ± 150	Jeffrey (2005), Zuraina *et al.* (1998b)	10,007 ± 227	Beta-109140
Song Keplek	8.132608S, 111.002214E	East Java	8870 ± 210	Simanjuntak and Asikin (2004), Forestier (2007: 91)	9952 ± 299	None
Tanjung Pinang	*c.* 2.065208N, 128.514578E	North Malaku (Morotai Island)	8860 ± 110	Bellwood *et al.* (1998)	9485–9260[e]	ANU-7782
Buwawansi 1	0.016667S, 129.333333E	North Malaku (Gebe Island)	8550 ± 70	Bellwood *et al.* (1998)	9182–8955[e]	Wk-4628
Song Terus	*c.* 8.160003S, 111.013525E	East Java	8350 ± 100	Sémah *et al.* (2004)	9362 ± 125	Beta-124008
Gua Lawa	*c.* 7.808175N, 111.367386E	East Java	8190 ± 170	Simanjuntak and Asikin (2004)	9141 ± 232	Unknown
Guri Cave	*c.* 9.274253N, 117.984761E	Palawan (Lipuut Point)	8130 ± 180	Fox (1978), Pawlik (pers. comm.)	Uncalibrated	AERIK-20
Niah Caves (Area D)	3.813908N, 113.781414E	Borneo (Sarawak)	8005 ± 50	Barton *et al.* (2009)	8915 ± 103	OxA-11864
			7948 ± 39	Barton *et al.* (2009)	8793 ± 109	OxA-18358
Agop Atas (MAD 1)	*c.* 4.719803N, 118.164086E	Borneo (Sabah)	7920 ± 370	Bellwood (1988)	8792 ± 370	ANU-2397
Guri Cave	*c.* 9.274253N, 117.984761E	Palawan (Lipuut Point)	7890 ± 90	Fox (1978), Pawlik (pers. comm.)	Uncalibrated	AERIK-21
Agop Atas (MAD 1/28)	*c.* 4.719803N, 118.164086E	Borneo (Sabah)	7690 ± 100	Bellwood (1988)	Uncalibrated	ANU-3087

(continued)

Pleistocene/ Holocene Epoch	Site (Island Southeast Asia)	Decimal Lat. and Long.	Location	Uncalibrate Date (years bp)	Reference(s)	Calibrated Date (years BP)	Lab Code
	Niah Caves (Area A)	3.813908N, 113.781414E	Borneo (Sarawak)	7606 ± 35	Pritchard *et al.* (2009)	8400 ± 20	OxA-15161
	Paso	*c.* 1.206269N, 124.854392E	Sulawesi (Minahasa Peninsula)	7530 ± 450	Bellwood (1976)	8366 ± 478	ANU-1517
	Agop Atas (MAD 1)	*c.* 4.719803N, 118.164086E	Borneo (Sabah)	7390 ± 270	Bellwood (1988)	8205 ± 264	ANU-2398
	Paso	*c.* 1.206269N, 124.854392E	Sulawesi (Minahasa Peninsula)	7360 ± 310	Bellwood (1976)	8178 ± 305	ANU-1518
	Hinai/Sukajadi Pasar III	*c.* 3.838378N, 98.415514E	Sumatra (North Sumatra)	7340 ± 360	Edwards McKinnon (1991)	8162 ± 356	SUA-1107
	Lie Siri	*c.* 8.435489S, 126.361992E	East Timor (Baucau)	7270 ± 160	Glover (1986)	8088 ± 160	ANU-236
	Leang Sarru	*c.* 3.932464N, 126.701619E	Sulawesi (Talaud Islands)	7660 ± 40	Ono *et al.* (2010)	8033–8144[f]	TERRA-070407a05
	Ulu Leang 1	*c.* 4.989058S, 119.668239E	Sulawesi (South Peninsula)	7170 ± 650	Olsen and Glover (2004)	8019 ± 660	ANU-606
	Duryong Cave	*c.* 9.303197N, 118.074022E	Palawan (nr. Lipuut Point)	7000 ± 250	Fox (1970), Pawlik (pers. comm.)	7834 ± 231	UCLA-288
	Liang Nabulei Lisa	6.001867S, 134.56045E	Aru Islands (Kobroor)	6970 ± 160	O'Connor *et al.* (2006)	7754–7167[a]	ANU-10905
	Golo Cave	*c.* 0.008975S, 129.391692E	North Malaku (Gebe Island)	7400 ± 10	Szabó *et al.* (2007)	7965–7743[b]	ANU-9449
	Kria Cave	1.251975S, 132.333511E	New Guinea (Bird's Head)	6900 ± 80	Pasveer (2004)	7930–7590[c]	OxA-6043
	Uai Bobo 2	*c.* 8.636519S, 126.433997E	East Timor (Hatu Ariana)	7010 ± 125	Glover (1986)	7838 ± 123	ANU-328

	Kria Cave	1.251975S, 132.333511E	New Guinea (Bird's Head)	6760 ± 50	Pasveer (2004)	7680–7510[c]	GrA-9103
	Lie Siri	*c.* 8.435489S, 126.361992E	East Timor (Baucau)	6635 ± 140	Glover (1986)	7516 ± 110	ANU-171
Mid-Holocene	Song Terus	*c.* 8.160003S, 111.013525E	East Java	6390 ± 80	Sémah *et al.* (2004)	7316 ± 80	Beta-124009
	Um Kapat Papo	0.166667S, 129.566667E	North Malaku (Gebe Island)	6670 ± 60	Bellwood *et al.* (1998)	7198–7017[e]	ANU-9318
	Kria Cave	1.251975S, 132.333511E	New Guinea (Bird's Head)	5940 ± 50	Pasveer (2004)	6890–6660	GrA-9102
	Ille Cave	11.196111N, 119.505278E	Palawan (Sibaltan Bay)	5769 ± 37	Lewis *et al.* (2008)	6565 ± 57	OxA-16095
	Niah Caves (Gan Kira)	3.810333N, 113.785436E	Borneo (Sarawak)	5710 ± 80	Hunt and Rushworth (2005)	6494 ± 95	Beta-193909
	Uai Bobo 2	*c.* 8.636519S, 126.433997E	East Timor (Hatu Ariana)	5520 ± 60	Glover (1986)	6307 ± 48	ANU-187
	Daeo 2	*c.* 2.058892N, 128.478525E	North Malaku (Morotai Island)	5530 ± 70	Bellwood *et al.* (1998)	6450–6283[e]	ANU-9452
	Banyan Valley Cave	*c.* 19.466667N, 98.233333E	Thailand (Mae Hong Son)	5360 ± 120	Reynolds (1992)	6145 ± 141	GaK-4341
	Song Keplek	8.132608S, 111.002214E	East Java	5174 ± 160	Forestier (2007: 91)	5930 ± 183	Unknown
	Song Gupuh	8.161111S, 110.979939E		5109 ± 36	Morwood *et al.* (2008)	5820 ± 70[d]	Wk-18547
				4946 ± 35	Morwood *et al.* (2008)	5630 ± 30[d]	Wk 18546
	Leang Tuwo Mane'e	*c.* 4.505278N, 126.815408E	Sulawesi (Talaud Islands)	4860 ± 130	Bellwood (1976)	Uncalibrated	ANU-1717
	Kria Cave	1.251975S, 132.333511E	Indonesian Papua	4370 ± 50	Pasveer (2004)	5210–4840[c]	GrA-9100
	Guri Cave	*c.* 9.274253N, 117.984761E	Palawan (Lipuut Point)	4220 ± 140	Fox (1978), Pawlik (pers. comm.)	4769 ± 189	AERIK-19

(continued)

Pleistocene/ Holocene Epoch	Site (Island Southeast Asia)	Decimal Lat. and Long.	Location	Uncalibrate Date (years bp)	Reference(s)	Calibrated Date (years BP)	Lab Code
	Gua Braholo	*c.*8.081664N, 110.754894E	East Java	4120 ± 100	Simanjuntak and Asikin (2004)	4640 ± 155	None
Late Holocene	Niah Caves (Lobang Hangus)	3.813228N, 113.785544E	Borneo (Sarawak)	3945 ± 40	Piper and Rabett (2009)	4404 ± 50	OxA-13940
	Uai Bobo 2	*c.* 8.636519S, 126.433997E	East Timor (Hatu Ariana)	3740 ± 90	Glover (1986)	4096 ± 132	ANU-239
	Ulu Leang 1	*c.* 4.989058S, 119.668239E	Sulawesi (South Peninsula)	4390 ± 90	Olsen and Glover (2004)	3840 ± 172	PRL-230
	Uai Bobo 1	*c.* 8.636519S, 126.433997E	East Timor (Hatu Ariana)	3470 ± 90	Glover (1986)	3734 ± 118	ANU-414
	Golo Cave	*c.* 0.008975S, 129.391692E	North Malaku (Gebe Island)	3230 ± 80	Szabó *et al.* (2007)	3691–3244[b]	ANU-9448
	Song Keplek	8.132608S, 111.002214E	East Java	3260 ± 110	Simanjuntak and Asikin (2004), Forestier (2007: 91)	3484 ± 122	None
	Toé Cave	1.284889S, 132.218361E	New Guinea (Bird's Head)	2930 ± 65	Pasveer (2004)	3320–2870[c]	OxA-6201
	Gua Lawa	*c.* 7.808175N, 111.367386E	East Java	3040 ± 90	Simanjuntak and Asikin (2004)	3242 ± 123	Unknown
	Tanjung Pinang	*c.* 2.065208N, 128.514578E	North Malaku (Morotai Island)	3390 ± 70	Bellwood *et al.* (1998)	3209–2959[e]	ANU-7778
	Uai Bobo 1	*c.* 8.636519S, 126.433997E	East Timor (Hatu Ariana)	2190 ± 80	Glover (1986)	2194 ± 110	ANU-237

Pleistocene/ Holocene Epoch	Site (Mainland Southeast Asia)	Decimal Lat. and Long.	Location	Uncalibrate Date (years bp)	Reference(s)	Calibrated Date (years BP)	Lab Code
Pre–Last Glacial Maximum	Lang Rongrien	8.177375N, 98.879361E	Thailand (Krabi)	37,265 ± 1000	Anderson (1988, 1997)	42,358 ± 885	PITT-1248
				37,000 ± 1780	Anderson (1988, 1997)	42,108 ± 1589	SI-6819
				34,380 ± 2030	Anderson (1997)	39,691 ± 1938	PITT-1247
	Lang Kamnan	13.981111N, 99.42E	Thailand (Kanchanaburi)	30,880 ± 760	Shoocongdej (1996, 2000)	36,259 ± 766	GX-20068
	Hang Chổ	20.835278N, 105.652222E	Vietnam (Hoà Bình)	29,140 ± 200	Yi *et al.* (2008)	34,556 ± 247	SNU07–223
	Lang Rongrien	8.177375N, 98.879361E	Thailand (Krabi)	27,110 ± 615	Anderson (1997)	32,421 ± 673	SI-6816
	Moh Khiew	*c.* 8.162725N, 98.92345E	Thailand (Krabi)	25,800 ± 600	Matsumura and Pookajorn (2005)	30,989 ± 707 cal. BP	TK-933 Pr
Last Glacial Maximum	Hang Trống	20.250444N, 105.890111E	Vietnam (Ninh Bình)	20,537 ± 61	Rabett *et al.* (ms. in prep.)	24,438 ± 93	UBA-17272
	Mai Da Điều	*c.* 20.4N, 105.266667E	Vietnam (Thanh Hoá)	19,700 ± 150	Nguyen Gia Doi (2005)	23,534 ± 212	Bln 3542–86
	Hang Chổ	20.835278N, 105.652222E	Vietnam (Hoà Bình)	19,560 ± 220	Yi *et al.* (2008)	Uncalibrated	SNU 04-A027
	Xóm Trại	*c.* 20.533333N, 105.466667E		18,420 ± 150	Nguyen Viet (2000)	22,012 ± 211	Bln-3472
	Hang Chổ	20.835278N, 105.652222E		18,170 ± 160	Yi *et al.* (2008)	21,634 ± 268	SNU 04–233
	Lang Vanh	*c.* 20.483333N, 105.45E		16,470 ± 80	Nguyen Viet (2000)	Uncalibrated	Bln-3617
	Xóm Trại	*c.* 20.533333N, 105.466667E		16,130 ± 90	Nguyen Viet (2000)	19,259 ± 114	Bln-3042
	Xom Tre	*c.* 20.516667N, 105.466667E		16,040 ± 200	Nguyen Viet (2000)	Uncalibrated	Bln-3707

(*continued*)

Pleistocene/ Holocene Epoch	Site (Mainland Southeast Asia)	Decimal Lat. and Long.	Location	Uncalibrate Date (years bp)	Reference(s)	Calibrated Date (years BP)	Lab Code
Last Termination	Tham Khuong	*c.* 21.583333N, 103.316667E	Vietnam (Lai Chau)	15,800 ± 150	Ha Van Tan (1997)	Uncalibrated	HCMV-03/93
	Hang Trống	20.250444N, 105.890111E	Vietnam (Ninh Bình)	15,300 ± 38	Rabett *et al.* (ms. in prep.)	18,549 ± 62	UBA-14886
				14,878 ± 45	Rabett *et al.* (ms. in prep.)	17,860 ± 166	UBA-14885
	Gua Sagu	*c.* 3.980197N, 103.146114E	Peninsular Malaysia (Pahang)	14,410 ± 180	Zuraina (1998), Zuraina *et al.* (1998a)	16,991 ± 327	Beta-33181
	Tham Lod	*c.* 19.560519N, 98.278997E	Thailand (Mae Hong Son)	13,640 ± 80	Shoocongdej (2006)	15,877 ± 141	Beta-168224
	Gua Gunung Runtuh	5.1175N, 100.9675E	Peninsular Malaysia (Perak)	13,600 ± 120	Zuraina (1994)	Uncalibrated	Beta-38338
	Padah-lin Cave (1A)	*c.* 21.134169N, 96.357811E	Burma (South Shan State)	13,400 ± 200	U Aung Thaw (1971)	Uncalibrated	R-2547/5[B]
	Hang Trống	20.250444N, 105.890111E	Vietnam (Ninh Bình)	13,227 ± 35	Rabett *et al.* (ms. in prep.)	15,403 ± 112	UBA-14884
	Gua Sagu	*c.* 3.980197N, 103.146114E	Peninsular Malaysia (Pahang)	12,750 ± 160	Zuraina (1998), Zuraina *et al.* (1998a)	14,843 ± 225	Beta-42121
	Con Moong	*c.* 20.3N, 105.9E	Vietnam (Ninh Bình, Thanh Hoá, Hoà Bình)	12,350 ± 70	Nguyen Viet (2000)	Uncalibrated	Bln-3490
	Tham Lod	*c.* 19.560519N, 98.278997E	Thailand (Mae Hong Son)	12,100 ± 60	Shoocongdej (2006)	13,891 ± 70	Beta-168223
	Con Moong	*c.* 20.3N, 105.9E	Vietnam (Ninh Bình, Thanh Hoá, Hoà Bình)	11,755 ± 75	Kohl and Quitta (1978)	Uncalibrated	Bln-1713

	Hang Boi	20.258889N, 105.888056E	Vietnam (Ninh Bình)	11,741 ± 43	Rabett and Piper (2012)	13,601 ± 43	UBA-14887
	Spirit Cave	*c.* 19.567383N, 98.111597E	Thailand (Mae Hong Son)	11,350 ± 560	Gorman (1969, 1970)	13,208 ± 573	FSU-315
	Con Moong	*c.* 20.3N, 105.9E	Vietnam (Ninh Bình, Thanh Hoá, Hoà Bình)	11,090 ± 185	Kohl and Quitta (1978)	Uncalibrated	ZK-379
	Moh Khiew	*c.* 8.162725N, 98.92345E	Thailand (Krabi)	11,020 ± 150	Pookajorn *et al.* (1994)	12,893 ± 128	OAEP-1284
	Hang Trống	20.250444N, 105.890111E	Vietnam (Ninh Bình)	10,827 ± 32	Rabett *et al.* (ms. in prep.)	12,736 ± 41	UBA-9301
	Gua Tenggek	*c.* 4.014956N, 103.160394E	Peninsular Malaysia (Pahang)	10,660 ± 110	Zuraina (1998), Zuraina *et al.* (1998a)	12,601 ± 105	Beta-40969
				10,545 ± 80	Zuraina (1998), Zuraina *et al.* (1998a)	12,494 ± 103	Beta-40968
	Hang Boi	20.258889N, 105.888056E	Vietnam (Ninh Bình)	10,497 ± 45	Rabett *et al.* (2011)	12,447 ± 72	UBA-10167
				10,444 ± 45	Rabett *et al.* (2011)	12,362 ± 97	UBA-10168
	Hang Chổ	20.835278N, 105.652222E	Vietnam (Hoà Bình)	10,450 ± 250	Yi *et al.* (2008)	12,289 ± 378	SNU 03-156
	Hang Boi	20.258889N, 105.888056E	Vietnam (Ninh Bình)	10,362 ± 32	Rabett *et al.* (2009a)	12,182 ± 83	UBA-8373
	Con Moong	*c.* 20.3N, 105.9E	Vietnam (Ninh Bình, Thanh Hoá, Hoà Bình)	10,330 ± 70	Nguyen Viet (2000)	12,136 ± 132	Bln-3485
Early Holocene	Gua Gunung Runtuh	5.1175N, 100.9675E	Peninsular Malaysia (Perak)	10,120 ± 110	Zuraina (1994)	Uncalibrated	Beta-38394
	Hang Boi	20.258889N, 105.888056E	Vietnam (Ninh Bình)	9981 ± 45	Rabett *et al.* (2011)	11,400 ± 108	UBA-10166

(*continued*)

Pleistocene/ Holocene Epoch	Site (Mainland Southeast Asia)	Decimal Lat. and Long.	Location	Uncalibrate Date (years bp)	Reference(s)	Calibrated Date (years BP)	Lab Code
	Khao Talu	*c.* 14.004167N, 99.316667E	Thailand (Kanchanaburi)	9980 ± 530	Pookajorn (1985), Shoocongdej (1996)	11,525 ± 790	OAEP-055
	Con Moong	*c.* 20.3N, 105.9E	Vietnam (Ninh Bình, Thanh Hoá, Hoà Bình)	9905 ± 150	Nguyen Viet (2000)	11,351 ± 229	ZK- 340
	Hang Boi	20.258889N, 105.888056E	Vietnam (Ninh Bình)	9900 ± 50	Rabett *et al.* (2011)	11,280 ± 56	UBA-10165
				9863 ± 59	Rabett *et al.* (2009a)	11,236 ± 39	UBA-8372
	Lang Rongrien	8.177375N, 98.879361E	Thailand (Krabi)	9655 ± 90	Anderson (1997)	11,055 ± 158	SI-6817
	Gua Gunung Runtuh	5.1175N, 100.9675E	Peninsular Malaysia (Perak)	9460 ± 90	Zuraina (1994)	Uncalibrated	Beta-37818
	Hang Boi	20.258889N, 105.888056E	Vietnam (Ninh Bình)	9406 ± 44	Rabett *et al.* (2011)	10,631 ± 58	UBA-10163
				9397 ± 49	Rabett *et al.* (2009a)	10,620 ± 64	UBA-8371
	Laang Spean	*c.* 12.882878N, 102.933203E	Cambodia (Battambang)	8750 ± 900	Mourer (1977)	9860 ± 1154	LY-439
	Spirit Cave	*c.* 19.567383N, 98.111597E	Thailand (Mae Hong Son)	8550 ± 200	Gorman (1969, 1970)	9546 ± 226	GaK-1846
	Con Moong	*c.* 20.3N, 105.9E	Vietnam (Ninh Bình, Thanh Hoá, Hoà Bình)	8500 ± 60	Nguyen Viet (2000)	9500 ± 34	Bln-3482

	Hang Chổ	20.835278N, 105.652222E	Vietnam (Hoà Bình)	8430 ± 180	Yi *et al.* (2008)	Uncalibrated	SNU 04-A011
	Moh Khiew	*c.* 8.162725N, 98.92345E	Thailand (Krabi)	8420 ± 90	Pookajorn *et al.* (1994)	9439 ± 84	OAEP-1292
	Lang Kamnan	13.981111N, 99.42E	Thailand (Kanchanaburi)	8305 ± 90	Shoocongdej (1996, 2000)	9308 ± 129	OAEP-1178
	Spirit Cave	*c.* 19.567383N, 98.111597E	Thailand (Mae Hong Son)	8520 ± 290	Gorman (1969, 1970)	9520 ± 351	FSU-318
	Khao Talu	*c.* 14.004167N, 99.316667E	Thailand (Kanchanaburi)	8400 ± 640	Pookajorn (1985), Shoocongdej (1996)	9390 ± 788	OAEP-111
	Lang Rongrien	8.177375N, 98.879361E	Thailand (Krabi)	8300 ± 85	Anderson (1997)	9302 ± 125	SI-6215B
	Mai Da Điều	*c.* 20.4N, 105.266667E	Vietnam (Thanh Hoá)	8200 ± 70	Nguyen Gia Doi (2005)	8838 ± 142	Bln 3540–86
	Lang Rongrien	8.177375N, 98.879361E	Thailand (Krabi)	7765 ± 65	Anderson (1997)	8541 ± 66	SI-6213
	Padah-lin Cave (1A)	*c.* 21.134169N, 96.357811E	Burma (South Shan State)	7740 ± 125	U Aung Thaw (1971)	8528 ± 133	(R-2547/3)
	Sakai Cave	*c.* 7.32N, 99.80E	Thailand (Palian District, Trang)	7620 ± 160	Pookajorn *et al.* (1994)	8421 ± 151	OAEP-1364
	Lang Kamnan	13.981111N, 99.42E	Thailand (Kanchanaburi)	7540 ± 180	Shoocongdej (1996, 2000)	8347 ± 170	OAEP-1178
	Khao Talu	*c.* 14.004167N, 99.316667E	Thailand (Kanchanaburi)	7530 ± 530	Pookajorn (1985), Shoocongdej (1996)	8377 ± 565	OAEP-056
Mid-Holocene	Da But	*c.* 20.016667N, 105.716667E	Vietnam (Thanh Hoá)	6540 ± 60	Nguyen Viet (2005)	Uncalibrated	Bln-3509 II
	Gua Cha	*c.* 5.046686N, 101.770431E	Peninsular Malaysia (Kelantan)	6300 ± 170	Adi (1991)	7207 ± 190	ANU-2218

(*continued*)

Pleistocene/ Holocene Epoch	Site (Mainland Southeast Asia)	Decimal Lat. and Long.	Location	Uncalibrate Date (years bp)	Reference(s)	Calibrated Date (years BP)	Lab Code
	Gua Bintong/Bukit Chuping	6.496456N, 100.260931E	Peninsular Malaysia (Perlis)	5200 ± 200	Haile (1971)	5960 ± 224	Unknown
	Khao Talu	*c.* 14.004167N, 99.316667E	Thailand (Kanchanaburi)	5100 ± 500	Pookajorn (1985), Shoocongdej (1996)	5839 ± 573	OAEP-106
	Mai Da Điều	*c.* 20.4N, 105.266667E	Vietnam (Thanh Hoá)	4940 ± 90	Nguyen Gia Doi (2005)	Uncalibrated	NU-1119–95
	Gua Kechil	*c.* 3.833333N, 101.916667E	Peninsular Malaysia (Pahang)	4800 ± 800	Dunn (1964)	5457 ± 967	GX-0418
	Khao Talu	*c.* 14.004167N, 99.316667E	Thailand (Kanchanaburi)	4520 ± 430	Pookajorn (1985), Shoocongdej (1996)	5152 ± 550	OAEP-107
	Moh Khiew	*c.* 8.162725N, 98.92345E	Thailand (Krabi)	4240 ± 150	Pookajorn *et al.* (1996)	4795 ± 203	OAEP-1290
	Khao Talu	*c.* 14.004167N, 99.316667E	Thailand (Kanchanaburi)	4215 ± 95	Pookajorn (1985), Shoocongdej (1996)	4771 ± 126	OAEP-054
Late Holocene	Nong Nor	*c.* 13.478195N, 101.196601E	Thailand (Chonburi)	3980 ± 90	Higham *et al.* (1998: 39)	4440 ± 125	Wk-2029
	Khok Phanom Di	13.584725N, 101.140644E	Thailand (Chonburi)	3560 ± 60	Higham and Thosarat (2004a)	3852 ± 80	ANU-5493
	Khao Talu	*c.* 14.004167N, 99.316667E	Thailand (Kanchanaburi)	3420 ± 380	Pookajorn (1985), Shoocongdej (1996)	3696 ± 485	OAEP-053

Khok Phanom Di	13.584725N, 101.140644E	Thailand (Chonburi)	3310 ± 128	Higham and Thosarat (2004a)	3542 ± 149	ANU-5482
Ban Lum Khao	15.249125N, 102.342975E	Thailand (Nakhon Ratchasima)	3120 ± 50	Higham and Thosarat (2004b)	3346 ± 49	Wk-4511
Gua Cha	*c.* 5.046686N, 101.770431E	Peninsular Malaysia (Kelantan)	3020 ± 230	Adi (1991)	3197 ± 284	ANU-2217
Ban Lum Khao	15.249125N, 102.342975E	Thailand (Nakhon Ratchasima)	3000 ± 80	Higham and Thosarat (2004b)	3188 ± 120	Wk-4509
Gua Sagu	*c.* 3.980197N, 103.146114E	Peninsular Malaysia (Pahang)	2835 ± 70	Zuraina (1998), Zuraina *et al.* (1998a)	2941 ± 95	Beta-36341
Gua Gunung Runtuh	5.1175N, 100.9675E	Peninsular Malaysia (Perak)	2620 ± 80	Zuraina (1994)	Uncalibrated	Beta-37817
Padah-lin Cave (1A)	*c.* 21.134169N, 96.357811E	Burma (South Shan State)	1750 ± 81	U Aung Thaw (1971)	1663 ± 101	R-2547/1
Gua Sagu	*c.* 3.980197N, 103.146114E	Peninsular Malaysia (Pahang)	1240 ± 100	Zuraina (1998), Zuraina *et al.* (1998a)	1164 ± 100	Beta-32486

Note: Radiocarbon and calendar dates for Last Termination tool assemblages mentioned in the text. All calibrated dates are produced using the Fairbanks_0107 curve (Fairbanks *et al.*, 2005), aside from listed exceptions.

[a] Calibrated using the CALIB 3.4 curve.

[b] Calibrated using the CALIB 4.3 curve (Stuiver and Reimer, 1993; Suiver *et al.*, 1998).

[c] Calibrated using the IntCal04 curve.

[d] Calibrated using the CALIB 5.0.1 with marine 04 calibration data.

[e] Calibrated using CALIB 3.0 to 1 sigma.

[f] Calibrated with CalPal2007_Hulu.

Bibliography

Abbot, R.T. 1989. *Compendium of Landshells*. American Malacologists, Melbourne.

Abed, A.M., Yasin, S., Sadaqa, R., Al-Hawari, Z. 2008. The paleoclimate of the eastern desert of Jordan during Marine Isotope Stage 9. *Quaternary Research* 69: 458–68.

Adcock, G.J., Dennis, E.S., Easteal, S., Huttley, G.A., Jermiin, L.S., Peacock, W.J., Thorne, A. 2001. Mitochondrial DNA sequences in ancient Australians: implications for modern human origins. *Proceedings of the National Academy of Sciences of the United States of America* 98: 537–42.

Adi Taha. 1987. Archaeology in peninsular Malaysia: past, present and future. *Journal of Southeast Asian Studies* 18: 205–11.

Adi Taha. 1991. Gua Cha and the archaeology of the Orang Asli. *Indo-Pacific Prehistory Association Bulletin* 11: 363–72.

Adovasio, J.M., Soffer, O., Klíma, B. 1996. Upper Palaeolithic fibre technology: interlaced woven finds from Pavlov I, Czech Republic, *c.*26,000 years ago. *Antiquity* 70: 526–34.

Aiello, L.C. 2010. Five years of *Homo floresiensis*. *American Journal of Physical Anthropology* 142: 167–79.

Aigner, J.S. 1981. *Archaeological Remains in Pleistocene China*. C.H. Beck, Munich.

Akazawa, T., Aoki, K., Bar-Yosef, O. 1998. *Neandertals and Modern Humans in Western Asia*. Plenum Press, New York.

Allen, H.R. 1986. Holocene mangroves and middens in northern Australia and Southeast Asia. *Bulletin of the Indo-Pacific Prehistory Association* 7: 1–16.

Allen, J. 1998. Reinterpreting the 1969–1972 Willandra Lakes archaeological survey. *Archaeology in Oceania* 33: 207–20.

Allen, J. and O'Connell, J.F. 1995. Introduction. *Antiquity* 69: vii–ix.

Allen, J. and O'Connell, J.F. 2008. Getting from Sunda to Sahul. Clark, G., Leach, F., O'Connor, S. (eds.) *Islands of Inquiry: Colonisation, Seafaring and the Archaeology of Maritime Landscapes*. Terra Australis 29. Australian National University, Canberra, pp. 31–46.

Allen, J., Gosden, C., White, J.P. 1989. Human Pleistocene adaptations in the tropical island Pacific: recent evidence from New Ireland, a Greater Australian outlier. *Antiquity* 63: 548–61.

Alley, R.B., Shuman, C.A., Meese, D.A., Gow, A.J., Taylor, K.C., Cuffey, K.M., Fitzpatrick, J.J., *et al.* 1997. Visual-stratigraphic dating of the GISP2 ice core: basis, reproducibility, and application. *Journal of Geophysical Research* 102: 26367–81.

Alroy, J. 2000. New methods for quantifying macroevolutionary patterns and processes. *Paleobiology* 26: 707–33.

Alroy, J. 2001. A multispecies overkill simulation of the End-Pleistocene megafaunal mass extinction. *Science* 292: 1893–6.

Ambrose, S.H. 1998a. Late Pleistocene human population bottlenecks, volcanic winter, and differentiation of modern humans. *Journal of Human Evolution* 34: 623–51.

Ambrose, S.H. 1998b. Chronology of the Later Stone Age and food production in East Africa. *Journal of Archaeological Science* 25: 377–92.

Ammerman, A.J. and Cavalli-Sforza, L.L. 1984. *The Neolithic Transition and the Genetics of Populations in Europe*. Princeton University Press, Princeton.

An, Z. and Ho, C.K. 1989. New magnetostratigraphic dates of Lantian *Homo erectus*. *Quaternary Research* 32: 213–21.

An, Z., Kutzbach, J.E., Prell, W.L., Porter, S.C. 2001. Evolution of Asian monsoons and phased uplift of the Himalaya-Tibetan plateau since Late Miocene times. *Nature* 411: 62–6.

Anderson, A., Chappell, J., Gagan, M., Grove, R. 2005. Prehistoric maritime migration in the Pacific islands: an hypothesis of the ENSO forcing. *The Holocene* 16: 1–6.

Anderson, A. and O'Connor, S. 2008. Indo-Pacific migration and colonization – introduction. *Asian Perspectives* 47: 2–11.

Anderson, D.D. 1988. Excavations of a Pleistocene rockshelter in Krabi and the prehistory of Southern Thailand. Charoenwongsa, P. and Bronsen, B. (eds.) *Prehistoric Studies: The Stone and Metal Ages in Thailand*. Thai Antiquity Working Group, with the support of the John F. Kennedy Foundation of Thailand, Bangkok, pp. 43–59.

Anderson, D.D. 1990. *Lang Rongrien Rockshelter: A Pleistocene–Early Holocene Archaeological Site from Krabi, Southwestern Thailand*. University Museum Monograph 71. University Museum, University of Pennsylvania, Philadelphia.

Anderson, D.D. 1997. Cave archaeology in Southeast Asia. *Geoarchaeology: An International Journal* 12: 607–38.

Anderson, D.E., Goudie, A.S., Parker, A.G. 2007. *Global Environments through the Quaternary*. Oxford University Press, Oxford.

Anderson, D.G., Faught, M.K. 2000. Palaeoindian artefact distributions: evidence and implications. *Antiquity* 74: 507–13.

Anderson, D.G. and Gillam, J.C. 2000. Paleoindian colonization of the Americas: implications from an examination of physiography, demography, and artifact distribution. *American Antiquity* 65: 43–66.

Andrews, C.W. 1903. Note on a tooth of *Elephas namadicus*. Annandale, N. (ed.) *Fasciculi Malayenses: Anthropological and Zoological Results of an Expedition to Perak and the Siamese Malay States 1901–1902*. Longmans, Green, London.

Andrews, C.W. 1905. Fossil tooth of *Elephas namadicus* from Perak. *Journal of the Federated Malay States Museums* 1: 81–2.

Anisyutkin, N.K. and Timofeyev, V.I. 2006. The Palaeolithic flake industry of Vietnam. *Archaeology, Ethnology, and Anthropology in Eurasia* 27: 16–14.

Annandale, N. 1906. The introduction of the blow-gun to southern India. *Man* 14–16: 26.

Anshari, G., Kershaw, A.P., Van Der Kaars, S. 2001. A Late Pleistocene and Holocene pollen and charcoal record from peat swamp forest, Lake Sentarum Wildlife Reserve, West Kalimantan, Indonesia. *Palaeogeography, Palaeoclimatology, Palaeoecology* 171: 213–28.

Anshari, G., Kershaw, A.P., Van Der Kaars, S., Jacobsen, G. 2004. Environmental change and peatland forest dynamics in the Lake Sentarum area, West Kalimantan, Indonesia. *Journal of Quaternary Science* 19: 637–55.

Antón, S.C. 2003. Natural history of *Homo erectus*. *Yearbook of Physical Anthropology* 46: 126–70.

Argue, D., Donlon, D., Groves, C., Wright, R. 2006. *Homo floresiensis*: Microcephalic, pygmoid, *Australopithecus*, or *Homo*? *Journal of Human Evolution* 51: 360–74.

Armitage, S.J., Jasim, S.A., Marks, A.E., Parker, A.G., Usik, V.I., Uerpmann, H-P. 2011. The Southern Route "Out of Africa": evidence for an early expansion of modern humans into Arabia. *Science* 331: 453–6 (and suppl.).

Arnould, J. 2001. Reflections on human presence in space. *Acta Astronautica* 49: 489–94.

Arz, H.W., Pätzold, J., Wefer, G. 1998. Correlated millennial-scale changes in surface hydrography and terrigenous sediment yield inferred from last-glacial marine deposits off northeastern Brazil. *Quaternary Research* 50: 157–66.

Ash, J., Gallup, G.G., Jr. 2007. Paleoclimatic variation and brain expansion during human evolution. *Human Nature* 18: 109–24.

Ashkenazi, S., Motro, U., Goren-Inbar, N., Biton, R., Rabinovich, R. 2005. New morphometric parameters for assessment of body size in the fossil freshwater crab assemblage from the Acheulian site of Gesher Benot Ya'aqov, Israel. *Journal of Archaeological Science* 32: 675–89.

Ashton, N., Lewis, S.G., Parfitt, S.A., Penkman, K.E.H., Coope, G.R. 2008. New evidence for complex climate change in MIS 11 from Hoxne, Suffolk, UK. *Quaternary Science Reviews* 27: 652–68.

Atkinson, Q.D., Gray, R.D., Drummond, A.J. 2008. mtDNA variation predicts population size in humans and reveals a major Southern Asian chapter in human prehistory. *Molecular Biology and Evolution* 25: 468–74.

Aulagnier, S., Haffner, P., Mitchell-Jones, A.J., Moutou, F., Zima, J. 2009. *Mammals of Europe, North Africa and the Middle East.* A&C Black, London.

Baab, K.L. 2008. The taxonomic implications of cranial shape variation in *Homo erectus*. *Journal of Human Evolution* 54: 827–47.

Bacon, A-M., Demeter, F., Duringer, P., Rousse, S., Dodo, Y., Matsumura, H., Vu The Long, *et al.* 2006. Records of murine rodents (Mammalia, Rodentia) in the Pleistocene localities of Tan Vinh and Ma U'Oi (northern Vietnam) and their implications to past distribution. *Annales de Paléontologie* 92: 367–83.

Bacon, A-M., Demeter, F., Tougard, C., de Vos, J., Sayavongkhamdy, T., Antoine, P-O., Bouasisengpaseuth, B., Sichanthongtip, P. 2008. Redécouverte d'une faune pléistocène dans les remplissages karstiques de Tam Hang au Laos: premiers résultats. *Comptes Rendus Palevol* 7: 277–88 (in French with English summary).

Bacus, E.A. 2001. Southeast Asia, archaeology of. Smelser, N.J., Wright, J., Baltes, P.B. (eds.) *International Encyclopedia of the Social and Behavioral Sciences.* Elsevier Science, Oxford, pp. 14656–61.

Bae, K. 2010. Peopling of the Korean Peninsula. Norton, C.J. and Braun, D.R. (eds.) *Asian Paleoanthropology: From Africa to China and Beyond.* Springer, Dordrecht, pp. 181–90.

Bahuchet, S., McKey, D., de Garine, I. 1991. Wild yams revisited: is independence from agriculture possible for rain forest hunter-gatherers? *Human Ecology* 19: 213–43.

Bai, Y., Fang, X., Nie, J., Wang, Y., Wu, F. 2009. A preliminary reconstruction of the paleoecological and paleoclimatic history of the Chinese Loess Plateau from the application of biomarkers. *Palaeogeography, Palaeoclimatology, Palaeoecology* 271(1-2): 161–69.

Bailey, R.C., Head, G., Jenike, M., Owen, B., Rechtman, R., Zechenter, E. 1989. Hunting and gathering in tropical rain forest: is it possible? *American Anthropologist* 91: 59–82.

Balloux, F. 2009. The worm in the fruit of the mitochondrial DNA tree. *Heredity* 104: 419–20.

Balme, J. 1995. 30,000 years of fishery in western New South Wales. *Archaeology in Oceania* 30: 1–21.

Bardají, T., Goy, L.J., Zazo, C., Hillaire-Marcel, C., Dabrio, C.J., Cabero, A., Ghaleb, B., *et al.* 2009. Sea level and climate changes during OIS 5e in the Western Mediterranean. *Geomorphology* 104: 22–37.

Barkai, R., Gopher, A., Lauritzen, S.E., Frumkin, A. 2003. Uranium series dates from Qesem Cave, Israel, and the end of the Lower Palaeolithic. *Nature* 423: 977–9.

Barker, G., Lloyd-Smith, L., Barton, H., Cole, F., Hunt, C., Piper, P., Rabett, R., Paz, V., Szabó, K. 2011. Foraging-farming transitions at the Niah Caves, Sarawak, Borneo. *Antiquity* 85: 1–18.

Barker, G., Barton, H., Beavitt, P., Bird, M., Daly, P., Doherty, C., Gilbertson, D., Hunt, C., Krigbaum, J., Lewis, *et al.* 2002a. Prehistoric foragers and farmers in Southeast Asia: renewed investigations at Niah cave, Sarawak. *Proceedings of the Prehistoric Society* 68: 147–64.

Barker, G., Barton, H., Bird, M., Cole, F., Daly, P., Gilbertson, D., Hunt, C., Krigbaum, J., Lampert, C., Lewis, H., *et al.* 2002b. The Niah Cave Project: The third (2002) season of fieldwork. *Sarawak Museum Journal* 78: 87–178.

Barker, G., Barton, H., Bird, M., Daly, P., Datan, I., Dykes, A., Farr, L., Gilbertson, D.D., Harrisson, B., Hunt, C.O., *et al.* 2007. The 'human revolution' in tropical Southeast Asia: the antiquity and behaviour of anatomically modern humans at Niah Cave (Sarawak, Borneo). *Journal of Human Evolution* 52: 243–61.

Barker, G., Antoniadou, A., Armitage, S., Brooks, I., Candy, I., Connell, K., Douka, K., *et al.* 2010. The Cyrenaican Prehistory Project 2010: the fourth season of investigations of the Haua Fteah cave and its landscape, and further results from the 2007–2009 fieldwork. *Libyan Studies* 41: 63–88.

Barker, G., Gilbertson, D., Reynolds T. (eds.). Forthcoming. *Rainforest Foraging and Farming in Island Southeast Asia: The Archaeology and Environmental History of the Niah Caves, Sarawak*, Niah Cave Project, Vol. 1. McDonald Institute for Archaeological Research, Cambridge.

Barnabas, S., Shouche, Y., Suresh, C.G. 2005. High-resolution mtDNA studies of the Indian population: implications for Palaeolithic settlement of the Indian subcontinent. *Annals of Human Genetics* 70: 42–58.

Barsky, D., Celiberti, V., Cauche, D., Grégoire, S., Lebègue, F., de Lumley, H., Toro-Moyano, I. 2010. Raw material discernment and technological aspects of the Barranco León and Fuente Nueva 3 stone assemblages (Orce southern Spain). *Quaternary International* 223–4: 201–19.

Barton, C.M., Clark, G.A., Cohen, A.E. 1994. Art as information: explaining Upper Palaeolithic art in Western Europe. *World Archaeology* 26: 185–207.

Barton, H. 2005. The case for rainforest foragers: the starch record at Niah Cave, Sarawak. *Asian Perspectives* 44: 56–72.

Barton, H., Piper, P., Rabett, R., Reeds, I. 2009. Composite hunting technologies from the Terminal Pleistocene and Early Holocene, Niah Cave, Borneo. *Journal of Archaeological Science* 36: 1708–14.

Barton, M., Bean, N., Dunleavy, S., Gray, I., White, A. 2002. *Wild New World: Recreating Ice-Age North America*. BBC Worldwide, London.

Barstra, G-J. 1998. Short history of the archaeological exploration of the Maros caves in South Sulawesi. *Modern Quaternary Research in Southeast Asia* 15: 193–209.

Barstra, G.-J. 1992. Pacitan and Sangiran, and Java Man's tools. Bellwood, P., Datta, A., Chatterjee, P.G., Sen, A.K. (eds.) *Man and His Culture: A Resurgence*. Books and Books, New Delhi, pp. 93–103.

Barstra, G.-J. 1989. Late *Homo erectus* or Ngandong man of Java. *Palaeohistoria* 29: 1–7.

Barstra, G.-J. 1988. Ngandong man: age and artifacts. *Journal of Human Evolution* 17: 325–37.

Bar-Yosef, O. 1994. The Lower Paleolithic of the Near East. *Journal of World Prehistory* 8: 211–65.

Bar-Yosef, O. 2000. The Middle and Early Upper Palaeolithic in Southwest Asia and neighboring regions. Bar-Yosef, O. and Pilbeam, D. (eds.) *The Geography of Neanderthals and Modern Humans in Europe and the Greater Mediterranean*. Peabody Museum of Archaeology and Ethnology, Harvard University, Cambridge, pp. 107–56.

Bar-Yosef, O. 2002. The Upper Paleolithic Revolution. *Annual Review of Anthropology* 31: 363–93.

Bar-Yosef, O. 2007. The dispersal of modern humans in Eurasia: a cultural interpretation. Mellars, P., Stringer, S., Bar-Yosef, O., Boyle, K. (eds.) *Rethinking the Human Revolution: New Behavioural and Biological Perspectives on the Origins and Dispersal of Modern Humans*. McDonald Institute Monographs, Cambridge, pp. 207–17.

Bar-Yosef, O., Eren, M.I., Yuan, J., Cohen, D.J., Li, Y. Forthcoming. Were bamboo tools made in prehistoric Southeast Asia? An experimental view from South China. *Quaternary International.*

Baskaran, M., Maratheb, A.R., Rajagurub, S.N., Somayajulu, B.L.K. 1986. Geochronology of Palaeolithic cultures in the Hiran Valley, Saurashtra, India. *Journal of Archaeological Science* 13: 505–14.

Bassinot, F.C., Labeyrie, L.D., Vincent, E., Quidelleur, X., Shackleton, N.J., Lancelot, Y. 1994. The astronomical theory of climate and the age of the Brunhes-Matuyama magnetic reversal. *Earth and Planetary Science Letters* 126: 91–108.

Bates, T.S., Lamb, B.K., Guenther, A., Dignon, J., Stoiber, R.E. 1992. Sulfur emissions to the atmosphere from natural sources. *Journal of Atmospheric Chemistry* 14: 315–37.

Baumann, K.-H., Lackschewitz, K.S., Mangerud, J., Spielhagen, R.F., Wolf-Welling, T.C.W., Henrich, R., Kassens, H. 1995. Reflection of Scandinavian ice sheet fluctuations in Norwegian Sea sediments during the past 150,000 years. *Quaternary Research* 43: 185–97.

Bautista, A. 1999. Development of zooarchaeology for Southeast Asia. Paper presented at the International Colloquium on Archaeology in Southeast Asia in the 3rd Millennium, 27–29 September, Penang, Malaysia.

Beaton, J.M. 1991. Colonizing continents: some problems from Australia and the Americas. Dillehay, T.D. and Meltzer, D.J. (eds.) *The First Americans: Search and Research*. CRC Press, Boca Raton, pp. 209–30.

Beck, C., Jones, G.T. 2010. Clovis and the Western Stemmed: population migration and the meeting of two technologies in the intermountain West. *American Antiquity* 75(1): 81–116.

Bellwood, P. 1976. Archaeological research in Minahasa and Talaud Islands, northern Indonesia. *Asian Perspectives* 19: 240–87.

Bellwood, P. 1979. *Man's Conquest of the Pacific: The Prehistory of Southeast Asia and Oceania*. Collins, Auckland.

Bellwood, P. 1984. Archaeological research in the Madai-Baturong region, Sabah. *Bulletin of the Indo-Pacific Prehistory Association* 5: 38–54.

Bellwood, P. 1987. The prehistory of Island Southeast Asia: a multi-disciplinary review of recent research. *Journal of World Prehistory* 1: 71–124.

Bellwood, P. (ed.). 1988. *Archaeological Research in South-eastern Sabah*. Sabah Museum Monograph 2, Kota Kinabalu.

Bellwood, P. 1997. *Prehistory of the Indo-Malaysian Archipelago*. 2nd ed. University of Hawaii Press, Honolulu.

Bellwood, P. 1998. Human dispersals and colonizations in prehistory – the Southeast Asian data and their implications. Omoto, K. and Tobias, P.V. (eds.) *The Origins and Past of Modern Humans – Towards Reconciliation*. World Scientific, Singapore, pp. 188–205.

Bellwood, P. 2006. *First Farmers: The Origins of Agricultural Societies*. Blackwell, Oxford.

Bellwood, P. 2007. *Prehistory of the Indo-Malay Archipelago*. 3rd ed. Australian National University E Press, Canberra.

Bellwood, P. Nitihaminoto, G., Irwin, G., Gunadi, Waluyo, A., Tanudirjo, D. 1998. 35,000 years of prehistory in the northern Moluccas. *Modern Quaternary Research in Southeast Asia* 15: 233–75.

Bennett, E., Nyaoi A.J., Jephte Sompud. 2000. Saving Borneo's bacon: The sustainability of hunting in Sarawak and Sabah. Robinson, J.G. and Bennett, E.L. (eds.) *Hunting for Sustainability in Tropical Forests*. Columbia, New York, pp. 305–24.

Berger, J-F and Guilaine, J. 2009. The 8200 cal BP abrupt environmental change and the Neolithic transition: a Mediterranean perspective. *Quaternary International* 200: 31–49.

Bettis, E.A., III, Milius, A.K., Carpenter, S.J., Larick, R., Yahdi Zaim, Yan Rizal, Ciochon, R.L., *et al.* 2009. Way out of Africa: Early Pleistocene paleoenvironments inhabited by *Homo erectus* in Sangiran, Java. *Journal of Human Evolution* 56: 11–24.

Binford, L. 1979. Organisation and formation processes: looking at curated technologies. *Journal of Anthropological Research* 35: 255–73.

Binford, L. 1980. Willow smoke and dog's tails: hunter-gatherer settlement systems and archaeological site formation. *American Antiquity* 45: 4–20.

Binford, L.R. 1981. *Bones: Ancient Men, Modern Myths*. Academic Press, New York.

Binford, L.R. 1985. Human ancestors: changing views of their behavior. *Journal of Anthropological Archaeology* 4: 292–327.

Binford, L.R. and Ho, C.K. 1985. Taphonomy at a distance: Zhoukoudian, 'The Cave Home of Beijing Man'? *Current Anthropology* 26: 413–42.

Bird, D.W., Richardson, J.L., Veth, P.M., Barham, A.J. 2002. Explaining shellfish variability in middens on the Meriam Islands, Torres Strait, Australia. *Journal of Archaeological Science* 29: 457–69.

Bird, M.I., Turney, C.S.M., Fifield, L.K., Jones, R., Ayliffe, L.K., Palmer, A., Cresswell, R., Robertson, S. 2002. Radiocarbon analysis of the early archaeological site of Nauwalabila I, Arnhem Land, Australia: implications for sample suitability and stratigraphic integrity. *Quaternary Science Reviews* 21: 1061–75.

Bird, M., Taylor, D., Hunt, C. 2005. Palaeoenvironments of insular Southeast Asia during the last glacial period: a savanna corridor in Sundaland? *Quaternary Science Reviews* 24: 2228–42.

Bird, M., Boobyer, E.M., Bryant, C., Lewis, H.A., Paz, V., Edryd Stephens, W. 2007. A long record of environmental change from bat guano deposits in Makangit Cave, Palawan, Philippines. *Earth and Environmental Science Transactions of the Royal Society of Edinburgh*, 98: 59–69.

Bird-David, N. 1990. The giving environment: another perspective on the economic system of gatherer-hunters. *Current Anthropology* 31: 189–96.

Birdsell, J.B. 1977. The recalibration of a paradigm for the first peopling of Greater Australia. Allen, J., Golson, J., Jones, R. (eds.) *Sunda and Sahul: Prehistoric Studies in Southeast Asia, Melanesia and Australia*. Academic Press, London, pp. 113–67.

Birdsell, J.B. 1979. A reassessment of the age, sex, and population affinities of the Niah cranium. *American Journal of Physical Anthropology* 50: 419.

Bischoff, J.L., Williams, R.W., Rosenbauer, R.J., Aramburu, A., Arsuaga, J.L., García, N., Cuenca-Bescós, G. 2007. High-resolution U-series dates from the Sima de los Huesos hominids yields $600^{+\infty}/_{-66}$ kyrs: implications for the evolution of the early Neanderthal lineage. *Journal of Archaeological Science* 34: 763–70.

Björck, S., Walker, M.J.C., Cwynar, L.C., Johnsen, S., Knudsen, K.-L., Lowe, J.J., Wohlfarth, B., and INTIMATE Members. 1998. An event stratigraphy for the Last Termination in the North Atlantic region based on the Greenland ice-core record: a proposal by the INTIMATE group. *Journal of Quaternary Science* 13: 283–92.

Black, D. 1926. Tertiary man in Asia—the Chou Kou Tien Discovery. *Science* 64: 586–87.

Black, D. 1934. The Croonian lecture: on the discovery, morphology, and environment of *Sinanthropus pekinensis*. *Philosophical Transactions of the Royal Society of London, Series B* 223: 57–120.

Blackwood, B. 1950. *The Technology of a Modern Stone Age People in New Guinea*. Pitt-Rivers Museum Occasional Paper on Technology 3. Oxford University Press, Oxford.

Blades, B.S. 1999. Aurignacian lithic economy and early modern human mobility: new perspectives from classic sites in the Vézère valley of France. *Journal of Human Evolution* 37: 91–120.

Blain, H.-A., Bailon, S., Cuenca-Bescós, G., Arsuaga, J.L., Bermúdez de Castro, J.M., Carbonell, E. 2009. Long-term climate record inferred from early-middle Pleistocene amphibian and squamate reptile assemblages at the Gran Dolina Cave, Atapuerca, Spain. *Journal of Human Evolution* 56: 55–65.

Blanck, T., McCord, W.P., Minh, L. 2006. On the variability of *Curoa trifasciata* (Bell, 1825). *Frankfurt Contributions to Natural History* 31, Edition Chimaira, Frankfurt am Main.

Blanton, D.B. 2003. The weather is fine, wish you were here, because I'm the last one alive: 'learning' the environment in the English New World colonies. Rockman, M. and Steele, J. (eds.) *Colonisation of Unfamiliar Landscapes: The Archaeology of Adaptation*. Routledge, London, pp. 190–200.

Blockley, S.P.E., Blockley, M., Donahue, E., Lane, S., Lowe, J., Pollard, M. 2006. The chronology of abrupt climate change and Late Upper Palaeolithic human adaptation in Europe. *Journal of Quaternary Science* 21: 575–84.

Boaz, N.T., Ciochon, R.L., Xu, Q., Liu, J. 2000. Large mammalian carnivores as a taphonomic factor in the bone accumulation at Zhoukoudian. *Acta Anthropologica Sinica* 19(Suppl.): 224–34.

Boaz, N.T., Ciochon, R.L., Xu, Q., Liu, J. 2004. Mapping and taphonomic analysis of the *Homo erectus* loci at Locality 1 Zhoukoudian, China. *Journal of Human Evolution* 46: 519–49.

Bobe, R., Behrensmeyer, A.K. 2004. The expansion of grassland ecosystems in Africa in relation to mammalian evolution and the origin of the genus *Homo*. *Palaeogeography, Palaeoclimatology, Palaeoecology* 207: 399–420.

Bocherens, H., Billiou, D., Mariotti, A., Toussaint, M., Patou-Mathis, M., Bonjean, D., Otte, M. 2001. New isotopic evidence for dietary habits of Neandertals from Belgium. *Journal of Human Evolution* 40: 497–505.

Bocherens, H., Drucker, D.G., Billiou, D., Patou-Mathis, M., Vandermeersch, B. 2005. Isotopic evidence for diet and subsistence pattern of the Saint-Césaire I Neanderthal: review and use of a multi-source mixing model. *Journal of Human Evolution* 49: 71–87.

Bond, G., Heinrich, H., Broecker, W., Labeyrie, L., McManus, J., Andrews, J., Huon, S., *et al.* 1992. Evidence for massive discharges of icebergs into the North Atlantic Ocean during the last glacial period. *Nature* 360: 245–9.

Bond, G., Broecker, W., Johnsen, S., McManus, J., Labeyrie, L., Jouzel, J., Bonani, G. 1993. Correlations between climate records from North Atlantic sediments and Greenland ice. *Nature* 365: 144–7.

Bond, G.C. and Lotti, R. 1995. Iceberg discharges into the North Atlantic on millennial time scales during the Last Glaciation. *Science* 267: 1005–10.

Bond, G., Showers, W., Cheseby, M., Lotti, R., Almasi, P., deMenocal, P., Priore, P., *et al.* 1997. A pervasive millennial-scale cycle in North Atlantic Holocene and glacial climates. *Science* 278: 1257–66.

Bonin, F., Devaux, B., Dupré, A. 2006. *Turtles of the World*. Pritchard, P.C.H. (trans.). A&C Black, London.

Boslough, M.B.E., Crawford, D.A. 2008. Low-altitude airbursts and the impact threat. *International Journal of Impact Engineering* 35: 1441–8.

Bouzouggara, A., Barton, N., Vanhaeren, M., d'Errico, F., Collcutt, S., Higham, T., Hodge, E., *et al.* 2007. 82,000-year-old shell beads from North Africa and implications for the origins of modern human behavior. *Proceedings of the National Academy of Sciences of the United States of America* 104: 9964–9.

Bowdler, S. 1990. Peopling Australasia: the 'coastal colonization' hypothesis re-examined. Mellars, P. (ed.) *The Emergence of Modern Humans: An Archaeological Perspective.* Cornell University Press, Ithaca, pp. 327–43.

Bowdler, S. 2006. The Hoabinhian: early evidence for SE Asian trade networks? Bacus, E.A., Glover, I.C., Piggott, V.C. (eds.) *Uncovering Southeast Asia's Past.* National University of Singapore Press, Singapore, pp. 355–59.

Bowler, J.M. 1998. Willandra Lakes revisited: environmental framework for human occupation. *Archaeology in Oceania* 33: 120–55.

Bowler, J.M., Johnston, H., Olley, J.M., Prescottk, J.R., Roberts, R.G., Shawcross, W., Spooner, N.A. 2003. New ages for human occupation and climatic change at Lake Mungo, Australia. *Nature* 421: 837–40.

Boyle, K. 2007. Changing biodiversity and complexity across the Middle–Upper Palaeolithic Transition. Mellars, P., Stringer, S., Bar-Yosef, O., Boyle, K. (eds.) *Rethinking the Human Revolution: New Behavioural and Biological Perspectives on the Origins and Dispersal of Modern Humans.* McDonald Institute Monographs, Cambridge, pp. 303–14.

Boyle, M., Burkitt, M.C., Field, H., Garrod, D., Oakley, K. 1963. Recollections of the Abbé Breuil. *Antiquity* 37: 12–18.

Brantingham, P.J. 1998. Mobility, competition, and Plio-Pleistocene hominid foraging groups. *Journal of Archaeological Method and Theory* 5: 57–98.

Brantingham, P.J., Krivoshapkin, A.I., Li, J., Tserendagva, T. 2001. The Initial Upper Palaeolithic in northeast Asia. *Current Anthropology* 42: 735–46.

Brantingham, P.J., Kuhn, S.L., Kerry, K.W. 2004. *The Early Upper Palaeolithic Beyond Western Europe.* University of California Press, London.

Bräuer, G., Yokoyama, Y., Falguères, C., Mbua, E. 1997. Human origins backdated. *Nature* 386: 337–8.

Breuil, H. 1912. Les Subdivisions du paleolithique superieur et leur signification. Congrès International d'anthropologie et d'Archéologie Préhistorique. *Compte Rendu* 14 (Geneva): 165–238.

Brewer, S., Guiot, J., Sánchez-Goñi, S.F., Klotz, S. 2008. The climate in Europe during the Eemian: a multi-method approach using pollen data. *Quaternary Science Reviews* 27(25–6): 2303–15.

Britton, K., Grimes, V., Niven, L., Steele, T.E., McPherron, S., Soressi, M., Kelly, T.E., *et al.* 2011. Strontium isotope evidence for migration in late Pleistocene Rangifer: implications for Neanderthal hunting strategies at the Middle Palaeolithic site of Jonzac, France. *Journal of Human Evolution* 61: 176–85.

Broecker, W.S., Thurber, D.L., Goddard, J., Ku, T.-L., Matthews, R.K., Mesolella, K.J. 1968. Milankovitch hypothesis supported by precise dating of coral reefs and deep-sea sediments. *Science* 159: 297–300.

Broecker, W.S. 1994. Massive iceberg discharges as triggers for global climate change. *Nature* 372: 421–4.

Broecker, W.S. 1995. Chaotic climate. *Scientific American* 273: 62–8.

Broecker, W. 1996. Glacial climate in the tropics. *Science* 272: 1902–4.

Bronson, B. and Asmar, T. 1975. Prehistoric investigations at Tianko Panjang Cave, Sumatra. *Asian Perspectives* 18: 128–45.

Brook, B.W., Bowman, D.M.J.S., Burney, D.A., Flannery, T.F., Gagan, M.K., Gillespie, R., Johnson, C.N., *et al.* 2007. Would the Australian megafauna have become extinct if humans had never colonised the continent? *Quaternary Science Reviews* 26: 560–64.

Brooks, A.S., Helgren, D.M., Cramer, J.S., Franklin, A., Hornyak, W., Keating, J.M., Klein, R.G., *et al.* 1995. Dating and context of three Middle Stone Age sites with bone points in the Upper Semliki Valley, Zaire. *Science* 268: 548–53.

Brosius, J.P. 1991. Foraging in tropical rain forests: the case of the Penan of Sarawak, East Malaysia (Borneo). *Human Ecology* 19: 123–50.

Brosius, J.P. 1993. Contrasting subsistence ecologies among Penan foragers, Sarawak (East Malaysia). Hladik, C.M., Hladik, A., Linares, O.F., Pagezy, H., Semple, A., Hadley, M. (eds.) *Tropical Forests, People and Food.* Man and the Biosphere 13. UNESCO/Parthanon, Paris, pp. 515–22.

Brosius, J.P. 2007. Prior transcripts, divergent paths: resistance and acquiescence to logging in Sarawak, East Malaysia. Sercombe, P., Sellato, B. (eds.) *Beyond the Green Myth: Borneo's Hunter-Gatherers in the Twenty-First Century.* Nias Press, Copenhagen, pp. 289–333.

Brothwell, D.R. 1960. Upper Pleistocene skull from Niah Caves. *Sarawak Museum Journal* 9: 323–49.

Brown, C.H. 1985. Mode of subsistence and folk biological taxonomy. *Current Anthropology* 26: 43–64.

Brown, P., Sutikna, T., Morwood, M.J., Soejono, R.P., Jatmiko, Wayhu Saptomo, E. Rokus Awe Due. 2004. A new small-bodied hominin from the Late Pleistocene of Flores, Indonesia. *Nature* 431: 1055–91.

Brumm, A. and Moore, M.W. 2005. Symbolic revolutions and the Australian archaeological record. *Cambridge Archaeological Journal* 15: 157–75.

Brumm, A., Aziz, F., van den Bergh, G.D., Morwood, M.J., Moore, M.W., Kurniawan, I., Hobbs, D.R., Fullagar, R. 2006. Early stone technology at Flores and its implications for *Homo Floresiensis. Nature* 441: 624–28.

Brumm, A., Jensen, G.M., van den Bergh, G.D., Morwood, M.J., Kurniawan, I., Aziz, F., Storey, M. 2010. Hominins on Flores, Indonesia, by one million years ago. *Nature* 464: 748–52.

Bui Vinh. 1991. The Da But culture in the Stone Age of Vietnam. *Bulletin of the Indo-Pacific Prehistory Association* 10: 127–31.

Bulbeck, D. 1982. A re-evaluation of possible evolutionary processes in Southeast Asia since the late Pleistocene. *Bulletin of the Indo-Pacific Prehistory Association* 3: 1–21.

Bulbeck, D.F. 2003. Hunter-gatherer occupation of the Malay Peninsula from the Ice Age to the Iron Age. Mercador, J. (ed.) *Under the Canopy: The Archaeology of Tropical Rainforests.* Rutgers University Press, New Brunswick, pp. 119–60.

Bulbeck, D. 2004. Divided in space, united in time: the Holocene prehistory of South Sulawesi. *Modern Quaternary Research in Southeast Asia* 18: 129–66.

Bulbeck, D. 2006. Economic and Technological Change during the Middle and Late Holocene in the Lamoncong Highlands, South Sulawesi, Indonesia. Bacus, E.A., Glover, I.C., Piggott, V.C. (eds.) *Uncovering Southeast Asia's Past.* National University of Singapore Press, pp. 393–410.

Bulbeck, D.F. 2008. An integrated perspective on the Austronesian diaspora: the switch from cereal agriculture to maritime foraging in the colonisation of island Southeast Asia. *Australian Archaeology* 67: 31–51.

Bulbeck, D., Pasqua, M., Di Lello, A. 2000. Culture history of the Toalean of South Sulawesi, Indonesia. *Asian Perspectives* 39: 71–108.

Bulbeck, D., Sumantri, I., Hiscock, P. 2004. Leang Sakapao 1, a second dated Pleistocene site from South Sulawesi. *Modern Quaternary Research in Southeast Asia* 18: 111–28.

Bulmer, S. 2005. Reflections in stone: axes and the beginnings of agriculture in the Central Highlands of New Guinea. Pawley, A., Attenborough, R., Golsen, J., Hide, R. (eds.) *Papuan Pasts: Cultural, Linguistic and Biological Histories of Papuan-Speaking Peoples.* Australian National University, Canberra, pp. 387–450.

Bunn, H.T., Bartman, L.E., Kroll, E.M. 1988. Variability in bone assemblage formation from Hadza hunting, scavenging and carcass processing. *Journal of Anthropological Archaeology* 7: 412–57.

Burleigh, R. 1981. Radiocarbon dating of freshwater shells from Leang Burung 2: Part 1. *Modern Quaternary Research in Southeast Asia* 6: 51–52.

Burroughs, W.J. 2005. *Climate Change in Prehistory: The End of the Reign of Chaos.* Cambridge University Press, Cambridge.

Bush, A.B.G. and Fairbanks, R.G. 2003. Exposing the Sunda Shelf: tropical responses to eustatic sea level change. *Journal of Geophysical Research* 108: 1–10.

Cabrera, V., Pike-Tay, A., Bernaldo de Quirós, F. 2006. Trends in Middle Palaeolithic settlement in Cantabrian Spain: the late Mousterian at Castillo cave. Victoria Cabrera, Obra selecta. *Zona Arqueológica* 6: 175–87.

Caldecott, J. and Caldecott, S. 1985. A horde of pork. *New Scientist* 15: 32–35.

Cameron, D., Patnaik, R., Sahni, A. 2004. The phylogenetic significance of the Middle Pleistocene Narmada hominin cranium from Central India. *International Journal of Osteoarchaeology* 14: 419–47.

Cann, R.L., Stoneking, M., Wilson, A.C. 1987. Mitochondrial DNA and human evolution. *Nature* 325: 31–6.

Cannon, C.H., Morley, R.J., Bush, A.B.G. 2009. The current refugial rainforests of Sundaland are unrepresentative of their biogeographic past and highly vulnerable to disturbance. *Proceedings of the National Academy of Sciences of the United States of America* 106: 11188–93.

Carbonell, E., Bermúdez de Castro, J.M., Parés, J.M., Pérez-González, A., Cuenca-Bescós, G., Ollé, A., Mosquera, M., *et al.* 2008a. The first hominin of Europe. *Nature* 452: 465–70.

Carbonell, E., Mosquera, M., Rodríquez, X.P., Bermúdez de Castro, J.M., Burjachs, F., Rosell, J., Sala, R., Vallverdú, X. 2008b. Eurasian gates: the earliest human dispersals. *Journal of Anthropological Research* 64: 195–228.

Carto, S.L., Weaver, A.J., Hetherington, R., Yin, L., Wiebe, E.C. 2009. Out of Africa and into an ice age: on the role of global climate change in the late Pleistocene migration of early modern humans out of Africa. *Journal of Human Evolution* 56: 139–51.

Cavalli-Sforza, L.L., Piazza, A., Menozzi, P., Mountain, J. 1988. Reconstruction of human evolution: bringing together genetic, archaeological, and linguistic data. *Proceedings of the National Academy of Sciences of the United States of America* 85: 6002–6.

Ceron, J. n.d. The Macrobotanical remains in Hang Boi and Hang Trống cave Sites. Report prepared for the Tràng An Archaeological Project.

Chagnon, N.A. 1992. *Yanomamo*. 4th ed. Case Studies in Cultural Anthropology Series. Spindler, G. and Spindler, L. (eds.). Harcourt Brace Jovanovich College Publishers, Fort Worth.

Charles, E., Piper, P., Lewis, H., Paz, V., Ochoa, J., Ronquillo, W. Forthcoming. Late Quaternary sea level changes and the Palaeohistory of Palawan Island, Philippines: mapping palaeogeographic changes using open source GIS.

Chan, H. 2007. History and the Punan Vuhang: response to economic and resource tenure change. Sercombe, P. and Sellato, B. (eds.) *Beyond the Green Myth: Borneo's Hunter-Gatherers in the Twenty-First Century*. Nias Press, Copenhagen, pp. 192–226.

Chapman, C.R. and Morrison, D. 1994. Impacts on the Earth by asteroids and comets: assessing the hazard. *Nature* 367: 33–40.

Chapman, V. 1986. Inter-site variability in southwest Sulawesi: results of the 1969 Australian-Indonesian Archaeological Expedition. *Archaeology in Oceania* 21: 76–84.

Chappell, J. 1991. Late Quaternary environmental changes in eastern and central Australia, and their climatic interpretation. *Quaternary Science Reviews* 10: 377–90.

Chappell, J. and Shackleton, N.J. 1986. Oxygen istopes and sea level. *Nature* 324: 137–40.

Chase, P.G. 1989. How different was Middle Palaeolithic subsistence? A zooarchaeological perspective on the Middle to Upper Palaeolithic transition. Mellars, P. and Stringer, C. (eds.) *The Human Revolution*. Edinburgh University Press, Edinburgh, pp. 321–37.

Chase, P.G., Debénath, A., Dibble, H.L., McPherron, S.P., Schwarcz, H.P., Stafford, T.W., Jr., Tournepiche, J.-F. 2007. New dates for the Fontéchevade (Charente, France) *Homo* remains. *Journal of Human Evolution* 52: 217–21.

Chauhan, P.R. 2008. Large mammal fossil occurrences and associated archaeological evidence in Pleistocene contexts of peninsular India and Sri Lanka. *Quaternary International* 192: 20–42.

Chazan, M. 1995. Conceptions of time and the development of Paleolithic chronology. *American Anthropologist* 97: 457–67.

Chen, C.-H., Lee, M.-Y., Iizuka, Y., Dehn, J., Wei, K.-Y., Carey, S. 2004. First Toba supereruption: comment and reply. *Geology*. Online Forum, June.

Chesner, C.A. 1998. Petrogenesis of the Toba Tuffs, Sumatra, Indonesia. *Journal of Petrology* 39: 397–438.

Chesner, C.A., Rose, W.I., Deino, A., Drake, R., Westgate, J.A. 1991. Eruptive history of Earth's largest Quaternary caldera (Toba, Indonesia) clarified. *Geology* 19: 200–3.

Chia, S. 1997. Prehistoric pottery sources and technology in Peninsular Malaysia based on compositional and morphological studies. Monograph. *Malaysia Museums Journal* 33, Kuala Lumpur.

Chitkament, T. 2006–7. *Lithic Analysis of Moh Khiew Rockshelter (Locality I) in Krabi River Valley, Krabi Province, Southwestern Thailand*. Erasmus mundus Master in Quaternary and Prehistory, Universitat Rovira I Virgili.

Chlachula, J. 2001. Pleistocene climate change, natural environments and Palaeolithic occupation of the Altai area, west-central Siberia. *Quaternary International* 80–1: 131–67.

Chung, S.K. and Yin, E.H. 1977. Regional geology: Peninsular Malaysia. Chung, S.K. (ed.) *Kementerian Perusahaan Utama, Lapuran Tahunan Kajibumi Malayia (Ministry of Primary Industries, Annual Report Geological Survey Malaysia)*, Government Printing Office, Kuching, pp. 64–76.

Ciochon, R.L., Olsen, J.W. 1991. Palaeoanthropological and archaeological discoveries from Lang Trang caves: a new Middle Pleistocene hominid site from Northern Vietnam. *Bulletin of the Indo-Pacific Prehistory Association* 10: 59–73.

Ciochon, R., Vu The Long, Larick, R., Gonzáles, L., Grün, R., De Vos, J., Yonge, C., Taylor, L., Yoshida, H., Reagan, M. 1996. Dated co-occurrence of *Homo erectus* and *Gigantopithecus* from Tham Khuyen Cave, Vietnam. *Proceedings of the National Academy of Sciences of the United States of America* 93: 3016–20.

Ciochon, R.L., Bettis, E.A., III. 2009. Asian *Homo erectus* converges in time. *Nature* 458: 153–4.

Ciochon, R. 2009. The mystery ape of Pleistocene Asia. *Nature* 459: 910–11.

Clark, G. 1971. *World Prehistory: A New Outline*. Cambridge University Press, Cambridge.

Clark, G.A. 1997a. Through a glass darkly: conceptual issues in modern human origins research. Clark, G.A. and Willermet, C.M. (eds.) *Conceptual Issues in Modern Human Origins Research*. Aldine de Gryter, New York, pp. 60–76.

Clark, G.A. 1997b. The Middle-Upper Paleolithic transition in Europe: an American perspective. *Norwegian Archaeological Review* 30: 25–53.

Clark, G.A. 2006. Modern human origins research in Iberia: a critique of the indigenist model. *Zona Arqueológica* 7: 348–57.

Clark, G.A. and Lindly, J.M. 1989. The case of continuity: observations on the biocultural transition in Europe and Western Asia. Mellars, P. and Stringer, C. (eds.) *The Human Revolution*. Edinburgh University Press, Edinburgh, pp. 626–76.

Clark, J.D., Beyene, Y., Wolde Gabriel, G., Hartk, W.K., Renne, P.R., Gilbert, H., Defleurq, A., *et al*. 2003. Stratigraphic, chronological and behavioural contexts of Pleistocene *Homo sapiens* from Middle Awash, Ethiopia. *Nature* 423: 747–52.

Clark, J.G.D. 1954. *Excavations at Star Carr: An Early Mesolithic Site at Seamer Near Scarborough, Yorkshire*. Cambridge University Press, Cambridge.

Clark, J.G.D. 1999. Dorothy Garrod. Murray, T. (ed.) *Encyclopedia of Archaeology: The Great Archaeologists* (Pt. 1). ABC-CLIO, Oxford, pp. 401–12.

Clark, P.U., Alley, R.B., Keigwin, L.D., Licciardi, J.M., Johnsen, S.J., Wang, H. 1996. Origin of the first global meltwater pulse following the last glacial maximum. *Paleoceanography* 11: 563–77.

Clark, P.U., Dyke, A.S., Shakun, J.D., Carlson, A.E., Clark, J., Wohlfarth, B., Mitrovica, J.X., Hostetler, S.W., McCabe, A.M. 2009. The Last Glacial Maximum. *Science* 325: 710–14.

Claassen, C. 1998. *Shells*. Cambridge University Press, Cambridge.

Cohen, T.J., Nanson, G.C., Jansen, J.D., Jones, B.G., Jacobs, Z., Treble, P., Price, D.M., May, J-H., Smith, A.M., Ayliffe, L.K., Hellstrom, J.C. 2011. Continental aridification and the vanishing of Australia's megalakes. *Geology* 39: 167–70.

Colani, M. 1930. Recherches sur le Préhistorique Indochinos. *Extrait du Bulletin de l'Ecole Française d'Extrême-Orient* XXX(3–4).

Colinvaux, P.A. and Bush, M.B. 1991. The rain-forest ecosystem as a resource for hunting and gathering. *American Anthropologist* 93: 153–60.

Collings, H.D. 1936. Report of an archaeological excavation in Kedah, Malay Peninsula. *Bulletin of the Raffles Museum, Series B* 1: 5–16.

Collings, H.D. 1937. An excavation at Bukit Chuping, Perlis. *Bulletin Raffles Museum, Series B* 1: 94–119.

Collings, H.D. 1938. Pleistocene site in the Malay Peninsula. *Nature* 142: 575–6.

Comas, D., Plaza, S., Wells, R.S., Yuldaseva, N., Lao, O., Calafell, F., Bertranpetit, J. 2004. Admixture, migrations, and dispersals in Central Asia: evidence from maternal DNA lineages. *European Journal of Human Genetics* 12: 495–504.

Conard, N.J. and Bolus, M. 2003. Radiocarbon dating the appearance of modern humans and timing of cultural innovations in Europe: new results and new challenges. *Journal of Human Evolution* 44: 331–71.

Conard, N.J., Grootes, P.M., Smith, F.H. 2004. Unexpectedly recent dates for human remains from Vogelherd. *Nature* 430: 198–201.

Conard, N.J. 2003. Palaeolithic ivory sculptures from southwestern Germany and the origins of figurative art. *Nature* 426: 830–32.

Conkey, M. 1997. Beyond art and between the caves: thinking about context in the interpretative process. Conkey, M., Soffer, O., Stratmann, D., Jablonski, N. (eds.) *Beyond Art: Pleistocene Image and Symbol*. Wattis Symposium Series in Anthropology, Memoirs of the California Academy of Sciences 23, University of California Press, San Fransisco, pp. 343–67.

Cook, E.J. 2009. A record of late Quaternary environments at lunette-lakes Bolac and Turangmoroke, Western Victoria, Australia, based on pollen and a range of non-pollen palynomorphs. *Review of Palaeobotany and Palynology* 153: 185–224.

Cook, S.F. 1946. A reconsideration of shellmounds with respect to population and nutrition. *American Antiquity* 12: 50–53.

Corlett, R.T. 2007. The impact of hunting on the mammalian fauna of tropical Asian forests. *Biotropica* 39: 292–303.

Corvinus, G. 2004. *Homo erectus* in East and Southeast Asia, and the questions of the age of the species and its association with stone artefacts, with special attention to handaxe-like tools. *Quaternary International* 117: 141–51.

Cosford, J., Qing, H., Yuan, D., Zhang, M., Holmden, C., Patterson, W., Hai, C. 2008. Millennial-scale variability in the Asian monsoon: evidence from oxygen isotope records from stalagmites in southeastern China. *Palaeogeography, Palaeoclimatology, Palaeoecology* 266: 3–12.

Cosgrove, R. 1995. Late Pleistocene behavioural variation and time trends: the case from Tasmania. *Archaeology in Oceania* 30: 83–104.

Cosgrove, R. and Pike-Tay, A. 2004. The Middle Palaeolithic and Late Pleistocene Tasmania hunting behaviour: a reconsideration of the attributes of modern human behaviour. *International Journal of Osteoarchaeology* 14: 321–32.

Cosgrove, R., Field, J., Ferrier, Å. 2007. The archaeology of Australia's tropical rainforests. *Palaeogeography, Palaeoclimatology, Palaeoecology* 251: 150–73.

Covert, H.H., Trinh Dzanh, Do Duc Quang, Le Quoc Thanh, Nguyen Manh Phi. 2002. New mammalian fossils from Ninh Binh Province, Viet Nam. *Journal of Geology, Series B* 19–20: 109–12.

Cox, M.P. 2008. Accuracy of molecular dating with the rho statistic: deviations from coalescent expectations under a range of demographic models. *Human Biology* 80: 335–57.

Cranbrook, Earl of, and Labang, D. 2003. Bearded pigs (*Sus barbatus*): tooth-wear and aging wild populations in Sarawak. *Sarawak Museum Journal* 79: 163–82.

Cranbrook, Earl of, and Piper, P.J. 2007. The Javan Rhinoceros *Rhinoceros sondaicus* in Borneo. *Raffles Bulletin of Zoology* 55: 229–32.

Cranbrook, Earl of, and Piper, P.J. 2009. Borneo Records of Malay Tapir, *Tapirus indicus* Desmarest: A zooarchaeological and historical review. *International Journal of Osteoarchaeology* 19: 491–507.

Crassard, R. 2009. The Middle Palaeolithic of Arabia: the view from the Hadramawt region of Yemen. Petraglia, M.D. and Rose, J.I. (eds.) *The Evolution of Human Populations in Arabia*. Springer, Dordrecht, pp. 151–68.

Cruciani, F., La Fratta, R., Santolamazza, P., Sellitto, D., Pascone, R., Moral, P., Watson, E., Guida, V., *et al.* 2004. Phylogeographic analysis of haplogroup E3b (E-M215) Y chromosomes reveals multiple migratory events within and out of Africa. *American Journal of Human Genetics* 74: 1014–22.

Crutzen, P.J. 2002. Geology of mankind. *Nature* 415: 23.

Cucchi, T., Dobney, K., Fujita, M. 2009. New insights into pig taxonomy, domestication and human dispersal in Island South East Asia through molar shape analysis: the *Sus* remains from Niah, and Lobang Kudih caves in Sarawak. *International Journal of Osteoarchaeology* 19: 508–30.

Curran, L.M. and Leighton, M. 2000. Vertebrate responses to spatiotemporal variation in seed production of mast-fruiting Dipterocarpaceae. *Ecological Monographs* 70: 101–28.

Czarnetzki, A., Jakob, T., Pusch, C.M. 2003. Palaeopathological and variant conditions of the *Homo heidelbergensis* type specimen (Mauer, Germany). *Journal of Human Evolution* 44: 479–95.

Dalén, L., Orlando, L., Shapiro, B., Brandström Durling, M., Quam, R., Gilbert, M.T.P., Díez Fernández-Lomana, J.C.D., *et al.* Forthcoming. Partial genetic turnover in Neandertals: continuity in the east and population replacement in the west. *Molecular Biology and Evolution*.

Daly, P. 1969. Approaches to faunal analysis in archaeology. *American Antiquity* 34: 146–53.

Dang, N.T. and Ho, T.H. 2003. Hai loai cua moi thuoc ho Potamidae o mein nam Viet Nam [Two new potamid crab species of Potamidae from the southern part of Vietnam]. *Journal of Biology, National Centre for Natural Science and Technology of Vietnam* 25: 7–13.

Dansgaard, W., Clausen, H.B., Gundestrup, N., Hammer, C.U., Johnsen, S.F., Kristinsdottir, P.M., Reeh, N. 1982. A New Greenland deep ice core. *Science* 218: 1273–77.

Dansgaard, W., White, J.W.C., Johnsen, S.J. 1989. The abrupt termination of the Younger Dryas climate event. *Nature* 339: 532–4.

Dansgaard, W., Johnsen, S.J., Clausen, H.B., Dahl-Jensen, D., Gundestrup, N.S., Hammer, C.U., Hviderg, C.S., *et al.* 1993. Evidence for general instability of past climate from a 250-kyr ice-core record. *Nature* 364: 218–20.

Datan, I. 1993. *Archaeological Excavations at Gua Sireh (Serian) and Lubang Angin (Gunung Mulu National Park) Sarawak, Malaysia. Sarawak Museum Journal* (Special Monograph 6).

Davidson, I. and Nobel, W. 1992. Why the first colonisation of the Australian region is the earliest evidence of modern human behaviour. *Archaeology in Oceania* 27: 113–19.

Davidson, J.H.C.S. 1979. Archaeology in Northern Viet-Nam since 1954. Smith, R.B. and Watson, W. (eds.) *Early Southeast Asia: Essays in Archaeology, History and Historical Geography*. Oxford University Press, New York, pp. 98–124.

Davis, R.S. 1987. Regional perspectives on the Soviet Central Asian Palaeolithic. Soffer, O. (ed.) *The Pleistocene Old World: Regional Perspectives*. Plenum Press, New York, pp. 121–33.

Davis, R.S., Ranov, V.A. 1980. On the Central Asian Paleolithic: reply to recent comments. *Current Anthropology* 21: 809–10.

Davis, R.S. and Ranov, V.A. 1999. Recent work on the Palaeolithic of Central Asia. *Evolutionary Anthropology* 7: 86–93.

Davies, W. 2001. A very model of a modern human industry: new perspectives on the origins and spread of the Aurignacian in Europe. *Proceedings of the Prehistoric Society* 67: 195–217.

Davies, W. 2007. Re-evaluating the Aurignacian as an expression of modern human mobility and dispersal. Mellars, P., Boyle, K., Bar-Yosef, O., Stringer, C. (eds.) *Rethinking the Human Revolution*. McDonald Institute Monographs, Cambridge, pp. 263–74.

de Queiroz, A. 2005. The resurrection of oceanic dispersal in historical biogeography. *Trends in Ecology and Evolution* 20: 68–73.

Dean, J.S. 2000. Complexity theory and sociocultural change in the American Southwest. McIntosh, R.J., Tainter, J.A., Keech McIntosh, S. (eds.) *The Way the Wind Blows: Climate, History and Human Action*. Columbia University Press, New York, pp. 89–118.

Delgado, R.A., and Van Schaik, C.P. 2000. The behavioral ecology and conservation of the orangutan (*Pongo pygmaeus*): A tale of two islands. *Evolutionary Anthropology* 9: 201–18.

deMenocal, P.B. 2004. African climate change and faunal evolution during the Pliocene-Pleistocene. *Earth and Planetary Science Letters* 220: 3–24.

deMenocal, P.B. 2011. Climate and human evolution. *Science* 331: 540–42.

Demeter, F., Bacon, A-M., Kim Thuy Nguyen, Vu The Long, Matsumura, H., Ha Huu Nga, Schuster, M., Mai Huong Nguyen, Coppens, Y. 2004. An archaic *Homo* molar from northern Vietnam. *Current Anthropology* 45: 535–60.

Demeter, F., Bacon, A-M., Kim Thuy Nguyen, Vu The Long, Duringer, P., Roussé, S., Coppens, Y. Matsumura, *et al.* 2005. Discovery of a second human molar and cranium fragment in the late Middle to Late Pleistocene cave of Ma U'Oi (Northern Vietnam). *Journal of Human Evolution* 48: 393–402.

Deng, C., Wei, Q., Zhu, R., Wang, H., Zhang, R., Ao, H., Chang, L., Pan, Y. 2006. Magnetostratigraphic age of the Xiantai Paleolithic site in the Nihewan Basin and implications for early human colonization of Northeast Asia. *Earth and Planetary Science Letters* 244: 336–48.

Denison, S.M., Maslin, M.A., Boot, T.C., Pancost, R.D., Ettwein, V.J. 2006. Precession-forced changes in South West African vegetation during Marine Isotope Stages 101–100 (~2.56–2.51 Ma). *Palaeogeography, Palaeoclimatology, Palaeoecology* 220: 375–86.

Dennell, R. 1990. Progressive gradualism, imperialism and academic fashion: Lower Palaeolithic archaeology in the 20th century. *Antiquity* 64: 549–58.

Dennell, R. 2003. Dispersal and colonisation. Long and short chronologies: how continuous is the Early Pleistocene record for hominids outside East Africa? *Journal of Human Evolution* 45: 421–40.

Dennell, R. 2004. Hominin dispersals and Asian biogeography during the Lower and Early Middle Pleistocene *c.* 2.0–0.5 Mya. *Asian Perspectives* 43: 205–26.

Dennell, R. 2005. The Solo (Ngandong) *Homo erectus* assemblage: a taphonomic assessment. *Oceania* 40: 81–90.

Dennell, R.W. 2010. The colonization of 'Savannahstan': issues of timing(s) and patterns of dispersal across Asia in the Late Pliocene and Early Pleistocene. Norton, C.J. and Braun, D.R. (eds.) *Asian Paleoanthropology: From Africa to China and Beyond*. Springer, Dordrecht, pp. 7–30.

Dennell, R. 2009. *The Palaeolithic Settlement of Asia*. Cambridge University Press, Cambridge.

Dennell, R.W., Rendell, H., Hailwood, E. 1988. Early tool-making in Asia: two-million-year-old artefacts in Pakistan. *Antiquity* 62: 98–106.

Dennell, R.W., Rendell, H.M., Mohammad Halim, Moth, E. 1992. A 45,000-year-old open-air paleolithic site at Riwat, northern Pakistan. *Journal of Field Archaeology* 19: 17–33.

Dennell, R. and Roebroeks, W. 2005. An Asian perspective on early human dispersal from Africa. *Nature* 438: 1099–1104.

Denton, G.H., Anderson, R.F., Toggweiler, J.R., Edwards, R.L., Schaefer, J.M., Putnam, A.E. 2010. The last glacial termination. *Science* 328: 1652–6.

Derenko, M.V., Grzybowski, T., Malyarchuk, B.A., Dambueva, I.K., Denisova, G.A., Czarny, J., Dorzhu, C.M., *et al.* 2003. Diversity in mitochondrial DNA lineages in South Siberia. *Annals of Human Genetics* 67: 391–411.

Derev'anko, A.P., Markin, S. 1998. The Middle and Upper Palaeolithic of western and central Siberia. Derev'anko, A.P., Shimkin, D.B., Powers, W.R. (eds.) *The Palaeolithic of Siberia: New Discoveries and Interpretations*. University of Illinois Press, Urbana, pp. 79–84.

Derev'anko, A.P., Shimkin, D.B., Powers, W.R. 1998. *The Palaeolithic of Siberia: New Discoveries and Interpretations*. University of Illinois Press, Urbana.

Derev'anko, A.P., Postnov, A.V., Rybin, E.P., Kuzmin, Y.V., Keates, S.G. 2005. The Pleistocene peopling of Siberia: a review of environmental and behavioural aspects. *Bulletin of the Indo-Pacific Prehistory Association* 25: 57–68.

Derev'anko, A.P., Petrin, V.T., Rybin, E.P. 2000. The Kara-Bom site and the characteristics of the Middle-Upper Palaeolithic transition in the Altai. *Archaeology, Ethnology and Anthropology of Eurasia* 2: 33–52.

Derev'anko, A.P., Postnov, A.V., Rybin, E.P., Kuzmin, Y.V., Keates, S.G. 2005. The Pleistocene peopling of Siberia: a review of environmental and behavioural aspects. *Bulletin of the Indo-Pacific Prehistory Association* 25: 57–68.

d'Errico, F. and Blackwell, L. (eds.). 2005. *From Tools to Symbols: From Early Hominids to Modern Humans*. Wits University Press, Johannesburg.

d'Errico, F. and Sánchez Goñi, M.F. 2003. Neandertal extinction and the millennial scale climatic variability of OIS 3. *Quaternary Science Reviews* 22: 769–88.

d'Errico, F., Henshilwood, C., Vanhaeren, M., van Niekerk, K. 2005. *Nassarius kraussianus* shell beads from Blombos Cave: evidence for symbolic behaviour in the Middle Stone Age. *Journal of Human Evolution* 48: 3–24.

d'Errico, F. and Henshilwood, C.S. 2007. Additional evidence for bone technology in the southern African Middle Stone Age. *Journal of Human Evolution* 52: 142–63.

de Terra, H. 1938. Preliminary report on recent geological and archaeological discoveries relating to early man in Southeast Asia. *Proceedings of the National Academy of Sciences* 24: 407–13.

de Terra, H. and Movius, H.L., Jr. 1943. Research on early man in Burma, with supplementary reports upon the Pleistocene vertebrates and mollusks of the region, and Pleistocene geology and early man in Java. *Transactions of the American Philosophical Society* 32: 263–464.

Détroit, F., Dizon, E., Falguères, C., Hameau, S., Ronquillo, W., Sémah, F. 2004. Upper Pleistocene *Homo sapiens* from the Tabon cave (Palawan, the Philippines): description and dating of new discoveries. *Comptes Rendus Palevol* 3: 705–12.

de Vries, H., Waterbolk, H.T. 1958. Groningen radiocarbon dates III. *Science* 128: 1550–6.

Dickinson, W.R. 2003. Impact of Mid-Holocene hydro-isostatic highstand in regional sea level on habitability of islands in Pacific Oceania. *Journal of Coastal Research* 19: 489–502.
Diehl, J.F., Onstott, T.C., Chesner, C.A., Knight, M.D. 1987. No short reversals of Brunhes age recorded in the Toba tuffs, north Sumatra, Indonesia. *Geophysical Research Letters* 14: 753–6.
Dincause, D.F. 1993. Fluted points in the Eastern forests. Soffer, O. and Praslov, N.D. (eds.) *From Kostenki to Clovis: Upper Palaeolithic-Palaeo-Indian Adaptations*. Plenum Press, New York, pp. 279–91.
Dixon, E.J. 2001. Human colonization of the Americas: timing, technology and process. *Quaternary Science Reviews* 20: 277–99.
Dizon, E., Détroit, F., Sémah, F., Falguères, C., Hameau, S., Ronquillo, W., Cabanis, E. 2002. Notes on the morphology and age of the Tabon Cave fossil *Homo sapiens*. *Current Anthropology* 43: 660–66.
Dong, W., Jin, C.-Z., Xu, Q.-Q., Liu, J.-Y., Tong, H.-W., Zheng, L.-T. 2000. A comparative analysis on the mammalian faunas associated with *Homo erectus* in China. *Acta Anthropologica Sinica* 19(Suppl.): 246–56.
Douka, K. 2009. Radiocarbon dating the Middle to Upper Palaeolithic Transition. Seminar presentation to the Palaeolithic-Mesolithic Discussion Group, Department of Archaeology, University of Cambridge.
Dove, M.R. 1993. The responses of Dayak and bearded pig to mast-fruiting in Kalimantan: an analysis of nature-culture analogies. Hladik, C.M., Hladik, A., Linares, O.F., Pagezy, H., Semple, A., Hadley, M. (eds.) *Tropical Forests, People and Food*. Man and the Biosphere 13. UNESCO/Parthanon, Paris, pp. 113–23.
Dowsett, H., Thompson, R., Barron, J., Cronin, T., Fleming, F., Ishman, S., Poore, R., Willard, D., Holtz, T., Jr. 1994. Joint investigations of the Middle Pliocene climate I: PRISM paleoenvironmental reconstructions. *Global and Planetary Change* 9: 169–95.
Drake, N.A., Blench, R.M., Armitage, S.J., Bristow, C.S., White, K.H. 2011. Ancient watercourses and biogeography of the Sahara explain the peopling of the desert. *Proceedings of the National Academy of Sciences of the United States of America* 108: 458–62.
Dubois, E. 1896. On *Pithecanthropus erectus*: a transitional form between man and the apes. *Journal of the Anthropological Institute of Great Britain and Ireland* 25: 240–55.
Dunbar, R. 2000. Climate variability during the Holocene: an update. McIntosh, R.J., Tainter, J.A. and Keech McIntosh, S. (eds.) *The Way the Wind Blows: Climate, History and Human Action*. Columbia University Press, New York, pp. 45–88.
Dunn, F.L. 1964. Excavations at Gua Kechil, Pahang. *Journal of the Malay Branch of the Royal Asiatic Society* 37: 87–124.
Dunn, F.L. 1966. Radiocarbon dating of the Malayan Neolithic. *Proceedings of the Prehistoric Society* 32: 352–53.
Dunn, F.L. 1975. Rain-forest collectors and traders: a study of resource utilization in modern and ancient Malaya. *Monograph of the Malaysian Branch of the Royal Asiatic Society* 5, Kuala Lumpur.
Dupree, L., Angel, J.L., Brill, R.H., Caley, E.R., Davis, R.S., Kolb, C.C., Marshack, A., Perkins, D., Jr., Solem, A. 1972. Prehistoric research in Afghanistan (1959–1966). *Transactions of the American Philosophical Society* 62: 1–84.
Earl, G.W. 1863. On the shell-mounds of Province Wellesley, in the Malay Peninsula. *Transactions of the Ethnological Society of London* 2: 119–29.
Echo-Hawk, R.C. 2000. Ancient history in the New World: integrating oral traditions and the archaeological record in deep time. *American Antiquity* 65: 267–90.
Eckhardt, R.B., Henneberg, M. 2010. LB1 from Liang Bua, Flores: Craniofacial asymmetry confirmed, plagiocephaly diagnosis dubious. *American Journal of Physical Anthropology* 143: 331–4.
Eder, J.F. 1984. The impact of subsistence change on mobility and settlement pattern in a tropical forest foraging economy: some implications for archaeology. *American Anthropologist* 86: 837–53.
Eder, J.F. 1988. Batak foraging camps today: a window on the history of a hunting-gathering economy. *Human Ecology* 16: 35–55.
Edwards McKinnon, E. 1991. The Hoabinhian in the Wampu/Lau Biang Valley of northeastern Sumatra: an update. *Bulletin of the Indo-Pacific Prehistory Association* 10: 132–42.
Ehlers, J. and Gibbard, P.L. 2007. The extent and chronology of Cenozoic Global Glaciation. *Quaternary International* 164–5: 6–20.

Endicott, K. 1984. The economy of the Batek of Malaysia: annual and historical perspectives. *Research in Economic Anthropology* 6: 29–52.

Endicott, K. and Bellwood, P. 1991. The possibility of independent foraging in the rain forest of peninsular Malaysia. *Human Ecology* 19: 151–85.

Endicott, P., Gilbert, M.T.P., Stringer, C., Lalueza-Fox, C., Willerslev, E., Hansen, A.J., Cooper, A. 2003. The genetic origin of the Andaman Islanders. *American Journal of Human Genetics* 72: 178–84.

Endicott, P., Metspalu, M., Stringer, C., Macaulay, V., Coopers, A., Sanchez, J.J. 2006. Multiplexed SNP typing of ancient DNA clarifies the origin of Andaman mtDNA haplogroups amongst South Asian tribal populations. *PLOS One* 1: 1–8.

Endicott, P., Ho, S.Y.W., Metspalu, M., Stringer, C. 2009. Evaluating the mitochondrial timescale of human evolution. *Trends in Ecology and Evolution* 24: 515–21.

Engelhardt, R.A. and Rogers, P. 1997. Maritime adaptive strategies in post-Pleistocene Southeast Asia: an ethnoarchaeological model for the nature and distribution of archaeological sites. *Bulletin of the Indo-Pacific Prehistory Association* 16: 177–92.

EPICA Community Members. 2004. Eight glacial cycles from an Antarctic ice core. *Nature* 429: 623–8.

EPICA Community Members. 2006. One-to-one coupling of glacial climate variability in Greenland and Antarctica. *Nature* 444: 195–98.

Erdbrink, D.P. 1954. Mesolithic remains of the Sampung stage in Java: some remarks and additions. *Southwestern Journal of Anthropology* 10: 294–303.

Erlandson, J.M. 2001. The archaeology of aquatic adaptations: paradigms for a new millennium. *Journal of Archaeological Research* 9: 287–350.

Erlandson, J.M., Braje, T.J. 2011. From Asia to the Americas by boat? Paleogeography, paleoecology, and stemmed points of the northwest Pacific. *Quaternary International* 239: 28–37.

Erlandson, J.M., and Moss, M.L. 1996. The Pleistocene-Holocene transition along the Pacific coast of North America. Straus, L.G., Eriksen, B.V., Erlandson, J.M., Yesner, D.R. (eds.) *Humans at the End of the Ice Age: The Archaeology of the Pleistocene-Holocene Transition*. Plenum, New York, pp. 277–301.

Evans, I.H.N. 1918. Preliminary report on cave exploration, near Lenggong, Upper Perak. *Journal of the Federated Malay States Museums* 7: 227–34.

Evans, I.H.N. 1920. Cave-dwellers in Pahang. *Journal of the Federated Malay States Museums* 9: 37–52.

Evans, I.H.N. 1928. Further excavations at Gunong Pondok. *Journal of the Federated Malay States Museums* 12: 161–62.

Evans, I.H.N. 1929. On ancient kitchen-middens in province Wellesley. *Journal of the Federated Malay States Museums* 15: 15–18.

Evans, I.H.N. 1931. A search for antiquaries in Kedah and Perlis. *Journal of the Federated Malay States Museums* 15: 43–50.

Evans, I.H.N. 1937. *The Negritos of Malaya*. Cambridge University Press, Cambridge.

Everden, N. 1993. *The Natural Alien: Humankind and the Environment*. 2nd ed. University of Toronto Press, Toronto.

Excoffier, L. and Schneider, S. 1999. Why hunter-gatherer populations do not show signs of Pleistocene demographic expansions. *Proceedings of the National Academy of Sciences of the United States of America* 103: 10597–602.

Fairbanks, R.G., Mortlocka, R.A., Tzu-Chien Chiua, Li Caoa, Kaplana, A., Guilderson, T.P., Fairbanks, *et al.* 2005. Radiocarbon calibration curve spanning 0 to 50,000 years BP based on paired 230Th/ 234U/ 238U and 14C dates on pristine corals. *Quaternary Science Reviews* 24: 1781–96.

Fairbridge, R.W. 1972. Climatology of a Glacial Cycle. *Quaternary Research* 2: 283–302.

Fagundes, N.J.R., Kanitz, R., Eckert, R., Valls, A.C.S., Bogo, M.R, Salzano, F.M., Smith, D.G., *et al.* 2008. Mitochondrial population genomics supports a single pre-Clovis origin with a coastal route for the peopling of the Americas. *American Journal of Human Genetics* 82: 583–92.

Falk, D., Hildebolt, C., Smith, K., Morwood, M.J., Sutikna, T., Brown, P., Jatmiko, Saptomo W.E., Brunsden, B., Prior, F. 2005. The brain of LB1, *Homo floresiensis*. *Science* 308: 242–5.

Farbstein, R.A. 2009. *Pavlovian portable art: socio-technical process, aesthetic context*. Unpublished PhD dissertation, Department of Archaeology, University of Cambridge.

Feathers, J.K. 2002. Luminescence dating in less than ideal conditions: case studies from Klasies River main site and Duinefontein, South Africa. *Journal of Archaeological Science* 29: 177–94.

Feathers, J.K., Migliorini, E. 2001. Luminescence dating at Katanda – a reassessment. *Quaternary Science Reviews* 20: 961–6.

Féblot-Augustins, J. 2009. Revisiting European Upper Palaeolithic raw material transfers: the demise of the cultural ecological paradigm? Adams, B. and Blades, B.S. (eds.) *Lithic Materials and Paleolithic Societies*. Blackwell, Oxford, pp. 25–46.

Feibel, C. 2004. Quaternary lake margins of the Levant Rift Valley. Goren-Inbar, N. and Speth, J.D. (eds.) *Human Paleoecology in the Levantine Corridor*. Oxbow Books, Oxford, pp. 21–36.

Feibel, C.S., Brown, F.H., McDougall, I. 1989. Stratigraphic context of fossil hominids from the Omo Group deposits: Northern Turkana Basin, Kenya and Ethiopia. *American Journal of Physical Anthropology* 78: 595–622.

Fernández, M.H. 2006. Rodent paleofaunas as indicators of climatic change in Europe during the last 125,000 years. *Quaternary Research* 65: 308–23.

Field, J.S., Petraglia, M.D., Lahr, M.M. 2007. The southern dispersal hypothesis and the South Asian archaeological record: examination of dispersal routes through GIS analysis. *Journal of Anthropological Archaeology* 26: 88–108.

Field, M.H., Huntley, B., Müller, H. 1994. Eemian climate fluctuations observed in a European pollen record. *Nature* 371: 779–83.

Finlayson, C. 2004. *Neanderthals and Modern Humans: An Ecological and Evolutionary Perspective*. Cambridge University Press, Cambridge.

Finlayson, C. and Carrión, J.S. 2007. Rapid ecological turnover and its impact on Neanderthal and other human populations. *TRENDS in Ecology and Evolution* 22: 213–22.

Fix, A.G. 1995. Malayan paleosociology: implications for patterns of genetic variation among the Orang Asli. *American Anthropologist* 97: 313–23.

Fix, A.G. 1999. *Migration and Colonization in Human Microevolution*. Cambridge University Press, Cambridge.

Fix, A.G. 2002. Colonization models and initial genetic diversity in the Americas. *Human Biology* 74: 1–10.

Flenley, J.R. 1997. The Quaternary in the tropics: an introduction. *Journal of Quaternary Science* 12: 345–6.

Folco, L., Glass, B.P., D'Orazio, M., Rochette, P. 2010. A common volatilization trend in Transantarctic Mountain and Australasian microtektites: Implications for their formation model and parent crater location. *Earth and Planetary Science Letters* 293: 135–39.

Foley, R.A. and Lahr, M.M. 1997. Mode 3 technologies and the evolution of modern humans. *Cambridge Archaeological Journal* 7: 3–36.

Ford, J.A. 1959. Eskimo prehistory in the vicinity of Point Barrow, Alaska. *Anthropological Papers American Museum of Natural History* 47(1).

Formicola, V. and Buzhilova, A.P. 2004. Double child burial from Sunghir (Russia): pathology and inferences for Upper Palaeolithic funerary practices. *American Journal of Physical Anthropology* 124: 189–98.

Forestier, H. 2007. *Ribuan Gunung, Ributan Alat Batu: Prasejarah Song Keplek, Gunung Sewa, Jawa Timur*. Published by KPG (Kepustakaan Populer Gramedia), École française d'Extrême-Orient, Institut de Recherche pour le Développement, Pusat Penelitian dan Pengembangan Arkeologi Nasional, and Forum Jakarta-Paris (in Bahasa Indonesia).

Forestier, H., Driwantoro, D., Guillaud, D., Budiman, Siregar, D. 2006. New data for the prehistoric chronology of South Sumatra. Simanjuntak, T., Hisyam, M., Prasetyo, B., Nastiti, T.S. (eds.) *Archaeology: Indonesian Perspective: R.P. Soejono's Festschrift*. Indonesian Institute of Sciences, LIPI Press, Jakarta, pp. 177–92.

Forster, P. and Matsumura, S. 2005. Did early humans go north or south? *Science* 308: 965–66.

Fox, R. 1970. *The Tabon Caves: Archaeological Explorations on Palawan Island, Philippines*. National Museum, Manila.

Fox, R. 1978. The Philippine Palaeolithic. Ikawa-Smith, F. (ed.) *Early Palaeolithic in South and Southeast Asia*. Mouton, The Hague, pp. 59–85.

Fox, W. 1990. *Towards a Transpersonal Ecology*. Shambala, Boston.

Friedlaender, J., Schurr, T., Gentz, F., Koki, G., Friedlaender, F., Horvat, G., Babb, P., *et al.* 2005. Expanding Southwest Pacific mitochondrial haplogroups P and Q. *Molecular Biology Evolution* 22: 1506–17.

Fronval, T. and Jansen, E. 1996. Rapid changes in ocean circulation and heat flux in the Nordic seas during the last interglacial period. *Nature* 383: 806–10.

Fuhrer, K., Neftel, A., Anklin, M., Maggi, V. 1993. Continuous measurements of hydrogen peroxide, formaldehyde, calcium and ammonium concentrations along the new GRIP ice core from summit, central Greenland. *Atmospheric Environment* 27A: 1873–80.

Fullagar, R. and Field, J. 1997. Pleistocene seed-grinding implements from the Australian arid zone. *Antiquity* 71: 300–7.

Gabunia, L., de Lumley, M-A., Vekua, A., Lordkipanidze, D., de Lumley, H. 2002. Découverte d'un nouvel hominidé à Dmanissi (Transcaucasie, Géorgie). *Comptes Rendus Palevol* 1: 243–53 (in French, with English summary).

Gabunia, L., Vekua, A., Lordkipanidze, D., Swisher, C.C., III, Ferring, R., Justus, A., Nioradze, M., *et al.* 2000. Earliest Pleistocene hominid cranial remains from Dmanisi, Republic of Georgia: taxonomy, geological setting, and age. *Science* 288: 1019–25.

Gabunia, L. and Vekua, A. 1995. A Plio-Pleistocene hominid from Dmanisi, East Georgia, Caucasus. *Nature* 373: 509–12.

Gaboardi, M., Deng, T.T., Wang, Y. 2005. Middle Pleistocene climate and habitat change at Zhoukoudian, China, from the carbon and oxygen isotopic record from herbivore tooth enamel. *Quaternary Research* 63: 329–38.

Gagan, M.K., Ayliffe, L.K., Hopley, D., Cali, J.A., Mortimer, G.E., Chappell, J., McCulloch, M.T., Head, M.J. 1998. Temperature and surface-ocean water balance of the Mid-Holocene tropical western Pacific. *Science* 279: 1014–18.

Gamble, C. 1982. Interaction and alliance in Paleolithic society. *Man* 17: 92–107.

Gamble, C. 1993. *Timewalkers: The Prehistory of Global Colonization.* Alan Sutton, Stroud, UK.

Gamble, C. 1986. *The Palaeolithic Settlement of Europe.* Cambridge University Press, Cambridge.

Gamble, C. 1991. The social context for European Palaeolithic art. *Proceedings of the Prehistoric Society* 57: 3–15.

Gamble, C. 1998. Palaeolithic society and the release from proximity: a network approach to intimate relations. *World Archaeology* 29: 426–49.

Gamble, C. 1999. *The Palaeolithic Societies of Europe.* Cambridge University Press, Cambridge.

Gamble, C. and Soffer, O. (eds.). 1989a. *The World at 18000 BP: Vol. 1. High Latitudes.* Routledge, London.

Gamble, C. and Soffer, O. (eds.). 1989b. *The World at 18000 BP: Vol. 2. Low Latitudes.* Routledge, London.

Ganopolski, A. and Rahmstorf, S., 2001. Rapid changes of glacial climate simulated in a coupled climate model. *Nature* 409: 154–58.

Garrod, D., Dudley Buxton, L.H., Elliot Smith, G., Bate, D.M.A., Spiller, R.C., Hinton, M.A.C., Fischer, P. 1928. Excavation of a Mousterian rock-shelter at Devil's Tower, Gibraltar. *Journal of the Royal Anthropological Institute of Great Britain and Ireland* 58: 33–113.

Garrod, D. and Bate, D.M.A. 1937. *The Stone Age of Mount Carmel: Excavations at Wady El-Mughara.* Vol. 1. Oxford Univeristy Press, Oxford.

Garrod, D.A.E. 1956. Acheuléo-Jabrudien et Pré-Aurignacien de la Grotte de Taboun (Mont Carmel): étude stratigraphique et chronologique. *Quaternaria* 3: 39–59.

Garvey, J. 2007. Surviving an Ice Age: the zooarchaeological record from southwestern Tasmania. *Palaios* 22: 583–5.

Gaudzinski, S. 2004. Subsistence patterns of Early Pleistocene hominids in the Levant-taphonomic evidence from the 'Ubeidiya Formation (Israel). *Journal of Archaeological Science* 31: 65–75.

Gebauer, H.D. and Price, L. 1995. Gua Tempurung: a tunnel cave in the marble hills (fengcong type karst) of Kinta Valley, Perak, West Malaysia. *Journal of the Malay Branch of the Royal Asiatic Society* 68: 29–52.

Gell, A. 1995. The language of the forest: landscape and phonological iconism in Umeba. Hirsch, E. and O'Hanlon, M. (eds.) *The Anthropology of Landscape: Perspectives on Place and Space.* Clarendon Press, Oxford, pp. 232–54.

G.E.S. (Anonymous). 1934. Davidson Black. 1884–1934. *Obituary Notices of Fellows of the Royal Society* 1: 360–5.

Geyh, M.A., Kudrass, H.R., Streif, H. 1979. Sea-level changes during the late Pleistocene and Holocene in the Strait of Malacca. *Nature* 278: 441–3.

Giaccio, B., Isaia, R., Fedele, F.G., Di Canzio, E., Hoffecker, J., Ronchitelli, A., Sinitsyn, A.A., Anikovich, M., Lisitsyn, S.N., Popov, V.V. 2008. The Campanian Ignimbrite and Codola tephra layers: two temporal/stratigraphic markers for the Early Upper Palaeolithic in southern Italy and eastern Europe. *Journal of Volcanology and Geothermal Research* 177: 208–26.

Gibbard, P.L. 2003. Definition of the Middle–Upper Pleistocene boundary. *Global and Planetary Change* 36: 201–8.

Gibbard, P. and van Kolfschoten, T. 2005. The Pleistocene and Holocene epochs. Gradstein, F.M., Ogg, J.G., Smith, A.G. (eds.) *A Geologic Time Scale*. Cambridge University Press, Cambridge, pp. 441–52.

Gibbard, P. and Head, M.J. 2008. Formal request to ICS that the base of the Quaternary System/Period be lowered to the Gelasian Stage GSSP (at 2.588 Ma), and that the base of the Pleistocene Series be lowered to the same position. Letter submitted to the International Commission on Stratigraphy, Subcommission on Quaternary Stratigraphy, http://www.stratigraphy.org/forum/viewtopic.php?f=6&t=21.

Gibbard, P.L., Head, M.J., Walker, M.J. 2010. Submission on Quaternary Stratigraphy. Formal ratification of the Quaternary System/Period and the Pleistocene Series/Epoch with a base at 2.58 Ma. *Journal of Quaternary Science* 25: 96–102.

Gibbard, P.L., Smith, A.G., Zalasiewicz, J.A., Barry, T.L., Cantrill, D., Coe, A.L., Cope, J.C.W., *et al*. 2005. What status for the Quaternary? *Boreas* 34: 1–6.

Gillespie, R. and Brook, B.W. 2006. Is there a Pleistocene archaeological site at Cuddie Springs? *Archaeology in Oceania* 41: 1–11.

Glass, B.P. and Koeberl, C. 2006. Australasian microtektites and associated impact ejecta in the South China Sea and the Middle Pleistocene supereruption of Toba. *Meteoritics and Planetary Science* 41: 305–26.

Glover, I.C. 1976. Ulu Leang Cave, Maros: a preliminary sequence of post-Pleistocene cultural development in South Sulawesi. *Archipel* 11: 113–54.

Glover, I.C. 1977. The Late Stone Age in eastern Indonesia. *World Archaeology* 9: 42–61.

Glover, I.C. 1981. Leang Burung 2: an Upper Palaeolithic rock shelter in south Sulawesi, Indonesia. *Modern Quaternary Research in Southeast Asia* 6: 1–38.

Glover, I.C. 1986. *Archaeology in Eastern Timor, 1966–67*. Terra Austalis 11. Australian National University, Canberra.

Glover, I.C. and Presland, G. 1985. Microliths in Indonesian flaked stone industries. Misra, V.N. and Bellwood, P. (eds.) *Recent Advances in Indo-Pacific Prehistory: Proceedings of the International Symposium, Poona, Dec. 19–21st 1978*. E.J. Brill, Leiden, pp. 185–95.

Goebel, T. 2002. The "microblade adaptation" and recolonization of Siberia during the Late Upper Pleistocene. Elston, R.G. and Kuhn S.L. (eds.) *Thinking Small: Global Perspectives on Microlithization*. Archeological Papers of the American Anthropological Association Number 12, pp. 117–31.

Goebel, T. 2004. The Early Upper Palaeolithic of Siberia. Brantingham, P.J., Kuhn, S.L. and Kerry, K.W. (eds.) *The Early Upper Palaeolithic Beyond Western Europe*. University of California Press, London, pp. 162–95.

Goldberg, P., Weiner, S., Bar-Yosef, O., Xu, Q., Liu, J. 2001. Site formation processes at Zhoukoudian, China. *Journal of Human Evolution* 41: 483–530.

Goldstein, D.B., Linares, A.R., Cavalli-Sforza, L.L., Feldman, M.W. 1995. Genetic absolute dating based on microsatellites and the origin of modern humans. *Proceedings of the National Academy of Sciences of the United States of America* 92: 6723–7.

Gonzalez, C. and Dupont, L.M. 2009. Tropical salt marsh succession as sea-level indicator during Heinrich events. *Quaternary Science Reviews* 28: 939–46.

Gorecki, P., Grant, M., O'Connor, S., Veth, P. 1997. The morphology, function and antiquity of Australian grinding implements. *Archaeology in Oceania* 32: 141–50.

Goren-Inbar, N., Feibel, C.S., Verosub, K.L., Melamed, Y., Kislev, M.E., Tchernov, E., Saragusti, I. 2000. Pleistocene milestones on the Out-of-Africa Corridor at Gesher Benot Ya'aqov, Israel. *Science* 289: 944–7.

Goren-Inbar, N., Werker, E., Feibel, C.S. 2002. *The Acheulian Site of Gesher Benot Ya'aqov, Israel: The Wood Assemblage*. Oxbow Books, Oxford.

Goren-Inbar, N., Alperson, N., Kislev, M.E., Simchoni, O., Melamed, Y., Ben-Nun, A., Werker, E. 2004. Evidence of hominin control of fire at Gesher Benot Ya'aqov, Israel. *Science* 304: 725–27.

Gorman, C.F. 1969. Hoabinhian: a pebble-tool complex with early plant associations in Southeast Asia. *Science* 163: 671–3.

Gorman, C.F. 1970. Excavations at Spirit Cave, northern Thailand. *Asian Perspectives* 13: 79–107.

Gorman, C.F. 1971. The Hoabinhian and after: subsistence patterns in Southeast Asia during the Late Pleistocene and early Recent periods. *World Archaeology* 2: 300–20.

Gorman, C.F. 1977. A priori models and Thai prehistory: a reconsideration of the beginnings of agriculture in Southeast Asia. Reed, C.A. (ed.) *Origins of Agriculture*. Mouton, The Hague, pp. 321–55.

Grassman, L., Jr. 1999. Ecology and behavior of the Indochinese leopard in Kaeng Krachan National Park, Thailand. *Natural History Bulletin of the Siam Society* 47: 77–93.

Gravina, B., Mellars, P., Bronk Ramsey, C. 2005. Radiocarbon dating of interstratified Neanderthal and early modern human occupations at the Chatelperronian type-site. *Nature* 438: 51–56.

Grayson, D.K. and Delpech, F. 2002. Specialized early Upper Palaeolithic hunters in southwestern France? *Journal of Archaeological Science* 29: 1439–49.

Grayson, D.K. and Meltzer, D.J. 2003. A requiem for North American overkill. *Journal of Archaeological Science* 30: 585–93.

Green, R.E., Krause, J., Ptak, S.E., Briggs, A.W., Ronan, M.T., Simons, J.F., Lei, D., *et al.* 2006. Analysis of one million base pairs of Neanderthal DNA. *Nature* 444: 330–36.

Green, R.E., Malaspinas, A.-S., Krause, J., Briggs, A.W., Johnson, P.L.F., Uhler, C., Meyer, M., *et al.* 2008. A complete Neandertal mitochondrial genome sequence determined by high-throughput sequencing. *Cell* 134: 416–26.

Green, R.E., Krause, J., Briggs, A.W., Maricic, T., Stenzel, U., Kircher, M., Patterson, N., *et al.* 2010. A draft sequence of the Neandertal genome. *Science* 328: 710–22.

Griffin, P.B. 1985. A contemporary view of the shift from hunting to horticulture: the Agta case. Misra, V.N. and Bellwood, P. (eds.) *Recent Advances in Indo-Pacific Prehistory: Proceedings of the International Symposium held at Poona December 19th–21st*, E.J. Brill, Leiden, pp. 349–52.

Griffin, P.B. 1989. Hunting, farming and sedentism in a rainforest foraging society. Kent, S. (ed.) *Farmers as Hunters: The Implications of Our Sedentism*. Cambridge University Press, Cambridge, pp. 70–80.

Griffin, P.B. 1997. Technology and variation in arrow design among the Agta of northeastern Luzon. Knecht, H. (ed.) *Projectile Technology*. Plenum Press, New York, pp. 267–86.

Griffin, P.B. and Griffin, M.B. 2000. Agta hunting and sustainability of resource use in Northeastern Luzon, Philippines. Robinson, J.G. and Bennett, E.L. (eds.) *Hunting for Sustainability in Tropical Forests*. New York, Columbia, pp. 325–35.

Griffin, P.B. and Solheim, W.G., II. 1990. Ethnoarchaeological research in Asia. *Asian Perspectives* 28: 145–62.

GRIP Members. 1993. Climate instability during the last interglacial period recorded in the GRIP ice core. *Nature* 364: 204–7.

Grootes, P.M., Stuiver, M., White, J.W.C., Johnsen, S., Jouzel, J. 1993. Comparison of oxygen isotope records from the GISP2 and GRIP Greenland ice cores. *Nature* 366: 552–54.

Grote, A.R. 1877. On the peopling of America. *American Naturalist* 11: 221–26.

Groube, L., Chappel, J., Muke, J., Price, D. 1986. A 40,000 year-old human occupation site at Huon Peninsula, Papua New Guinea. *Nature* 324: 453–55.

Groube, L. 1989. The taming of the rainforests: a model for Late Pleistocene forest exploitation in New Guinea. Harris, D. and Hillman, G. (eds.) *Foraging and Farming: The Evolution of Plant Exploitation*. Unwin Hyman, London, pp. 292–304.

Groves, C.P. and Grubb, P. 1987. Relationships of living deer. Wemmer, C.M. (ed.) *Biology and Management of the Cervidae*. Smithsonian Institution Press, Washington, DC, pp. 21–59.

Grün, R. and Stringer, C. 2000. Tabun revisited: revised ESR chronology and new ESR and U-series analyses of dental material from Tabun C1. *Journal of Human Evolution* 39: 601–12.

Grün, R., Stringer, C., McDermott, F., Nathan, R., Porat, N., Robertson, S., Taylor, L., Mortimer, G., Eggins, S., McCulloch, M. 2005. U-series and ESR analyses of bones and teeth relating to the human burials from Skhul. *Journal of Human Evolution* 49: 316–34.

Grün, R., Eggins, S., Aubert, M., Spooner, N., Pike, A.W.G., Müller, W. 2010. ESR and U-series analyses of faunal material from Cuddie Springs, NSW, Australia: implications for the timing of the extinction of the Australian megafauna. *Quaternary Science Reviews* 29: 596–610.

Guillaud, D. (ed.). 2006. *Menyelusuri Sungai, Merunut Waktu: Penelitian Arkeologi di Sumatera Selatan*. IRD Editions, Jakarta.

Guillou, H., Singer, B.S., Laj, C., Kissel, C., Scaillet, S., Jicha, B.R. 2004. On the age of the Laschamp geomagnetic excursion. *Earth and Planetary Science Letters* 227: 331–43.

Gunn, A. 2003. Voles, lemmings and caribou – population cycles revisited? *Rangifer* 14: 105–11.

Guthrie, R.D. 1983. Osseous projectile points: biological considerations affecting raw material selection and design among Paleolithic and Paleoindian peoples. Clutton-Brock, J. and Grigson, C. (eds.) *Animals and Archaeology 1: Hunters and Their Prey*. British Archaeological Reports, International Series 163, Oxford, pp. 273–94.

Guthrie, D. 1989. Woolly arguments against the Mammoth Steppe – a new look at the palynological data. *Review of Archaeology* 10: 16–34.

Guthrie, D., and van Kolfschoten, T. 2000. Neither warm and moist, nor cold and arid: the ecology of the Mid Upper Palaeolithic. Roebroeks, W., Mussi, M., Svoboda, J., Fennema, K. (eds.) *Hunters of the Golden Age: The Mid Upper Palaeolithic of Eurasia 30,000–20,000 BP*. Analecta Praehistorica Leidensia 31. University of Leiden Press, Leiden, pp. 13–30.

Habgood, P.J. and Franklin, N.R. 2008. The revolution that didn't arrive: a review of Pleistocene Sahul. *Journal of Human Evolution* 55: 187–222.

Haidle, M.N. and Pawlik, A.F. 2010. Pleistocene modernity: an exclusively Afro-European issue: an introduction to Session A1. *Bulletin of the Indo-Pacific Prehistory Association* 30: 3–8.

Haile, N.S. 1971. Quaternary shorelines in West Malaysia and adjacent parts of the Sunda Shelf. *Quaternaria* 15: 333–43.

Hameau, S., Falguères, C., Bahain, J.J., Sémah, F., Sémah, A.M., Dolo, J.M. 2007. ESR dating in Song Terus cave (East Java, Indonesia). *Quaternary Geochronology* 2: 398–402.

Hames, R.B. 1979. A comparison of the efficiencies of the shotgun and the bow in neotropical forest hunting. *Human Ecology* 7: 219–52.

Hamid Mohd Isa. 2007. *Bengkel alat batu zaman Pleistosen pertengahan (70,000 tahan dahulu) di Kota Tampan, Lenggong, Perak*. Unpublished master's thesis, Pusat Penyelidikan Arkeologi Malaysia, Universisi Sains Malaysia (in Malay).

Ha Huu Nga. 1998. Relationships between the Soi Nhu and Ha Long cultures. Paper presented at the 16th Congress of the Indo-Pacific Prehistory Association, 1–7 July, Melaka.

Hagemoen, R.I.M. and Reimers, E. 2002. Reindeer summer activity pattern in relation to weather and insect harassment. *Journal of Animal Ecology* 71: 883–92.

Haines, P.W., Howard, K.T., Alib, J.R., Burretta, C.F., Sangad Bunopas. 2004. Flood deposits penecontemporaneous with ~0.8 Ma tektite fall in NE Thailand: impact-induced environmental effects? *Earth and Planetary Science Letters* 225: 19–28.

Han, D. and Xu, C. 1985. Pleistocene mammalian faunas of China. Wu, R. and Olsen, R.W. (eds.) *Palaeoanthropology and Palaeolithic Archaeology in the People's Republic of China*. Academic Press, Orlando, pp. 267–89.

Hanebuth, T., Stattegger, K., Grootes, P.M. 2000. Rapid flooding of the Sunda Shelf: a late-glacial sea-level record. *Science* 288: 1033–35.

Hanebuth, T.J.J., Saito, Y., Tanabe, S., Quang Lan Vu, Quang Toan Ngo. 2006. Sea levels during late marine isotope stage 3 (or older?) reported from the Red River delta (northern Vietnam) and adjacent regions. *Quaternary International* 145–6: 119–34.

Hansen, J.M. 1991. *Excavations at Franchthi Cave: Palaeoethnobotany of Franchthi Cave*. Fascicle 7. Indiana University Press, Bloomington.

Harland, J.F., Barrett, J.H., Carrott, J., Dobney, K., Jaques, D. 2003. The York System: an integrated zooarchaeological database for research and teaching. *Internet Archaeology* 13, http://intarch.ac.uk/journal/issue13/harland_index.html.

Harpending, H.C. and Relethford, J.H. 1997. Population perspectives on human origins research. Clark, G.A. and Willermet, C.M. (eds.) *Conceptual Issues in Modern Human Origins Research.* Aldine de Gruyter, New York, pp. 361–68.

Harpending, H.C., Sherry, S.T., Rogers, A.R., Stoneking, M. 1993. The genetic structure of ancient human populations. *Current Anthropology* 34: 483–96.

Harris, N. 2006. The elevation history of the Tibetan Plateau and its implications for the Asian monsoon. *Palaeogeography, Palaeoclimatology, Palaeoecology* 241: 4–15.

Harrisson, T. 1957. The great cave of Niah: a preliminary report on Bornean prehistory. *Man* 57: 161–66.

Harrisson, T. 1966. Lobang Angus, a frequentation cave at Niah – I. *Sarawak Museum Journal* 28-29: 217–23.

Harrisson, T. and Medway, Lord. 1962. A first classification of prehistoric bone and tooth artifacts (based on material from Niah Great Cave). *Sarawak Museum Journal* 19-20: 335–62.

Haslam, M., Korisettar, R., Petraglia, M., Smith, T., Shipton, C., Ditchfield, R. 2010. In Foote's steps: the history, significance and recent archaeological investigation of the Billa Surgam Caves in southern India. *South Asian Studies* 26: 1–19.

Ha Van Tan. 1976. The Hoabınhian in the context of Viet Nam. *Vietnamese Studies* 46: 127–97.

Ha Van Tan. 1997. The Hoabinhian and before. *Bulletin of the Indo-Pacific Prehistory Association* 16: 35–41.

Ha Van Tan. 1999. Prehistoric archaeological research in Vietnam today, paper presented at the International Colloquium on Archaeology in Southeast Asia in the 3rd Millennium, Penang, Malaysia.

Hays, J.D., Imbrie, J., Shackleton, N.J. 1976. Variations in the Earth's orbit: pacemaker of the Ice Ages. *Science* 194: 1121–32.

Hayward, M.W., Henschell, P., O'Brien, J., Hofmeyr, M., Balme, G., Kerley, G.I.H. 2006. Prey preferences of the leopard *Panthera pardus*. *Journal of Zoology, London* 270: 298–313.

Headland, T.N. 1987. The wild yam question: how well could independent hunter-gatherers live in a tropical rain forest ecosystem? *Human Ecology* 15: 463–91.

Heaney, L.R. 1991. A synopsis of climate and vegetational change in Southeast Asia. *Climate Change* 19: 53–61.

Heine-Geldern, R. 1936. Prehistoric research in Indonesia. Repr. in *Annual Bibliography of Indian Archaeology* IX: 26–38.

Heine-Geldern, R. 1945. *Prehistoric Research in the Netherlands Indies.* Southeast Asia Institute, New York.

Heinrich, H. 1988. Origin and consequences of cyclic ice rafting in the northeast Atlantic Ocean during the past 130,000 years. *Quaternary Research* 29: 142–52.

Heinrich, J. 2004. Demography and cultural evolution: how adaptive cultural processes can produce maladaptive losses – the Tasmanian case. *American Antiquity* 69: 197–214.

Henshilwood, C.S. 2007. Fully symbolic *sapiens* behaviour: innovations in the Middle Stone Age at Blombos Cave, South Africa. Mellars, P., Stringer, S., Bar-Yosef, O., Boyle, K. (eds.) *Rethinking the Human Revolution: New Behavioural and Biological Perspectives on the Origins and Dispersal of Modern Humans.* McDonald Institute Monographs, Cambridge, pp. 123–32.

Henshilwood, C. and Marean, C. 2003. The origin of modern human behaviour. *Current Anthropology* 44: 627–51.

Henshilwood, C.S., d'Errico, F., Yates, R., Jacobs, Z., Tribolo, C., Duller, G.A.T., Mercier, N., *et al.* 2002. Emergence of modern human behavior: Middle Stone Age engravings from South Africa. *Science* 295: 1278–80.

Henshilwood, C. and Sealy, J. 1997. Bone artefacts from the Middle Stone Age at Blombos Cave, Southern Cape, South Africa. *Current Anthropology* 38: 890–95.

Henshilwood, C.S., Sealy, J.C., Yates, R., Cruz-Uribe, K., Goldberg, P., Grine, F.E., Klein, R.G., Poggenpoel, C., van Niekerk, K., Watts, I. 2001. Blombos Cave, Southern Cape, South Africa: preliminary report on the 1992–1999 excavations of the Middle Stone Age levels. *Journal of Archaeological Science* 28: 421–48.

Hertler, C. and Volmer, R. 2008. Assessing prey competition in fossil carnivore communities – a scenario for prey competition and its evolutionary consequences for tigers in Pleistocene Java. *Palaeogeography, Palaeoclimatology, Palaeoecology* 257: 67–80.

Hetherington, R. and Reid, R.G.B. 2010. *The Climate Connection: Climate Change and Modern Human Evolution.* Cambridge University Press, Cambridge.

Hewitt, G. 2003. Ice ages, species distributions and evolution. Rothschild, L.J. and Lister, A.M. (eds.) *Evolution on Planet Earth: The Impact of the Physical Environment.* Academic Press, London, pp. 339–61.

Hienz, R.D., Brady, J.V., Hursh, S.R., Banner, M.J., Gasior, E.D., Spence, K.R. 2007. Communication constraints, indexical countermeasures, and crew configuration effects in simulated space-dwelling groups. *Acta Astronautica* 60: 362–78.

Higham, C.F.W. 1993. The bone, antler and turtle carapace technology. Higham, C.F.W. and Thosarat, R. (eds.) *The Excavation of Khok Phanom Di: A Prehistoric Site in Central Thailand: Vol. III: The Material Culture Part 1.* Society of Antiquaries, Thames and Hudson, London, pp. 1–44.

Higham, C.F.W. and Maloney, B. 1989. Coastal adaptation, sedentism, and domestication: a model for socio-economic intensification in prehistoric Southeast Asia. Harris, D.R. and Hillman, G.C. (eds.) *Foraging and Farming: The Evolution of Plant Exploitation.* Unwin Hyman, London, pp. 650–66.

Higham, C.F.W. and Thosarat, R. 1998. *Prehistoric Thailand.* Thames and Hudson, London.

Higham, C.F.W. and Thosarat, R. 2004a. *The Excavation of Khok Phanom Di: Volume VII: Summary and Conclusions.* Society of Antiquaries of London, London.

Higham, C.F.W. and Thosarat, R. (eds.) 2004b. *The Excavation of Ban Lum Khao: Vol. 1. The Origins of the Civilization of Angkor Project.* Fine Arts Department, Bangkok.

Higham, C.F.W., Fordyce, R.E., O'Reilly, D.J.W. 1998. The faunal remains and worked bone. Higham, C.W.F. and Thosarat, R. (eds.) *The Excavation of Nong Nor: A Prehistoric Site in Central Thailand.* Oxbow Books, Oxford.

Higham, T., Bronk Ramsey, C., Karavanić, I., Smith, F.H., Trinkaus, E. 2006. Revised direct radiocarbon dating of the Vindija G1 Upper Paleolithic Neandertals. *Proceedings of the National Academy of Sciences of the United States of America* 103: 553–57.

Hill, C., Soares, P., Mormina, M., Macaulay, V., Meehan, W., Blackburn, J., Clarke, D., *et al.* 2006. Phylogeography and ethnogenesis of Aboriginal Southeast Asians. *Molecular Biology Evolution* 23: 2480–91.

Hill, C., Soares, P., Mormina, M., Macaulay, V., Clarke, D., Blumbach, P.B., Vizuete-Forster, M., *et al.* 2007. A mitochondrial stratigraphy for Island Southeast Asia. *American Journal of Human Genetics* 80: 29–43.

Hiscock, P. 2006. Artefacts on Aru: evaluating the technological sequences. O'Connor, S., Spriggs, M. and Veth, P. (eds.) *The Archaeology of the Aru Islands, Eastern Indonesia.* Terra Australis 22. Australian National University, Canberra, pp. 205–34.

Hoáng Xuân Chinh. 1991. Faunal and cultural changes from Pleistocene to Holocene in Vietnam. *Bulletin of the Indo-Pacific Prehistory Association* 10: 74–8.

Hoffman, P.F. and Schrag, D.P. 2002. The snowball Earth hypothesis: testing the limits of global change. *Terra Nova* 14: 129–55.

Hoffman, P.F. and Schrag, D.P. 2000. The Snowball Earth. *Scientific American* 282: 68–75.

Holdaway, S. and Porch, N. 1995. Cyclical patterns in the Pleistocene occupation of Southwest Tasmania. *Archaeology in Oceania* 30: 74–82.

Hooijer, D.A. 1962. Report upon a collection of Pleistocene mammals from tin-bearing deposits in a limestone cave near Ipoh, Kinta Valley, Perak. *Federation Museums Journal* VII: 1–5.

Hooijer, D.A. 1963. Further Hell mammals from Niah. *Sarawak Museum Journal* 11: 196–200.

Hope, G., Kershaw, A.P., van der Kaars, S., Sun, X., Liew, P-M., Heusser, L.E., Takahara, H., *et al.* 2004. History of vegetation and habitat change in the Austral-Asian region. *Quaternary International* 118–119: 103–126.

Hori, K., Tanabe, S., Saito, Y., Haruyama, S., Viet Nguyen, Kitamura, A. 2005. Delta initiation and Holocene sea-level change: example from the Song Hong (Red River) delta, Vietnam. *Sedimentary Geology* 164: 237–49.

Hou, Y., Potts, R., Yuan, B., Guo, Z., Deino, A., Wang, W., Clark, J., Xie, G., Huang, W. 2000. Mid-Pleistocene Acheulean-like stone technology of the Bose Basin, south China. *Science* 287: 1622–26.

Housley, R.A., Gamble, C.S., Street, M., Pettitt, P. 1997. Radiocarbon evidence for the Late glacial human recolonisation of Northern Europe. *Proceedings of the Prehistoric Society* 63: 25–54.

Huang, C.C., Pang, J., Zhao, J. 2000. Chinese loess and the evolution of the east Asian monsoon. *Progress in Physical Geography* 24: 75–96.

Huang, W., Ciochon, R., Gu, Y., Larick, R., Fang, Q., Schwarcz, H., Yonge, C., de Vos, J., Ring, W. 1995. Early *Homo* associated with artefacts from Asia. *Nature* 378: 275–78.

Huber, C., Leuenberger, M., Spahni, R., Flückiger, J., Schwander, J., Stocker, T.F., Johnsen, S., Landais, A., Jouzel, J. 2006. Isotope calibrated Greenland temperature record over Marine Isotope Stage 3 and its relation to CH_4. *Earth and Planetary Science Letters* 243: 504–19.

Hudjashov, G., Kivisild, T., Underhill, P.A., Endicott, P., Sanchez, J.J., Lin, A.A., Sheng, P., *et al.* 2007. Revealing the prehistoric settlement of Australia by Y chromosome and mtDNA analysis. *Proceedings of the National Academy of Sciences of the United States of America* 104: 8726–30.

Huffman, O.F., Zaim, Y., Kappelman, J., Ruez, D.R., Jr., de Vos, J., Rizal, Y., Aziz, F., Hertler, C. 2006. Relocation of the 1936 Mojokerto skull discovery site near Perning, East Java. *Journal of Human Evolution* 50: 431–51.

Hughen, K., Lehman, S., Southon, J., Overpeck, J., Marchal, O., Herring, C., Turnbull, J. 2004. 14C activity and global carbon cycle changes over the past 50,000 years. *Science* 303: 202–7.

Hunt, C.O. and Rushworth, G. 2005. Cultivation and human impact at 6000 cal yr B.P. in tropical lowland 4 forest at Niah, Sarawak, Malaysian Borneo. *Quaternary Research* 64: 460–68.

Hunt, C.O., Gilbertson, D.D., Rushworth, G. 2007. Modern humans in Sarawak, Malaysian Borneo, during Oxygen Isotope Stage 3: palaeoenvironmental evidence from the Great Cave of Niah. *Journal of Archaeological Science* 34: 1953–69.

Hunt, C.O., Gilbertson, D.D., Rushworth, G. 2012. A 50,000-year record of late Pleistocene tropical vegetation and human impact in lowland Borneo. *Quaternary Science Reviews* 37: 61–80.

Huntley, B. 1991. How plants respond to climate change: migration rates, individualism and the consequences for plant communities. *Annals of Botany* 67(Suppl. 1): 15–22.

Huntley, B. and Webb, T., III. 1989. Migration: species' response to climatic variations caused by changes in the Earth's orbit. *Journal of Biogeography* 16: 5–19.

Hutterer, K.L. 1976. An evolutionary approach to the Southeast Asian cultural sequence. *Current Anthropology* 17: 221–42.

Hyodo, M., Nakaya, H., Urabe, A., Saegusa, H., Xue, S., Yin, J., Ji, X. 2002 Paleomagnetic dates of hominid remains from Yuanmou, China, and other Asian sites. *Journal of Human Evolution* 43: 27–41.

Ingman, M. and Gyllensten, U. 2003. Mitochondrial genome variation and evolutionary history of Australian and New Guinean aborigines. *Genome Research* 13: 1600–6.

Ingold, T. 1993. The temporality of landscape. *World Archaeology* 25: 152–74.

Institution of Mechanical Engineers. 2009. *Climate Change: Adapting to the Inevitable.* Adaptation Report. 13 March. London.

Intergovernmental Panel on Climate Change. 2007. *Climate Change: The Physical Science Basis (Summary for Policy-makers). Contribution of Working Group I to the Fourth Assessment Report of the IPCC.* http://www.ipcc.ch/.

Jacob, T. 1973. Palaeoanthropological discoveries in Indonesia with special reference to the finds of the last two decades. *Journal of Human Evolution* 2: 473–85.

Jacob, T., Indriati, E., Soejono, R.P., Hsü, K., Frayer, D.W., Eckhardt, R.B., Kuperavage, A.J., *et al.* 2006. Pygmoid Australomelanesian *Homo sapiens* skeletal remains from Liang Bua, Flores: population affinities and pathological abnormalities. *Proceedings of the National Academy of Sciences of the United States of America* 103: 1421–6.

Jacobi, R.M., Higham, T.F.G., Bronk Ramsey, C. 2006. AMS radiocarbon dating of Middle and Upper Palaeolithic bone in the British Isles: improved reliability using ultrafiltration. *Journal of Quaternary Science* 21: 557–73.

Jacobs, Z., Duller, G.A.T., Wintle, A.G., Henshilwood, C.S. 2006. Extending the chronology of deposits at Blombos Cave, South Africa, back to 140 ka using optical dating of single and multiple grains of quartz. *Journal of Human Evolution* 51: 255–73.

Jacobs, Z., Roberts, R.G., Galbraith, R.F., Deacon, H.J., Grün, R., Mackay, A., Mitchell, P., *et al.* 2008. Ages for the Middle Stone Age of Southern Africa: implications for human behavior and dispersal. *Science* 322: 733–5.

James, H.V.A., Petraglia, M.D. 2005. Modern human origins and the evolution of behavior in the later Pleistocene record of South Asia. *Current Anthropology* 46: S3–S27.

Janis, C. 2003. Tectonics, climate change, and the evolution of mammalian ecosystems. Rothschild, L.J., Lister, A.M. (eds.) *Evolution on Planet Earth: The Impact of the Physical Environment*. Academic Press, London, pp. 319–37.

Jeffrey Abdullah. 2005. Human teeth of the Palaeolithic period from Gua Balamangan, Sabah. Zuraina Majid (ed.) *The Perak Man and Other Prehistoric Skeletons of Malaysia*. Penerbit Universiti Sains Malaysia, Pinang, pp. 229–37.

Jiménez-Espejo, F.J., Martínez-Ruiz, F., Finlayson, C., Paytan, A., Sakamoto, T., Ortega-Huertas, M., Finlayson, G., *et al.* 2007. Climate forcing and Neanderthal extinction in Southern Iberia: insights from a multiproxy marine record. *Quaternary Science Reviews* 26: 836–52.

Jochim, M.A. 1983. Paleolithic art in ecological perspective. Bailey, G. (ed.) *Hunter-Gatherer Economy in Prehistory*, Cambridge University Press, Cambridge, pp. 212–19.

Johns, T. and Kubo, I. 1988. A survey of traditional methods employed for the detoxification of plant foods. *Journal of Ethnobiology* 8: 81–129.

Johnsen, S.J., Dansgaard, W., Clausen, H.B., Langway, C.C., Jr. 1972. Oxygen isotope profiles through the Antarctic and Greenland ice-sheets. *Nature* 235: 429–34.

Johnsen, S.J., Clausen, H.B., Dansgaard, W., Fuhrer, K., Gundestrup, N., Hammer, C.U., Iversen, P., *et al.* 1992. Irregular glacial interstadial recorded in a new Greenland ice core. *Nature* 359: 312–13.

Johnsen, S.J., Dahl-Jensen, D., Gundestrup, N., Steffensen, J.P, Clausen, H.B., Miller, H., Masson-Delmotte, V., *et al.* 2001. Oxygen isotope and palaeotemperature records from six Greenland ice-core stations: Camp Century, Dye-3, GRIP, GISP2, Renland and NorthGRIP. *Journal of Quaternary Science* 16: 299–307.

Johnson, B.J., Miller, G.H., Fogel, M.L., Magee, J.W., Gagan, M.K., Chivas, A.R. 1999. 65,000 years of vegetation change in Central Australia and the Australian summer monsoon. *Science* 284: 1150–52.

Johnson, E. 1985. Current developments in bone technology. *Advances in Archaeological Method and Theory* 8: 157–235.

Joly, K., Jandt, R.R., Meyers, C.R., Cole, M.J. 2007. Changes in vegetative cover on Western Arctic Herd winter range from 1981 to 2005: potential effects of grazing and climate change. *Rangifer* 17: 199–207.

Jones, R. 1989. East of Wallace's Line: issues and problems in the colonization of the Australian continent. Mellars, P. and Stringer, C. (eds.) *The Human Revolution*. Edinburgh University Press, Edinburgh, pp. 743–82.

Jones, S.C. and Pal, J.N. 2009. The Palaeolithic of the Middle Son valley, north-central India: changes in hominin lithic technology and behaviour during the Upper Pleistocene. *Journal of Anthropological Archaeology* 28: 323–41.

Jourdan, C. 2001. Creolization: sociocultural aspects. Smelser, N.J. and Baltes, P.B. (eds.) *International Encyclopedia of the Social and Behavioral Sciences*. Elsevier Science, Oxford, pp. 2903–6.

Jouzel, J., Masson-Delmotte, V., Cattani, O., Dreyfus, G., Falourd, S., Hoffmann, G., Minster, B., *et al.* 2007. Orbital and millennial Antarctic climate variability over the past 800,000 years. *Science* 317: 793–96.

Jungers, W.L., Larson, S.G., Harcourt-Smith, W., Morwood, M.J., Sutikna, T., Rokhus Awe Due, Djubiantono, T. 2009. Descriptions of the lower limb skeleton of *Homo floresiensis*. *Journal of Human Evolution* 57: 538–54.

Jungers, W.L., Harcourt-Smith, W.E.H., Wunderlich, R.E., Tocheri, M.W., Larson, S.G., Sutikna, T., Rokhus Awe Due, Morwood, M.J. 2009. The foot of *Homo floresiensis*. *Nature* 459: 81–4.

Kamminga, J. 2007. Reinterpretation of the Hoabinhian at Sai Yok, Thailand. *Archaeological Heritage of Malaysia* 1: 40–60.

Kar, A., Singhvi, A.K., Rajaguru, S.N., Juyal, N., Thomas, J.V., Banerjee, D., Dhir, R.P. 2001. Reconstruction of the late Quaternary environment of the lower Luni plains, Thar Desert, India. *Journal of Quaternary Science* 16: 61–68.

Kaifu, Y., Aziz, F., Indriati, E., Jacob, T., Kurniawan, I., Baba, H. 2008. Cranial morphology of Javanese *Homo erectus*: new evidence for continuous evolution, specialization, and terminal extinction. *Journal of Human Evolution* 55: 551–80.

Karkanas, P., Koumouzelis, M., Kozlowski, J.K., Sitlivy, V., Sobczyk, K., Berna, F., Weiner, S. 2004. The earliest evidence for clay hearths: Aurignacian features in Klisoura Cave 1, southern Greece. *Antiquity* 301: 513–25.

Keates, S. 2002. The Movius Line: fact or fiction? *Bulletin of the Indo-Pacific Prehistory Association* 22: 17–24.

Keates, S.G. 2010. Evidence for the earliest Pleistocene hominid activity in the Nihewan Basin of northern China. *Quaternary International* 223–4: 408–17.

Keivany, Y. and Nelson, J.S. 2000. Taxonomic review of the genus *pungitius*, Ninespine sticklebacks (Gasterosteidae). *Cybium* 24: 107–22.

Kellman, M. and Tackaberry, R. 1997. *Tropical Environments: The Functioning and Management of Tropical Ecosystems*. Routledge, London.

Kelly, R.L. 1995. *The Foraging Spectrum: Diversity in Hunter-Gatherer Lifeways*. Smithsonian Institution Press, Washington, DC.

Kelly, R.L. 2003. Colonization of new land by hunter-gatherers: expectations and implications based on ethnographic data. Rockman, M. and Steele, J. (eds.) *Colonization of Unfamiliar Landscapes: The Archaeology of Adaptation*. Routledge, London, pp. 44–58.

Kelly, R.L, and Todd, L.C. 1988. Coming into the country: early Paleoindian hunting and mobility. *American Antiquity* 53: 231–44.

Kennedy, K.A.R. 1979. The deep skull of Niah: an assessment of twenty years of speculation concerning its evolutionary significance. *Asian Perspectives* 20: 32–50.

Kennedy, K.A.R. and Deraniyagala, S.U. 1989. Fossil remains of 28,000-year old hominids from Sri Lanka. *Current Anthropology* 30: 394–9.

Kennedy, K.A.R. and Zahorsky, J.L. 1997. Trends in prehistoric technology and biological adaptations: new evidence from Pleistocene deposits at Fa Hien cave, Sri Lanka. Allchin, R. and Allchin B. (eds.) *South Asian Archaeology 1995*, vol. 2, Oxford-IBH, New Delhi, pp. 839–53.

Kennedy, K.A.R., Deraniyagala, S.U., Roertgen, W.J., Chiment, J., Disotell, T. 1987. Upper Pleistocene fossil hominids from Sri Lanka. *American Journal of Physical Anthropology* 72: 441–61.

Kienast, M., Steinke, S., Stattegger, K., Calvert, S.E. 2001. Synchronous Tropical South China Sea SST change and Greenland warming during deglaciation. *Science* 291: 2132–34.

Kienast, M., Hanebuth, T.J.J., Pelejero, C., Steinke, S. 2003. Synchroneity of meltwater pulse 1a and the Bølling warming: new evidence from the South China Sea. *Geology* 31: 67–70.

Kijngam, A. 1991. The remains of fish, crabs and turtles. Higham, C.F.W. and Bannanurag, R. (eds.) *The Excavation of Khok Phanom Di: Vol. 2. The Biological Report (Part 1)*. Society of Antiquaries of London, pp. 223–30.

Ki-Kydd, K., Piper, P. 2004. Identification of morphological variation in the humeri of Bornean primates and its application to zooarchaeology. *Archaeofauna* 13: 85–95.

Kimura, Y. 2002. Examining time trends in the Oldowan technology at Beds I and II, Olduvai Gorge. *Journal of Human Evolution* 43: 291–321.

King, V. 1995. Tropical rainforests and indigenous peoples: symbiosis and exploitation. *Sarawak Museum Journal* 69: 1–22.

Kirch, P.V. 1980. The archaeological study of adaptation: theoretical and methodological issues. *Advances in Archaeological Method and Theory* 3: 101–56.

Kissel, C., Laj, C., Labeyrie, L., Dokken, T., Voelker, A., Blamart, D. 1999. Rapid climatic variations during marine isotopic stage 3: magnetic analysis of sediments from Nordic Seas and North Atlantic. *Earth and Planetary Science Letters* 171: 489–502.

Kitchener, A.C., Beaumont, M.A., Richardson, D. 2006. Geographical variation in the clouded leopard, *Neofelis nebulosa*, reveals two species. *Current Biology* 16: 2377–83.

Kivisild, T., Rootsi, S., Metspalu, M., Mastana, S., Kaldma, K., Parik, J., Metspalu, E., *et al.* 2003. The genetic heritage of the earliest settlers persists both in Indian tribal and caste populations. *American Journal of Human Genetics* 72: 313–32.

Kivisild, T. 2007. Complete MtDNA sequences – quest on 'Out-of-Africa' route completed? Mellars, P., Stringer, S., Bar-Yosef, O., Boyle, K. (eds.) *Rethinking the Human Revolution: New Behavioural and*

Biological Perspectives on the Origins and Dispersal of Modern Humans. McDonald Institute Monographs, Cambridge, pp. 21–32.

Klein, R.G. 1999. *The Human Career: Human Biological and Cultural Origins*. 2nd ed. University of Chicago Press, Chicago.

Koeberl, C., Glass, B.P., Keates, S.G. (Comment); Potts, R., Huang, W., Hou, Y., Deino, A., Yuan, B., Guo, Z., Clark, J. (Response). 2000. Tektites and the age paradox in Mid-Pleistocene China. *Science* 289: 507a.

Kohl, G. and Quitta, H. 1978. Paleolithic and Hoabinhian. *Radiocarbon* 20: 388.

Kolb, C.C. 1980. More on the Central Asian Paleolithic. *Current Anthropology* 21: 130–1.

Kotlia, B.S. and Joshi, M. 2008. Reconstruction of Late Pleistocene palaeoecology of the Upper Narmada valley (Central India) using fossil communities. *Palaeoworld* 17: 153–9.

Koumouzelis, M., Ginter, B., Kozlowski, J.K., Pawlikowski, M., Bar-Yosef, O., Albert, R.M., Litynska-Zajac, M., *et al*. 2001. The early Upper Palaeolithic in Greece: the excavations in Klisoura Cave. *Journal of Archaeological Science* 28: 515–39.

Kramer, A., Djubiantono, T., Aziz, F., Bogard, J.S., Weeks, R.A., Weinand, D.C., Hames, W.E., *et al*. 2005. The first hominid fossil recovered from West Java, Indonesia. *Journal of Human Evolution* 48: 661–67.

Kramer, A. 1993. Human taxonomic diversity in the Pleistocene: does *Homo erectus* represent multiple hominid species? *American Journal of Physical Anthropology* 91: 161–71.

Krause, J., Orlando, L., Serre, D., Viola, B., Prüfer, K., Richards, M.P., Hublin, J.-J., Hänni, C., Derevianko, A.P., Pääbo, S. 2007. Neanderthals in Central Asia and Siberia. *Nature* 449: 902–4.

Krause, J., Fu, Q., Good, J.M., Viola, B., Shunkov, M.V., Derevianko, A.P., Pääbo, S. 2010. The complete mitochondrial DNA genome of an unknown hominin from southern Siberia. *Nature* 464: 894–97.

Krigbaum, J. 2001. *Human paleodiet in tropical Southeast Asia: isotopic evidence from Niah Cave and Gua Cha*. Unpublished PhD thesis, Department of Anthropology, New York University.

Krigbaum, J. and Datan, I. 1999. The Deep Skull of Niah: Borneo's first people. *Borneo* 5: 12–17.

Krigbaum, J. and Datan, I. 2005. The deep skull and associated human remains from Niah Cave. Zuraina Majid (ed.) *The Perak Man and Other Prehistoric Skeletons of Malaysia*. Universiti Sains Malaysia Pusat Penyelidikan Arkeologi Malaysia, Penang, pp. 131–54.

Kuhn, S. and Stiner, M. 1998. The earliest Aurignacian of Riparo Mochi (Liguria, Italy). *Current Anthropology* 39(2, Suppl.): S175–89.

Kunz, J., Bollinger, K., Jessberger, E.K., Storzer, D. 1995. Ages of Australasian tektites (abstract). *Lunar and Planetary Science*, 26th: 809–10.

Kvavadze, E., Bar-Yosef, O., Belfer-Cohen, A., Boaretto, E., Jakeli, N., Matskevich, Z., Meshveliani, T. 2009. 30,000-year-old wild flax fibers. *Science* 325: 1359.

Labeyrie, L., Vidal, L., Cortijo, E., Paterne, M., Arnold, M., Duplessy, J.C., Vautravers, M., *et al*. 1995. Surface and deep hydrology of the Northern Atlantic Ocean during the past 150,000 years. *Philosophical Transactions of the Royal Society: Biological Sciences* 348: 255–64.

Lahr, M., Foley, R. 1994. Multiple dispersals and modern human origins. *Evolutionary Anthropology* 3: 48–60.

Lahr, M.M. and Foley, R.A. 1998. Towards a theory of modern human origins: geography, demography, and diversity in recent human evolution. *Yearbook of Physical Anthropology* 41: 137–76.

Laland, K.N. and Brown, G.R. 2006. Niche construction, human behavior, and the adaptive-lag hypothesis. *Evolutionary Anthropology* 15: 95–104.

Lamb, A. 1959. Malaya: prehistory, fieldwork. *Asian Perspectives* 3: 32–35.

Lambeck, K. and Chappell, J. 2001. Sea level change through the last glacial cycle. *Science* 292: 679–86.

Lambeck, K., Esat, T.M., Potter, E.-K. 2002a. Links between climate and sea levels for the past three million years. *Nature* 419: 199–206.

Lambeck, K., Yokoyama, Y., Purcell, T. 2002b. Into and out of the Last Glacial Maximum: sea-level change during Oxygen Isotope Stages 3 and 2. *Quaternary Science Reviews* 21: 343–60.

Lampert, C.D., Glover, I.C., Hedges, R.E.M., Heron, C.P., Higham, T.F.G., Stern, B., Shoocongdej, R., Thompson, G.B. 2003. Dating resin coating on pottery: the Spirit Cave early ceramic dates revised. *Antiquity* 77: 126–33.

Lampert, C.D., Glover, I.C., Hedges, R.E.M., Heron, C.P., Higham, T.F.G., Stern, B., Shoocongdej, S., Thompson, G.B. 2004. Response to J.C. White. *Antiquity* 78: 186–87.

Lampert, R.J. 1966. An excavation at Durras North, New South Wales. *Archaeology and Physical Anthropology in Oceania* 1: 83–118.

Langbroek, M. and Roebroeks, W. 2000. Extraterrestrial evidence on the age of the hominids from Java. *Journal of Human Evolution* 38: 595–600.

Langley, M.C., Clarkson, C., Ulm, S. 2011. From small holes to grand narratives: the impact of taphonomy and sample size on the modernity debate in Australia and New Guinea. *Journal of Human Evolution* 61: 197–208.

Langub, J. 1988. The Penan strategy. Denslow, J.S. and Padoch, C. (eds.) *People of the Tropical Rainforest*. University of California Press, Berkeley, pp. 207–8.

Langway, C.C., Jr. 2008. The history of early polar ice cores. *Cold Regions Science and Technology* 52: 101–17.

Larick, R., Ciochon, R.L., Yahdi Zaim, Sudijono, Suminto, Rizal, Y., Aziz, F., Reagan, M., Heizler, M. 2001. Early Pleistocene ^{40}Ar/^{39}Ar ages for Bapang Formation hominins, Central Jawa, Indonesia. *Proceedings of the National Academy of Sciences of the United States of America* 98: 4866–71.

Larson, S.G., Jungers, W.L., Morwood, M.J., Sutikna, T., Jatmiko, Wahyu Saptomo, E., Rokus Awe Due, Djubiantono, T. 2007. *Homo floresiensis* and the evolution of the hominin shoulder. *Journal of Human Evolution* 53: 718–31.

Larson, S.G., Jungers, W.L., Tocheri, M.W., Orr, C.M., Morwood, M.J., Sutikna, T., Rokhus Due Awe, Djubiantono, T. 2009. Descriptions of the upper limb skeleton of *Homo floresiensis*. *Journal of Human Evolution* 57: 555–70.

Latinis, D.K. 2000. The development of subsistence systems models for island Southeast Asia band Near Oceania: the nature and role of arboriculture and arboreal-based economics. *World Archaeology* 32: 41–67.

Launius, R.D. and McCurdy, H.E. 2007. Robots and humans in space flight: technology, evolution and interplanetary travel. *Technology in Society* 29: 271–81.

Leavesley, M.G., Bird, M.I., Fifield, L.K., Hausladen, P.A., Santos, G.M., di Tada, M.L. 2002. Buang Merabak: early evidence for human occupation in the Bismarck Archipelago, Papua New Guinea. *Australian Archaeology* 54: 55–56.

Lee, H. 2009. Seminar presentation to the Palaeolithic-Mesolithic Discussion Group, Department of Archaeology, University of Cambridge.

Lee, M.-Y., Chen, C.-H., Wei, K.-Y., Iizuka, Y., Carey, S. 2004. First Toba supereruption revival. *Geology* 32: 61–4.

Lee, S-H., Wolpoff, M.H. 2003. The pattern of evolution in Pleistocene human brain size. *Paleobiology* 29: 186–96.

Leighton, M. and Leighton, D.R. 1983. Vertebrate responses to fruiting seasonality within a Bornean rain forest. Sutton, S.L., Whitmore, T.C., Chadwick, A.C. (eds.) *Tropical Rainforest: Ecology and Management*. Blackwell, Oxford, pp. 181–96.

Lekagul, B. and McNeely, J.A. 1988. *Mammals of Thailand*. Association for the Conservation of Wildlife, Bangkok.

Leong, S.H. 1999. Archaeology in Malaysia since the mid-seventies. Paper presented at the International Colloquium on Archaeology in Southeast Asia in the 3rd Millennium, Penang, Malaysia.

Leuschner, D.C. and Sirocko, F. 2000. The low-latitude monsoon climate during Dansgaard-Oeschger cycles and Heinrich events. *Quaternary Science Reviews* 19: 243–54.

Lewis, H., Johnson, K., Ronquillo, W. 2007–8. Preliminary results of speleothem dating from Tabon Cave, Palawan, Philippines: moisture increase at the Last Glacial Maximum. *Hukay* 12: 35–50.

Lewis, H., Paz, V., Lara, M., Barton, H., Piper, P., Ochoa, J., Vitales, T., *et al.* 2008. Terminal Pleistocene to mid-Holocene occupation and an early cremation burial at Ille Cave, Palawan, Philippines. *Antiquity* 82: 318–35.

Lewis-Williams, D. 2002. *The Mind in the Cave*. Thames and Hudson, London.

Li, Y., Yang, J., Xia, Z., Mo, D. 1998. Tectonic geomorphology in the Shanxi Graben System, northern China. *Geomorphology* 23: 77–89.

Liat, L.B., Das, I. 1999. *Turtles of Borneo and Peninsula Malaysia*. Natural History Publications (Borneo), Kota Kinabalu.

Lieberman, D.E. 2005. Further fossils from Flores. *Nature* 437: 957–58.

Lieberman, D.E. 2009. *Homo floresiensis* from head to toe. *Nature* 459: 41–42.

Lieberman, D.E. and Shea, J.J. 1994. Behavioral differences between archaic and modern humans in the Levantine Mousterian. *American Anthropologist* 96: 300–32.

Liu, T. and Ding, Z. 1998. Chinese loess and the paleomonsoon. *Annual Review of Earth Planetary Science* 26: 111–45.

Liu, Z.C. 1985. Sequence of sediments at Locality 1 in Zhoukoudian and correlation with loess stratigraphy in northern China and with the chronology of deep-sea cores. *Quaternary Research* 23: 139–53.

Liu, W., Jin, C.-Z., Zhang, Y.-Q., Cai, Y.-J., Xing, S., Wu, X.-J., Cheng, H., *et al.* 2010. Human remains from Zhirendong, South China, and modern human emergence in East Asia. *Proceedings of the National Academy of Sciences of the United States of America* 107: 19201–206.

López-González, F., Grandal-d'Anglade, A., Vidal-Romaní, J.R. 2006. Deciphering bone depositional sequences in caves through the study of manganese coatings. *Journal of Archaeological Science* 33: 707–17.

Lordkipanidze, D., Jashashvili, T., Vekua, A., Ponce de León, M.S., Zollikofer, C.P.E., Rightmire, G.P., Pontzer, H., *et al.* 2007. Postcranial evidence from early *Homo* from Dmanisi, Georgia. *Nature* 449: 305–10 (and Suppl.).

Loutre, M.F. and Berger, A. 2003. Marine Isotope Stage 11 as an analogue for the present interglacial. *Global and Planetary Change* 36: 209–17.

Louys, J., Curnoe, D., Tong, H. 2007. Characteristics of Pleistocene megafauna extinctions in Southeast Asia. *Palaeogeography, Palaeoclimatology, Palaeoecology* 243: 152–73.

Lowe, J.J., Rasmussen, S.O., Björck, S., Hoek, W.Z., Steffensen, J.P., Walker, M.J.C., Yu, Z.C. and, INTIMATE Group. 2008. Synchronisation of palaeoenvironmental events in the North Atlantic region during the Last Termination: a revised protocol recommended by the INTIMATE group. *Quaternary Science Reviews* 27: 6–17.

Loy, T.H., Spriggs, M., Wickler, S. 1992. Direct evidence for human use of plants 28,000 years ago: starch residues on stone artefacts from the northern Solomon Islands. *Antiquity* 66: 898–912.

Lu, H., van Huissteden, K., An, Z., Nugteren, G., Vandenberghe, J. 1999. East Asia winter monsoon variations on a millennial time-scale before the last glacial–interglacial cycle. *Journal of Quaternary Science* 14: 101–10.

Lu, H., van Huissteden, K., Zhou, J., Vandenberghe, J., Liu, X., An, Z. 2000. Variability of East Asian winter monsoon in Quaternary climatic extremes in north China. *Quaternary Research* 54: 321–7.

Lubell, D. 2004a. Prehistoric edible land snails in the circum-Mediterranean: the archaeological evidence. Brugal, J.-P. and Desse, J. (eds.) *Petits Animaux et Sociétés Humaines*. CÉPAM, CNRS, Sophia Antipolis, France, pp. 77–98.

Lubell, D. 2004b. Are land snails a signature for the Mesolithic–Neolithic transition in the circum-Mediterranean? Budja, M. (ed.) *The Neolithization of Eurasia – Paradigms, Models and Concepts Involved, Documenta Praehistorica* XXVII, pp. 1–24.

Lupo, K.D. 1994. Butchery marks and carcass acquisition strategies: distinguishing hunting from scavenging in archaeological contexts. *Journal of Archaeological Science* 21: 827–37.

Lüthi, D., Le Floch, M., Bereiter, B., Blunier, T., Barnola, J-M., Siegenthaler, U., Raynaud, D., *et al.* 2008. High-resolution carbon dioxide concentration record 650,000–800,000 years before present. *Nature* 453: 379–82.

Lyman, R.L. 1994. *Vertebrate Taphonomy: Cambridge Manuals in Archaeology*. Cambridge University Press, Cambridge.

Lyman, R.L. 2008. *Quantitative Paleozoology*. Cambridge University Press, Cambridge.

Macaulay, V., Hill, C., Achilli, A., Rengo, C., Clarke, D., Meehan, W., Blackburn, J., *et al.* 2005. Single, rapid coastal settlement of Asia revealed by analysis of complete mitochondrial genomes. *Science* 308: 1034–36 (and Suppl.).

MacDonald, D.H. 1998. Subsistence, sex, and cultural transmission in Folsom culture. *Journal of Anthropological Archaeology* 17: 217–39.

Macintosh, N.W.G. 1978. The Tabon cave mandible (with appendix: dental features of the Tabon mandible, by B.C.W. Barker). *Archaeology and Physical Anthropology in Oceania* 13: 143–66.

MacKinnon, K., Hatta, G., Halim, H., Mangalik, A. 1997. *The Ecology of Kalimantan: The Ecology of Indonesia Series*. Vol. III. Oxford University Press, Oxford.

Maeda, Y., Siringan, F., Omura, A., Berdin, R., Hosono, Y., Atsumi, S., Nakamura, T. 2004. Higher-than-present Holocene mean sea levels in Ilocos, Palawan and Samar, Philippines. *Quaternary International* 115–16: 15–26.

Magee, J. 1988. Oriented soil samples from the Tingkayu Basin. Bellwood, P. (ed.) *Archaeological Research in South-eastern Sabah*. Sabah Museum Monograph 2, Kota Kinabalu, pp. 31–37.

Mallol, C. 2006. What's in a beach? Soil micromorphology of sediments from the Lower Palaeolithic site of 'Ubeidiya, Israel. *Journal of Human Evolution* 51: 185–206.

Maloney, B.K. and McCormac, F.G. 1995. A 30,000-year pollen and radiocarbon record from highland Sumatra as evidence for climate change. *Radiocarbon* 37: 181–90.

Mandryk, C.A.S., Josenhans, H., Fedje, D.W., Mathewes, R.W. 2001. Late Quaternary paleoenvironments of northwestern North America: implications for inland versus coastal migration routes. *Quaternary Science Reviews* 20: 301–14.

Markova, A.K., Puzachenko, A.Y., van Kolfschoten, T. 2010. The North Eurasian mammal assemblages during the end of MIS 3 (Brianskian–Late Karginian–Denekamp Interstadial). *Quaternary International* 212: 149–58.

Marean, C.W. 2007. Heading north: an Africanist perspective on the replacement of Neanderthals by modern humans. Mellars, P., Stringer, S., Bar-Yosef, O., Boyle, K. (eds.) *Rethinking the Human Revolution: New Behavioural and Biological Perspectives on the Origins and Dispersal of Modern Humans*. McDonald Institute Monographs, Cambridge, pp. 367–79.

Marín Arroyo, A.B., Landete Ruiz, M.D., Vidal Bernabeu, G., Seva Román, R., González Morales, M.R., Straus, L.G. 2008. Archaeological implications of human-derived manganese coatings: a study of blackened bones in El Mirón Cave, Cantabrian Spain. *Journal of Archaeological Science* 35: 801–13.

Marks, A. 1990. The Middle and Upper Palaeolithic of the Near-East and the Nile Valley: the problem of cultural transformations. Mellars, P. (ed.) *The Emergence of Modern Humans: An Archaeological Perspective*. Cornell University Press, Ithaca, pp. 56–80.

Marliac, A. and Simanjuntak, T. 1996. Preliminary report on the site of Song Gentong Kabupaten Tulungagung, East Java (Indonesia). Communication to the 6th International Congress of European Association of Southeast Asian Archaeologists, International Institute for Asian Studies, Leiden, 2–6 September.

Marlow, L. 2007. The hominid dispersal into Early and Middle Pleistocene Europe: an approach from biogeography. Unpublished doctoral thesis, Department of Archaeology, University of Cambridge.

Márquez, S., Mowbray, K., Sawyer, G.J., Jacob, T., Silvers, A. 2001. New fossil hominid Calvaria from Indonesia – Sambungmacan 3. *The Anatomical Record* 262: 344–68.

Martin, P.S. 1973. The discovery of America. *Science* 179: 969–74.

Martínez-Navarro, B., Turq, A., Agustí Ballester, J., Oms, O. 1997. Fuente Nueva-3 (Orce, Granada, Spain) and the first human occupation of Europe. *Journal of Human Evolution* 33: 611–20.

Marwick, B. 2009. Biogeography of Middle Pleistocene hominins in mainland Southeast Asia: a review of current evidence. *Quaternary International* 202: 51–58.

Marwick, B. 2008. What attributes are important for the measurement of assemblage reduction intensity? Results from an experimental stone artefact assemblage with relevance to the Hoabinhian of mainland Southeast Asia. *Journal of Archaeological Science* 35: 1189–1200.

Matsumura, H., and Pookajorn, S. 2005. A morphometric analysis of the Late Pleistocene human skeleton from the Moh Khiew Cave in Thailand. *Journal of Comparative Human Biology* 56: 93–118.

Matsumura, H. and Zuraina Majid. 1999. Metric analysis of an Early Holocene human skeleton from Gua Gunung Runtuh, Malaysia. *American Journal of Physical Anthropology* 109: 327–40.

Matthews, J.M. 1966. A review of the 'Hoabinhian in Indo-China'. *Asian Perspectives* 9: 86–95.

Mayewski, P.A., Meeker, L.D., Whitlow, S., Twickler, M.S., Morrison, M.C., Alley, R.B., Bloomfield, P., Taylor, K. 1993. The atmosphere during the Younger Dryas. *Science* 261: 195–97.

Mayr, E. 1947. Ecological factors in speciation. *Evolution* 1: 263–88.

Mazák, J.H. and Groves, C.P. 2006. A taxonomic revision of the tigers (*Panthera tigris*) of Southeast Asia. *Mammalian Biology* 71: 268–87.

McBrearty, S. 2007. Down with the Revolution. Mellars, P., Stringer, S., Bar-Yosef, O., Boyle, K. (eds.) *Rethinking the Human Revolution: New Behavioural and Biological Perspectives on the Origins and Dispersal of Modern Humans*. McDonald Institute Monographs, Cambridge, pp. 133–51.

McBrearty, S. and Brooks, A.S. 2000. The revolution that wasn't: a new interpretation of the origins of modern human behaviour. *Journal of Human Evolution* 39: 453–63.

McCarthy, F.D. 1940. Comparison of the prehistory of Australia with that of Indo-China, the Malay Peninsula and the Netherlands East Indies. Chasen, F.N. and Tweedie, M.W.F. (eds.) *Proceedings of the Third Congress of Prehistorians of the Far East*, Singapore, 24–30 January, 1938.

McCown, T.D. 1937. Mugharet Es-Skhul. Description and excavations. Garrod, D.A., and Bate, D.M.A. (eds.) *The Stone Age of Mount Carmel. Volume I: Excavations at the Wady El-Mughara*. Clarendon Press, Oxford, pp. 91–107.

McDermott, F., Grün, R., Stringer, C.B., Hawkesworth, C.J. 1993. Mass-spectrometric U-series dates for Israeli Neanderthal/early modern hominid sites. *Nature* 363: 252–55.

McDougall, I., Brown, F.H., Fleagle, J.G. 2005. Stratigraphic placement and age of modern humans from Kibish, Ethiopia. *Nature* 433: 733–36.

Medway, Lord. 1966. Animal remains from Lobang Angus, Niah. *Sarawak Museum Journal* 28-29: 185–216.

Meehan, B. 1982. *Shell Bed to Shell Midden*. Australian Institute of Aboriginal Studies, Canberra.

Meese, D.A., Gow, A.J., Alley, R.B., Zielinski, Z.A., Grootes, P.M., Ram, M., Taylor, K.C., Mayewski, P.A., Bolzan, J.F. 1997. The Greenland Ice Sheet Project 2 depth-age scale: methods and results. *Journal of Geophysical Research* 102: 26411–23.

Médail, F. and Diadema, K. 2009. Glacial refugia influence plant diversity patterns in the Mediterranean Basin. *Journal of Biogeography* 36: 1333–45.

Meijaard, E. 2004. Biogeographic history of the Javan leopard *Panthera pardus* based on a craniometric analysis. *Journal of Mammalogy* 85: 302–10.

Meijaard, E., Sheil, D., Nasi, R., Augeri, D., Rosenbaum, B., Iskandar, D., Titiek Setyawati, *et al.* 2005. *Life after Logging: Reconciling Wildlife Conservation and Production Forestry in Indonesian Borneo*. CIFOR and UNESCO, Jakarta.

Mellars, P. 1973. The character of the Middle-Upper Palaeolithic transition in southwest France. Renfrew, C. (ed.) *The Explanation of Culture Change*. Duckworth, London, pp. 255–76.

Mellars, P. 1989. Technological changes and the Middle-Upper Palaeolithic transition: economics, social and cognitive perspectives. Mellars, P. and Stringer, C. (eds.) *The Human Revolution*. Edinburgh University Press, Edinburgh, pp. 338–65.

Mellars, P. 1990. *The Emergence of Modern Humans: An Archaeological Perspective*. Cornell University Press, Ithaca, NY.

Mellars, P. 1993. Archaeology and the population-dispersal hypothesis of modern human origins in Europe. Aitkin, M.J., Stringer, C.B., Mellars, P.A. (eds.) *The Origins of Modern Humans and the Impact of Chronometric Dating*. Princeton University Press, Princeton, NJ, pp. 196–213.

Mellars, P. 1994. The Upper Palaeolithic revolution. Cunliffe, B. (ed) The Oxford Illustrated Prehistory of Europe. Oxford University Press, Oxford, pp. 42–78.

Mellars, P. 1996. *The Neanderthal Legacy: An Archaeological Perspective from Western Europe*. Princeton University Press, Princeton, NJ.

Mellars, P. 2004a. Neanderthals and the modern human colonization of Europe. *Nature* 432: 461–65.

Mellars, P.A. 2004b. Reindeer specialization in the early Upper Palaeolithic: the evidence from south west France. *Journal of Archaeological Science* 31: 613–17.

Mellars, P. 2005. The impossible coincidence. A single-species model of the origins of modern human behaviour in Europe. *Evolutionary Anthropology* 14: 12–27.

Mellars, P. 2006a. Going East: new genetic and archaeological perspectives on the modern human colonization of Eurasia. *Science* 313: 796–800.

Mellars, P. 2006b. The ecological basis of Upper Palaeolithic cave art. *Zona Arqueológica* 7: 2–11.

Mellars, P. 2006c. Why did modern human populations disperse from Africa *ca.* 60,000 years ago? A new model. *Proceedings of the National Academy of Sciences of the United States of America* 103: 9381–86.

Mellars, P. and Stringer, C. (eds.). 1989. *The Human Revolution: Behavioural and Biological Perspectives in the Origins of Modern Humans*. Edinburgh University Press, Edinburgh.

Mellars, P., Stringer, S., Bar-Yosef, O., Boyle, K. (eds.). 2007a. *Rethinking the Human Revolution: New Behavioural and Biological Perspectives on the Origins and Dispersal of Modern Humans*. McDonald Institute Monographs, Cambridge.

Mellars, P., Gravina, B., Bronk Ramsey, C. 2007b. Confirmation of Neanderthal/modern human interstratifications at the Chatelperron type-site. *Proceedings of the National Academy of Sciences of the United States of America* 104: 3657–62.

Mercader, J. 2003. *Under the Canopy: The Archaeology of Tropical Rain Forests*. Rutgers University Press, New Brunswick, NJ.

Mercier, N., Valladas, H., Bar-Yosef, O., Vandermeersch, B., Stringer, C., Joron, J.-L. 1993. The thermoluminescence date for the Mousterian burial of Es-Skhul, Mt. Carmel. *Journal of Archaeological Science* 20: 169–74.

Merriwether, D.A., Hodgson, J.A., Friedlaender, F.R., Allaby, R., Cerchio, S., Koki, G., Friedlaender, J.S. 2005. Ancient mitochondrial M haplogroups identified in the Southwest Pacific. *Proceedings of the National Academy of Sciences of the United States of America* 102: 13034–39.

Micheels, A. and Montenari, M. 2008. A snowball Earth versus a slushball Earth: results from Neoproterozoic climate modelling sensitivity experiments. *Geosphere* 4: 401–10.

Mijares, A.S.B. 2005. The archaeology of Peñablanca Caves sites, Northern Luzon, Philippines. *Journal of Austonesian Studies* 1: 65–93.

Mijares, A.S.B. 2008. The Late Pleistocene to Early Holocene foragers of Northern Luzon. *Bulletin of the Indo-Pacific Prehistory Association* 28. 99 107.

Mijares, A.S., Détroit, F., Piper, P., Grün, R., Bellwood, P., Aubert, M., Champion, G., Cuevas, N., DeLeon, A., Dizon, E. 2010. New evidence for a 67,000-year-old human presence at Callao Cave, Luzon, Philippines. *Journal of Human Evolution* 59: 123–32.

Millard, A.R. and Pike, A.W.G. 1999. Uranium-series dating of the Tabun Neanderthal: a cautionary note. *Journal of Human Evolution* 36: 581–85.

Miller, J.M.A. 2000. Craniofacial variation in *Homo habilis*: an analysis of the evidence for multiple species. *American Journal of Physical Anthroplogy* 112: 103–28.

Minc, L.D. 1986. Scarcity and survival: the role of oral tradition in mediating subsistence crises. *Journal of Anthropological Archaeology* 5: 39–113.

Miracle, P.T. 1998. The spread of modernity in Palaeolithic Europe. Omoto, K. and Tobias, P.V. (eds.) *The Origins and Past of Modern Humans – Towards Reconciliation*. World Scientific, Singapore, pp. 171–83.

Misra, V.N. 2001. Prehistoric human colonization of India. *Journal of Biosciences* 26: 491–531.

Mithen, S. 1996. *The Prehistory of the Mind*. Thames and Hudson, London.

Mocellin, J.S.P. and Suedfeld, P. 1991. Voices from the ice: diaries of polar explorers. *Environment and Behavior* 23: 704–22.

Mohd Mokhtar Saidin. 2000. Bukit Jawa, Lenggong, a Middle Pleistocene site in Malaysia? Paper presented at the International Symposium on Geology, Bandung, Indonesia.

Mohd Mokhtar Saidin. 2006a. Bukit Bunuh, Lenggong, Malaysia: new evidence of Late Pleistocene culture in Malaysia and Southeast Asia. Bacus, E.A., Glover, I.C., Piggott, V.C. (eds.) *Uncovering Southeast Asia's Past*. National University Press, Singapore, pp. 60–64.

Mohd Mokhtar Saidin. 2006b. Bukit Bunuh, Lenggong, Perak: its contribution to the Palaeolithic culture. Abd. Aziz Tajuddin and Ong, L. K. (eds.) *Fundamental Research at USM 2002–2005* 3: 3–11.

Mohd Mokhtar Saidin. 2007a. The application of fission-track (FT) and optically stimulated luminescence (OSL) dating to Bukit Bunuh, Lenggong, Perak. *Archaeological Heritage of Malaysia* 1: 73–80.

Mohd Mokhtar Saidin. 2007b. Palaeolithic chronology in Malaysia. Paper presented at Human Origins Patrimony Studies in Southeast Asia (HOPSEA), 10–12 December, Paris.

Molengraaff, G.A.F. and Weber, M. 1921. On the relation between the pleistocene glacial period and the origin of the Sunda sea (Java and South China-sea), and its influence on the distribution of coral reefs and on the land- and freshwater fauna. *Netherlands Academy of Arts and Sciences Proceedings* 23: 395–439.

Molodkov, A., Bolikhovskaya, N. 2009. Climate change dynamics in Northern Eurasia over the last 200 ka: evidence from mollusc-based ESR-chronostratigraphy and vegetation successions of the loess palaeosol records. *Quaternary International* 201: 67–76.

Mona, S., Grunz, K.E., Brauer, S., Pakendorf, B., Castrí, L., Sudoyo, H., Marzuki, S., *et al.* 2009. Genetic admixture history of Eastern Indonesia as revealed by Y-chromosome and mitochondrial DNA analysis. *Molecular Biology and Evolution* 6: 1865–77.

Moncel, M.-H. 2010. Oldest human expansions in Eurasia: favouring and limiting factors. *Quaternary International* 223–4: 1–9.

Monecke, K., Finger, W., Klarer, D., Kongko, W., McAdoo, B.G., Moore, A.L., Sudrajat, S.U. 2005. A 1,000-year sediment record of tsunami recurrence in northern Sumatra. *Nature* 455: 1232–34.

Mook, W.G. 1981. Radiocarbon dating of freshwater shells from Leang Burung 2: part 2. *Modern Quaternary Research in Southeast Asia* 6: 53–54.

Moore, M.W. and Brumm, A. 2007. Stone artifacts and hominins in island Southeast Asia: new insights from Flores, eastern Indonesia. *Journal of Human Evolution* 52: 85–102.

Moore, M.W., Sutikna, T., Jatmiko, Morwood, M.J., Brumm, A. 2009. Continuities in stone flaking technology at Liang Bua, Flores, Indonesia. *Journal of Human Evolution* 57: 503–26.

Morrogh-Bernard, H., Husson, S., Page, S.E., Rieley, J.O. 2003. Population status of the Bornean orangutan (*Pongo pygmaeus*) in Sebangau peat samp forest, Central Kalimantan, Indonesia. *Biological Conservation* 110: 141–52.

Morwood, M.J. and Hobbs, D.R. 1995. Themes in the prehistory of tropical Australia. *Antiquity* 69: 747–68.

Morwood, M.J. and Jungers, W.L. 2009. Conclusions: implications of the Liang Bua excavations for hominin evolution and biogeography. *Journal of Human Evolution* 57: 640–8.

Morwood, M.J., O'Sullivan, P.B., Aziz, F., Raza, A. 1998. Fission-track ages of stone tools and fossils on the east Indonesian island of Flores. *Nature* 392: 173–9.

Morwood, M., O'Sullivan, P., Susanto, E.E., Aziz, F. 2003. Revised age for Mojokerto 1, an early *Homo erectus* cranium from East Java, Indonesia. *Australian Archaeology* 57: 1–4.

Morwood, M.J., Soejono, R.P., Roberts, R.G., Sutikna, T., Turney, C.S.M., Westaway, K.E., Rink, W.J., *et al.* 2004. Archaeology and age of a new hominin from Flores in eastern Indonesia (and Suppl.). *Nature* 431: 1087–91.

Morwood, M.J., Brown, P., Jatmiko, Sutikna, T., Wahyu Saptomo, E., Westaway, K.E., Rokus Awe Due, *et al.* 2005. Further evidence for small-bodied hominins from the Late Pleistocene of Flores, Indonesia. *Nature* 437: 1012–17.

Morwood, M.J., Sutikna, T., Saptomo, E.W., Westaway, K.E., Jatmiko, Rokus Awe Due, Moore, M.W., *et al.* 2008. Climate, people and faunal succession on Java, Indonesia: evidence from Song Gupuh. *Journal of Archaeological Science* 35: 1776–89.

Mountain, J.L. and Cavalli-Sforza, L.L. 1997. Multilocus genotypes, a tree of individuals, and human evolutionary history. *American Journal of Human Genetics* 61: 705–18.

Mourer, R. 1977. Laang Spean and the prehistory of Cambodia. *Modern Quaternary Research in Southeast Asia* 3: 29–56.

Mourer, C. and Mourer, R. 1970. The prehistoric industry of Laang Spean, province of Battambang, Cambodia. *Archaeology and Phyisical Anthropology in Oceania* 5: 128–46.

Mourer, C. and Mourer, R. 1971. Prehistoric research in Cambodia during the last ten years. *Asian Perspectives* 14: 35–42.

Movius, H.L., Jr. 1938. Stone Age archaeology of Burma. Section in a paper by de Terra, H., Preliminary report on recent geological and archaeological discoveries relating to early man in Southeast Asia. *Proceedings of the National Academy of Sciences of the United States of America* 24: 407–13.

Movius, H.L., Jr. 1944. Early man and Pleistocene stratigraphy in Southern and Eastern Asia. *Papers of the Peabody Museum of American Archaeology and Ethnology* 19: 1–125.

Movius, H.L., Jr. 1948. The Lower Palaeolithic cultures of Southern and Eastern Asia. *Transactions of the American Philosophical Society* 38: 329–420.

Moyle, R.G., Filardib, C.E., Smith, C.E., Diamond, J. 2009. Explosive Pleistocene diversification and hemispheric expansion of a "great speciator". *Proceedings of the National Academy of Sciences of the United States of America* 106: 1863–68.

Mudar, K., Anderson, D. 2007. New evidence for Southeast Asian Pleistocene foraging economies: faunal remains from the early levels of Lang Rongrien rockshelter, Krabi, Thailand. *Asian Perspectives* 46: 298–334.

Mulvaney, J. and Soejono, R.P. 1970. The Australian-Indonesian archaeological expedition to Sulawesi. *Asian Perspectives* 13: 163–77.

Mulvaney, J. and Soejono, R.P. 1971. Archaeology in Sulawesi, Indonesia. *Antiquity* 45: 26–33.

Mulvaney, J. and Kamminga, J. 1999. *Prehistory of Australia.* Smithsonian Institute Press, Washington.

Murphy, K.L., Rygalov, V.Y., Johnson, S.B. 2009. Minimal support technology and in situ resource utilization for risk management of planetary spaceflight missions. *Advances in Space Research* 43: 1275–84.

Myers, T.P., Voorhies, M.R., Corner, R.G. 1980. Spiral fractures and bone pseudotools at paleontological sites. *American Antiquity* 45: 483–90.

Nadel, D., Danin, A., Werker, E., Schick, T., Kislev, E.M., Stewart, K. 1994. 19,000-year-old twisted fibers from Ohalo II. *Current Anthropology* 35: 451–8.

Nadel, D., Carmi, I., Segal, D. 1995. Radiocarbon dating of Ohalo II: archaeological and methodological implications. *Journal of Archaeological Science* 22: 811–22.

Nadel, D., Weiss, E., Simchoni, O., Tsatskin, A., Danin, A., Kislev, M. 2004. Stone Age hut in Israel yields world's oldest evidence of bedding. *Proceedings of the National Academy of Sciences of the United States of America* 101: 6821–6.

Nakagawa, T., Kitagawa, H., Yasuda, Y., Tarasov, P.E., Gotanda, K., Sawai, Y. 2005. Pollen/event stratigraphy of the varved sediment of Lake Suigetsu, central Japan from 15,701 to 10,217 SG vyr BP (Suigetsu varve years before present): description, interpretation, and correlation with other regions. *Quaternary Science Reviews* 24: 1691–1701.

Narbonne, G.M., Gehling, J.G. 2003. Life after snowball: The oldest complex Ediacaran fossils. *Geology* 31: 27–30.

Newbold, T.J. 1844. Notes on the Osseous Breccia and deposit in the caves of Billasurgam, southern India. *Journal of the Asiatic Society of Bengal* 13: 610–11.

Newcomer, M. 1987. Appendix 3: Study and replication of bone tools from Ksar Akil, Republic of Lebanon. Bergman, C. (ed.) *Ksar Akil, Lebanon: A Technological and Typological Analysis of the Later Palaeolithic Levels of Ksar Akil: Vol. 2. Levels XIII–VI.* British Archaeological Reports, International Series S329, London, pp. 284–307.

NGRIP Members. 2004. High-resolution record of northern hemisphere climate extending into the last interglacial period. *Nature* 431: 147–51.

Ng, P.K.L. and Kosuge, T. 1995. On a new *Somanniathelphusa* Bott, 1968, from Vietnam (Crustacea: Decapoda: Brachyura: Parathelphusidae). *Proceedings of the Biological Society of Washington* 108: 61–67.

Nguyen Gia Doi. 2005. Results of recent research into the lithic industries from Late Pleistocene/Early Holocene sites in northern Vietnam. *Bulletin of the Indo-Pacific Prehistory Association* 25: 95–97.

Nguyen Khac Su. 2008. Palaeolithic archaeology in Vietnam, Laos and Cambodia. *Vietnam Archaeology* 3: 7–17.

Nguyen Lan Cuong. 2007. Unique burial practice by ancient cavemen of the Hoa Binh civilisation in Vietnam. *Anthropologischer Anzeiger* 65: 129–35.

Nguyen Ngoc Thach. 2005. *Shells of Vietnam.* Conch Books, Hackenheim.

Nguyêñ Văn Lũ, Nguyêñ Cao Tâń. 2002. Nhuńg phât hun mở̀x vé khaở cồ huc năm.

Nguyen Viet. 2000. Homeland of the Hoabinhan in Vietnam. Paper presented at the Association of Southeast Asian Archaeologists in Europe, Sarteano, Italy.

Nguyen Viet. 2004. Hoabhinhian food strategy in Viet Nam. Paz, V. (ed.) *Southeast Asian Archaeology: Wilhem G. Solheim II Festschrift.* University of the Philippines Press, Manila, pp. 442–62.

Nguyen Viet. 2005. The Da But culture: evidence for cultural development in Vietnam during the middle Holocene. *Bulletin of the Indo-Pacific Prehistory Association* 25: 89–93.

Nguyen Viet. 2008. Hoabinhian macrobotanical remains from archaeological sites in Vietnam: indicators of climate changes from the late Pleistocene to the early Holocene. *Bulletin of the Indo-Pacific Prehistory Association* 28: 80–3.

Noone, H.D. 1936. Report on the settlements and welfare of the Ple-Temiar Senoi of the Perak-Kelantan Watershed. *Journal of Federated Malay States Museums* 19: 24–8.

Norton, C.J. and Braun, D.R. 2010. *Asian Paleoanthropology: From Africa to China and Beyond.* Springer, Dordrecht.

Nosil, P., Vines, T.H., Funk, D.J. 2005. Perspective: reproductive isolation caused by natural selection against immigrants from divergent habitats. *Evolution* 59: 705–19.

Ochoa, J. 2008. *Terrestrial vertebrates from Ille Cave, northern Palawan, Philippines: subsistence and palaeoecology in the Terminal Pleistocene to the Holocene.* Unpublished Master's thesis, University of the Philippines, Manila.

O'Connell, J.F. and Allen, J. 2004. Dating the colonization of Sahul (Pleistocene Australia–New Guinea): a review of recent research. *Journal of Archaeological Science* 31: 835–53.

O'Connell, J.F. and Allen, H.J. 2007. Pre-LGM Sahul (Pleistocene Australia–New Guinea) and the archaeology of early modern humans. Mellars, P., Stringer, S., Bar-Yosef, O., Boyle, K. (eds.) *Rethinking the Human Revolution: New Behavioural and Biological Perspectives on the Origins and Dispersal of Modern Humans.* McDonald Institute Monographs, Cambridge, pp. 395–410.

O'Connor, S. 2006. Unpacking the Island Southeast Asian Neolithic cultural package, and finding local complexity. Bacus, E.A., Glover, I.C., Piggott, V.C. (eds.) *Uncovering Southeast Asia's Past.* National University of Singapore Press, Singapore, pp. 74–87.

O'Connor, S. 2007. New evidence from East Timor contributes to our understanding of earliest modern human colonisation east of the Sunda Shelf. *Antiquity* 81: 523–35.

O'Connor, S., Spriggs, M., Veth, P. 2002. Direct dating of shell beads from Lene Hara Cave, East Timor. *Australian Archaeology* 55: 18–21.

O'Connor, S. and Veth, P. 2005. Early Holocene shell fish hooks from Lene Hara Cave, East Timor establish complex fishing technology was in use in Island South East Asia five thousand years before Austronesian settlement. *Antiquity* 79: 249–56.

O'Connor, S., Aplin, K., Szabó, K., Pasveer, J., Veth, P., Spriggs, M. 2006. Liang Lemdubu: a Pleistocene cave site in the Aru Islands. O'Connor, S., Spriggs, M. and Veth, P. (eds.) *The Archaeology of the Aru Islands, Eastern Indonesia. Terra Australis* 22. Australian National University, Canberra pp. 171–204.

Okladnikov, A.P. 1978. The Palaeolithic of Mongolia. Ikawa-Smith, F. (ed.) *Early Palaeolithic in South and East Asia.* Mouton, The Hague, pp. 317–25.

Okladnikov, A.P. and Pospelova, G.A. 1982. Ulalinka, the oldest Palaeolithic site in Siberia. *Current Anthropology* 23: 710–12.

Olsen, J.W., Ciochon, R.L. 1990. A review of evidence for postulated Middle Pleistocene occupations in Viet Nam. *Journal of Human Evolution* 19: 761–88.

Olsen, S. 1989. On distinguishing natural from cultural damage on archaeological antler. *Journal of Archaeological Science* 16: 125–35.

Olsen, S.L. and Glover, I.C. 2004. The bone industry of Ulu Leang 1 and Leang Burung 1 rockshelters, Sulawesi, Indonesia, in its regional context. *Modern Quaternary Research in Southeast Asia* 18: 273–99.

Olsen, S.L. and Shipman, P. 1988. Surface modification on bone: trampling versus butchery. *Journal of Archaeological Science* 15: 535–53.

Ono, R., Soegondho, S., Yoneda, M. 2010. Changing marine exploitation during Late Pleistocene in northern Wallacea: shell remains from Leang Sarru rockshelter in Talaud Islands. *Asian Perspectives* 48: 318–41.

Oota, H., Pakendorf, B., Weiss, G., von Haeseler, A., Pookajorn, S., Settheetham-Ishida, W., Tiwawech, D., Ishida, T., Stoneking, M. 2005. Recent origin and cultural reversion of a hunter-gatherer group. *PLOS Biology* 3: 536–42.

Oppenheimer, C. 2001. Limited global change due to the largest known Quaternary eruption, Toba ~74kyr BP? *Quaternary Science Reviews* 21: 1593–1609.

Oppenheimer, S. 2004. *Out of Eden: The Peopling of the World.* Constable, London.

Oppenheimer, S. 2009. The great arc of dispersal of modern humans: Africa to Australia. *Quaternary International* 202: 2–13.

Oppenheimer, S. 1998. *Eden in the East: The Drowned Continent of Southeast Asia.* Weidenfeld and Nicolson, London.

Osborne, A.H., Vance, D., Rohling, E.J., Barton, N., Rogerson, M., Fello, N. 2008. A humid corridor across the Sahara for the migration of early modern humans out of Africa 120,000 years ago. *Proceedings of the National Academy of Sciences of the United States of America* 105: 16444–7.

Otte, M. 2007. Arguments for population movement of anatomically modern humans from Central Asia to Europe. Mellars, P., Stringer, S., Bar-Yosef, O., Boyle, K. (eds.) *Rethinking the Human Revolution: New Behavioural and Biological Perspectives on the Origins and Dispersal of Modern Humans.* McDonald Institute Monographs, Cambridge, pp. 359–66.

Otte, M., Biglari, F., Flas, D., Shidrang, S., Zwyns, N., Mashkour, M., Naderi, R., *et al.* 2007. The Aurignacian in the Zagros region: new research at Yafteh Cave, Lorestan, Iran. *Antiquity* 81: 82–96.

Palombo, M.R. and Mussi, M. 2006. Large mammal guilds at the time of the first human colonization of Europe: the case of the Italian Pleistocene record. *Quaternary International* 149: 94–103.

Pan, Y. and Zong, G. 1991. Mammalian faunas. Qian, F., Zhou, G.X. (eds.) *Quaternary Geology and Paleoanthropology of Yuanmou, Yunnan, China.* Science Press, Beijing. Selected trans. by W. Downs, 1993, pp. 94–102.

Panagopoulou, E., Karkanas, P., Tsartsidou, G., Kotjabopoulou, E., Harvati, K., Ntinou, M. 2002–4. Late Pleistocene archaeological and fossil human evidence from Lakonis Cave, southern Greece. *Journal of Field Archaeology* 29: 323–49.

Pappu, S., Gunnell, Y., Taieb, M., Brugal, J.-P., Touchard, Y. 2003. Excavations at the Palaeolithic site of Attirampakkam, South India: preliminary findings. *Current Anthropology* 44: 591–98.

Parfitt, S.A., Barendregt, R.W., Breda, M., Candy, I., Collins, M.J., Coope, G.R., Durbidge, P., *et al.* 2005. The earliest record of human activity in northern Europe. *Nature* 438: 1008–12.

Parfitt, S.A., Ashton, N.M., Lewis, S.G., Abel, R.L., Coope, G.R., Field, M.H., Gale, R., *et al.* 2010. Early Pleistocene human occupation at the edge of the boreal zone in northwest Europe. *Nature* 466: 229–33.

Parker, A.G. 2009. Pleistocene climate change in Arabia: developing a framework for hominin dispersal over the last 350 ka. Petraglia, M.D. and Rose, J.I. (eds.) 2009. *The Evolution of Human Populations in Arabia.* Springer, Dordrecht, pp. 39–49.

Partin, J.W., Cobb, K.M., Adkins, J.F., Clark, B., Fernandez, D.P. 2007. Millennial-scale trends in west Pacific warm pool hydrology since the Last Glacial Maximum. *Nature* 449: 453–55.

Pasqua, M. and Bulbeck, D. 1998. A technological interpretation of the Toalian, South Sulawesi. *Modern Quaternary Research in Southeast Asia* 15: 211–32.

Pasveer, J.M. 1998. Kria Cave: an 8000-year occupation sequence from the Bird's Head of Irian Jaya. *Modern Quaternary Research in Southeast Asia* 15: 67–89.

Pasveer, J.M. 2004. The Djief Hunters: 26,000 years of rainforest exploitation on the Bird's Head of Papua, Indonesia. *Modern Quaternary Research in Southeast Asia* 17.

Pasveer, J.M. 2006. Bone artefacts from Liang Lemdubu and Liang Nabulei Lisa, Aru Islands. O'Connor, S., Spriggs, M. and Veth, P. (eds.) *The Archaeology of the Aru Islands, Eastern Indonesia. Terra Australis* 22. Australian National University, Canberra pp. 235–54.

Pasveer, J.M., Clarke, S.J., Miller, G.H. 2002. Late Pleistocene occupation of inland rainforest, Bird's Head, Papua. *Archaeology in Oceania* 37: 92–95.

Pasveer, J.M. and Bellwood, P. 2004. Prehistoric bone artefacts from the Northern Moluccas. Keates, S. and Pasveer, J. (eds.) *Quaternary Research in Indonesia,* Lisse, A.A. Balkema, pp. 301–59.

Patnaik, R., Chauhan, P.R., Rao, M.R., Blackwell, B.A.B., Skinner, A.R., Ashok Sahni. 2009. New geochronological, paleoclimatological, and archaeological data from the Narmada Valley hominin locality, central India. *Journal of Human Evolution* 56: 114–33.

Patole-Edoumba, E. 2009. A typo-technological definition of Tabonian industries. *Bulletin of the Indo-Pacific Prehistory Association* 29: 21–25.

Pawlik, A.F. 2004. The Palaeolithic site of Arubo 1 in Central Luzon, Philippines. *Bulletin of the Indo-Pacific Prehistory Association* 24: 3–12.

Payne, J. and Francis, C. 1998. *A Field Guide to the Mammals of Borneo.* Sabah Society, Kota Kinabalu.

Peacock, B.A.V. and Dunn, F.L. 1968. Recent archaeological discoveries in Malaysia, 1967 West Malaysia. *Journal of the Malaysian Branch of the Royal Asiatic Society* 41: 171–79.

Perera, H.N. 2010. *Prehistoric Sri Lanka: late Pleistocene rockshelters and an open-air site.* British Archaeological Reports, International Series 2142, Oxford.

Perera, N., Kourampas, N., Simpson, I.A., Deraniyagala, S.U., Bulbeck, D., Kamminga, J., Perera, J., Fuller, D.Q., Szabó, K., Oliveira, N.V. 2011. People of the ancient rainforest: Late Pleistocene foragers at the Batadomba-lena rockshelter, Sri Lanka. *Journal of Human Evolution* 61: 254–69.

Peresani, M., Fiore, I., Gala, M., Romandini, M., Tagliacozzo, A. 2011. Late Neandertals and the intentional removal of feathers as evidenced from bird bone taphonomy at Fumane Cave 44 ky B.P., Italy. *Proceedings of the National Academy of Sciences of the United States of America* 108: 3888–93.

Perlès, C. 1999. Long-term perspective on the occupation of the Franchthi Cave: continuity and discontinuity. Bailey, G.N., Adam, E., Panagopoulou, E., Perlès, C., Zachos, K. (eds.) *The Palaeolithic*

Archaeology of Greece and Adjacent Areas, Proceedings of the ICOPAG Conference, Ioannina, British School at Athens Studies 3, Athens, pp. 310–18.

Perlès, C. 2000. Greece, 30,000–20,000 bp. Roebroeks, W., Mussi, M., Svoboda, J., Fennema, K. (eds.) *Hunters of the Golden Age: The Mid Upper Palaeolithic of Eurasia 30,000–20,000 BP*. Analecta Praehistorica Leidensia 31. University of Leiden Press, Leiden, pp. 375–97.

Petersen, R.M. 1969. Würm II Climate at Niah Cave. *Sarawak Museum Journal* 17: 67–79.

Peterson, C.E., Shen, C., Chen, C., Chen, W., Tang, Y. 2003. Taphonomy of an Early Pleistocene archaeofauna from Xiaochangliang, Nihewan Basin, north China. Shen, C. and Keates, S. (eds.) *Current Research in Chinese Pleistocene Archaeology*. BAR International Series 1179. Archaeopress, Oxford, pp. 83–98.

Petit, J.R., Jouzel, J., Raynaud, D., Barkov, N.I., Barnola, J.-M., Basile, I., Bender, M., *et al.* 1999. Climate and atmospheric history of the past 420,000 years from the Vostok ice core, Antarctica. *Nature* 399: 429–36.

Petraglia, M.D. 2007. Mind the gap: factoring the Arabian Peninsula and the Indian Subcontinent into Out of Africa models. Mellars, P., Stringer, S., Bar-Yosef, O., Boyle, K. (eds.) *Rethinking the Human Revolution: New Behavioural and Biological Perspectives on the Origins and Dispersal of Modern Humans*. McDonald Institute Monographs, Cambridge, pp. 411–24.

Petraglia, M.D. and Alsharekh, A. 2002. The Middle Palaeolithic of Arabia: implications for modern human origins, behaviour and dispersals. *Antiquity* 77: 671–84.

Petraglia, M.D. and Rose, J.I. (eds.). 2009. *The Evolution of Human Populations in Arabia*. Springer, Dordrecht.

Petraglia, M., Korisettar, R., Boivin, N., Clarkson, C., Ditchfield, P., Jones, S., Koshy, J., *et al.* 2007. Middle Paleolithic assemblages from the Indian Subcontinent before and after the Toba super-eruption. *Science* 317: 114–16.

Petraglia, M.D., Drake, N., Alsharekh, A. 2009. Acheulean landscapes and large cutting tools assemblages in the Arabian Peninsula. Petraglia, M.D. and Rose, J.I. (eds.) *The Evolution of Human Populations in Arabia*. Springer, Dordrecht, pp. 103–16.

Pettitt, P.B. and Bader, N.O. 2000. Direct AMS radiocarbon dates for the Sunghir mid Upper Palaeolithic burials. *Antiquity* 74: 269–70.

Pettitt, P., Bahn, P., Züchner, C. 2009. The Chauvet conundrum: are claims for the 'birthplace of art' premature? Bahn, P.G. (ed.) *An Enquiring Mind: Studies in Honor of Alexander Marshack*. Oxbow Books, Oxford, pp. 253–78.

Pfeffer, P. and Caldecott, J. 1986. The bearded pig (*Sus barbatus*) in East Kalimantan and Sarawak. *Journal of the Malay Branch of the Royal Asiatic Society* 59: 81–100.

Pham Huy Thong. 1980. Con Moong Cave: a noteworthy archaeological discovery in Vietnam. *Asian Perspectives* 23: 17–21.

Pham Huy Thong, Hoang Xuán Chinh, Nguyên Khán Su. 1990. *Hang Con Moong*. Hà Nôi.

Pickering, T.R. 2002. Reconsideration of criteria for differentiating faunal assemblages accumulated by hyenas and hominids. *International Journal of Osteoarchaeology* 12: 127–41.

Pike-Tay, A. and Cosgrove, R. 2006. 'Mousterians in the Antipodes'? Exploring 'modern human behaviour' in Late Pleistocene Tasmania. *Zona Arqueológica* 7: 337–46.

Piper, P.J. and Cranbrook, Earl of. 2007. Niah Cave bone IX: first Pleistocene record of Hose's pygmy shrew Suncus hosei. *Sarawak Museum Journal* 84: 227–32.

Piper, P.J., Ochoa, J., Lewis, H., Paz, V., Ronquillo, W.P. 2008. The first evidence for the past presence of the tiger *Panthera tigris* (L.) on the island of Palawan, Philippines: extinction in an island population. *Palaeogeography, Palaeoclimatology, Palaeoecology* 264: 123–7.

Piper, P.J., Ochoa, J., Robles, E.C., Lewis, H., Paz, V. 2011. Palaeozoology of Palawan Island, Philippines. *Quaternary International* 233: 142–58.

Piper, P. and Rabett, R. 2009. Hunting in a tropical rainforest: evidence from the Late Pleistocene and Early Holocene at Lobang Hangus, Niah Cave, Borneo. *International Journal of Osteoarchaeology* 19: 551–65.

Piper, P. and Rabett, R. Forthcoming-a. Vertebrate fauna from the Niah Caves. Barker, G., Gilbertson, D., Reynolds T. (eds.) *Rainforest Foraging and Farming in Island Southeast Asia: The Archaeology and Environmental History of the Niah Caves, Sarawak, Niah Cave Project*. Vol. 2. McDonald Institute for Archaeological Research, Cambridge.

Piper, P. and Rabett, R. Forthcoming-b. Late Pleistocene subsistence strategies in Island Southeast Asia and its implications for understanding the development of modern human behaviour. Dennell, R. and Porr, M. (eds.) *Southern Asia, Australia, and the Search for Human Origins*. Cambridge University Press, Cambridge.

Piper, P.J., Rabett, R.J., Cranbrook, Earl of. 2007a. New discoveries of an extinct Giant Pangolin (*Manis* cf. *palaeojavanica* Dubois) at Niah Cave, Sarawak, Borneo: biogeography, palaeoecology and taxonomic relationships. *Sarawak Museum Journal* 84: 205–26.

Piper, P., Cranbrook, Earl of, Rabett, R. 2007b. Confirmation of the presence of the tiger *Panthera tigris* (L.) in Late Pleistocene and Holocene Borneo. *Malayan Nature Journal* 59: 259–67.

Piper, P.J., Rabett, R.J., Kurui, E. 2008. Using community composition and structural variation in terminal Pleistocene vertebrate assemblages to identify human hunting behaviour at Niah Caves, Borneo. *Bulletin of the Indo-Pacific Prehistory Association* 28: 88–98.

Pokines, J. and Krupa, M. 1997. Self-barbed antler spearpoints and evidence of fishing in the Late Upper Palaeolithic of Cantabrian Spain. Knecht, H. (ed.) *Projectile Technology*. Plenum Press, New York, pp. 241–62.

Politis, G.G. 1996. Moving to produce: Nukak mobility and settlement patterns in Amazonia. *World Archaeology* 27: 492–511.

Politis, G.G. 1999. Plant exploitation among the Nukak hunter-gatherers of Amazonia: between ecology and ideology. Gosden, C. and Hather, J. (eds.) *The Prehistory of Food: Appetites for Change*. Routledge, London, pp. 99–125.

Politis, G.G. and Messineo, P.G. 2008. The Campo Laborde site: new evidence for the Holocene survival of Pleistocene megafauna in the Argentine Pampas. *Quaternary International* 191: 98–114.

Politis, G.G. and Saunders, N.J. 2002. Archaeological correlates of ideological activity: food-taboos and spirit-animals in an Amazonian hunter-gatherer society. Miracle, P. and Milner N. (eds.) *Consuming Passions and Patterns of Consumption*. McDonald Institute for Archaeological Research, Cambridge, pp. 113–30.

Pookajorn, S. 1984. *The Hoabinhian of Mainland Southeast Asia: New Data from Recent Thai Excavation in the Ban Kao area*. Paper 16. Thai Khadi Research Institute, Thammasat University.

Pookajorn, S. 1985. Ethnoarchaeology with the Phi Tong Luang (Mlabrai): forest hunters of northern Thailand. *World Archaeology* 17: 206–21.

Pookajorn, S. 1996. Human activities and environmental changes during the Late Pleistocene to Middle Holocene in South Thailand and Southeast Asia. Straus, L.G., Erikson, B.V., Erlandson, J.M. (eds.) *Humans at the End of the Ice Age: The Archaeology of the Pleistocene–Holocene Transition*. Plenum Press, New York, pp. 201–13.

Pookajorn, S., Sinsakul, S., Chaimanee, Y., Ukrit, A., Eiamsiri, P., Songkhla, B., Thongpakdee, A., *et al.* 1994. *Final Report of Excavations at Moh Khiew Cave, Krabi Province; Sakai Cave, Trang Province and Ethnoarchaeological Research of Hunter-Gatherer Group, Socall Mani, or Sakai, or Orang Asli at Trang Province: The Hoabinhian Research Project in Thailand*. Vol. 2. Faculty of Archaeology, Silpakorn University, Bangkok.

Pope, G.G., Frayer, D.W., Liangchareon, M., Kulasing, P., Nakabanlang, S. 1981. Palaeoanthropological investigations of the Thai-American expedition in northern Thailand (1978–1980): an interim report. *Asian Perspectives* 21: 147–63.

Pope, G.G., Barr, S., Macdonald, A., Nakabanlang, S. 1986. Earliest radiometrically dated artifacts from Southeast Asia. *Current Anthropology* 27: 275–9.

Pope, G.G. 1988. Recent advances in Far Eastern palaeoanthropology. *Annual Review of Anthropology* 17: 43–77.

Pope, S.T. 1962. *Bows and Arrows*. University of California Press, Berkeley.

Posey, D. 1985. Native and indigenous guidelines for new Amazonian development strategies: understanding biological diversity through ethnoecology. Hemming, J. (ed.) *Change in the Amazon Basin*. Manchester University Press, Manchester, pp. 156–81.

Potts, R. 1996a. *Humanity's Descent: The Consequences of Ecological Instability*. Avon, New York.

Potts, R. 1996b. Evolution and climate variability. *Science* 273: 922–3.

Potts, R. 1998a. Environmental hypotheses of hominin evolution. *Yearbook of Physical Anthropology* 41: 93–136.

Potts, R. 1998b. Variability selection in hominin evolution. *Evolutionary Anthropology* 7: 81–96.

Potts, R. 2003. Environmental variability and its impact on adaptive evolution, with special reference to human origins. Rothschild, L.J. and Lister, A.M. (eds.) *Evolution on Planet Earth: The Impact of the Physical Environment.* Academic Press, London, pp. 363–78.

Prange, M., Lohmann, G., Romanova, V., Butzin, M. 2004. Modelling tempo-spatial signatures of Heinrich Events: influence of the climatic background state. *Quaternary Science Reviews* 23: 521–27.

Prasad, M.S., Mahale, V.P., Kodagali, V.N. 2007. New sites of Australasian microtektites in the central Indian Ocean: implications for the location and size of source crater. *Journal of Geophysical Research* 112, E06007.

Price, G.J., Zhao, J.-X., Feng, Y.-X., Hocknull, S.A. 2009. New U/Th ages for Pleistocene megafauna deposits of southeastern Queensland, Australia. *Journal of Asian Earth Sciences* 34: 190–7.

Pritchard, P.C.H. 1979. *Encyclopedia of Turtles.* T.F.H. Publications, Neptune, New Jersey.

Pritchard, P., Rabett, R., Piper, P. 2009. Distinguishing species of geoemydid and trionychid turtles from shell fragments: evidence from the Pleistocene at Niah Cave, Sarawak, Borneo. *International Journal of Osteoarchaeology* 19: 531–50.

Prokopenko, A.A., Williams, D.F., Kuzmin, M.I., Karabanov, E.B., Khursevich, G.K., Peck, J.A. 2002. Muted climate variations in continental Siberia during the mid-Pleistocene epoch. *Nature* 418: 65–68.

Pureepatpong, N. 2006. Recent investigation of early people (Late Pleistocene to Early Holocene) from Ban Rai and Tham Lod rock shelter sites, Pang Mapha district, Mae Hong Son province, northwestern Thailand. Bacus, E.A., Glover, I.C., Piggott, V.C. (eds.) *Uncovering Southeast Asia's Past.* National University of Singapore Press, Singapore, pp. 38–45.

Puri, R.K. 2005a. *Deadly Dances in the Bornean Rainforest: Hunting Knowledge of the Penan Benalui.* KITLV Press, Leiden.

Puri, R.K. 2005b. Postabandonment ecology of Penan forest camps: anthropological and ethnobotanical approaches to the history of a rainforest valley in East Kalimantan. Dove, M., Sajise, P.E. and Doolittle, A.A. (eds.) *Conserving Nature in Culture: Case Studies from Southeast Asia.* Monograph 54. Yale Southeast Asian Studies, New Haven, pp. 25–82.

Pyatt, F.B., Wilson, B., Barker, G.W. 2005. The chemistry of tree resins and ancient rock paintings in the Niah Caves, Sarawak (Borneo): some evidence of rain forest management by early human populations. *Journal of Archaeological Science* 32: 897–901.

Pyatt, F.B., Barker, G.W., Rabett, R.J., Szabó, K., Wilson, B. 2010. Analytical examination of animal remains from Borneo: the painting of bone and shell. *Journal of Archaeological Science* 37: 2102–5.

Qian, F., Zhou, G.X. (eds.). 1991. *Quaternary Geology and Paleoanthropology of Yuanmou, Yunnan, China.* Science Press, Beijing. (Selected trans. W. Downs, North Arizona University, 1993).

Qin, H. 1997. Morphological, allozymic, and karyotypic distinctions between *Neritina* (*Dostia*) *violacea* and *N.* (*D.*) *cornucopia* (Gastropoda: Neritoidea). *Journal of Zoology, London* 241: 343–69.

Rabett, R.J. n.d.-a. Classifying prehistoric bone technology for functionality and biography.

Rabett, R.J. n.d.-b. Faunal analysis of the late- to post-Pleistocene cave sites of Gua Sagu and Gua Tenggek, Pahang, Malaysia. Report submitted to the Pusat Penyelidikan Arkeologi Malaysia, Universiti Sains Malaysia.

Rabett, R.J. 2002. Bone tool technology and subsistence variability in prehistoric Southeast Asia. Unpublished doctoral thesis, Department of Archaeology, University of Cambridge.

Rabett, R.J. 2004a. The bone artefacts. Higham, C.F.W. and Thosarat, R. (eds.) *The Excavation of Ban Lum Khao: Vol. 1. The Origins of the Civilization of Angkor Project.* Fine Arts Department, Bangkok, pp. 197–201.

Rabett, R.J. 2004b. The ones that come ready-made: the use of *Sus* tusks as tools at prehistoric coastal sites in Malaysia. *Archaeofauna* 13: 131–43.

Rabett, R.J. 2005. The early exploitation of Southeast Asian mangroves: bone technology from caves and open sites. *Asian Perspectives* 44: 154–79.

Rabett, R.J. 2007. Vertebrate faunal remains from Gua Cha: new material found in association with recently re-discovered human remains from the 1954 Sieveking excavation. *Archaeological Heritage of Malaysia* 1: 61–72.

Rabett, R.J. 2008. Analysis of faunal remains: bone tool analysis. Pearsall, D.M. (ed.) *Encyclopaedia of Archaeology.* Vol. 2. Elsevier, New York, pp. 931–36.

Rabett, R.J. 2011. Techno-modes, techno-facies and palaeo-cultures: change and continuity in the Pleistocene of Southeast, Central and North Asia. Roberts, B., Vander Linden, M. (eds.) *Investigating Archaeological Cultures: Material Culture, Variability and Transmission*. Springer, New York, pp. 97–136.

Rabett, R. and Barker, G. 2007. Through the looking glass: new evidence on the presence and behaviour of Late Pleistocene humans at Niah Cave, Sarawak, Borneo. Mellars, P., Stringer, S., Bar-Yosef, O., Boyle, K. (eds.) *Rethinking the Human Revolution: New Behavioural and Biological Perspectives on the Origins and Dispersal of Modern Humans*. McDonald Institute Monographs, Cambridge, pp. 383–94.

Rabett, R., Barker, G. 2010. Late Pleistocene and Early Holocene forager mobility in Southeast Asia. Bellina, B., Wisseman Christie, J., Bacus, L., Pryce, T.O. (eds.) *50 Years of Archaeology in Southeast Asia: Essays in Honour of Ian Glover*. River Books, Bangkok, pp. 66–77.

Rabett, R. and Piper, P. Forthcoming. Eating your tools: early butchery and modification of primate bones in tropical Southeast Asia. Seetah, K., and Gravina, B. (eds.) *Bones for Tools, Tools for Bones*. McDonald Institute Monographs, Cambridge.

Rabett, R.J., Piper, P.J. 2012. The emergence of bone technologies at the end of the Pleistocene in Southeast Asia: regional and evolutionary implications. *Cambridge Archaeological Journal* 22: 37–56.

Rabett, R., Piper, P., Barker, G. 2006. Bones from Hell: preliminary results of new work on the Harrisson faunal assemblage from the deepest part of the Niah Cave, Sarawak. Bacus, E.A., Glover, I.C., Piggott, V.C. (eds.) *Uncovering Southeast Asia's Past*. National University of Singapore Press, Singapore, pp. 46–59.

Rabett, R., Barker, G., Hunt, C.O., Naruse, T., Piper, P., Raddatz, E., Reynolds, T., *et al.* 2009a. The Tràng An project: Late- to Post-Pleistocene settlement of the lower Song Hong valley, North Vietnam. *Journal of the Royal Asiatic Society* 19: 83–109.

Rabett, R., Stimpson, C., Nguyệñ Van Son, Nguyêñ Cao Tâñ, Đỗ Van Cừơng, Ngô Thi Huy, Nguyêñ Van Thai, Đỗ Thị Tuyền. 2009b. Archaeological survey in the Tràng An eco-resort, Ninh Bình, North Vietnam: a brief report. *Archaeological Heritage of Malaysia* 2: 117–27.

Rabett, R., Appleby, J., Blyth, A., Farr, L., Gallou, A., Giffiths, T., Hawkes, J., *et al.* 2011. Inland shell midden site-formation: investigation into a late Pleistocene to early Holocene midden from Tràng An, northern Vietnam. *Quaternary International* 239: 153–69.

Rabinovich, R., Gaudzinski-Windheuser, S., Goren-Inbar, N. 2008. Systematic butchering of fallow deer (*Dama*) at the early Middle Pleistocene Acheulian site of Gesher Benot Ya'aqov (Israel). *Journal of Human Evolution* 54: 134–49.

Radcliffe-Brown, A.P. 1948. *The Andaman Islanders*. Free Press, Glencoe, IL.

Raddatz, E.R. 2006. *Bamboo technology in the Palaeolithic of Southeast Asia*. Unpublished MPhil dissertation, Department of Archaeology, University of Cambridge.

Rahmstorf, S. 2002. Ocean circulation and climate during the past 120,000 years. *Nature* 419: 207–14.

Ramakrishnan, U., Richard, G.C., Pelkey, N.W. 1999. Tiger decline caused by the reduction of large ungulate prey: evidence from a study of leopard diets in southern India. *Biological Conservation* 89: 113–20.

Rambo, T. 1985. Primitive polluters: Semang impact on the Malaysian tropical rain forest ecosystem. *Anthropological Papers Museum of Anthropology University of Michigan* 76: 29–104.

Ranere, A.J., and Cooke, R.G. 2003. Late glacial and early Holocene occupation of Central American tropical forests. Mercader, J. (ed.) *Under the Canopy: The Archaeology of Tropical Rain Forests*. Rutgers University Press, New Brunswick, pp. 201–48.

Ranov, V. 1995. The 'Loessic Palaeolithic' in south Tadjikistan, Central Asia: its industries, chronology and correlation. *Quaternary Science Reviews* 14: 731–45.

Ranov, V.A., Carbonell, E., Rodriguez, X.P. 1995. Kuldara: earliest human occupation in Central Asia in its Afro-Asian context. *Current Anthropology* 36: 337–46.

Ranov, V.A. and Davis, R.S. 1979. Toward a new outline of the Soviet Asian Palaeolithic. *Current Anthropology* 20: 249–70.

Rasmussen, M., Guo, X., Wang, Y., Lohmueller, K.E., Rasmussen, S., Albrechtsen, A., Skotte, L., *et al.* 2011. An Aboriginal Australian genome reveals separate human dispersals into Asia. *Science* 334: 94–8.

Rasmussen, S.O., Andersen, K.K., Svensson, A.M., Steffensen, J.P., Vinther, B.M., Clausen, H.B., Siggaard-Andersen, M.-L., *et al.* 2006. A new Greenland ice core chronology for the last glacial termination. *Journal of Geophysical Research* 111, D06102.

Rasmussen, S.O., Seierstad, I.K., Andersen, K.K., Bigler, M., Dahl-Jensen, D., Johnsen, S.J. 2008. Synchronization of the NGRIP, GRIP, and GISP2 ice cores across MIS2 and palaeoclimatic implications. *Quaternary Science Reviews* 27: 18–28.

Rawlinson, S.E. and Bell, M.A. 1982. A stickleback fish (*Pungitius*) from the Neogene Sterling Formation, Kenai Peninsula, Alaska. *Journal of Paleontology* 56: 583–8.

Reich, D., Green, R.E., Kircher, M., Krause, J., Patterson, N., Durand, E.Y., Viola, B. *et al.* 2010. Genetic history of an archaic hominin group from Denisova Cave in Siberia. *Nature* 468: 1053–1060.

Reich, D., Patterson, N., Kircher, M., Delfin, F., Nandineni, M.R., Pugach, I., Ko, A. M-S., *et al.* 2011. Denisova admixture and the first modern human dispersals into Southeast Asia and Oceania. *The American Journal of Human Genetics* 89: 516–528.

Reimer, P.J., Baillie, M.G.L., Bard, E., Bayliss, A., Beck, J.W., Bertrand, C.J.H., Blackwell, P.G., *et al.* 2004. INTCAL04 terrestrial radiocarbon age calibration, 0–26 cal kyr BP. *Radiocarbon* 46: 1029–58.

Relethford, J.H. 1995. Genetics and modern human origins. *Evolutionary Anthropology* 4: 53–63.

Relethford, J.H. 2001. Absence of regional affinities of Neandertal DNA with living humans does not reject multiregional evolution. *American Journal of Physical Anthropology* 115: 95–8.

Relethford, J.H. 2008. Genetic evidence and the modern human origins debate. *Heredity* 100: 555–63.

Rendell, H. and Dennell, R.W. 1985. Dated Lower Palaeolithic artefacts from northern Pakistan. *Current Anthropology* 26: 393.

Renfrew, C. and Bahn, P.G. 1991. *Archaeology: Theory Methods and Practice.* 1st ed. Thames and Hudson, London.

Reynolds, T.E.G. 1989. Techno-typology in Thailand: a case study of the Tham Khao Khi Chan. *Indo-Pacific Prehistoric Association* 9: 33–43.

Reynolds, T.E.G. 1990. The Hoabinhian: a review. Barnes, G.L. (ed.) *Bibliographic Review of Far Eastern Archaeology.* Oxbow Books, Oxford, pp. 1–30.

Reynolds, T.E.G. 1992. Excavations at Banyan Valley Cave, northern Thailand: a report on the 1972 seasson. *Asian Perspectives* 31: 77–97.

Reynolds, T.E.G. 1993. Problems in the Stone Age of South East Asia. *Proceedings of the Prehistoric Society* 59: 1–15.

Reynolds, T.E.G. 2007. Problems of the Stone Age of South-east Asia revisited. *Proceedings of the Prehistoric Society* 73: 39–58.

Richardson, J.B., III. 1998a. Archaeological evidence for the origins of El Niño and its impact upon pre-Columbian Peruvian civilizations. Research seminar presented at the Department of Archaeology, University of Cambridge.

Richardson, J.B., III. 1998b. From mastodons and mangroves to desert: Late Pleistocene to Mid-Holocene climate and cultural adaptations on the Peruvian north coast. A presentation at the FERCO International Conference on Climate and Culture at 3,000 B.C., University of Maine at Orono.

Richardson, J.A. 1947. An outline of the geomorphological evolution of British Malaya. *Geological Magazine* 84: 129–44.

Riddington, R. 1983. From artifice to artefact: stages in the industrialization of a northern hunting people. *Journal of Canadian Studies* 18: 55–66.

Rigaud, J., and Simek, J. 1990. The Last Pleniglacial in the South of France. Soffer, O., Gamble, C. (eds.) *The world at 18000 B.P.* Vol. 1. Unwin Hyman, London, pp. 69–86.

Rightmire, P.G., Lordkipanidze, D., Vekua, A. 2006. Anatomical descriptions, comparative studies and evolutionary significance of the hominin skulls from Dmanisi, Republic of Georgia. *Journal of Human Evolution* 50: 115–41.

Rightmire, P.G. 2001. Patterns of hominid evolution and dispersal in the Middle Pleistocene. *Quaternary International* 75: 77–84.

Rival, L. 1993. The growth of family trees: understanding Huaorani perceptions of the forest. *Man* 28: 635–52.

Rival, L. 1996. Blowpipes and spears: the social significance of Huaorani technological choices. Descola, P. and Palsson, G. (eds.) *Nature and Society: Anthropological Perspectives*. Routledge, London, pp. 145–64.

Roberts, M.B., Stringer, C.B., Parfitt, S.A. 1990. A hominid tibia from Middle Pleistocene sediments at Boxgrove, UK. *Nature* 369: 311–13.

Roberts, R.G., Jones, R., Smith, M.A. 1990. Thermoluminescence dating of a 50,000-year-old human occupation site in northern Australia. *Nature* 345: 153–6.

Roberts, R.G., Jones, R., Spooner, N.A., Head, M.J., Murray, A.S., Smith, M.A. 1994. The human colonisation of Australia: optical dates of 53,000 and 60,000 years bracket human arrival at Deaf Adder Gorge, Northern Territory. *Quaternary Geochronology (Quaternary Science Reviews)* 13: 575–83.

Roberts, R.G., Flannery, T.F., Ayliffe, L.K., Yoshida, H., Olley, J.M., Prideaux, G.J., Laslett, G.M., *et al.* 2001. New ages for the last Australian megafauna: continent-wide extinction about 46,000 years ago. *Science* 292: 1888–92.

Roberts, R.G., Morwood, M.J., Westaway, K.E. 2005. Illuminating Southeast Asian prehistory: new archaeological and paleoanthropological frontiers for luminescence dating. *Asian Perspectives* 44: 294–319.

Robinson, J.G. and Bennett, E.L. 2000. Carrying capacity limits to sustainable hunting in tropical forests. Robinson, J.G. and Bennett, E.L. (eds.) *Hunting for Sustainability in Tropical Forests*. Columbia University Press, New York, pp. 13–30.

Rockman, M. 2003. Knowledge and learning in the archaeology of colonization. Rockman, M. and Steele, J. (eds.) *Colonization of Unfamiliar Landscapes: The Archaeology of Adaptation*. Routledge, London, pp. 3–24.

Rockman, M. 2008. So the weather's different here too? Application of the archaeological model of landscape learning to questions of past and future climate change. Paper precirculated for the Sixth World Archaeological Congress, Dublin.

Roe, G. 2009. On the interpretation of Chinese loess as a paleoclimate indicator. *Quaternary Research* 71: 150–61.

Roebroeks, W. 2003. Landscape learning and the earliest peopling of Europe. Rockman, M. and Steele, J. (eds.) *Colonization of Unfamiliar Landscapes: The Archaeology of Adaptation*. Routledge, London, pp. 99–115.

Roebroeks, W. 2001. Hominid behaviour and the earliest occupation of Europe: an exploration. *Journal of Human Evolution* 41: 437–61.

Rolland, N. 2002. The initial hominid colonization of Asia: a survey of anthropic evidence from biogeographic and ecological perspectives. *Bulletin of the Indo-Pacific Prehistory Association* 22: 3–15.

Ros Fatihah Muhammad and Yeap, E.B. 2000. Proposed conservation of Badak Cave C, Lenggong as vertebrate fossil site extraordinary. Annual Geological Conference 2000. *Geological Society of Malaysia*, pp. 189–96.

Ros Fatihah Muhammad, Yoshida, D., Tani, A. Smart, P.L. 2002. Implications of electron spin resonance and Uranium-series dating techniques on speleothems in the Kinta and Lenggong Valleys, West Malaysia. *Advances in ESR Applications* 18: 19–26.

Rose, J.I. 2004. New evidence for the expansion of an Upper Pleistocene population out of East Africa, from the site of Station One, northern Sudan. *Cambridge Archaeological Journal* 14: 205–16.

Rose, J.I. and Petraglia, M.D. 2009. Tracking the origin and evolution of human populations in Arabia. Petraglia, M.D. and Rose, J.I. (eds.) *The Evolution of Human Populations in Arabia*. Springer, Dordrecht, pp. 1–12.

Roucoux, K.H., Tzedakis, P.C., de Abreu, L., Shackleton, N.J. 2006. Climate and vegetation changes 180,000 to 345,000 years ago recorded in a deep-sea core off Portugal. *Earth and Planetary Science Letters* 249: 307–25.

Rousseau, D.D., Kuklac, G., McManus, J. 2006. What is what in the ice and the ocean? *Quaternary Science Reviews* 25: 2025–30.

Roy, K., Valentine, J.W., Jablonski, D., Kidwell, S.M. 1996. Scales of climatic variability and time averaging in Pleistocene biotas: implications for ecology and evolution. *TREE* 11: 458–63.

Rule, S., Brook, B.W., Haberle, S.G., Turney, C.S.M., Kershaw, A.P., Johnson, C.N. 2012. The aftermath of megafaunal extinction: ecosystem transformation in Pleistocene Australia. *Science* 335: 1483–85.

Rybin, E.P. 2005. Land use and settlement patterns in the mountain belt of south Siberia: mobility strategies and the emergence of 'cultural geography' during the Middle-to-Upper Palaeolithic transition. *Indo-Pacific Prehistoric Association Bulletin* 25: 79–87.

Sarasin, F. 1933. Historical researches in Siam. *Journal of the Siam Society* 26: 171–202.

Sathiamurthy, E. and Voris, H.K. 2006. Maps of Holocene sea level transgression and submerged lakes on the Sunda Shelf. *Natural History Journal of Chulalongkorn University (Suppl.)* 2: 1–44.

Satterthwait, L.D. 1986. Aboriginal Australian net hunting. *Mankind* 16: 31–48.

Schemske, D.W. 2000. Understanding the Origin of Species. *Evolution* 54: 1069–73.

Schemske, D.W. 2010. Adaptation and the Origin of Species. *The American Naturalist* 176(S1): S4–S25.

Schick, K.D. and Toth, N. 1993. *Making Silent Stones Speak.* Weidenfeld and Nicolson, London.

Schick, K., Toth, N., Wei, Q., Clark, J.D., Etler, D. 1991. Archaeological perspectives in the Nihewan Basin, China. *Journal of Human Evolution* 21: 13–26.

Schimanski, A. and Stattegger, K. 2005. Deglacial and Holocene evolution of the Vietnam shelf: stratigraphy, sediments and sea-level change. *Marine Geology* 214: 365–87.

Schulz, H., von Rad, U., Erlenkeuser, H. 1998. Correlation between Arabian Sea and Greenland climate oscillations of the past 110,000 years. *Nature* 393: 54–57.

Schurr, T.G. and Sherry, S.T. 2004. Mitochondrial DNA and Y chromosome diversity and the peopling of the Americas: evolutionary and demographic evidence. *American Journal of Human Biology* 16: 420–39.

Schwarcz, H.P., Simpson, J.J., Stringer, C.B. 1998. Neanderthal skeleton from Tabun: U-series data by gamma-ray spectrometry. *Journal of Human Evolution* 35: 635–45.

Schwartz, J.H. and Tattersall, I. 1996. Whose teeth? *Nature* 381: 201–2.

Scrivenor, J.B. 1928. *The Geology of the Malayan Ore Deposits.* Macmillan, London.

Scrivenor, J.B. 1941. Geological research in the Malay peninsula and archipelago. *Geological Magazine* 78: 125–50.

Sellato, B. 2007. Resourceful children of the forest: the Kalimantan Punan through the twentieth century. Sercombe, P. and Sellato, B. (eds.) *Beyond the Green Myth: Borneo's Hunter-Gatherers in the Twenty-First Century.* Nias Press, Copenhagen, pp. 61–90.

Sellato, B. and Sercombe, P.G. 2007. Borneo, hunter-gatherers and change. Sercombe, P. and Sellato, B. (eds.) *Beyond the Green Myth: Borneo's Hunter-Gatherers in the Twenty-First Century.* Nias Press, Copenhagen, pp. 1–49.

Sémah, A-M. and Sémah, F. 2012. The rain forest in Java through the Quaternary and its relationships with humans (adaptation, exploitation and impact on the forest). *Quaternary International* 249: 120–28.

Sémah, F., Sémah, A.-M., Simanjuntak, T. 2003. More than a million years of human occupation in insular Southeast Asia: the early archaeology of eastern Java. Mercador, J. (ed.) *Under the Canopy: The Archaeology of Tropical Rainforests.* Rutgers University Press, New Brunswick, pp. 161–90.

Sémah, F., Sémah, A.-M., Falguères, C., Détroit, F., Gallet, X., Hameau, S., Moigne, A.-M., Simanjuntak, H.T. 2004. The significance of the Punung karstic area (eastern Java) for the chronology of the Javanese Palaeolithic, with special reference to Song Terus Cave. *Modern Quaternary Research in Southeast Asia* 18: 45–62.

Semaw, S. 2000. The world's oldest stone artefacts from Gona, Ethiopia: their implications for understanding stone technology and patterns of human evolution between 2.6–1.5 million years ago. *Journal of Archaeological Science* 27: 1197–1214.

Sepulchre, P., Ramstein, G., Kageyama, M., Vanhaeren, M., Krinner, G., Sánchez-Goñi, M.-F., d'Errico, F. 2007. H4 abrupt event and late Neanderthal presence in Iberia. *Earth and Planetary Science Letters* 258: 283–92.

Shackleton, N.J. 1987. Oxygen isotopes, ice volume and sea level. *Quaternary Science Reviews* 6: 183–90.

Shackleton, N.J., An, Z., Dodonov, A.E., Gavin, J., Kukla, G.J., Ranov, V.A., Zhou, L.P. 1995. Accumulation rate of loess in Tadjikistan and China: relationship with glocal ice volume cycles. *Quaternary Proceedings* 4: 1–6.

Shackleton, N.J., Fernanda Sánchez-Goñi, M., Pailler, D., Lancelot, Y. 2002. Marine Isotope Substage 5e and the Eemian Interglacial. *Global and Planetary Change* 36: 151–55.

Shackleton, N.J., Fairbanks, R.G., Chiu, T-C., Parrenin, F. 2004. Absolute calibration of the Greenland time scale: implications for Antarctic time scales and for Δ14C. *Quaternary Science Reviews* 23: 1513–22.

Shane, P., Self, S., Blake, S., Rampino, M.R. 2004. First Toba supereruption: comment and reply. *Geology*, June.

Shchelinsky, V.E., Dodonov, A.E., Baigusheva, V.S., Kulakov, S.A., Simakova, A.N., Tesakov, A.S., Titov, V.V. 2010. Early Palaeolithic sites on the Taman Peninsula (Southern Azov Sea region, Russia): Bogatyri/Sinyaya Balka and Rodniki. *Quaternary International* 223–4: 28–35.

Shea, J.J. 2007. The boulevard of broken dreams: evolutionary discontinuity in the Late Pleistocene Levant. Mellars, P., Stringer, S., Bar-Yosef, O., Boyle, K. (eds.) *Rethinking the Human Revolution: New Behavioural and Biological Perspectives on the Origins and Dispersal of Modern Humans*. McDonald Institute Monographs, Cambridge, pp. 219–32.

Shea, J.J. 2008. Transitions or turnovers? Climatically-forced extinctions of *Homo sapiens* and Neanderthals in the east Mediterranean Levant. *Quaternary Science Reviews* 27: 2253–70.

Shea, J.J. 2011. *Homo sapiens* is as *Homo sapiens* was. *Current Anthropology* 52: 1–35.

Shen, C., Gao, X., Wei, Qi. 2010. The earliest hominin occupations in the Nihewan basin of northern China: recent progress from field investigations. Norton, C.J. and Braun, D.R. (eds.) *Asian Paleoanthropology: From Africa to China and Beyond*. Springer, Dordrecht, pp. 169–80.

Shen, G., Gao, X., Gao, B., Granger, D.E. 2009. Age of Zhoukoudian *Homo erectus* determined with 26Al/10Be burial dating. *Nature* 458: 198–200.

Shen, G., Wang, W., Wang, Q., Zhao, J., Collerson, K., Zhou, C., Tobias, P.V. 2002. U-Series dating of Liujiang hominid site in Guangxi, southern China. *Journal of Human Evolution* 43: 817–29.

Shennan, S. 2001. Demography and cultural innovation: a model and its implications for the emergence of modern human culture. *Cambridge Archaeological Journal* 11: 5–16.

Shimanski, A. and Stattegger, K. 2005. Deglacial and Holocene evolution of the Vietnam shelf: stratigraphy, sediments and sea-level change. *Marine Geology* 214: 365–87.

Shpakova, E.G. and Derev'anko, A.P. 2000. The interpretation of odontological features of Pleistocene human remains from the Altai. *Archaeology, Ethnology and Anthropology of Eurasia* 1: 125–38.

Shoocongdej, R. 1996. *Forager mobility organization in seasonal tropical environments: a view from Lang Kamnan Cave, western Thailand*, Unpublished PhD dissertation, Department of Anthropology, University of Michigan.

Shoocongdej, R. 2000. Forager mobility organization in seasonal tropical environments of western Thailand. *World Archaeology* 32: 14–40.

Shoocongdej, R. 2006. Late Pleistocene activities at the Tham Lod rockshelter in Highland Pang Mapha, Mae Hong Son province, northwestern Thailand. Bacus, E.A., Glover, I.C., Piggott, V.C. (eds.) *Uncovering Southeast Asia's Past*. National University of Singapore Press, Singapore, pp. 22–37.

Shunkov, M. 2005. The characteristics of the Altai (Russia) Middle Palaeolithic in regional context. *Bulletin of the Indo-Pacific Prehistory Association* 25: 69–77.

Shutler, R., Jr., Head, J.M., Donahue, D.J., Jull, A.J.T., Barbetti, M.F., Matsu'ura, S., de Vos, J., Storm, P. 2004. AMS radiocarbon dates on bone from cave sites in southeast Java, Indonesia, including Wajak. *Modern Quaternary Research in Southeast Asia* 18: 89–93.

Sieveking, G.De.G. 1954. Excavations at Gua Cha, Kelantan 1954, Part 1. *Federation Museums Journal* 1–2: 75–138.

Sillitoe, P. 1988. *Made in Niugini: Technology in the Highlands of Papua New Guinea*. British Museums Publications, London, in association with the Univeristy of Durham Publications Board.

Simanjuntak, T. 2000. New discoveries at Braholo Cave, western Gunung Sewu (Indonesia). *Proceedings of the 7th International Conference EurASEAA 1998*, pp. 219–26.

Simanjuntak, T. and Asikin, I.N. 2004. Early Holocene human settlement in eastern Java. *Bulletin of the Indo-Pacific Prehistory Association* 24: 13–19.

Simanjuntak, T., Sémah, F., Gaillard, C. 2010. The palaeolithic in Indonesia: nature and chronology. *Quaternary International* 223–4: 418–21.

Simmons, T., Falsetti, A.B., Smith, F.H. 1991. Frontal bone morphometrics of southwest Asian Pleistocene hominids. *Journal of Human Evolution* 20: 249–269.

Simpson, I.A., Kourampas, N., Perera, H.N. 2008. Bellan-bandi Palassa, Sri Lanka: formation processes of a Mesolithic open air site identified through thin section micromorphology. *Archaeologia—Journal of Archaeology* 4: 3–18.

Siong, K.T. and Clements, R. 2008. Taxonomy and distribution of the Neritidae (Mollusca: Gastropoda) in Singapore. *Zoological Studies* 47: 481–94.

Sirocko, F., Seelos, K., Schaber, K., Rein, B., Dreher, F., Diehl, M., Lehne, R., Jäger, K., Krbetschek, M., Degering, D. 2005. A late Eemian aridity pulse in central Europe during the last glacial inception. *Nature* 436: 833–36.

Skeat, W.W. 1902. The Wild Tribes of the Malay Peninsula. *Journal of the Anthropological Institute of Great Britain and Ireland* 32: 124–41.

Sloan, C. 1975. A study of the Punan Busang, III: Punan hunting methods. *Malayan Nature Journal* 28(3–4): 146–51.

Smit, J., van Bijden, A.J.M., Troelstra, S.R. 1991. Analysis of the Australasian microtektite event, the Toba Lake event, and the Cretaceous/Paleogene boundary, eastern Indian Ocean. Weissel, J. and Peirce, J., *et al.* (eds.) *Proceedings of the Ocean Drilling Program, Scientific Results*. Vol. 121. Ocean Drilling Program, College Station, pp. 489–503.

Smith, C.I., Chamberlain, A.T., Riley, M.S., Stringer, C., Collins, M.J. 2003. The thermal history of human fossils and the likelihood of successful DNA amplification. *Journal of Human Evolution* 45: 203–17.

Smith, F.H., Janković, I., Karavanić, I. 2005. The assimilation model, modern human origins in Europe, and the extinction of Neandertals. *Quaternary International* 137: 7–19.

Smith, P.J. 2009. *A 'Splendid Idiosyncrasy': Prehistory at Cambridge 1915–50*. British Archaeological Reports, British Series 485, Oxford

Smith, P.J. n.d. From 'small, dark and alive' to 'cripplingly shy': Dorothy Garrod as the first woman professor at Cambridge. http://www.arch.cam.ac.uk/~pjs1011/Pams.html.

Smith, T.M., Tafforeau, P., Reid, D.J., Grün, R., Eggins, S., Boutakiout, M., Hublin, J-J. 2007. Earliest evidence of modern human life history in North African early *Homo sapiens*. *Proceedings of the National Academy of Sciences of the United States of America* 104: 6128–33.

Soares, P., Trejaut, J.A., Loo, J.-H., Hill, C., Mormina, M., Lee, C.-L., Chen, Y.-M., *et al.* 2008. Climate change and postglacial human dispersals in Southeast Asia. *Molecular Biology Evolution* 25: 1209–18.

Sobel, J.M., Chen, G.F., Watt, L.R., Schemske, D.W. 2010. The biology of speciation. *Evolution* 64: 295–315.

Soffer, O. 1989. The Middle to Upper Palaeolithic transition on the Russian Plain. Mellars, P. and Stringer, C. (eds.) *The Human Revolution*. Edinburgh University Press, Edinburgh, pp. 714–42.

Soffer, O. 1997. The mutability of Upper Palaeolithic art in Central and Eastern Europe: patterning and significance. Conkey, M., Soffer, O., Stratmann, D., Jablonski, N. (eds.) *Beyond Art: Pleistocene Image and Symbol*. Wattis Symposium Series in Anthropology. Memoirs of the California Academy of Sciences 23. California University Press, San Francisco, pp. 239–61.

Soffer, O., Adovasio, J.M., Illingworth, J.S., Amirkhanov, H.A., Praslov, N.D., Street, M. 2000. Palaeolithic perishables made permanent. *Antiquity* 74: 812–21.

Solheim, W. 1983. Archaeological research in Sarawak, past and future. *Sarawak Museum Journal* 53: 35–58.

Solheim, W. 1994. Comments of recent conferences: the Hoabinhian 60 years after Madeline Colani. *Southeast Asian Archaeology International Newsletter* 4: 9–12.

Sonakia, A. and de Lumley, H. 2006. Narmada *Homo erectus* – a possible ancestor of the modern Indian. *Comptes Rendus Palevol* 5: 353–57.

Sørensen, P. 1988. *Archaeological Excavations in Thailand: Surface Finds and Minor Excavations*. Scandinavian Institute of Asian Studies Occasional Papers 1. Curzon Press, London.

Sørensen, P. 1999. The Ban Kao culture of Thailand and Malaysia: its origin, the present geographical extension, its chronology and relative chronological subphases. Paper presented at the International Colloquium on Archaeology in Southeast Asia in the 3rd Millennium, Penang, Malaysia.

Sørensen, M.A.P. and Hatting, T. 1966. *Archaeological Excavations in Thailand: Vol. II. Ban Kao*. Thai-Danish Prehistoric Expedition 1960–2, Munksgaard, Copenhagen.

Spahni, R., Chappellaz, J., Stocker, T.F., Loulergue, L., Hausammann, G., Kawamura, K., Flückiger, J., *et al.* 2005. Atmospheric methane and nitrous oxide of the Late Pleistocene from Antarctic ice cores. *Science* 310: 1317–21.

Spoor, F., Leakey, M.G., Gathogo, P.N., Brown, F.H., Antón, S.C., McDougall, I., Kiarie, C., Manthi, F.K., Leakey, L.N. 2007. Implications of new early Homo fossils from Ileret, east of Lake Turkana, Kenya. *Nature* 448: 688–91.

Spötl, C., Scholz, D., Mangini, A. 2008. A terrestrial U/Th-dated stable isotope record of the Penultimate Interglacial. *Earth and Planetary Science Letters* 276: 283–92.

Stark, M. 2004. Pre-Angkorian and Angkorian Cambodia. Glover, I. and Bellwood, P. (eds.) *Southeast Asia: From Prehistory to History*. Routledge Curzon, London, pp. 89–119.

Stauffer, P.H. 1973. Cenozoic. Gobbett, D.J. and Hutchison, C.S. (eds.) *Geology of the Malay Peninsula (West Malaysia and Singapore)*. John Wiley, New York, pp. 143–76.

Steele, J. and Rockman, M. 2003. Where do we go from here? Modelling the decision-making process during exploratory dispersal. Rockman, M. and Steele, J. (eds.) *Colonization of Unfamiliar Landscapes: The Archaeology of Adaptation*. Routledge, London, pp. 130–43.

Steffensen, J.P., Andersen, K.K., Bigler, M., Clausen, H.B., Dahl-Jensen, D., Fischer, H., Goto-Azuma, K., *et al.* 2008. High-resolution Greenland ice core data show abrupt climate change happens in few years. *Science* 321: 680–4.

Stein, J.K. 1992. *Deciphering a Shell Midden*. Academic Press, San Diego.

Steinke, S., Kienast, M., Pflaumann, U., Weinelt, M., Stattegger, K. 2001. A high-resolution sea-surface temperature record from the tropical South China Sea (16,500–3000 yr B.P.). *Quaternary Research* 55: 352–62.

Sterling, E.J., Hurley, M.M., Le Duc Minh. 2006. *Vietnam: A Natural History*. Yale University Press, New Haven.

Stewart, B. 2009. Digging the high life: new excavations at Melikane Rockshelter, Lesotho. Paper presented at the McDonald Institute for Archaeological Research, University of Cambridge.

Stewart, B.A., Dewar, G.I., Morley, M.W., Inglis, R.H., Wheeler, M., Jacobs, Z., Roberts, R.G. Forthcoming. Afromontane foragers of the Late Pleistocene: Site formation, chronology and occupational pulsing at Melikane Rockshelter, Lesotho. *Quaternary International*.

Stiles, D. 1992. The hunter-gatherer 'revisionist' debate. *Anthropology Today* 8(2): 13–17.

Stimpson, C. 2009. Raptor and owl bone from Niah: identifications and morphological variation in the humerus and tarsometatarsus of selected raptors. *International Journal of Osteoarchaeology* 19: 476–90.

Stimpson, C. Forthcoming. Local scale, proxy evidence for the presence of closed canopy forest in north-western Borneo in the late Pleistocene: bones of strategy I bats from the archaeological record of the Great Cave of Niah, Sarawak. *Palaeogeography, Palaeoclimatology, Palaeoecology*.

Stiner, M. 2002. Carnivory, coevolution, and the geographic spread of the genus *Homo*. *Journal of Archaeological Research* 10: 1–63.

Stirling, C.H., Esat, T.M., Lambeck, K., McCulloch, M.T. 1998. Timing and duration of the Last Interglacial: evidence for a restricted interval of widespread coral reef growth. *Earth and Planetary Science Letters* 160: 745–62.

Stokes, C.R., Clark, C.D., Darby, D.A., Hodgson, D.A. 2005. Late Pleistocene ice export events into the Arctic Ocean from the M'Clure Strait Ice Stream, Canadian Arctic Archipelago. *Global and Planetary Change* 49: 139–62.

Stoneking, M. and Delfin, F. 2010. The human genetic history of East Asia: weaving a complex tapestry. *Current Biology* 20: R188–93.

Storm, P. 1995. The evolutionary significance of the Wajak skulls. *Scripta Geologica* 110: 1–247.

Storm, P. 2001. The evolution of humans in Australasia from an environmental perspective. *Palaeogeography, Palaeoclimatology, Palaeoecology* 171: 363–83.

Storm, P. 2012. A carnivorous niche for Java Man? A preliminary consideration of the abundance of fossils in Middle Pleistocene Java. *Comptes Rendus Palevol* 11(2–3): 191–202.

Storm, P., Aziz, F., de Vos, J., Kosasih, Dikdik, Baskoro, Sinung, Ngaliman, van den Hoek Ostende, L.W. 2005. Late Pleistocene *Homo sapiens* in a tropical rainforest fauna in East Java. *Journal of Human Evolution* 49: 536–45.

Stringer, C. 2000. Coasting out of Africa. *Nature* 405: 24–7.

Stringer, C. 2003. Out of Ethiopia. *Nature* 423: 692–95.

Stringer, C. 2007. The origin and dispersal of *Homo sapiens*: our current state of knowledge. Mellars, P., Stringer, S., Bar-Yosef, O., Boyle, K. (eds.) *Rethinking the Human Revolution: New Behavioural and Biological Perspectives on the Origins and Dispersal of Modern Humans*. McDonald Institute Monographs, Cambridge, pp. 15–20.

Stringer, C.B., Grün, R., Schwarcz, H.P., Goldberg, P. 1989. ESR dates for the hominid burial site of Es Skhul in Israel. *Nature* 338: 756–58.

Stringer, C.B. and Hublin, J-J. 1999. New age estimates for the Swanscombe hominid, and their significance for human evolution. *Journal of Human Evolution* 37: 873–77.

Stuebing, R.B., Ghazally Ismail, Ling Hau Ching. 1994. The distribution and abundance of the Indo-Pacific crocodile *Crocodylus porosus* Schneider in the Klias River, Sabah, East Malaysia. *Biological Conservation* 69: 1–7.

Stuijts, I-L.M. 1993. *Late Pleistocene and Holocene vegetation of West Java, Indonesia*. Modern Quaternary Research in Southeast Asia 12, A A Balkema Publishers, Amsterdam.

Stuiver, M. and Reimer, P.J. 1993. Extended 14C data base and revised CALIB 3.014C age calibration program. *Radiocarbon* 35: 215–30.

Stuiver, M., Reimer, P.J., Bard, E., Beck, J.W., Burr, G.S., Hughen, K.A., Kromer, B., McCormac, G., van der Plicht, J., Spurk, M. 1998. INTCAL98 radiocarbon age calibration, 24000–0 *cal BP. Radiocarbon* 40: 1041–83.

Summerhayes, G.R., Leavesley, M., Fairbairn, A., Mandui, H., Field, J., Ford, A., Fullagar, R. 2011. Human adaptation and plant use in highland New Guinea 49,000 to 44,000 years ago. *Science* 330: 78–81.

Sun, X., Li, X. 1999. A pollen record of the last 37 ka in deep sea core 17940 from the northern slope of the South China Sea. *Marine Geology* 156: 227–44.

Sun, X., Li, X., Luo, Y., Chen, X. 2000. The vegetation and climate at the last glaciation on the emerged continental shelf of the South China Sea. *Palaeogeography, Palaeoclimatology, Palaeoecology* 160: 301–16.

Suntharalingam, T., Teoh, L.H. and van de Meene, E.A. 1977. A systematic investigation of the Quaternary deposits in the coastal plain of Taiping, Perak. Chung, S.K. (ed) *Kementerian Perusahaan Utama, Lapuran Tahunan Kajibumi Malayia* (Ministry of Primary Industries, Annual Report Geological Survey Malaysia), Government Printing Office, Kuching, pp. 116–26.

Svensson, A., Andersen, K.K., Bigler, M., Clausen, H.B., Dahl-Jensen, D., Davies, S.M., Johnsen, S.J., *et al.* 2008. A 60 000 year Greenland stratigraphic ice core chronology. *Climate of the Past* 4: 47–57.

Svoboda, J. 1994. The Pavlov site, Czech Republic: lithic evidence from the Upper Palaeolithic. *Journal of Field Archaeology* 21: 69–81.

Svoboda, J., Péan, S., Wojtal, P. 2005. Mammoth bone deposits and subsistence practices during Mid-Upper Palaeolithic in Central Europe: three cases from Moravia and Poland. *Quaternary International* 126–8: 209–21.

Swann, G.E.A., Mackay, A.W., Leng, M.J., Demory, F. 2005. Climatic change in Central Asia during MIS 3/2: a case study using biological responses from Lake Baikal. *Global and Planetary Change* 46: 235–53.

Swindler, D.R. 2002. *Primate Dentition: An Introduction to the Teeth of Non-human Primates*. Cambridge University Press, Cambridge.

Swindler, D.R. and Erwin, J. 1986. *Comparative Primate Biology: Vol. 1. Systematics, Evolution and Anatomy*. Alan R. Liss, New York.

Swisher, C.C., III, Curtis, G.H., Jacob, T., Getty, A.G., Suprijo, A., Widiasmoro. 1994. Age of the earliest known hominds in Java, Indonesia. *Science* 263: 1118–21.

Swisher, C.C., III, Rink, W.J, Antón, S.C., Schwarcz, H.P., Curtis, G.H., Suprijo, A., Widiasmoro. 1996. Latest *Homo erectus* of Java: potential contemporaneity with *Homo sapiens* in Southeast Asia. *Science* 274: 1870–4.

Szabó, K., Ramirez, H., Anderson, A., Bellwood, P. 2003. Prehistoric subsistence strategies on the Batanes Islands, northern Philippines. *Bulletin of the Indo-Pacific Prehistory Association* 23: 163–71.

Szabó, K., Brumm, A., Bellwood, P. 2007. Shell artefact production at 32,000–28,000 BP in Island Southeast Asia: thinking across media? *Current Anthropology* 48: 701–23.

Tabbada, K.A., Trejaut, J., Loo, J.-H., Chen, Y.-M., Lin, M., Mirazón-Lahr, M., Kivisild, T., Corazon De Ungria, M.A. 2010. Philippine mitochondrial DNA diversity: a populated viaduct between Taiwan and Indonesia? *Molecular Biology and Evolution* 27: 21–31.

Takahata, N., Lee, S.-H., Satta, Y. 2001. Testing multiregionality of modern human origins. *Molecular Biology Evolution* 18: 172–83.

Tanabe, S., Kazuaki, H., Saito, Y., Shigeko, H., Van Phai Vu, Kitamura, A. 2003a. Song Hong (Red River) delta evolution related to millennium-scale Holocene sea-level changes. *Quaternary Science Reviews* 22: 2345–61.

Tanabe, S., Hori, K., Saito, Y., Haruyama, S., Le Quoc Doanh, Sato, Y., Hiraide, S. 2003b. Sedimentary facies and radiocarbon dates of the Nam Dinh-1 core from the Song Hong (Red River) delta, Vietnam. *Asian Earth Sciences* 21: 503–13.

Tanabe, S., Saito, Y., Vu, Q.L., Hanebuth, T.J.J., Kitamura, A., Ngo, Q.T. 2006. Holocene evolution of the Song Hong (Red River) delta system, northern Vietnam. *Sedimentary Geology* 187: 29–61.

Tang, Y., You, Y., Li, Y. 1981. Several Paleolithic and Early Pleistocene mammalian localities in Yangyuan and Yuxian Counties, Hebei Province. *Vertebrata PalAsiatica* 19: 256–67 (trans. W. Downs, 1992).

Tappen, M. 1994. Bone weathering in the tropical rain forest. *Journal of Archaeological Science* 21: 667–73.

Tarasov, L. and Peltier, W.R. 2006. A calibrated deglacial drainage chronology for the North American continent: evidence of an Arctic trigger for the Younger Dryas. *Quaternary Science Reviews* 25: 659–88.

Taylor, K.C., Lamorey, G.W., Doyle, G.A., Alley, R.B., Grootes, P.M., Mayewski, P.A., White, J.W.C., Barlow, L.K. 1993. The 'flickering switch' of late Pleistocene climate change. *Nature* 361: 432–36.

Tchernov, E. 1988. The biogeographical history of the southern Levant. Yom-Tov, Y. and Tchernov, E. (eds.) *The Zoogeography of Israel: The Distribution and Abundance at a Zoogeographical Crossroad.* Dr. W. Junk, Dordrecht, pp. 159–250.

Tchernov, E. 1998. The faunal sequence of the Southwest Asian Middle Palaeolithic in relation to hominid dispersals. Akazawa, T., Aoki, K., Bar-Yosef, O. (eds.) *Neandertals and Modern Humans in Western Asia.* Plemum Press, New York, pp. 77–90.

Templeton, A.R. 2007. Genetics and recent human evolution. *Evolution* 61: 1507–19.

Thangaraj, K., Chaubey, G., Kivisild, T., Reddy, A.G., Singh, V.K., Rasalkar, A.A., Singh, L. 2005. Reconstructing the origin of Andaman Islanders. *Science* 308: 996.

Theissen, K.M., Dunbar, R.B., Cooper, A.K., Mucciarone, D.A., Hoffmann, D. 2003. The Pleistocene evolution of the East Antarctic Ice Sheet in the Prydz bay region: stable isotopic evidence from ODP Site 1167. *Global and Planetary Change* 39: 227–56.

Thiel, B. 1990. Excavations at Musang Cave, northeast Luzon, Philippines. *Asian Perspectives* 28: 229–64.

Thompson, L.G., Yao, T., Davis, M.E., Henderson, K.A., Mosley-Thompson, E., Lin, P.-N., Beer, J., Synal, H.-A., Cole-Dai, J., Bolzan, J.F. 1997. Tropical climate instability: the last glacial cycle from a Qinghai-Tibetan ice core. *Science* 276: 1821–25.

Thomson, D.F. 1936. Notes on some bone and stone implements from North Queensland. *Miscellaneous Papers Journal of the Royal Anthropological Institute* 66: 71–81.

Tianyuan, L. and Etler, D. 1992. New Middle Pleistocene hominid crania from Yunxian in China. *Nature* 357: 404–7.

Tjia, H.D. 1996. Sea-level changes in the tectonically stable Malay-Thai peninsula. *Quaternary International* 31: 95–101.

Tolan-Smith, C. 2003. The social context of landscape learning and the late glacial–early postglacial recolonization of the British Isles. Rockman, M. and Steele, J. (eds.) *Colonization of Unfamiliar Landscapes: The Archaeology of Adaptation.* Routledge, London, pp. 116–29.

Tong, H. 2006. Composition des faunes de mammifères quaternaries en Chine selon un gradient Nord-Sud. *L'Anthropologie* 110: 870–87 (in French with English abstract).

Tong, H. 2007. Occurrences of warm-adapted mammals in North China over the Quaternary Period and their paleo-environmental significance. *Science in China Series D, Earth Sciences* 50: 1327–40.

Tostevin, G.B. 2007. Social intimacy, artefact visibility and acculturation models of Neanderthal–modern human interaction. Mellars, P., Boyle, K., Bar-Yosef, O., Stringer, C. (eds.) *Rethinking the Human Revolution.* McDonald Institute Monographs, Cambridge, pp. 341–57.

Tougard, C. 2001. Biogeography and migration routes of large mammal faunas in South-east Asia during the Late Middle Pleistocene: focus on the fossil and extant faunas of Thailand. *Palaeogeography, Palaeoclimatology, Palaeoecology* 168: 337–58.

Tougard, C., Montuire, S. 2006. Pleistocene paleoenvironmental reconstructions and mammalian evolution in South-East Asia: focus on fossil faunas from Thailand. *Quaternary Science Reviews* 25: 126–41.

Townsend, P.K. 1974. Sago production in a New Guinea economy. *Human Ecology* 2: 217–36.

Townsend, P.K. 1990. On the possibility/impossibility of tropical forest hunting and gathering. *American Anthropologist* 92: 745–47.

Treerayapiwat, C. 2005. Patterns of habitation and burial in the Ban Rai rock shelter, northwestern Thailand. *Asian Perspectives* 44: 231–45.

Trình Năng Chung. 2008. Research on Hoà Bình culture in the three Indochina countries. *Vietnam Archaeology* 3: 19–32.

Trinkaus, E. 2005. Early modern humans. *Annual Review of Anthropology* 34: 207–30.

Trinkaus, E., Moldovan, O., Milota, S., Bilgar, A., Sarcina, L., Athreya, S., Bailey, S.E., *et al.* 2003. An early modern human from the Pestera eu Oase, Romania. *Proceedings of the National Academy of Sciences of the United States of America* 100: 11231–6.

Tsuji, Y., Matsutomi, H., Imamura, F., Takeo, M., Kawata, Y., Matsuyama, M., Takahashi, T., Sunarjo, Harjadi, P. 1995. Damage to coastal villages due to the 1992 Flores Island earthquake tsunami. *Pure and Applied Geophysics* 44: 481–524.

Tudhope, A.W., Chilcott, C.P., McCulloch, M.T., Cook, E.R., Chappell, J., Ellam, R.M., Lea, D.W., Lough, J.M., Shimmield, G.B. 2001. Variability in the El Niño Southern Oscillation through a glacial-interglacial cycle. *Science* 291: 1511–17.

Turner, A. 1992. Large carnivores and earliest European hominids: changing determinants of resource availability during the Lower and Middle Pleistocene. *Journal of Human Evolution* 22: 109–26.

Turner, A. 1984. Hominids and fellow travellers: human migration into high latitudes as part of a large mammal community. Foley, R. (ed.) *Hominid Evolution and Community Ecology.* Academic Press, London, pp. 193–218.

Turner, A. and Antón, M. 1996. The giant hyaena, *Pachycrocuta Brevirostris* (Mammalia, Carnivora, Hyaenidae). *Geobios* 29: 455–68.

Turney, C.S.M. and Brown, H. 2007. Catastrophic early Holocene sea level rise, human migration and the Neolithic transition in Europe. *Quaternary Science Reviews* 26: 2036–41.

Turney, C.S.M., Flannery, T.F., Roberts, R.G., Reid, C., Fifield, L.K., Higham, T.F.G., Jacobs, Z., *et al.* 2008. Late-surviving megafauna in Tasmania, Australia, implicate human involvement in their extinction. *Proceedings of the National Academy of Sciences of the United States of America* 105: 12150–53 (and Suppl.).

Turney, C.S.M., Bird, M.I., Fifield, L.K., Roberts, R.G., Smith, M., Dortch, C.E., Grün, R., *et al.* 2001. Early human occupation at Devil's Lair, southwestern Australia 50,000 years ago. *Quaternary Research* 55: 3–13.

Turq, A., Martínez-Navarro, B., Palmquist, P., Arribas, A., Agustí, J., Rodriguez Vidal, J. 1996. Le Plio-Pléistocène de la region d'Orce, province de Grenade, Espagne: Bilan et perspectives de recherches. *Paleo* 8: 161–204.

Tweedie, M.W.F. 1936. Report on cave excavations carried out in Bukit Chintamani, near Bentong, Pahang. *Bulletin of the Raffles Museum, Series B* 1: 17–25.

Tweedie, M.W.F. 1937. Minor excavations carried out in caves in Pahang and Johore. *Bulletin of the Raffles Museum, Series B* 3: 154–55.

Tweedie, M.W.F. 1953. The Stone Age in Malaya. *Journal of the Malay Branch of the Royal Asiatic Society* 26: 1–90.

Tweedie, M.W.F. 1957. *Prehistoric Malaya.* Donald Moore, Singapore.

U Aung Thaw. 1971. The 'Neolithic' culture of Padah-lin Caves. *Asian Perspectives* 14: 123–33.

Urquhart, D. 1989. Biology. E. Hall (ed.) *People and Caribou in the Northwest Territories.* Department of Renewable Resources, Government of the Northwest Territories: Yellowknife, pp. 67–73.

Ursin, H., Bergan, T., Collet, J., Endresen, I.M., Lugg, D.J., Maki, P., Matre, R., *et al.* 1991. Psychobiological studies of individuals in small, isolated groups in the Antarctic and space analogues. *Environment and Behavior* 23: 766–81.

Urushibara-Yoshino, K. and Yoshino, M. 1997. Palaeoenvironmental change in Java island and its surrounding area. *Journal of Quaternary Science* 12: 435–42.

Vaks, A., Bar-Matthews, M., Ayalon, A., Matthews, A., Halicz, L., Frumkin, A. 2007. Desert speleothems reveal climatic window for African exodus of early modern humans. *Geology* 35: 831–34.

Valladas, H., Reyss, J.L., Joron, J.L., Valladas, G., Bar-Yosef, O., Vandermeersch, B. 1988. Thermoluminescence dating of Mousterian 'Proto-Cro-Magnon' remains from Israel and the origin of modern man. *Nature* 331: 614–16.

Valladas, H., Tisnérat-Laborde, N., Cachier, H., Arnold, M., Bernaldo de Quirós, F., Cabrera-Valdés, V., Clottes, J., *et al.* 2001a. Radiocarbon AMA dates for Paleolithic cave paintings. *Radiocarbon* 43: 977–86.

Valladas, H., Clottes, J., Geneste, J.-M., Garcia, M.A., Arnold, M., Cachier, H., Tisnérat-Laborde, N. 2001b. Evolution of prehistoric cave art. *Nature* 413: 479.

Valladas, H., Cachier, H., Maurice, P., Bernaldo de Quiró, F., Clottes, J., Cabrera Valdés, V., Uzquiano, P., Arnold, M. 2006. Direct radiocarbon dates for prehistoric paintings at the Altamira, El Castillo and Niaux caves. Victoria Cabrera, Obra selecta. *Zona Arqueológica* 6: 129–31.

van den Bergh, G. 1999. *The Late Neogene elephantoid-bearing faunas of Indonesia and their palaeozoogeographic implications. A study of the terrestrial faunal succession of Sulawesi, Flores and Java, including evidence for early hominid dispersal east of Wallace's Line.* Scripta Geologica 117 (Special issue).

van den Bergh, G.D., de Vos, J., Sondaar, P.Y., Aziz, F. 1996. Pleistocene zoogeographic evolution of Java (Indonesia) and glacio-eustatic sea level fluctuations: a background for the presence of *Homo*. *Bulletin of the Indo-Pacific Prehistory Association* 14: 7–21.

van den Bergh, G.D., deVos, J., Sondaar, P.Y. 2001. The late Quaternary palaeogeography of mammal evolution in the Indonesian Archipelago. *Palaeogeography, Palaeoclimatology, Palaeoecology* 171: 385–408.

van den Bergh, G.D., Meijer, H.J.M., Rokus Awe Due, Morwood, M.J., Szabó, K., van den Hoek Ostende, L.W., Sutikna, T., Saptomo, E.W., Piper, P.J., Dobney, K.M. 2009. The Liang Bua faunal remains: a 95 kyr. sequence from Flores, East Indonesia. *Journal of Human Evolution* 57: 527–37.

van der Kaars, W.A. and Dam, M.A.C. 1995. A 135,000-year record of vegetational and climatic change from the Bandung area, West-Java, Indonesia. *Palaeogeography, Palaeoclimatology, Palaeoecology* 117: 55–72.

van Es, L.J.C. 1930. The prehistoric remains in the Sampoeng cave, Residency of Ponorogo, Java. *Proceedings of the Fourth Pacific Science Congress, Java, 1929: Vol. III. Biological Papers*. Batavia-Bandoeng, pp. 329–40.

Vanhaeren, M., d'Errico, F., Stringer, C., James, S.L., Todd, J.A., Mienis, H.K. 2006. Middle Paleolithic shell beads in Israel and Algeria. *Science* 312: 1785–88.

Vanhaeren, M., d'Errico, F., Billy, I., Grousset, F. 2004. Tracing the source of Upper Palaeolithic shell beads by strontium isotope dating. *Journal of Archaeological Science* 31: 1481–8.

van Heekeren, H.R. 1957. *The Stone Age of Indonesia*. Nijhof, The Hague.

van Heekeren, H.R. 1972. *The Stone Age of Indonesia*. Nijhof, The Hague.

van Heekeren, H.R. and Knuth, Count E. 1967. *Archaeological Excavations in Thailand: Vol. 1. Sai-Yok*. Munksgaard, Copenhagen.

Vanoverbergh, M. 1925. Negritos of northern Luzon, chapter III: Material life. *Anthropos* 20: 399–415.

van Stein Callenfels, P.V. and Evans, I.H.N. 1928. Report of cave excavations in Perak. *Journal of the Federated Malay States Museums* XII: 145–59.

van Stein Callenfels, P.V. 1936. An excavation of three kitchen middens at Guar Kepah, Province Wellesley, Straits Settlements. *Bulletin of the Raffles Museum, Series B* 1: 27–37.

van Stein Callenfels, P.V. and Noone, H.D. 1940. Report on an excavation in the rock-shelter Gol Ba'it, near Sungai Siput (Perak). Tweedie, M.W.F. (ed.) *Proceedings of the Third Congress of Prehistorians of the Far East*. Singapore, 1938, pp. 119–25.

Vekua, A., Lordkipanidze, D., Rightmire, G.P., Agusti, J., Ferring, R., Maisuradze, G., Mouskhelishvili, A., *et al.* 2002. A new skull of early *Homo* from Dmanisi, Georgia. *Science* 297: 85–89.

Velasco, J.A.F-T. 1989. Thoughts on the transition from Magdalenian to the Azilian in Cantabria: evidence from the Cueva de Los Azules, Asturias. Bonsell, C. (ed.) *The Mesolithic in Europe*. John Donald, Edinburgh, pp. 582–8.

Velat Bujeng. 2007. The evidence of bone tools used at Lobang Batu Puteh, Bukit Sarang, Bintulu, Sarawak in late Neolithic culture. *Jurnal Arkeologi Malaysia* 20: 84–102.

Verleyen, E., Hodgson, D.A., Milne, G.A., Sabbe, K., Vyverman, W. 2005. Relative sea-level history from the Lambert Glacier region, East Antarctica, and its relation to deglaciation and Holocene glacier readvance. *Quaternary Research* 63: 45–52.

Vermeersch, P.M. 2005. European population changes during Marine Isotope Stages 2 and 3. *Quaternary International* 137: 77–85.

Verna, C. and d'Errico, F. 2011. The earliest evidence for the use of human bone as a tool. *Journal of Human Evolution* 60: 145–57.

Verpoorte, A. 2009. Limiting factors on early modern human dispersals: the human biogeography of Late Pleniglacial Europe. *Quaternary International* 201: 77–85.

Verstappen, H.T. 1975. On palaeo climates and landform development in Malesia. *Modern Quaternary Research in Southeast Asia* 1: 3–36.

Vidal, L., Schneider, R.R., Marchal, O., Bickert, T., Stocker, T.F., Wefer, G. 1999. Link between the North and South Atlantic during the Heinrich events of the last glacial period. *Climate Dynamics* 15: 909–19.

Vishnyatsky, L.B. 1999. The Palaeolithic of Central Asia. *Journal of World Prehistory* 13: 69–122.

Vishnyatsky, L.B. 2004. The Middle-Upper Palaeolithic interface in former Soviet Central Asia. Brantingham, P.J., Kuhn, S.L. and Kerry, K.W. (eds.) *The Early Upper Palaeolithic Beyond Western Europe*. University of California Press, London, pp. 151–61.

Vislobokova, I. 2009. A new species of Megacerini (Cervidae, Artiodactyla) from the Late Miocene of Taralyk-Cher, Tuva (Russia), and remarks on the relationships of the group. *Geobios* 42: 397–410.

Voeks, R.A. 2007. Penan ethnobotany: subsistence strategy and breadth of knowledge. Sercombe, P. and Sellato, B. (eds.) *Beyond the Green Myth: Borneo's Hunter-Gatherers in the Twenty-First Century*. Nias Press, Copenhagen, pp. 262–88.

Voelker, A.H.L. and Workshop Participants. 2002. Global distribution of centennial-scale records for Marine Isotope Stage (MIS) 3: a database. *Quaternary Science Reviews* 21: 1185–1212.

Vogel, J.C. and Waterbolk, H.T. 1963. Groningen radiocarbon dates IV. *Radiocarbon* 5: 163–202.

Voris, H.K. 2000. Maps of Pleistocene sea levels in Southeast Asia: shorelines, river systems and time durations. *Journal of Biogeography* 27: 1153–67.

Vu The Long, deVos, J., Ciochon, R.L. 1996. The fossil mammal fauna of the Lang Trang caves, Vietnam, compared with Southeast Asian fossil and recent mammal faunas: the geographical implications. *Bulletin of the Indo-Pacific Prehistory Association* 14: 101–9.

Walker, A. and Shipman, P. 1996. *The Wisdom of Bones*. Weidenfeld and Nicolson, London.

Walker, D. and Sieveking, A. de G. 1962. The Palaeolithic industry of Kota Tampan, Perak, Malaya. *Proceedings of the Prehistoric Society* 5: 103–39.

Walker, M. 2005. *Quaternary Dating Methods*. John Wiley, Chichester.

Walker, M.J.C., Björck, S., Lowe, J.J., Cwynar, L.C., Johnsen, S., Knudsen, K.L., Wohlfarth, B., INTIMATE Group. 1999. Isotopic 'events' in the GRIP ice core: a stratotype for the Late Pleistocene. *Quaternary Science Reviews* 18: 1143–50.

Walker, M., Johnsen, S., Rasmussen, S.O., Popp, T., Steffensen, J.-P., Gibbard, P., Hoek, W. *et al.* 2009. Formal definition and dating of the GSSP (Global Stratotype Section and Point) for the base of the Holocene using the Greenland NGRIP ice core, and selected auxiliary records. *Journal of Quaternary Science* 24: 3–17.

Walter, R.C., Buffler, R.T., Bruggemann, H., Guillaume, M.M.M., Berhe, S.M., Negassi, B., Libsekal, Y., *et al.* 2000. Early human occupation of the Red Sea coast of Eritrea during the last interglacial. *Nature* 405: 65–69.

Wang, H.-W., Mitra, B., Chaudhuri, T.K., Palanichamy, M.G., Kong, Q.-P., Zhang, Y.-P. 2011. Mitochondrial DNA evidence supports northeast Indian origin of the aboriginal Andamanese in the Late Paleolithic. *Journal of Genetics and Genomics* 38: 117–22.

Wang, P., Tian, J., Cheng, X., Xu, J. 2003. Carbon reservoir changes preceded major ice-sheet expansion at the mid-Brunhes event. *Geology* 31: 239–42.

Wang, W., Potts, R., Hou, Y., Chen, Y., Wu, H., Yuan, B., Huang, W. 2005. Early Pleistocene hominid teeth recovered in Mohui cave in Bubing Basin, Guangxi, South China. *Chinese Science Bulletin* 50: 2777–82.

Wang, W., Potts, R., Baoyin, Y., Huang, W., Cheng, H., Edwards, R.L., Ditchfield, P. 2007. Sequence of mammalian fossils, including hominoid teeth, from the Bubing Basin caves, South China. *Journal of Human Evolution* 52: 370–9.

Wasson, J.T. 2003. Large aerial bursts: an important class of terrestrial accretionary events. *Astrobiology* 3: 163–79.

Watanabe, H. 1985. The chopper-chopping tool complex of Eastern Asia: an ethnoarchaeological-ecological re-examination. *Journal of Anthropological Archaeology* 4: 1–18.

Watanabe, O., Jouzel, J., Johnsen, S., Parrenin, F., Shojik, H., Yoshida, N. 2003. Homogeneous climate variability across East Antarctica over the past three glacial cycles. *Nature* 422: 509–12.

Waters, M.R. and Stafford, T.W., Jr. 2007. Redefining the age of Clovis: implications for the peopling of the Americas. *Science* 315: 1122–26.

Webb, C. and Allen, J. 1990. A functional analysis of Pleistocene bone tools from two sites in Southwest Tasmania. *Archaeology in Oceania* 25: 75–78.

Webb, S.G. 2006. *The First Boat People*. Cambridge University Press, Cambridge.

Webb, W.F. 1948. *Foreign Land and Fresh Water Shells*. St. Petersburg Printing, FL.

Webster, J. 2001. Creolizing the Roman provinces. *American Journal of Archaeology* 105: 209–25.

Wei, G.-B., Lister, A.M. 2005. Significance of the dating of the Majuangou site for understanding Eurasian mammoth evolution. *Vertebrata PalAsiatica* 43: 243–4 (in Chinese and English).

Wei, K.-Y., Chiu, T.-C., Chen, Y.-G. 2003. Toward establishing a maritime proxy record of the East Asian summer monsoons for the late Quaternary. *Marine Geology* 201: 67–79.

Wei, Q. 2001. Some archeological sites from the Early Pleistocene in China. *Indo-Pacific Prehistory Association Bulletin* 21: 148–51.

Weiss, E., Kislev, M.E., Simchoni, O., Nadel, D. 2004. Small-grained wild grasses as staple food at the 23,000-year-old site of Ohalo II, Israel. *Economic Botany* 58: S125–34.

Wells, R.S., Yuldasheva, N., Ruzibakiev, R., Underhill, P.A., Evseeva, I., Blue-Smith, J., Li, J., *et al.* 2001. The Eurasian heartland: a continental perspective on Y-chromosome diversity. *Proceedings of the National Academy of Sciences of the United States of America* 98: 10244–49.

Westaway, K., Morwood, M.J., Roberts, R.G., Rokus, A.D., Zhao, J.-X., Storm, P., Aziz, F., *et al.* 2007. Age and biostratigraphic significance of the Punung Rainforest Fauna, East Java, Indonesia, and implications for *Pongo* and *Homo*. *Journal of Human Evolution* 53: 709–17.

Westgate, J.A., Shane, P.A.R., Pearce, N.J.G., Perkins, W.T., Korisettar, R., Chesner, C.A., Williams, M.A.J., Acharyya, S.K. 1998. All Toba Tephra occurrences across Peninsular India belong to the 75,000 yr B.P. eruption. *Quaternary Research* 50: 107–12.

Westley, K. and Dix, J. 2006. Coastal environments and their role in prehistoric migrations. *Journal of Maritime Archaeology* 1: 9–28.

Whatrup, C.S. 1982. The analysis, description and interpretation of the vertebrate faunal remains from the cave site of Ulu Leang 1, Sulawesi with special reference to the pigs. Institute of Archaeology, London.

White, C. and Peterson, N. 1969. Ethnographic interpretations of the prehistory of western Arnhem Land. *Southwestern Journal of Anthropology* 25: 45–67.

White, J.C. 2004. Comment on dates from a resin-coated sherd from Spirit Cave, Thailand. *Antiquity* 78: 184–6.

White, J.C. and Gorman, C. F. 2004. Pattern in 'amorphous' industries: the Hoabinhian viewed through a lithic reduction sequence. Paz, V. (ed.) *Southeast Asian Archaeology: Wilhem G. Solheim II Festschrift*. University of the Philippines Press, Manila, pp. 411–43.

White, J.P., Crook, K.A.W., Ruxton, B.P. 1970. Kosipe: a Late Pleistocene site in the Papuan Highlands. *Proceedings of the Prehistoric Society* 36: 152–70.

White, R. 1989. Production complexity and standardisation in Early Aurignacian bead and pendant manufacture: evolutionary implications. Mellars, P. and Stringer, C. (eds.) *The Human Revolution*. Edinburgh University Press, Edinburgh, pp. 366–90.

White, R. 2007. Systems of personal ornamentation in the Early Upper Palaeolithic: methodological challenges and new observations. Mellars, P., Stringer, S., Bar-Yosef, O., Boyle, K. (eds.) *Rethinking the Human Revolution: New Behavioural and Biological Perspectives on the Origins and Dispersal of Modern Humans*. McDonald Institute Monographs, Cambridge, pp. 287–302.

White, T. 1952. Observations on the butchery technique of some aboriginal peoples. *American Antiquity* 17: 337–38.

White, T.D., Asfaw, B., DeGusta, D., Gilbert, H., Richards, G.D., Suwa, G., Howell, F.C. 2003. Pleistocene *Homo sapiens* from Middle Awash, Ethiopia. *Nature* 423: 742–47.

Wickler, S. and Spriggs, M. 1988. Pleistocene human occupation of the Solomon Islands, Melanesia. *Antiquity* 62: 703–6.

Widmer, R.J. 1989. Archaeological research strategies in the investigation of shell-bearing sites, a Florida perspective. Paper presented at the Annual Meeting of the Society for American Archaeology, Atlanta.

Wild, E.M., Teschler-Nicola, M., Kutschera, W., Steier, P., Trinkaus, E., Wanek, W. 2005. Direct dating of Early Upper Palaeolithic human remains from Mladeč. *Nature* 435: 332–35.

Williams-Hunt, P.D.R. 1952. *An Introduction to the Malayan Aborigines*. Government Press, Kuala Lumpur.

Winterhalder, B. 1980. Environmental analysis in human evolution and adaptation research. *Human Ecology* 8: 135–70.

Wobst, H.M. 1976. Locational relationships in Palaeolithic society. *Journal of Human Evolution* 5: 49–58.

Wolff, E.W., Chappellaz, J., Blunier, T., Rasmussen, S.O., Svensson, A. 2010. Millennial-scale variability during the last glacial: the ice core record. *Quaternary Science Reviews* 29: 2828–38.

Wood, B. and Turner, A. 1995. Out of Africa and into Asia. *Nature* 378: 239–40.

Woodroffe, C.D. 1990. The impact of sea-level rise on mangrove shorelines. *Progress in Physical Geography* 14: 483–520.

Woodroffe, C.D., Chappell, J., Thom, B.G. 1988. Shell middens in the context of estuarine development: South Alligator River, Northern Territory. *Archaeology in Oceania* 23: 95–103.

Worm, H-U. 1997. A link between geomagnetic reversals and events and glaciations. *Earth and Planetary Science Letters* 147: 55–61.

Wray, L. 1897. The cave dwellers of Perak. *Journal of the Anthropological Institute of Great Britain and Ireland* 26: 36–47.

Wray, L. 1905. Further notes of the cave dwellers of Perak. *Journal of the Federated Malay States Museums* 1: 13–15.

Wroe, S. and Field, J. 2006. A review of the evidence for a human role in the extinction of Australian megafauna and an alternative interpretation. *Quaternary Science Reviews* 25: 2692–2703.

Wu, R. and Dong, X. 1985. *Homo erectus* in China. Wu, R. and Olsen, R.W. (eds.) *Palaeoanthropology and Palaeolithic Archaeology in the People's Republic of China*. Academic Press, Orlando, pp. 79–89.

Wu, R. and Olsen, R.W. 1985. *Palaeoanthropology and Palaeolithic Archaeology in the People's Republic of China*. Academic Press, Orlando.

Wu, X. 2004. On the origin of modern humans in China. *Quaternary International* 117: 131–40.

Wurster, C.M., Bird, M.I., Bull, I.D., Creed, F., Bryant, C., Dungait, J.A.J., Paz, V. 2010. Forest contraction in north equatorial Southeast Asia during the Last Glacial Period. *Proceedings of the National Academy of Sciences of the United States of America* 107: 15508–11.

Xia, Z., Kong, X., Jiang, X., Cheng, H. 2007. Precise dating of East-Asian-Monsoon D/O events during 95–56 ka BP: based on stalagmite data from Shanbao Cave at Shennongjia, China. *Science in China Series D, Earth Sciences* 50: 228–35.

Yeo, D.C.J. and Ng, P.K.L. 2007. On the genus 'Potamon' and allies in Indochina (Crustacea: Decapoda: Brachyura: Potamidae). *Raffles Bulletin of Zoology* 16: 273–308.

Yi, S. and Clark, G. 1983. Observations on the Lower Palaeolithic of Northeast Asia. *Current Anthropology* 24: 181–202

Yi, S., Lee, J.-J., Kim, S., Yoo, Y., Kim, D. 2008. New data on the Hoabinhian: investigations at Hang Cho cave, northern Vietnam. *Bulletin of the Indo-Pacific Prehistory Association* 28: 73–79.

Yokoyama, Y., Falguères, C., Sémah, F., Jacob, T., Grün, R. 2008. Gamma-ray spectrometric dating of late *Homo erectus* skulls from Ngandong and Sambungmacan, Central Java, Indonesia. *Journal of Human Evolution* 55: 274–77.

Yost, J. and Kelley, P. 1983. Shotguns, blowpipes and spears: an analysis of technological efficiency. Hames, R. and Vickers, W. (eds.) *Adaptive Responses of Native Amazonians*. Academic Press, New York, pp. 189–224.

Yotova, V., Lefebvre, J-F., Moreau, C., Gbeha, E., Hovhannesyan, K., Bourgeois, S., Bédarida, S., *et al.* 2011. An X-linked haplotype of Neandertal origin is present among all non-African populations. *Molecular Biology and Evolution* 28: 1957–62.

Zachos, J., Pagani, M., Sloan, S., Thomas, E., Billups, K. 2001. Trends, rhythms, and aberrations in global climate 65 Ma to present. *Science* 292: 686–93.

Zaliznyak, L. 1998. The ethnographic record, and structural changes in the prehistoric hunter-gatherer economy of boreal Europe. Zvelebil, M., Dennell, R., Domańska, L. (eds.) *Harvesting the Sea, Farming the Forest.* Sheffield Academic Press, Sheffield, pp. 45–51.

Zhang, S. 1985. The Early Palaeolithic of China. Wu, R. and Olsen, R.W. (eds.) *Palaeoanthropology and Palaeolithic Archaeology in the People's Republic of China.* Academic Press, Orlando, pp. 147–86.

Zhou, C., Liu, Z., Wang, Y., Huang, Q. 2000. Climatic cycles investigated by sediment analysis in Peking Man's Cave, Zhoukoudian, China. *Journal of Archaeological Science* 27: 101–9.

Zhu, R.X., Hoffman, K.A., Potts, R., Deng, C.L., Pan, Y.X., Guo, B., Shi, C.D., *et al.* 2001. Earliest presence of humans in northeast Asia. *Nature* 413: 413–17.

Zhu, R.X., An, Z., Potts, R., Hoffman, K.A. 2003. Magnetostratigraphic dating of early humans in China. *Earth Science Reviews* 61: 341–59.

Zhu, R.X., Potts, R., Xie, F., Hoffman, K.A., Deng, C.L., Shi, C.D., Pan, Y.X., *et al.* 2004. New evidence on the earliest human presence at high northern latitudes in northeast Asia. *Nature* 431: 559–62.

Zhu, R.X., Potts, R., Pan, Y.X., Yao, H.T., Lü, L.Q., Zhao, X., Gao, X., Chen, L.W., Gao, F., Deng, C.L. 2008. Early evidence of the genus *Homo* in East Asia. *Journal of Human Evolution* 55: 1075–85.

Zielinski, G.A., Mayewski, P.A., Meeker, L.D., Whitlow, S., Twickler, M.S. 1996. A 110,000-yr record of explosive volcanism from the GISP2 (Greenland) ice core. *Quaternary Research* 45: 109–18.

Zilhão, J., Angelucci, D.E., Badal-García, E., d'Errico, F., Daniel, F., Dayet, L., Douka, K., *et al.* 2010. Symbolic use of marine shells and mineral pigments by Iberian Neandertals. *Proceedings of the National Academy of Sciences of the United States of America* 107: 1023–28.

Zöller, L. 2000. Chronology of upper Pleistocene 'red silts' in the Siwalik system and constraints for the timing of the Upper Palaeolithic in Nepal. *Catena* 41: 229–35.

Zuraina Majid. 1982. The West Mouth, Niah, in the prehistory of Southeast Asia. *Sarawak Museum Journal* 31 (Special Issue).

Zuraina Majid. 1994. *The Excavations of Gua Gunung Runtuh and the Discovery of the Perak Man in Malaysia.* Department of Museums and Antiquity, Malaysia, Kuala Lumpur.

Zuraina Majid. 1990. The Tampanian problem resolved: archaeological evidence of a late Pleistocene lithic workshop. *Modern Quaternary Research in Southeast Asia* 11: 71–96.

Zuraina Majid. 1998. Radiocarbon dates and the cultural sequence in the Lenggong Valley and beyond. *Malaysia Museums Journal* 34: 241–9.

Zuraina Majid. 2003. *Archaeology in Malaysia.* Pusat Penyelidikan Arkeologi Malaysia, Pulau Pinang.

Zuraina Majid. 2005. *The Perak Man and Other Prehistoric Skeletons of Malaysia.* Penerbit Universiti Sains Malaysia, Pinang.

Zuraina Majid, Ang Bee Haut, Jeffrie Ignatius. 1998a. Late Pleistocene–Holocene sites in Pahang: excavations at Gua Sagu and Gua Tenggek. *Malaysia Museums Journal* 34: 65–115.

Zuraina Majid, Jeffrie Ignatius, Tjia, H.D., Koon, P. 1998b. Some interesting Late Pleistocene–Early Holocene finds from excavations in Balambangan Island, Sabah, Malaysia. *Sabah Society Journal* 15: 29–40.

Zuraina Majid, Mokhtar Saidin, Chia, M.S., Zolkurnian Hasan. 1994. Artifacts from the Gua Gunung Runtuh excavation. Zuraina Majid (ed.) *The Excavations of Gua Gunung Runtuh and the Discovery of the Perak Man in Malaysia.* Department of Museums and Antiquity, Kuala Lumpur, pp. 149–68.

Index

Abdur, 88
Abri Pataud, 76
Accipiter trivirgatus, cf., 219
Acclimatization, 2, 7, 268, 271
Acculturation, 70, 279, 288
Acheulean, 26, 27, 28, 29, 45, 47, 48, 51, 52, 58, 88
Acheulo-Yabrudian, 48
Adaptation
  high frequency processes, 286
  hominin adaptive trajectories, 7, 267, 268
  low frequency processes, 286–287
  tropical foragers (Southeast Asia), 283
  Variability selection hypothesis, 285–286
Additive strategies
  economic, 274, 280. *See* Strategy-switching (economic)
  technological, 165, 206, 283, 289
Agassi, Lake, 285
Ahmarian, 80
Ailuropoda melanoleuca fovealis, 35
Airstrip Mound site, 136
Altai Mountains, 50, 51, 94, 103
Altamira, 73
Amyda cartilaginea, 218, 230
Amyda sp., 37
Anderson, D.D., 111, 201
Anorrhinus galeritus, 219
Anthracoceros coronatus, 219
Anthracoceros malayanus, 219
Anthropocene, 8, 9, 274, 286, 289
Aq Kupruk, 104, 105
Arboreal-adapted taxa, 96, 110, 111, 113, 122, 151, 152, 155, 167, 209, 217, 219, 220, 227, 228, 230, 231, 240, 246, 250, 257, 258, 261, 262, 263, 265, 282
Arborophilia sp., 219
Arctictis binturong, 218, 229, 230, 231, 263
Arctogalidia trivirgata, 229
Arctonyx collaris, 241
Arcy-sur-Cure, 75
Arius sp., 219
Asian leaf turtle. *See Cyclemys dentata*
Asian soft-shell turtle. *See Amyda cartilaginea*
Asian wild dog. *See Cuon alipinus*
Assamese macaque. *See Macaca assamensis*
Athapaskan, 278
Atlantic thermohaline circulation (THC), 23–24
Attirampakkam, 106
Aurignacian, 69, 71, 72, 73, 76, 78, 102, 103, 268, 272
  Developed-, 280
  Proto-, 70, 78
Australo-Melanesian population, 109, 116
Australopithecines (robust), 286
Azilian, 74

Bacsonian, 188, 192, 194
Balobok rock-shelter, 159
Ban Don Mun, 54
Ban Lum Khao, 164, 165
Ban Mae Tha, 54
Ban Rai, 203
Banteng. *See Bos* cf. *javanicus*
Banyan Valley Cave, 201
Barranco León, 29
BAT 1, 173, 174
BAT 2, 173
Bat hawk. *See Macheiramphus alcinus*
Batadomba-lena, 107, 108
Batek, 274

Batu Gajah, finds from near, 55
Bau Du, 248
Bay cat. *See Felis badia*
Bear cat. *See Arctictis binturong*
Bearded pig 282. *See Sus barbatus*
Beli lena kitulgala, 107
Bellan-bandi Palassa, 107
Bellwood, P.S., 5, 122, 173, 174, 181
Bennett's Wallaby. *See Macropus rufogiseus*
Bent-winged bat. *See Miniopterus* sp.
Black eagle. *See Ictinaetus malayensis*, cf.
Black hornbill. *See Anthracoceros malayanus*
Black pond/marsh turtle. *See Siebenrockiella crassocollis*
Blombos Cave, 82, 83
Bobongara, 133, 136
Bogatyri/Sinyaya Balka, 29
Bolac, Lake, 131
Bond cycles, 22, 24
Bond events, 24
Bone technology. *See* Organic technology
Bornean gibbon. *See Hylobates* cf. *muelleri*
Bos cf, *javanicus*, 217, 220
Bose Basin, 39, 45, 59
Boxgrove, 49
Brahminy kite. *See Haliaster indus*
Breuil, H., 1, 88
Bubing Basin, 35, 45
Buceros cf, *rhinoceros*, 219
Buffy fish owl. *See Ketupa* cf. *ketupa*
Bui Ceri Uato, 168, 170
Bukit Bunuh, 59, 101, 112, 113, 199
Bukit Jawa, 55, 62, 63, 197
Burial, secondary, 194
Burials
  flexed, 194, 196
  Neolithic, 195
  secondary, 107, 173, 180, 194
Bushy crested hornbill. *See Anorrhinus galeritus*
Butchery, 258–259, 282
  châine operatoire, 228, 230
  cut-marks, 28, 30, 212, 216, 217, 223, 227, 228, 229, 231, 240, 245, 246, 252, 258, 259, 260, 262, 263
  Gesher Benot Ya'aqov, 47
  Gua Sagu and Gua Tenggek, 254, 261
  Hang Boi, 248
  impact scars, 216, 223
  Liang Bua, 125
  marrow extraction, 28, 30, 216, 223, 242
  Niah Caves, 157, 205, 212, 216, 218, 220, 227, 228, 229–230, 231, 262, 263
  scrape-marks, 30, 218
  stiff-joint butchery, 227
Buwawansi, 181

Callao Cave, 115, 116, 140
Cambrian Explosion, 8
Canis
  Canis cyonoides, 63
  Canis lupus, 63
  Canis lupus variabilis, 63
  Canis yuanmouensis, 37, 63
Carcass transportation. *See* Procurement
Carmel, Mt. *See* Wadi-el-Mughara ('Valley of Caves') and Wadi Kebara
Ceramic technology
  ceramic material sourcing, 276
  ceramic stratigraphic horizons, 128, 151, 181
  pottery sherds, 127, 173, 180, 181, 192, 198, 201, 202, 246, 249, 250, 251
  pre-ceramic stratigraphic horizons, 154, 160, 163, 166, 167, 175, 179, 180, 181, 198, 208
Cercopithecidae, 111, 121, 156, 157, 162, 167, 217, 219, 221, 227, 228, 240, 246, 263
Cervus unicolor, 113, 114, 217, 220, 240, 245, 254, 256, 260
Châtelperronian, 70, 75, 279, 280
Chauvet, Grotte, 73, 74, 75
Chenjiawo, 34–35, 36
Chinese three-striped box turtle. *See Cuora* cf. *trifasciata*
Chirki-on-Pravara, 52
Chongokni, 59
CLIMAP programme, 108
Climate
  climatic amelioration, 97, 267
  climatic fluctuations, 3, 12, 60, 66, 88, 249, 286, 287
  climatic instability, 6, 8, 62, 66, 87, 98, 140, 272, 282, 284, 288
  flickering switch, 20, 289
  rapid climate change, 6, 18–24, 73, 284, 289
Clouded leopard. *See Neofelis diardi*
Clovis, 134, 270
Coastal lagoons. *See* Ecological communities
Colani, M., 185, 186, 187, 190, 199
Colonization. *See* Transference, Diversification, Innovation Divergence and Creolization
  'abortive' colonization of the Levant, 92–93, 97
  acquired knowledge, 270
  immigrant inviability, 269
  in-situ resource utilizsation, 269
  island colonization (tsunami), 126–127
  landscape-learning, 270
  Later Upper Pleistocene Model, 68
  maladaptation during, 269, 271, 279, 280
  mega-patches, 270
  Multiple dispersal hypothesis, 104
  Northern Route out of Africa, 101
  Overkill hypothesis 137. *See* Extinction
  pioneer populations, 269, 278
  population contraction, 99, 271, 272
  population replacement, 48, 99, 267
  Population-dispersal hypothesis, 68, 80
  re-colonization, 45, 66, 204, 268, 285
  risk-reduction. *See* Risk
  seasonal forays, 278
  Southern Route hypothesis, 6, 85, 102, 139
  trans-location, 271
  transported knowledge, 270
  Wave of advance model, 106, 268
Common palm civet. *See Paradoxurus hermaphroditus*
Complexity (behavioural), 2, 47, 84, 108, 272, 280
Con Moong cave, 162, 188
Cosquer, La Grotte, 73
Cowtail stingray. *See Pastinachus sephen*

Creolization, 278–279
cultural 'hybridization' as alternative for, 279
Crested goshawk. *See Accipiter trivirgatus*, cf.
Crocodylia
Crocodylus cf, *porosus*, 230
Crocodylus ossifragus, 64
Crocodylus sp., 64
Garialis bengawanicus, 64
Cuddie Springs site, 134, 136
Cultural drift, 84
Cuon alipinus, 35
Cuon antiquus, 63
Cuon sp., 64
Cuora (=*Pyxidea*) *mouhotii*, 245
Cuora ambionensis, 261
Cuora cf, *trifasciata*, 245
Cyber-somatic devices, 287
Cyclemys dentata, 218, 245, 254, 257
Cyclemys oldhami, 242, 245
Cynailulus (=Acinonyx jubatus) sp., 63
Cynailulus (=Acinonyx), 37, 63
Cyprinidae sp., 219

Da But, 162, 190, 246, 249, 250
Da Phuc, 163
Dabutian, 190
Daeo village caves, 181
Dansgaard-Oeschger (D-O) cycles, 20–21
Dark-throated leaf turtle. *See Cyclemys oldhami*
Darra-i Kur, 104, 105, 107
Dasyatidae spp., 246
Dating methods
Amino Acid Racemization (AAR), 234, 249
Argon/Argon, 100
Electron Spin Resonance (ESR), 39, 52, 62, 91, 92
ESR/U-Series, 58, 81, 82, 91, 123, 124
Fairbanks_0107 calibration curve, 10
Fission-track dating, 40
Optically Stimulated Luminescence (OSL), 55, 81, 88, 100, 101, 122, 138
Palaeomagnetism, 26, 29, 42, 46, 50, 51
Radiocarbon dating 9, 10, 16, 107, 110, 114, 115, 117, 118, 120, 122, 124, 127, 129, 130, 143, 146, 158, 163, 187, 211, 220, 233, 234, 268, 270. *See* Appendix
Thermoluminescence (TL), 45, 50
Dawādmi, 27
Decapoda, freshwater
Potamidae, 238
Varunidae, 238
Villopotamon sp., 238, 239
Deglaciation, 8, 120, 146, 148, 151, 153, 205
Melt-water pulse-IA (mwp-IA), 143
Denisova Cave, 50, 51, 61, 62, 66, 94
Dennell, R., 5
Developed Oldowan, 28, 30, 60
Devil's Lair, the (Western Australia), 133
Dicerorhinus sumatrensis, 217
Dina, 51
Divergence (adaptive), 2, 133, 134, 269, 278–281
Divergence (climatic), 20
Divergence (genetic), 26, 49, 51, 102, 126
Diversification, 139, 206, 269, 272–274, 280, 290
Dmanisi, 5, 29–31, 140
Dogania subplana, 218, 254, 261
Douara Cave, 80
Dubois, M.E.F.T., 4, 40, 63
Dune, 16R, 93
Duryong Cave, 175
Dzudzuana Cave, 80

Early Upper Palaeolithic (EUP), 50, 69
East Indian box turtle. *See Cuora ambionensis*
Ecological communities
coastal lagoons, 22, 130, 246, 250, 282
faunal turn-over events, 41, 62–64, 65
Kedung Brubus fauna, 64
megafauna, 77, 136, 137, 282
Ngandong fauna, 54, 57, 58, 62, 64, 285
novel faunal associations, 98, 272
Punung fauna, 62, 96, 285
salt marsh, 246, 249, 250, 282
species diversity, 14, 61, 76, 78, 98, 273, 289
species' response to climate change, 98
Stegodon-Ailuropoda fauna, 35, 54
Trinil H.K, fauna, 63, 64
Villafranchian fauna, 29, 62
Eemian interglacial
concordance with MIS-5e, 85
Late Eemian Aridity Pulse (LEAP), 86
El Niño Southern Oscillation (ENSO), 273, 281, 286
Elaphas
Elaphas antiquus, 55
Elaphas maximus, 55
Elaphas namidicus, 55
Elaphas sp., 34, 35
Emiran, 80
Epigravettian, 78
Equus yunnanensis, 35, 37
Estuarine crocodile. *See* Crocodylia - *Crocodylus* cf. *porosus*
Eugeissona utilis (hill sago palm), 275
Evolved Pavlovian (Eastern Gravettian), 76
Exchange. *See* Procurement
Extinction, 33, 36, 38, 51, 100, 124, 137, 267, 269, 284, 285, 290
Eyre, Lake basin, 131

Fa Hien-lena 107, 108. *See* Hominin fossils
Facteur, Le, 76
Fauna, vertebrate. *See* individual taxa
Fay-Ne 1 at Jebel Faya, 88, 93
Felis badia, 217
Felis cf, *planiceps*, 229
Ferrassie, La, 76
Flat-headed cat. *See Felis* cf. *planiceps*
Foote, R.B., 4
Franchthi Cave, 78, 79
Fruit bat. *See Rousettus* sp.
Fuente Nueva 1, 29
Fumane Cave, 74

Gallus cf, *gallus*, 219

Garrod, D.A.E., 1, 88, 89, 90, 91
Geoemydidae (hard-shell turtles), 118, 218, 221, 245, 254, 262
Gesher Benot Ya'aqov, 46–47, 48, 60
Giant squirrel. *See Ratufa affinis*
Gibbon. *See Hylobates* sp.
Gigantopithecus blacki, 39
Glacials and Interglacials, 12–13
Glaciations
  Weichselian, 86
Global Stratotype Section and Point (GSSP), recognised, 9, 143
Glover, I.C., 168, 169, 170, 171, 172, 177, 178, 179
Gol Ba'it, 194
Golo Cave, 130, 160, 161, 181
Gongwangling, 35, 36
Gopeng Beds, the, 56
Gorman, C.F., 200, 201, 208, 209, 264
Gran Dolina, 29, 48
Gravettian, Eastern, 70, 76, 103
Greater mouse deer. *See Tragulus napu*
Greenland Interstadials
  GIS-1, 19, 20, 21, 143, 145, 146, 147, 150, 164, 205, 211, 221, 224, 231, 242, 258, 262, 263, 270, 282, 283
  GIS-11, 110
  GIS-12, 120
Greenland Stadials
  GS-1, 19, 20, 21, 86, 143, 147, 148, 205, 211, 242, 251, 263, 270
  GS-2, 19, 143
  GS-3, 19
Greenland Stadials and Interstadials, 18–20
Gua Badak, 55, 56
Gua Balambangan, 115–116, 158, 164, 174
Gua Bintong, 151, 163, 194
Gua Braholo, 153, 162, 167
Gua Cha, 195, 216, 257
Gua Gunung Runtuh, 101, 196–197
Gua Kechil, 164, 195
Gua Kerbau, 193, 194, 195
Gua Lawa, 150, 151, 162, 166, 167
Gua Naga, 56
Gua Pandan, 166
Gua Sagu, 210, 261, 262, 263
  faunal analysis from (Gua Sagu and Gua Tenggek), 251–261
  lithic technology from, 198–199
Gua Samang Buat, 122
Gua Sireh, 158, 164
Gua Tambun, 56
Gua Tenggek, 210
  faunal analysis from (Gua Sagu and Gua Tenggek), 255
  lithic technology from, 198, 199
Guak Kepah, 194
Gunung Sewa, 153, 167
Gunung Tempurung (and Gua Tempurung), 56, 57
Guri Cave, 175

Hadar, 65
Hagop Bilo, 153, 173, 184
Hairy-nosed otter. *See Lutra sumatrana*
Haliaster indus, 219
Hang Boi, 188–189, 210, 232, 233, 235, 253, 257, 258, 261, 263, 264, 273
  camp fire, modern, 238
  faunal analysis from, 232–251
  freshwater crabs, 239
  organic technology, 163
  perforated Neritidae shells, 239
Hang Chổ, 190
Hang Hum I and II, 109
Hang Trống, 163, 188, 189, 191, 262
Haplogroups
  A, 285
  B2, 285
  C, 103
  C1, 285
  D, 103
  D1, 285
  E1, 149, 150
  E1a, 149
  E1a1a, 149, 150
  E1b, 149
  E2, 149, 150
  E2a, 149
  E2b, 149
  G, 103
  L3, 97, 102, 104
  M, 97, 102, 104, 134
  M21, 276
  M31, 102
  M32, 102
  N, 102, 104
  Q, 134
  X2a, 285
Happisburgh, 29
Harrisson, T., 117, 118, 151, 231
  Sarawak Museum Archive of, 117, 123, 211
Hathnora, 51, 52, 107
Haua Fteah, 87
Heinrich Events, 21–23
Helarctos malayanus, 218
Heosemys spinosa, 218
Herto, upper fossiliferous deposits, 100
Himalayas, 31, 97
Hipposideros sp., 258
Hoabinhian, 112, 113, 114, 151, 163, 185–191, 192, 193–195, 197, 198, 199–202, 203, 206, 209, 236, 248, 250
Hog badger. *See Arctonyx collaris*
Holocene
  base of, 9
  Climatic Optimum, 260
Hominin fossils. *See* individual *Homo* species
  Batadomba-lena remains, 107
  Beli lena kitulgala remains, 107
  Bellan-bandi Palassa remains, 107
  C1 (Tabūn), 92
  Callao Cave remains, 116
  Darra-i Kur skull, 104
  Deep Skull, post cranial remains associated with, 118
  Deep Skull, the (Niah Caves), 101, 116–118, 119

Dmanisi hominins, 30–31
Fontéchevade hominin, 71
Irhoud 3 (Jebel Irhoud), 62, 87, 100, 103
KC4 (Kent's Cavern), 71
KNM-ER 26, 137
KNM-ER 26, 286
KNM-ER 26
KNM-ER, 42700, 26
KNM-ER, 42703, 26
KNM-WT, 15000, 26
LB1 (Liang Bua) 40, 124, 125. *See* Microcephalic, modern human; Liang Bua
LB-JR-8a (Liang Bua), 124
LBS4–32a (Liang Bua), 124
Liujing hominin (Tongtianyan Cave), 94, 96
Moh Khiew hominin, 109
Mungo III (Lake Mungo), 135, 139
Narmada specimens (Hathnora), 51, 62, 140
Ngandong hominins, 57, 58
Omo I and II hominins, 100
'Perak Man' (Gua Gunung Runtuh), 196
pigmented skull fragments (Niah Caves) 118. *See* Pigmentation
PU-198 (Punung), 96
Qafzeh hominins, 92
Skhūl hominins, 90–91, 92
Skhūl V, 90, 97
Sunghir I-III, 74–75
Tabon hominins, 114–115
TD6 (Gran Dolina), 48
Wajak, 4
YF-86B-N7-3 (Fa Hien-lena child), 93, 107
Zhirendong hominins, 95, 96
Hominoid fossils. *See Gigantopithecus blacki*
MH001 & MH0018 (Mohui Cave), 39
Yunxian crania, 96
Homo antecessor, 29, 48, 52
Homo erectus, 25–26, 31, 34, 35, 39, 40, 41, 45, 49, 51, 52–53, 54, 56–58, 60, 62–64, 65, 83, 94, 96, 109, 123, 126, 140, 267, 285
Homo (=Pithecanthropus) erectus, 4, 40
Homo erectus (=Sinanthropus pekinensis), 4
Homo ergaster, 25, 26, 31, 37, 39, 41
Homo floresiensis, 26, 40, 83, 115, 124, 125, 126, 127, 140, 167, 283, 286, 288
Homo georgicus, 31
Homo habilis, 26, 30, 31, 37, 39, 115
Homo heidelbergensis, 29, 48, 51, 52, 61, 106
Homo sapiens idaltu, 88
Homotherium ultimus, 63
Howieson's Poort, 81, 82, 107
Hoxne, 49
Human behaviour, modern, 2, 6, 69, 77, 82, 84, 130, 133, 140, 265, 266, 267, 288
Human Revolution in Europe, traits, 68–69
blade-based technologies, 69–70
mobility, 75–76
new lithic tool types, 70–71
non-lithic technological and representative media, 71–75
pace of material culture innovation, 77
population density, 76–77
subsistence, 80
Hunting strategies. *See* Procurement
Hyaenidae
Crocuta crocuta, 28
Crocuta crocuta ultima, 63
Crocuta sp., 64
Hyaena (=Pachycrocuta) brevirostris, 63, 64
Hyaena (=Pachycrocuta) licenti, 63
Hyaena sinensis, 63
Hyaena sp., 53, 62, 63
Pachycrocuta (=Hyaena) licenti, 37, 53
Hylobates cf, *muelleri*, 217, 228
Hylobates sp., 96, 219, 221
Hystricidae, 39, 219, 252

Ice cores (Antarctic)
Dome Fuji core, 18
EPICA, 18
EPICA (Dronning Maud Land core), 18
EPICA Dome C core, 15, 60
Vostok core, 15, 17, 18, 21
Vostok Station, 18
Vostok, Lake, 17
Ice cores (Greenland)
Camp Century, 15
GISP2, 15, 16, 17, 18, 19, 20, 125, 147
GRIP, 15, 17, 20, 62, 86, 87
NEEM, 15, 17
NGRIP, 15, 16, 20, 86, 87, 93
Renland cores, 15
Ice-rafted debris (IRD), 21, 22, 24
Ice-sheets
East Antarctic, 18, 143
Fennoscandian, 23, 98
Laurentide, 21, 143
Ictinaetus malayensis, cf., 219
Ille Cave, 159, 175, 176
Innovation, 1, 2, 68–77, 80, 138, 142, 151, 152, 155, 158, 164, 172, 185, 196, 206, 265, 266, 269, 274–275, 279, 280, 282, 284, 289
Inter-tropical Convergence Zone (ITCZ), 108, 286
Isampur locality, 52

Jalalpur, 51
Jamestown, Virginia, 268, 269, 271
Jebel Irhoud 62, 87. *See* Hominin fossils
Jerimalai, 128, 172, 184
Joe's Garden site, 136
Jungle fowl. *See Gallus* cf. *gallus*

Kalumpang sites, 194
Kao Pha Nam, 54
Kara Karmar, 104, 105
Karama, 50
Karatau I, 49
Katanda, 82
Keeled box turtle. *See Cuora (=Pyxidea) mouhotii*
Keewatin ice dome (Canada), 148
Keo Leng, 109
Keplek industry, 123, 167
Ketupa cf, *ketupa*, 219

Khao Talu, 201, 202
Khok Phanom Di, 148, 149, 158, 164
Kilu Cave, 136
Klasies River, 81
Klisoura Cave 1, 78
Kosipe, 134, 136
Kostenki I, 79
Kota Tampan, 4, 55
  KT1987, KT2005, 100, 109, 112, 196, 197, 199
Kria Cave, 160–161, 182–183
Ksar 'Akil, 81
Kul'dara, 31, 49
Kulturpumpe model, 73, 77
Kuprukian, 105
Kutikini Cave, 138

Laang Spean, 192
Lakhuti I, 49
Lakonis I, 70
Landscape-learning. *See* Colonization
Lang Bon, 162
Lang Kamnan, 201, 202, 203, 204
Lang Rongrien, 109–111, 112, 152, 185, 199, 201–202, 203, 204, 273
Lang Vanh, 187
Lascaux, 73
Last Glacial Maximum (LGM)
  definition, 9
Last Termination
  definition, 10
Late Pleistocene
  definition, 9
Lawin, 197
Leaf monkeys, 217, 228, 257
Leaf monkeys and macaques. *See* Cercopithecidae and individual species
Leang Burung 1, 128
Leang Burung 2, 128, 129
Leang Sakapao 1, 129
Leang Sarru, 180
Leang Tuwo Mane'e, 179–180, 184
Lene Hara, 127, 128, 172, 173, 184
Lenggong Valley, 55, 113, 122, 196, 197, 276
Leopard cat. *See Prionailurus bengalensis*
Lepus wongi, 35
Liang Bua, 123–127, 140, 167
Liang Lemdubu, 153, 161, 162, 183, 184
Liang Nebulei Lisa, 161
Lie Siri, 168, 169
Lithic raw materials
  basalt, 30, 115, 176, 198, 200
  calcite, 200
  chalcedony, 110, 202, 203
  chert, 45, 59, 76, 110, 111, 112, 115, 119, 122, 123, 124, 127, 129, 159, 167, 172, 174, 175, 176, 177, 180, 182, 183, 184, 192, 194, 197, 200, 202, 203
  flint, 29, 74, 92, 168
  granite, 197
  hornfels, 192
  iron-oxide, 197
  jasper, 94, 111, 119, 176, 202
  limestone, 29, 119, 123, 182, 184, 189, 191, 197, 198, 203
  obsidian, 76, 135, 165, 166, 167, 168, 176, 179, 180, 184
  quartz, 37, 45, 110, 112, 115, 119, 193, 197, 202, 203
  quartzite, 30, 35, 45, 50, 76, 111, 112, 182, 183, 192, 194, 197, 198, 200, 201, 202, 203
  sandstone, 45, 192, 197
  sandstone, metamorphic, 119
  schist, 193, 194
  shale, 203
  silicified limestone, 110, 166, 177, 183, 202
  silicified tuff, 182
  slate, 94, 194, 197, 198, 201
  suevite, 59, 112
Lithic technology
  adzes, 110, 136, 150, 169, 171, 197, 200
  arrowheads, bifacially flaked, 166
  arrowheads, hollow-base, 166, 177
  arrowheads, round-base, 166
  backed blades, 169
  bifacial implements, 28, 30, 31, 37, 59, 65, 110, 122, 125, 129, 192, 195, 197, 198, 199, 201, 202
  blade cores, 169, 172, 176
  blade technology, 69–70, 73, 80, 105, 113, 175, 180, 184, 206, 281
  blade/bladelet technology, 81
  blade-flakes, 187, 189, 191
  burins, 48, 70, 80, 94, 106, 119, 168, 169, 171, 172, 175, 189, 191
  Chopper-chopping tool schema, 59
  choppers, 28, 94, 115, 123, 186, 190, 192, 194, 197, 199, 201, 202
  composite technology, 81
  curation, 130, 172, 189
  débitage, 32, 47, 104, 105, 112, 115, 129, 130, 180
  decortification flakes, 203
  edge-gloss, 129, 167, 168, 169, 170, 171, 174, 176, 178, 184
  edge-ground pieces (proto-neoliths), 114, 186, 187, 188, 190, 193
  flake-based industries, 185, 192
  flake-blades, 104, 115, 165, 168, 170, 172, 174, 175, 176, 178, 179, 180, 184, 185, 186, 189, 206
  grinding stones, 150, 167, 173, 194, 200
  ground stone implements, 136, 183
  hafting, 136
  hammerstones, 115, 175, 176, 180, 182, 186, 197, 198, 201, 202, 203
  hard-hammer, freehand, 167
  micro-blades, 105, 106, 107, 124
  micro-cores, 105
  microliths, 107, 166
  multiplatform cores, 113, 123, 168, 169, 182
  naturally backed knives, 167
  nosed end-scrapers, 70, 168, 169, 171
  pebble tools, 31, 113, 122, 166, 173, 174, 185, 187, 188, 189, 190, 195, 197, 200, 201, 202, 204
  perforators, 124, 167
  polished stone, 136, 166, 173, 189, 200, 201
  pounders, 150, 168, 173, 194

prismatic cores, 70, 174, 184
pseudo Levallois points, 105, 129
re-tooling, 162, 189
retouched bladelets, 70
retouched blades, 70, 105, 171
retouched flakes, 94, 105, 113, 182, 183, 195, 200
scalar cores, 129
scrapers, 32, 35, 48, 70, 80, 93, 94, 104, 105, 113, 119, 135, 136, 165, 166, 167, 168, 169, 170, 172, 186, 190, 192, 201, 202
short axes, 186, 188, 192, 200, 201, 203
single platform cores, 189
soft-hammer reduction, 189, 190, 191
standardization, 45, 49, 105, 277, 282, 283, 289
Sumatralith, 186, 187, 188, 190, 192, 194, 195, 197, 199, 200, 201, 202, 203
tanged points, 169, 172, 184
thumb-nail scrapers, 136, 168, 171
Tingkayu bifaces, 130, 174, 206, 277
unifacial implements, 37, 114, 167, 186, 197, 198, 199, 202
unretouched pieces, 165, 180, 181
utilized flakes, 129, 167, 168, 169, 170, 174, 175, 176, 184, 201
waisted blades, 136
Liujiang. *See* Hominin fossils
Loess Plateau, Chinese, 27, 32, 34, 35, 41–42
Longgupo Cave, 35, 39
Long-tailed macaque. *See Macaca fascicularis*
Lower Palaeolithic, 5, 46, 65
Lower Pleistocene
definition, 9
Luochuan section, 34, 42
Lutra sumatrana, 218

Macaca
Macaca arctoides, 246, 247
Macaca assamensis, 245
Macaca fascicularis, 227, 229
Macaca mulatta, 245
Macaca nemestrina, 221, 227, 246, 257
Macaca sp., 113, 114, 162, 229
Macaca spp., 240
Macaques, 217, 227, 228
Macheiramphus alcinus, 219
Macropus rufogriseus, 138
Madai caves (MAD 1/28 and MAD 2), 158, 173–174
Magdalenian, 72, 73, 272
Mai Da Điều, 186, 190
Majuangou, 32, 33
Maladaptation. *See* Colonization
Malakunanja II, 133, 136
Malay civet. *See Viverra tangalunga*
Malayan flat-shell turtle. *See Notochelys platynota*
Malayan soft-shell turtle. *See Dogania subplana*
Malayan tapir. *See Tapirus indicus*
Mammuthus primigenius, 34
Mammuthus trogontherii, 34
Mangrove forest, 120, 146, 148, 223, 238, 246
exploitation of, 158, 282
growth of, 249
Mansuli site, 122
Marine cores
MD900963, 18
ODP, 17940, 143, 145
ODP 17954, 143
ODP 17961, 143
ODP 17964, 147
ODP 18252–3, 143
ODP 18287–3, 143, 148
ODP 43, 64
ODP 758B, 45
Marine oxygen isotope stage (MIS)
MIS-2, 19, 60, 87, 93, 131, 267, 284, 285
MIS-3, 21, 98, 108, 112, 122, 127, 131, 137, 267, 284
MIS-4, 109, 131, 285
MIS-5, 19, 86, 87, 92, 93, 139, 285
MIS-5a, 92
MIS-5b, 131
MIS-5c, 92
MIS-5d, 86, 93, 95, 97
MIS-5e, 56, 66, 85, 86, 87, 88, 92, 93, 95, 96, 137, 284, 285
MIS-5e, Event 1 & 2, 86
MIS-6, 48, 50, 51, 62, 86, 87, 88, 93, 95, 96, 99, 100, 286
MIS-7, 56, 61, 66
MIS-8, 48, 61
MIS-9, 18, 50
MIS-10, 61
MIS-11, 48, 49, 60, 61, 87, 284, 285
MIS-12, 60, 61
MIS-13, 49, 53, 61
MIS-14, 60, 61
MIS-15, 49, 53, 60
MIS-16, 60
MIS-17, 53, 60
MIS-18, 47, 60
MIS-19, 43, 53, 100
MIS-20, 43, 47, 100
Masked palm civet. *See Paguma larvata*
Mast-fruiting events, 273
Mata Menge, 57, 64, 123, 126
Matar, 58
Matenbek, 135
Matenkupkum, 135
Mauer, 48, 60
Mayr, E.W., 2
Mececyon (=Cuon) trinilensis, 63
Medway, Lord (Earl of Cranbrook), 151
Megantereon inexpectatus, 63
Megantereon nihowanensis, 63
Meilkane, 81
Melanesian population (modern), 196
Melayu Asli, 276
Mellars, P.A., 69, 84
Menez Dregan, 48
Mesolithic, 105, 163, 177, 186
Metal Age (Southeast Asia), 173, 175, 179, 180
Microcephalic, modern human, 4, 40, 126
Middle Palaeolithic, 30, 46, 69, 70, 72, 74, 75, 76, 77, 78, 79, 81, 93, 105, 106, 107, 140

Middle Pleistocene
definition, 9
Middle Son Valley, 52, 93
Middle Stone Age (MSA), 81, 82, 83, 88, 92, 93, 285, 288
Milankovich cycles, 10–12
Miniopterus sp., 258
Mladeč, 70, 71
Moh Khiew 110–111, 114, 163, 201, 202. *See* Hominin fossils
Mohui Cave, 39–40
Mojokerto, 26, 40
Mokhovo I, 50, 61
Mollusca, brackish water
Neritina (Dostia) cornucopia, 238
Neritina (Neripteron) violacea, 238
Neritina petitii, 220
Neritodryas dubia, 220
Neritodryas subsulcata, 220
Polymesoda erosa, 220
Mollusca, freshwater
Angulyagra (=Viviparus) polyzonata, 237
Angulyagra polyzonata, 237
Angulyagra sp., 114
Brotia (=Antimelania) costula, 114
Brotia (=Antimelania) siamensis, 114
Brotia perfectora, 129
Brotia sp., 196, 237
Brotia spinoza, 253, 254, 263
Corbicula, 34, 147, 162, 237, 249
Cristaria herculea, 237
Cuneopsis, 34
Lamprotula, 34
Limnea, 34
Pisidium, 34
Planorbis, 34
Unionidae, 237
Mollusca, marine
Cypraea sp., 238
Cyprea arabica, 248
Mollusca, terrestrial
Camaenidae, 236
Clausiliidae, 236
Cyclophorus dodrans, 114
Cyclophorus fulganatus, 114
Cyclophorus siamensis, 114, 188
Cyclophorus speciosus, 114
Cyclophorus spp., 236
Cyclophorus theodori, 236
Cyclophorus unicus, 236
Japonia?, 236
Leptopoma, 236
Pterocyclos, 236
Pupinidae, 236
Scabrina, 236
Streptaxidae, 236
Subulinidae, 236
Zonitidae, 236
Mongoloid Austronesian population, 149
Monitor lizards. *See Varanus* spp.
Monsoon
Asian monsoon, 42, 97
Summer monsoon, 42, 97, 98, 104, 132
Winter monsoon, 42, 97, 98
Mortuary profiles
age structure, 121, 227, 228, 231, 263
prime adult biased, 217
young adult biased, 231
Morwood, M.J., 124, 127
Mount Cripps cave, 137
Mouse deer. *See Tragulus* sp.
Mousterian, 46, 48, 50, 51, 62, 66, 70, 74, 80, 83, 85, 88, 90, 92, 94, 104, 105, 272, 280
Movius Line, the, 60
Movius, H.L, Jr., 5, 65, 83
Mungo, Lake 133. *See* Hominin fossils.; Willandra Lakes
Muntiacus, 55, 217, 241, 245
Muntjac (barking deer). *See Muntiacus*
Muong-Nong/Austalasian tektite strewn field (MN/AT), 43, 44–45, 267
Musang Cave, 159

Natural selection, 2, 66, 99
Nauwalabila I, 133, 136
Nawamoyn, 136
Neofelis diardi, 218, 229
Neogene, 9, 12
Neolithic, 117, 123, 127, 150, 165, 166, 175, 177, 179, 186, 195, 199, 201, 212, 251, 274, 276, 285, 288
Ngandong, 57, 58
Ngandong fauna. *See* Ecological communities
Ngebung 2, 58
Nguòm, 113, 114, 185, 190
Nguòmian, 113, 114, 201
Niah Caves, 101, 116–122, 151, 152, 153, 156, 157, 172–173, 205, 206, 211, 221, 257, 273, 277
Deep Skull. *See* Hominin fossils
Gan Kira, 223
Kain Hitam, 156
Lobang Hangus, faunal analysis from, 221–232, 258
removal of toxins (pit features), 123
West Mouth, Area A, faunal analysis from, 212–220, 258, 259
West Mouth, Area D, faunal analysis from, 220–221, 222
West Mouth, 'Hell' trench, 116, 117, 118, 119, 120, 121, 172, 205, 212
Niaux, 73
Nihewan Basin, 31, 34
Niujianbao, 35
North Atlantic conveyor 23, 24. *See* Atlantic thermohaline circulation (THC)
North Atlantic deep water current (NADW), 23, 24
Norton, C.J, and Braun, D.R., 5
Notochelys platynota, 218
Nyctereutes sinensis, 63

Ohalo II, 80
Okladnikov Cave, 51
Okladnikov, A.P., 82
Omo 64, 65. *See* Hominin fossils
Oral traditions, 287
Orangutan. *See Pongo pygmaeus*

Organic technology
antler, worked, 72, 80, 110
bamboo, 152, 176, 203, 209
blowpipes, 152, 228
'bone culture' hypothesis, 150, 151
bone points, 71, 82, 83, 121, 152, 154, 155, 163, 164, 178, 181, 188, 200, 220, 262
bone technology, 70, 72, 120, 150–165, 166, 177, 186, 205, 206, 220, 228, 246, 250, 261, 263, 265
bone technology, standardization, 154, 283
bone tool manufacturing debris, 154
bow and arrow, 152
catfish barbs, 157
composite technology, 158, 160, 161, 206, 277, 282
curation, bone technology, 157
edge-ground pieces, 150, 158, 159, 164
experimental studies, 151, 154, 157, 161
fletching, 152
gouges, 155, 156, 160
groove and snap technique, 110, 159
hafting, bone points, 152, 155, 158, 159, 160, 161
hafting, bone tool, 165
hafting, fibres, 157
needles, 158
perforators (awls), 71, 72, 150, 154, 159, 160, 161, 206
pig tusk, 120, 157, 159
resin, used in hafting, 157, 158
re-tooling, 164
Sampung bone industry, 151, 163
'spatulate' bone tools, 150, 159, 160, 161, 162, 177
stingray barbs, 157, 158, 164, 206, 264, 282
tanged arrowhead, 158
use-wear, bone points, 155, 159, 162
yam planters, 156
Ornamentation
beads, standardization, 74
body ornamentation, 1, 70, 72, 73, 74, 75, 80, 238
body ornamentation\t, 29
N. gibbosulus shell beads from Mugharet es-Skhūl, 81
perforated shell beads, 239
shell beads, perforated, 81, 108, 238
Oued Djebbana, 81

Pabbi Hills, 51
Pacitan, 55, 166, 167
Pacitanian, 165
Padah-Lin, 192
Paguma larvata, 217, 229
Pakefield, 29, 48, 60
Palaeo-cultures
definition, 280
quasi-stable nature of, 280
Palaeo-lake environments, 29, 32, 47, 55, 112, 113, 122, 131
Palaeolithic. *See* Lower, Middle, Upper Palaeolithic
use of term in this volume, 10
Palaeo-magnetic events
Laschamp, 9
Matuyama-Brunhes reversal, 9, 43, 47
Mid-Brunhes (MBE), 11, 61
Olduvai, 9
Panthera
Panthera cf, *pardus*, 37, 63, 64, 240, 241, 249, 258, 263
Panthera cf, *tigris*, 63
Panthera tigris, 37, 63, 64, 218, 241, 258
Panthera tigris sondaica, 64
Panthera youngi, 63
Paradoxurus hermaphroditus, 229
Parmerpar Meethaner, 136
Partridge. *See Arborophilia* sp.
Paso shell midden, 160, 179
Pastinachus sephen, 157, 158
Patne, 107, 108
Patpara, 93
Patte, M.E., 162, 185
Pavlov I, 79
Peñablanca caves, 115, 159, 176
Pestera cu Oase 1, 71
Phia Vai, 248
Pied hornbill. *See Anthracoceros coronatus*
Pigmentation
early use of at Niah Caves, 118, 119, 277
non-symbolic use of, 158
ochre, use of, 74, 82, 118, 135, 158
pigmented cowrie shell, 238
pigmented geoemydidae plastron (Niah Caves), 119
portrayal of early use, 72
skull fragments (Niah Caves), 118
tree-resin, 118
use of in Southeast Asia, 122, 130, 168, 289
Pig-tailed macaque. *See Macaca nemestrina*
Plant macro-molecules (lipids), study of, 250
Pleistocene Explosion of new life ways 290. *See* Cambrian Explosion
Pliocene epoch, 25, 42
Late Pliocene, 32, 34, 42
Plio-Pleistocene, 25, 29, 32, 34, 35, 41, 56, 65, 66
Pongo pygmaeus, 54, 96, 113, 114, 120, 217, 221, 227
Population genetics
aboriginal hair, DNA from, 102
bottle-neck events, 42, 99, 100, 101, 150, 205, 272, 285
coalescence, 10, 99, 102, 103, 135, 149
common ancestor, 26, 49, 51, 102, 276
daughter populations, 99, 101, 279
genetic drift, 49
genetic variability, 99, 149
haplogroups, individual. *See* Haplogroups
issues with, 103–104
larger long-term effective populations, 99
MtDNA, 49, 51, 85, 97, 99, 102, 103, 135, 149, 205, 276, 285
Orang Asli indigenous group (Peninsular Malaysia), 102, 109
Out of Africa II genetic model, 96, 98–99, 205
population generators, 102
Weak Garden of Eden hypothesis, 99
Y-chromosome, 85, 97, 99, 103, 104
ρ (rho) statistic, 102
Porcupine 241, 252, 259. *See* Hystricidae
Presbytis cristata, 221
Presbytis sp., 217, 262
Prionailurus bengalensis, 218, 229

Procurement 140. *See* Additive strategies. *See* Mortuary profiles
broad spectrum, 209, 264, 282
coastal foraging zone, 247
'dinnertime' camps, 137
embedded, 76
exchange, 76, 158, 195, 248, 250, 276, 278, 281
fishing, 128, 150, 155, 158, 164, 172, 173, 248, 261, 282
foraging radius, 247
hunting pressure, 137, 138, 217
hunting strategies, 121, 122, 152, 216, 219, 227, 257, 258, 261, 272, 273
logistical, 247, 278
mobility strategies 58, 134, 135, 208, 249, 250, 262, 264, 269, 276, 281, 283. *See* Human Revolution in Europe, traits.
prey selection, 228, 231, 264
removal of toxins (Niah pit features), 122, 123, 289
residential camps, 137, 138, 247, 248, 273, 278
sago, 121, 152, 209, 275
'schlepp' effect, 248
seasonality, 78, 110, 138, 209, 249, 250, 281
specialized, 77, 78, 138, 139, 272, 278, 280, 286, 290
subsistence strategies, 77, 92, 98, 114, 130, 139, 141, 207, 265, 283
swidden-based agriculture, 210, 274
transportation, 28, 217, 247, 248, 249, 251, 262, 263
traps, 111, 227, 229, 231, 246, 265
tropical foragers, 133, 142, 143, 151, 203, 208, 247, 250, 274, 276
yams, 209, 210
Pungitius sp., 34
Punung fauna. *See* Ecological communities
Python reticulatus, 218

Qafzeh, 92, 105
Quaternary
definition, 9

Rangifer tarandus, 76, 272
Rapid climate change. *See* Climate
Ratufa (=Sciurus) bicolor, 242
Ratufa affinis, 219, 221
Regionalism (adaptive), 7, 24, 65, 82, 288
Reindeer. *See Rangifer tarandus*
Reticulated python. *See Python reticulatus*
Rhesus macaque. *See Macaca mulatta*
Rhinoceros hornbill. *See Buceros* cf. *rhinoceros*
Rhinoceros sp., 55
Riparo Mochi, 70, 78
Risk, 209, 270, 271, 272, 273, 274, 275, 280
Riwat, 51, 106
Rodniki, 29
Roundleaf bat. *See Hipposideros* sp.
Rousettus sp., 235

Sa Huynh culture, 194
Sai Yok 1, 163
Sakai Cave, 163
Samarkandskaya, 94
Sambar deer. *See Cervus unicolor*
Sambungmacan, 58
Saptomo, Wayhy, 125
Savannah conditions, Pleistocene
Africa, 64, 132, 208, 286, 289
Central Asia, 29
'Savannahstan' model, 65
Southeast Asia, 108, 110, 144, 146, 204
Sea levels, 12, 57, 96, 97, 132, 142, 143, 148, 205, 220
coastal communities, 149, 248, 249
coastal inundation, 106, 248, 287
Seasonality. *See* Procurement
Sel'-Ungur, 49, 94
Semai, 276
Semang, 274, 276
Shanboa Cave, 42
Shark teeth (Gua Lawa), 167
Shell middens
BAT 1 (Borneo), 173
crushed shell deposits (Hang Boi), 234, 235
Da But (Vietnam), 162
Golo Cave (Gebe Island), 160
Hang Boi (Vietnam), 188, 234–236, 237
Hang Chổ (Vietnam), 191
Hang Trống (Vietnam), 163, 188
Ille Cave (Philippines), 159, 176
kitchen-middens, 185
MAD 2 (Borneo), 174
Paso (Sulawesi), 160
Province Wellesley (Peninsular Malaysia), 193
'shell-bearing midden camp', 250
Sukajadi Pasar III (Sumatra), 193
whole shell deposits (Hang Boi), 234, 235, 236
Shell technology
ostrich egg-shell beads, 108
shell adzes, 181
shell artefacts, 125, 127, 130, 177, 200
shell fishing hooks, 172, 173, 282
shell scrapers, 177, 188
standardization, 282, 283
Siebenrockiella crassocollis, 261
Silvered langur. *See Presbytis cristata*
Sima de los Huesos, 49
Sima del Elefante, 29
Site 9, 104
Slushball Earth, 8
Small-toothed palm civet. *See Arctogalidia trivirgata*
Snowball Earth, 8
Soa Basin, 125
Social systems
closed, 276, 282, 283, 289
differences in, 277
networks as buffer against scarcity, 273, 281
open, 276
Soi Nhu pebble and flake industry, 190
Son Vi, 187
Song Gupuh, 122, 123, 162
Song Keplek, 167
Song Perahu, 166
Song Terus, 123, 167
Sonvian, 112, 113, 187, 188

South Kov site, 136
Specialization, adaptive and social. *See* Procurement
Spiny hill turtle. *See Heosemys spinosa*
Spirit Cave, 199, 200–201, 202, 203, 208–209
Spotted Hyaena. *See* Hyaenidae, *Crocuta crocuta*
Stadials
  Bølling-Allerød 19, 20, 143. *See* GIS-1.
  temperature rise (Bølling-Allerød), 19
Stegodon orientalis, 35
Stegodon sp., 45, 56
Stegodon-Ailuropoda fauna. *See* Ecological communities
Steppe mammoth. *See Mammuthus trogontherii*
Still Bay Industry, 82
Stone tools. *See* Lithic technology
Strategy-switching (economic), 280, 283
Stump-tailed macaque. *See Macaca arctoides*
Subsistence strategies. *See* Procurement
Sukajadi Pasar III, 193
Sumatran rhinoceros. *See Dicerorhinus sumatrensis*
Sun bear. *See Helarctos malayanus*
Sunghir. *See* Hominin fossils
Sus barbatus, 121, 212, 215, 216, 230, 262, 273
Sus scrofa, 113
Sus sp., 56, 241, 245, 250, 259, 260
Swabian Jura sites
  Geißenklösterle, 72
  Hohle Fels, 72
  Hohlenstein-Stadel, 72
  Vogelherd, 70, 72, 73
Szeletian, 70, 279

Tabon, 114, 115, 116, 175–176, 185
Tabuhan industry, 123
Taforalt, 81
Tam Hang, 192
Tambora, volcano, 43
Tanjung Pinang, 180, 181
Taphonomy
  abrasion, bone, 33, 151, 212, 223, 252, 253, 260, 261
  anthropogenic processes, 28, 120, 234, 236
  butchery. *See* Butchery
  fluvial transportation, 28
  foot traffic (trampling), 34, 189, 212, 220, 223, 231, 234, 236, 252, 260
  fragmentation, bone, 218, 223, 234
  manganese oxide staining, 212, 252, 253, 260
  post-depositional processes, 40, 94, 129, 151, 174, 177, 212, 216, 234, 235, 236, 240, 253
  root-etching, 252
  weathering, bone, 30, 32, 212, 223, 252
Tapirus indicus, 217
Tapirus sinensis, 35
Techno-complex, Hoabinhian defined as, 186
Techno-modes
  Mode 1, 29, 37, 49, 62, 65, 83, 94, 135
  Mode 2, 26, 29, 45, 46, 65
  Mode 3, 46, 69, 70, 83, 92, 99
  Mode 4, 69, 70, 99, 100
Tektites. *See* Muong-Nong/Austalasian tektite strewn field (MN/AT)
Terra Amata, 48
Terrestrial cores
  Gunung Buda, stalagmite sequence (Borneo), 148
  Lake Pemerak (core HN3, cores A-C) (Kalimantan), 146
  Lake Suigetsu (Japan), 147
  Makangit Cave (Philippines), 146
  ND-1 (Vietnam), 147, 246, 250
  Pea Bullock (Sumatra), 148
  Situ Bayongbong (Java), 144, 145, 148
Tertiary, species survival from, 31
Terus industry, 123
Teshik-Tash, 51
Tham Khuong, 187
Tham Lod, 197, 201, 203
Tham Om, 109
Thar Desert, 93, 104
Tianko Panjang, 166
Tibetan Plateau, 4, 8, 97, 98, 104
Tiger. *See Panthera tigris*
Tingkayu, locales I and II, 122, 130, 174, 206, 277
Titans Shelter cave, 137
Toalean, 159–160, 176–180, 184, 185
  Maros points, 177, 178, 179, 184, 206
  Muduk bone bipoints, 177
  Pirri points, 177
Toba caldera, 42, 43
  Haranggoal Dacite Tuff (HDF), 43
  Middle Toba Tuff (MTT), 43
  Oldest Toba Tuff (OTT), 43, 44, 45, 63, 267
  Younger Toba Tuff (YTT), 43, 44, 55, 93, 100–101
Toé Cave, 161, 181, 182, 183
Tongtianyan Cave, 94
Tragulus napu, 217
Tragulus sp., 254
Tràng An. *See* Hang Boi and Hang Trống
Transference, 269–272, 280
Trans-somatic behaviour, 287
Tree shrew. *See Tupaia glis (=belangeri)*
Trinil, 4, 40
Trinil H.K, fauna. *See* Ecological communities
Trionychidae (soft-shell turtles), 218, 221, 254, 261, 262
Tupaia glis (=*belangeri*), 246
Turangmoroke, Lake, 131

Uai Bobo 1 and 2, 168–172
Ubeidya, 27, 52
Ulalinka, 50, 61
Ulu Leang 1, 128, 159, 177, 178, 179
Uluzzian, 70, 279
Um Kapat Papo, 181
Upper Palaeolithic, 6, 46, 48, 68, 69, 70, 71, 72, 74, 75, 76, 77, 78, 79, 80, 82, 84, 85, 94, 105, 106, 107, 108, 130, 140, 206, 266, 272, 277, 279, 283
Upper Pleistocene
  definition, 9
Ursus arctos, 63
Ursus cf, *spelaeus*, 63
Ursus thibetanus, 63
Ust'-Karakol, 94

van Es, L.J.C., 150
van Stein Callenfels, P.V., 150, 177, 186, 193

Varanus spp., 212, 218, 221, 230, 252
Variability Selection hypothesis. *See* Adaptation
Velika Pĕcina, 70
Vertessöllos, 48
Vilakuav site, 136
Villafranchian fauna. *See* Ecological communities
Viverra tangalunga, 218, 221
Vulpes cf, *chikushanensis*, 37

Wadi Fatimah, 27
Wadi Hadramawt, 46
Wadi Kebara
  Kebara Cave, 80
Wadi-el-Mughara ('Valley of Caves'), 88
  Mugharet el-Wad, 89, 90
  Mugharet es-Skhūl 81, 89, 90, 91, 92, 105. *See* Hominin fossils and Ornamentation
  Mugharet et-Tabūn 48, 89, 90. *See* Hominin fossils
Warreen, 137, 138
Wild dogs. *See Cuon* sp.
Willandra Lakes, 131, 137, 138
Wolo Sege, 41, 64, 126
Woolly mammoth. *See Mammuthus primigenius*
Wray, L, Jr., 4, 193

Xiaochangliang, 32–33, 34
Xifeng section, 42
Xóm Trại, 186, 187
Xom Tre, 187

Younger Dryas. *See* GS-1
Yuanmou, 35, 37, 38, 63
Yuanmou Basin, 26, 35, 37

Zamiatnin, Z.N., 83
Zaraisk, 79
Zhirendong. *See* Hominin fossils
Zhoukoudian, 4, 35, 45, 52, 53, 54, 60, 61, 63, 94
  Locality 1, 45, 52, 53, 54
Zuraina Majid, 196, 197, 253
Zuttiyeh Cave, 89